Understanding Physics

based on Fundamentals of Physics Sixth Edition, by Halliday, Resnick, and Walker

Part 2

Karen Cummings
Rensselaer Polytechnic Institute

Priscilla W. Laws
Dickinson College

Edward F. Redish
University of Maryland

Patrick J. Cooney
Millersville University

with additional members of the Activity Based Physics Group

David R. Sokoloff (University of Oregon)
Ronald K. Thornton (Tufts University)
Edwin F. Taylor (Massachusetts Institute of Technology)

John Wiley & Sons, Inc.
New York/Chichester/Weinheim/Brisbane
Singapore/Toronto

ACQUISITIONS EDITOR	Stuart Johnson
SENIOR PRODUCTION EDITOR	Elizabeth Swain
DEVELOPMENT EDITOR	Ellen Ford
SENIOR MARKETING MANAGER	Bob Smith
SENIOR DESIGNER	Madelyn Lesure
ILLUSTRATION EDITOR	Anna Melhorn
PHOTO EDITOR	Hilary Newman

This book was typeset by the authors and Progressive and was printed and bound by Bradford & Bigelow, Inc. The cover was printed by Phoenix.

This book is printed on acid-free paper. 

ISBN 0471-39429-7

Printed in the United States of America

10 9 8 7 6 5 4 3 2 1

This book is dedicated to Arnold Arons, whose pioneering work in physics education and reviews of early chapters have had a profound influence on our work.

TABLE OF CONTENTS

Chapter 20
The Kinetic Theory of Gases 20-1

Chapter 21
Entropy and the Second Law of Thermodynamics 21-1

Appendices

Preface

To the Student

The Nature of Physics

Welcome to the study of physics. The study of physics has been a part of the traditional education of scientists, engineers, and technologists for over a century. Why is this so? Why should you study physics and what's it all about? What distinguishes physics from the other sciences? It's easy to say what the other sciences study. Biology is about living things, and chemistry is about atoms and molecules. But physics is somehow different. Trying to stay in the same pattern, we could say that physics is about matter and energy. Because living things are matter and energy, and molecules are matter and energy, physics applies to both biological and chemical systems—just as it does to building bridges or electronic chips. In fact, physics is the most basic of the natural sciences. It sets the underlying rules both for the other sciences and for engineering.

Physics is also the process of learning about the physical world by finding ways to make sense of what we know through observation. As Nobel Laureate Richard Feynman[1] wrote about science, "*The test of all knowledge is experiment*. But what is the source of knowledge? Where do the laws that are to be tested come from?...Experiment, itself, helps to produce these laws, in the sense that it gives us hints. But also needed is imagination to create from these hints the great generalizations—to guess at the wonderful, simple, but very strange patterns beneath them all, and then to experiment to check again whether we have made the right guess." Progress in all of the natural sciences depends on this interaction between experiment and theory.

In learning physics you may be surprised to find that in some ways it seems much simpler than other courses like biology or chemistry. There are fewer things to consider and the systems we treat are simpler. If you were to write down all the equations that you have to know in a physics course and compare it to the number of organisms you are expected to be familiar with in a general biology class or to the number of chemical reactions you have to know in general chemistry, the physics number is relatively small. Also the situations seem simpler. A ball rolling down an inclined plane or a battery connected to a bulb are a lot simpler than a sea squirt or cyclohexane. But on the other hand, for many students, physics somehow seems harder than those other courses. What's going on? Although beginning students sometimes see physics as a jumble of separate equations to be memorized, this is not so. Most equations used in introductory physics courses can be derived from relatively few fundamental relationships.

In part, the reason physics seems hard is that we're not just trying to learn about the properties of specific systems. Very few people really care about balls rolling down inclined planes or simple battery and bulb circuits. What we are trying to learn about is the whole nature of the scientific process. How do we figure out scientific laws? What is the nature of measurement? When we propose a theory, what does it mean and when does it hold? How do we know what we know? These questions are critical to solving any real-world scientific problem such as creating a smaller computer chip that dissipates heat effectively or diagnosing what's really wrong with a patient.

---

[1]R. P. Feynman, *The Feynman Lectures on Physics,* Ch. 1, (Addison-Wesley, Reading, MA, 1964).

The Art of Idealization

In physics, we want to understand how things work. To do that, we start with the simplest possible system that shows a behavior. If we're studying motion, we start with a small, massive object whose structure we ignore at first. We pretend a baseball is just a tiny blob and figure out how it moves under the influence of a hit and of gravity—pretending that it is in a vacuum and that it never rotates or deforms. These are clearly not good assumptions for a real baseball! But they provide an excellent starting point for making sense of its basic motion. Over small distances (a few feet), and for reasonably low speeds (below about 20 miles/hour) the idealized description works very well. As you get up to higher speeds and distances, the effects of the air grow in importance—but now that you know the basic principles, you can put in additional physics that make the examples more realistic and extend the number of cases you can treat.

This approach of starting with the simplest cases, understanding them as completely as possible (inventing physical laws to describe the situations), and then extending our considerations a step at a time, has been a primary factor in building the powerful body of knowledge that is physics.

Following this procedure has led to striking and surprising results in contemporary physics theories that have improved our understanding of how the world works and have led to the development of applications that have transformed our world. The prediction that the apparent mass of an object would grow and its internal clock would slow down as it approached the speed of light has been strikingly confirmed over and over again. Our understanding of these strange effects makes the global positioning system possible. Our understanding of the quantum behavior of electrons inside atoms and molecules allows us to design new and ever smaller computer chips.

Yet to find the surprises in physics you don't need to wait till you study relativity or quantum mechanics (though they both are really interesting and lots of fun). Even the physics that you'll be studying here produces some really surprising insights into our everyday world. If you take a 1-inch ball made of wood and a similar ball made of lead, the lead ball may weigh 20 times as much as the wooden ball. Yet if you stand on a chair and drop the two of them together, they fall the 10 feet to the ground in almost exactly the same time. Even a Styrofoam ball, 5 times lighter than the wood, will hit only a tiny fraction of a second later than the lead ball. Why does that happen? Clearly, if you hold the lead ball in one hand and the wood in the other, gravity is pulling much harder on the lead than on the wood. Why doesn't the lead ball go faster?

Actually, you know the answer. You just have to put two insights together. Think about kicking a soccer ball and a bowling ball. If you give them the same kick (ouch!) the soccer ball will speed up a lot more than the bowling ball. So to get a heavy object moving (a bowling ball), you need more force than to get a light object (a soccer ball) moving. So why doesn't the lead ball fall more slowly than the wooden one? To make a heavy ball (the lead) fall as fast as a light ball (the wood) you need to pull on it with a bigger force. Well this is just what gravity does! Your hand tells you that gravity pulls harder on the lead ball. It turns out that the extra pull that gravity exerts on the heavy object is just what is needed to balance the tendency of a heavy object not to be moved as easily as a light object.

This book is full of other such examples. When an object is immersed in water, it seems to weigh less—and its weight reduction is equal to the weight of the water that it pushed out of the way. What could that water have to do with anything? That water is gone! When I connect two identical bulbs up to a battery, if I connect them in one way they'll both have the same brightness as a single bulb connected up. But, if I connect them in another way, they both get much dimmer. Huh? Why does that happen?

Pay attention to your observations and intuitions when reading this book. Sometimes they'll be right, sometimes they'll be partially right, and sometimes they'll be dead wrong. Not only will you learn more, you will begin to learn how to make sense of the physics world.

And remember: Physics is supposed to help you make sense of the physical world. If the physical phenomenon doesn't appear to make sense at first, you need to keep thinking about your ideas and what they imply. Keep asking questions and looking at experiments. Einstein said, "Physics is the refinement of common sense." The key here is on the word "refinement." Physics is more than common sense. It's common sense made careful and consistent by the continuous interaction of theory and experiment.

Using this Book as a Learning Tool

This textbook is only one of many resources that you will need to learn physics. It is critical that you observe physical phenomenon directly, do careful experiments, and understand the logical reasoning that is used to develop physics concepts and theories based on them. It is equally critical that you test and enrich your knowledge by applying it to problem solving. Problem solving involves making predictions about the behavior of physical systems using an appropriate combination of reasoning and an understanding of the equations that describe physical systems. Although we have tried in this alternate version of the textbook to present both the experimental results that support theories and some of the reasoning that has gone into the development of theories, you will understand the physics only when you make your own observations and are actively engaged in the reasoning process. We hope this book will help you enjoy the study of physics as much as we do.

To the Instructor

Welcome to *Understanding Physics*. This book is built on the foundations of Halliday, Resnik, and Walker's *Fundamentals of Physics.* You might well be asking yourself, "Why mess with success?" The textbook and its descendants, first written by Halliday and Resnick over 40 years ago, has been among the best-selling introductory physics texts since it appeared. It sets the standard against which all other texts are judged. That's why, when the Activity-Based Physics Group decided a new way of publishing physics materials was needed, we approached John Wiley & Sons.

Why a New Text?

We have chosen to create this alternative to the Sixth Edition of the popular HRW text for several reasons.

First, we responded to student complaints that textbooks are too fragmented. The tendency to insert sidebars, extra boxes, and examples breaks up the flow of expository material. We decided to adopt a more narrative style. For this reason the critical examples have been placed at the ends of chapters as Touchstone Examples. The rest have been moved to the problem book.

Second, we wanted to take advantage of what has been learned from scholarship in physics education about student learning difficulties. Thus, presentations that are known to be associated with common student confusions have been rewritten and clarified. Places where common student difficulties had previously been ignored or glossed over have been expanded or elaborated to help students over these traditional barriers.

Third, the topics have been rearranged somewhat (especially in the adoption of the *New Mechanics Sequence*[2]) to provide a more coherent learning path and story line. The story line is reinforced by the use of Reading Exercises that help students focus on thoughtful reading of the text sections in each chapter.

Fourth, the experimental evidence for many of the physical laws and relationships discussed in the narrative has been presented in graphical form. In almost all cases, the experiments described can be easily replicated by introductory physics students with computer-based data gathering tools commonly found in up-to-date physics labs.

Fifth, sections of the text have been reorganized and enhanced so that they reinforce the use of additional elements in a *Suite of Activity Based Physics Materials.*[3] Different Suite elements have been designed for use in lecture, laboratory, and recitation sessions. An electronic version of the text is under development that will allow students and instructors to link to related parts of the text as well as to appropriate Suite materials.

The Activity Based Physics Suite

In addition to this textbook, the *Activity Based Physics Suite* includes materials that can be used in lecture settings—the *Interactive Lecture Demonstration Series*[4] (currently under development). Suite materials that can be used in laboratory settings include the *Workshop Physics Activity Guide*[5] modules as well as the *Real Time Physics Laboratory* modules.[6] Other materials in the collection are suitable for use in recitation sessions such as the well-known University of Washington *Tutorials in Introductory Physics*[7] and a set of *Quantitative Tutorials*[8] developed at the University of Maryland. The student component of the *Activity Based Physics Suite* is rounded out with a set of collaborative problem solving materials, also developed at the University of Maryland.

---

[2]Priscilla W. Laws, "A New Order for Mechanics," *Proceedings of the Rensselaer Polytechnic Institute Conference on Introductory Physics Course*, 125-136, May 1993.

[3]http://physics.dickinson.edu/suite_prototype

[4]David R. Sokoloff and Ronald K. Thornton, "Using Interactive Lecture Demonstrations to Create an Active Learning Environment." *The Physics Teacher*, **35**, 340-347, September 1997.

[5]Priscilla W. Laws, *Workshop Physics Activity Guide*, Modules 1-4 w/ Appendices (John Wiley & Sons, New York, 1997).

[6]David R. Sokoloff, *RealTime Physics*, Modules 1-2, (John Wiley & Sons, New York, 1999).

[7]Lillian C. McDermott and Peter S. Shaffer, *Tutorials in Introductory Physics*, (Prentice-Hall, Upper Saddle River, NJ, 1998).

[8]E.F. Redish, "Implications of cognitive studies for teaching physics," *Am. J. Phys.*, **62**, 796-803, 1994.

The teacher's manual, *Teaching Physics with the Physics Suite*,[9] provides an overview of what has been learned about student difficulties with physics and a guide designed to aid instructors in the selection and integration of various *Suite* materials. Included as an Appendix to that volume is an *Action Research Kit* with conceptual and attitudinal surveys to help instructors gauge the effectiveness of their introductory physics teaching.

The Sixth Edition of HRW

Our modified and extended text is based on the Sixth Edition of Halliday, Resnick, and Walker.[10] Since this edition was just released in 2000, some instructors might be unaware of the many exciting changes that Jearl Walker has made in the Sixth Edition. Highlights of these changes which we have retained and embellished upon are worth mentioning:

Design Changes: more open format, reduction in the number of topics covered, and reduction in the number of sample problems included in the book.

Pedagogical Changes: more emphasis on reasoning, the statement of key ideas in sample problems, and the incorporation of more sample problems that stress applications.

A Final Word

Over the past decade we have learned how valuable it is for us as teachers to focus on what it is the students actually do to learn physics, and how valuable it can be for students to work with research-based materials that promote active learning. We hope you and your students find this book and some of the other Suite materials helpful in your quest to make the study of physics both exciting and understandable to your students.

Beta Version 1, July 2001

Karen Cummings (Rensselaer Polytechnical Institute)
Priscilla W. Laws (Dickinson College)
Edward F. Redish (University of Maryland)
Patrick J. Cooney (Millersville University)

with David R. Sokoloff (University of Oregon)
Ronald K. Thornton (Tufts University)

---

[9] E.F. Redish, *Teaching Physics with the Physics Suite*, (under preparation for publication by John Wiley & Sons).

[10] David Halliday, Robert Resnick, and Jearl Walker, *Fundamentals of Physics*, 6th Ed., (John Wiley and Sons, New York, 2001).

13 Equilibrium and Elasticity

Rock climbing may be the ultimate physics exam. Failure can mean death, and even "partial credit" can mean severe injury. For example, in a long chimney climb, in which your torso is pressed against one wall of a wide vertical fissure and your feet are pressed against the opposite wall, you need to rest occasionally or you will fall due to exhaustion. Here the exam consists of a single question: What can you do to relax your push on the walls in order to rest? If you relax without considering the physics, the walls will not hold you up.

What is the answer to this life-and-death, one-question exam?

The answer is in this chapter.

13-1 Equilibrium

Consider these objects: (1) a book resting on a table, (2) a hockey puck sliding across a frictionless surface with constant velocity, (3) the rotating blades of a ceiling fan, and (4) the wheel of a bicycle that is traveling along a straight path at constant speed. For each of these four objects:

1. The linear momentum $\vec{p}_{\text{com}}$ of its center of mass is constant.
2. Its angular momentum $\vec{L}_{\text{com}}$ about its center of mass, or about any other point, is also constant.

We say that such objects are in **equilibrium.** The two requirements for equilibrium are then

$$\vec{p}_{\text{com}} = \text{a constant} \quad \text{and} \quad \vec{L}_{\text{com}} = \text{a constant}. \tag{13-1}$$

Our concern in this chapter is with situations in which the constants in Eq. 13-1 are in fact zero; that is, we are concerned largely with objects that are not moving in any way—either in translation or in rotation—in the reference frame from which we observe them. Such objects are in **static equilibrium.** Of the four objects mentioned at the beginning of this section, only one—the book resting on the table—is in static equilibrium.

Fig. 13-1: A balancing rock near Petrified Forest National Park in Arizona. Although its perch seems precarious, the rock is in static equilibrium.

The balancing rock of Fig. 13-1 is another example of an object that, for the present at least, is in static equilibrium. It shares this property with countless other structures, such as cathedrals, houses, filing cabinets, and taco stands, that remain stationary over time.

As we discussed in Section 10-5, if a body returns to a state of static equilibrium after having been displaced from it by a force, the body is said to be in *stable* static equilibrium. A marble placed at the bottom of a hemispherical bowl is an example. However, if a small force can displace the body and end the equilibrium, the body is in *unstable* static equilibrium.

For example, suppose we balance a domino with the domino's center of mass vertically above the supporting edge as in Fig. 13-2*a*. The torque about the supporting edge due to the gravitational force $\vec{F}_g$ on the domino is zero, because the line of action of $\vec{F}_g$ is through that edge. Thus, the domino is in equilibrium. Of course, even a slight force on it due to some chance disturbance ends the equilibrium. As the line of action of $\vec{F}_g$ moves to one side of the supporting edge (as in Fig. 13-2*b*), the torque due to $\vec{F}_g$ increases the domino's rotation. Thus, the domino in Fig. 13-2*a* is in unstable static equilibrium.

The domino in Fig. 13-2*c* is not quite as unstable. To topple this domino, a force would have to rotate it through and then beyond the balance position of Fig. 13-2*a*, in which the center of mass is above a supporting edge. A slight force will not topple this domino, but a vigorous flick of the finger against the domino certainly will. (If we arrange a chain of such upright dominos, a finger flick against the first can cause the whole chain to fall.)

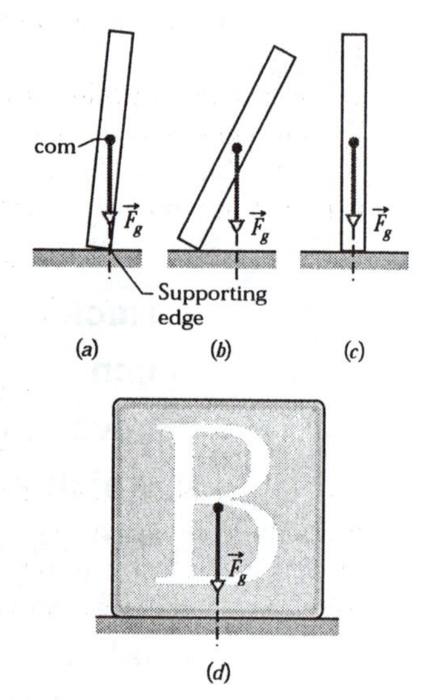

Fig. 13-2: (*a*) A domino balanced on one edge, with its center of mass vertically above that edge. The gravitational force $\vec{F}_g$ on the domino is directed through the supporting edge. (*b*) If the domino is rotated even slightly from the balanced orientation, then $\vec{F}_g$ causes a torque that increases the rotation. (*c*) A domino upright on a narrow side is somewhat more stable than the domino in (*a*). (*d*) A square block is even more stable.

The child's square block in Fig. 13-2*d* is even more stable because its center of mass would have to be moved even farther to get it to pass above a supporting edge. A flick of the finger may not topple the block. (This is why you never see a chain of toppling square blocks.) The worker in Fig. 13-3 is like both the domino and the square block: Parallel to the beam, his stance is wide and he is stable; perpendicular to the beam, his stance is narrow and he is unstable (and at the mercy of a chance gust of wind).

The analysis of static equilibrium is very important in engineering practice. The design engineer must isolate and identify all the external forces and torques that may act on a structure and, by good design and wise choice of materials, ensure that the structure will remain stable under these loads. Such analysis is necessary to ensure, for example, that bridges do not collapse under their traffic and wind loads, and that the landing gear of aircraft will survive the shock of rough landings.

13-2 The Requirements of Equilibrium

Fig. 13-3: A construction worker balanced above New York City is in static equilibrium but is more stable parallel to the beam than perpendicular to it.

The translational motion of a body is governed by Newton's Second Law in its linear momentum form, given by Eq. 8-23 as

$$\vec{F}_{\text{net}} = \frac{d\vec{p}}{dt}. \tag{13-2}$$

If the body is in translational equilibrium—that is, if $\vec{p}$ is a constant—then $d\vec{p}/dt = 0$ and we must have

$$\vec{F}_{\text{net}} = 0 \quad \text{(balance of forces).} \tag{13-3}$$

The rotational motion of a body is governed by Newton's Second Law in its angular momentum form, given by Eq. 12-28 as

$$\vec{\tau}_{\text{net}} = \frac{d\vec{L}}{dt}. \tag{13-4}$$

If the body is in rotational equilibrium—that is, if $\vec{L}$ is a constant—then $d\vec{L}/dt = 0$ and we must have

$$\vec{\tau}_{\text{net}} = 0 \quad \text{(balance of torques).} \tag{13-5}$$

Thus, the two requirements for a body to be in equilibrium are as follows:

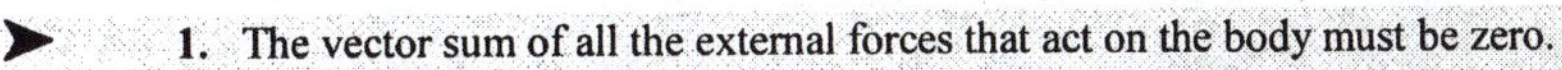

1. The vector sum of all the external forces that act on the body must be zero.
2. The vector sum of all the external torques that act on the body, measured about *any* possible point, must also be zero.

These requirements obviously hold for *static* equilibrium. They also hold for the more general equilibrium in which $\vec{p}$ and $\vec{L}$ are constant but not zero.

Equations 13-3 and 13-5, as vector equations, are each equivalent to three independent component equations, one for each direction of the coordinate axes:

Balance of forces	Balance of torques	
$F_{\text{net},x} = \Sigma F_x = 0$	$\tau_{\text{net},x} = \Sigma \tau_x = 0$	
$F_{\text{net},y} = \Sigma F_y = 0$	$\tau_{\text{net},y} = \Sigma \tau_y = 0$	(13-6)
$F_{\text{net},z} = \Sigma F_z = 0$	$\tau_{\text{net},z} = \Sigma \tau_z = 0$	

We shall simplify matters by considering only situations in which the forces that act on the body lie in the xy plane. This means that the only torques that can act on the body must tend to cause rotation around an axis parallel to the z axis. With this assumption, we eliminate one force equation and two torque equations from Eq. 13-6, leaving

$$F_{\text{net},x} = 0 \quad \text{(balance of forces),} \tag{13-7}$$

$$F_{\text{net},y} = 0 \quad \text{(balance of forces),} \tag{13-8}$$

$$\tau_{\text{net},z} = 0 \quad \text{(balance of torques).} \tag{13-9}$$

Here, $\tau_{\text{net},z}$ is the net torque that the external forces produce either about the z axis or about *any* axis parallel to it.

A hockey puck sliding at constant velocity over ice satisfies Eqs. 13-7, 13-8, and 13-9 and is thus in equilibrium *but not in static equilibrium.* For static equilibrium, the linear momentum $\vec{p}$ of the puck must be not only constant but also zero; the puck must be at rest on the ice. Thus, there is another requirement for static equilibrium:

➤**3.** The linear momentum $\vec{p}$ of the body must be zero.

READING EXERCISE 13-1: The figure gives six overhead views of a uniform rod on which two or more forces act perpendicularly to the rod. If the magnitudes of the forces are adjusted properly (but kept nonzero), in which situations can the rod be in static equilibrium?

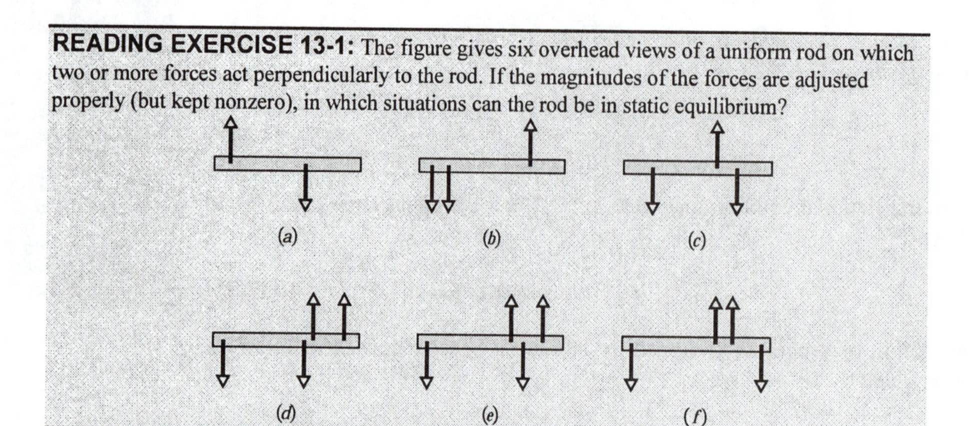

13-3 The Center of Gravity

The gravitational force on an extended body is the vector sum of the gravitational forces acting on the individual elements (the atoms) of the body. Instead of considering all those individual elements, we can say:

➤The gravitational force $\vec{F}_g$ on a body effectively acts at a single point, called the **center of gravity** (cog) of the body.

Here the word "effectively" means that if the forces on the individual elements were somehow turned off and force $\vec{F}_g$ at the center of gravity were turned on, the net force and the net torque (about any point) acting on the body would not change.

Until now, we have assumed that the gravitational force $\vec{F}_g$ acts at the center of mass (com) of the body. This is equivalent to assuming that the center of gravity is at the center of mass. Recall from Section 6-3 of Chapter 6 that, for a body of mass *M*, the force $\vec{F}_g$ is equal to $M\vec{a}_g$, where $\vec{a}_g$ is the acceleration that the force would produce if the body were to fall freely ($\vec{a}_g = \vec{g}$ where $\vec{g}$ is the gravitational force per unit mass). In the proof that follows, we show that

➤If $\vec{a}_g$ is the same for all elements of a body, then the body's center of gravity (cog) is coincident with the body's center of mass (com).

This is approximately true for everyday objects because $\vec{g}$ varies only a little along Earth's surface and decreases in magnitude only slightly with altitude. Thus, for objects like a mouse or a moose, we have been justified in assuming that the gravitational force acts at the center of mass. After the following proof, we shall resume that assumption.

Proof

First, we consider the individual elements of the body. Figure 13-4*a* shows an extended body, of mass M, and one of its elements, of mass m_i. A gravitational force $\vec{F}_{gi}$ acts on each such element and is equal to $m_i\vec{a}_{gi}$. The subscript on $\vec{a}_{gi}$ means $\vec{a}_{gi}$ is the gravitational acceleration *at the location of the ith mass element* (it can be different for other elements).

In Fig. 13-4*a*, each force $\vec{F}_{gi}$ produces a torque $\vec{\tau}_i$ on the element about the origin *O*, with moment arm x_i. Using Eq. 11-30 ($|\vec{\tau}| = r_\perp|\vec{F}|$), we can write the components of torque, along the *z* axis, τ_i

$$\tau_{i,z} = +x_i|\vec{F}_{gi}|. \qquad (13\text{-}10)$$

The net z component of torque on all the elements of the body is then

$$\tau_{\text{net},z} = \sum_i \tau_{i,z} = \sum_i x_i|\vec{F}_{gi}|. \qquad (13\text{-}11)$$

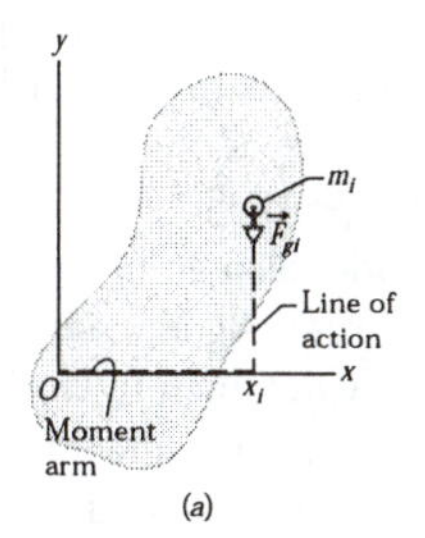

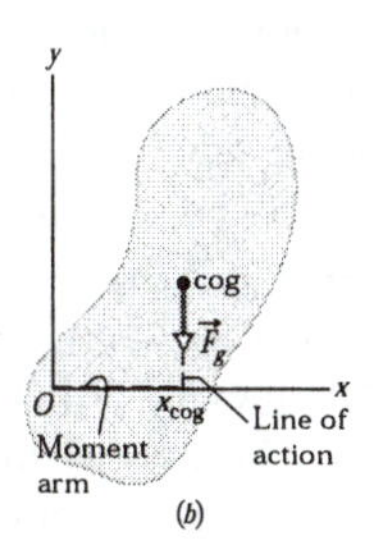

Fig. 13-4: (*a*) An element of mass m_i in an extended body. The gravitational force $\vec{F}_{gi}$ on it has moment arm x_i about the origin O of the coordinate system. (*b*) The gravitational force $\vec{F}_g$ on a body is said to act at the center of gravity (cog) of the body. Here it has moment arm x_{cog} about origin O.

Next, we consider the body as a whole. Figure 13-4*b* shows the gravitational force $\vec{F}_g$ acting at the body's center of gravity. This force produces a torque $\vec{\tau}$ on the body about O, with moment arm x_{cog}. Again using Eq. 11-31, we can write the z component of this torque as

$$\tau_z = x_{\text{cog}}|\vec{F}_g|. \qquad (13\text{-}12)$$

The gravitational force $\vec{F}_g$ on the body is equal to the sum of the gravitational forces $\vec{F}_{gi}$ on all its elements, so we can substitute $\Sigma\vec{F}_{gi}$ for $\vec{F}_g$ in Eq. 13-12 to write

$$\tau_{\text{net},z} = x_{\text{cog}}\Sigma|\vec{F}_{gi}| = x_{\text{cog}}|\Sigma\vec{F}_{gi}|. \qquad (13\text{-}13)$$

Now recall that the torque due to force $\vec{F}_g$ acting at the center of gravity is equal to the net torque due to all the forces $\vec{F}_{gi}$ acting on all the elements of the body. (That is how we defined the cog.) Therefore, $\tau_{\text{net},z}$ in Eq. 13-13 is equal to $\tau_{\text{net},z}$ in Eq. 13-11. Putting those two equations together, we can write

$$x_{\text{cog}}|\Sigma\vec{F}_{gi}| = \Sigma x_i|\vec{F}_{gi}|.$$

Substituting $m_i|\vec{a}_{gi}|$ for $\vec{F}_{gi}$ gives us

$$x_{\text{cog}}\Sigma m_i|\vec{a}_{gi}| = \Sigma x_i m_i|\vec{a}_{gi}|.$$

Now here is a key idea: If the accelerations $\vec{a}_{gi}$ at all the locations of the elements are the same, we can cancel g_i from this equation to write

$$x_{\text{cog}}\Sigma m_i = \Sigma x_i m_i. \qquad (13\text{-}14)$$

The sum Σm_i of the masses of all the elements is the mass M of the body. Therefore, we can rewrite Eq. 13-14 as

$$x_{\text{cog}} = \frac{1}{M}\Sigma x_i m_i. \qquad (13\text{-}15)$$

The right side of this equation gives the coordinate x_{com} of the body's center of mass (Eq. 8-9). We now have what we sought to prove:

$$x_{\text{cog}} = x_{\text{com}}. \qquad (13\text{-}16)$$

READING EXERCISE 13-2: Suppose that you skewer an apple with a thin rod, missing the apple's center of gravity. When you hold the rod horizontally and allow the apple to rotate freely, where does the center of gravity end up and why?

Touchstone Examples 13-3-1, 13-3-2, 13-3-3, and 13-3-4, at the end of this chapter, illustrate how to use what you learned in this section.

TE

13-4 Indeterminate Structures

For the problems of this chapter, we have only three independent equations at our disposal, usually two balance of forces equations and one balance of torques equation about a given rotation axis. Thus, if a problem has more than three unknowns, we cannot solve it.

It is easy to find such problems. In Touchstone Example 13-3-2, for example, we could have assumed that there is friction between the wall and the top of the ladder. Then there would have been a vertical frictional force acting where the ladder touches the wall, making a total of four unknown forces. With only three equations, we could not have solved this problem.

Consider also an unsymmetrically loaded car. What are the forces—all different—on the four tires? Again, we cannot find them because we have only three independent equations with which to work. Similarly, we can solve an equilibrium problem for a table with three legs but not for one with four legs. Problems like these, in which there are more unknowns than equations, are called **indeterminate.**

Yet solutions to indeterminate problems exist in the real world. If you rest the tires of the car on four platform scales, each scale will register a definite reading, the sum of the readings being the weight of the car. What is eluding us in our efforts to find the individual forces by solving equations?

The problem is that we have assumed—without making a great point of it—that the bodies to which we apply the equations of static equilibrium are perfectly rigid. By this we mean that they do not deform when forces are applied to them. Strictly, there are no such bodies. The tires of the car, for example, deform easily under load until the car settles into a position of static equilibrium.

We have all had experience with a wobbly restaurant table, which we usually level by putting folded paper under one of the legs. If a big enough elephant sat on such a table, however, you may be sure that if the table did not collapse, it would deform just like the tires of a car. Its legs would all touch the floor, the forces acting upward on the table legs would all assume definite (and different) values as in Fig. 13-5, and the table would no longer wobble. How do we find the values of those forces acting on the legs?

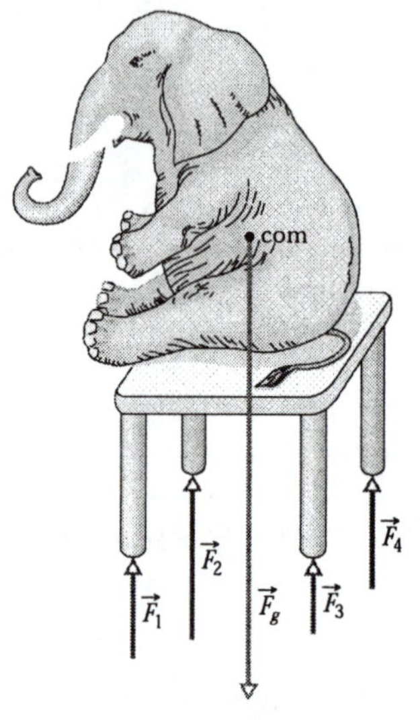

Fig. 13-5: The table is an indeterminate structure. The four forces on the table legs are different in magnitude and cannot be found from the laws of static equilibrium alone.

To solve such indeterminate equilibrium problems, we must supplement equilibrium equations with some knowledge of *elasticity,* the branch of physics and engineering that describes how real bodies deform when forces are applied to them. The next section provides an introduction to this subject.

READING EXERCISE 13-3: A horizontal uniform bar of weight 10 N is to hang from a ceiling by two wires that exert upward forces $\vec{F}_1$ and $\vec{F}_2$ on the bar. The figure shows four arrangements for the wires. Which arrangements, if any, are indeterminate (so that we cannot solve for numerical values of $\vec{F}_1$ and $\vec{F}_2$)?

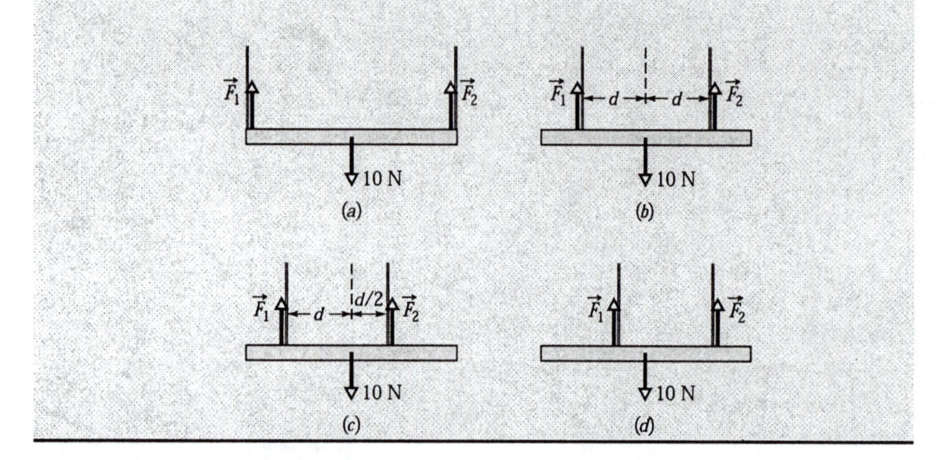

13-5 Elasticity

When a large number of atoms come together to form a metallic solid, such as an iron nail, they settle into equilibrium positions in a three-dimensional *lattice,* a repetitive arrangement in which each atom has a well-defined equilibrium distance from its nearest neighbors. The atoms are held together by interatomic forces that are modeled as tiny springs in Fig. 13-6. The lattice is remarkably rigid, which is another way of saying that the "interatomic springs" are extremely stiff. It is for this reason that we perceive many ordinary objects such as metal ladders, tables, and spoons as perfectly rigid. Of course, some ordinary objects, such as garden hoses or rubber gloves, do not strike us as rigid at all. The atoms that make up these objects *do not* form a rigid lattice like that of Fig. 13-6 but are aligned in long, flexible molecular chains, each chain being only loosely bound to its neighbors.

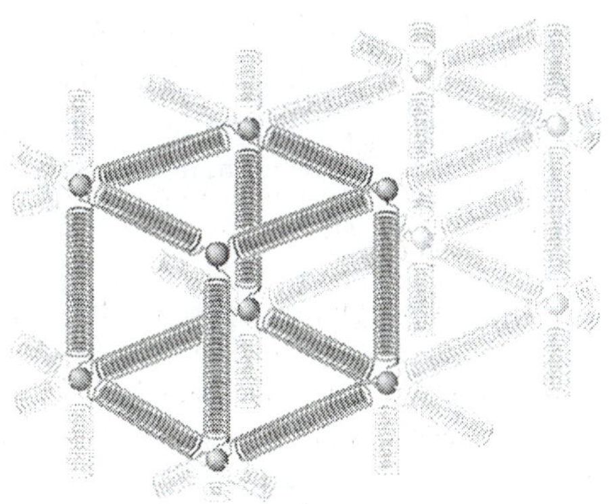

Fig. 13-6: The atoms of a metallic solid are distributed on a repetitive three-dimensional lattice. The springs represent interatomic forces.

All real "rigid" bodies are to some extent **elastic,** which means that we can change their dimensions slightly by pulling, pushing, twisting, or compressing them. To get a feeling for the orders of magnitude involved, consider a vertical steel rod 1 m long and 1 cm in diameter. If you hang a subcompact car from the end of such a rod, the rod will stretch, but only by about 0.5 mm, or 0.05%. Furthermore, the rod will return to its original length when the car is removed.

If you hang two cars from the rod, the rod will be permanently stretched and will not recover its original length when you remove the load. If you hang three cars from the rod, the rod will break. Just before rupture, the elongation of the rod will be less than $0.2\ \%$. Although deformations of this size seem small, they are important in engineering practice. (Whether a wing under load will stay on an airplane is obviously important.)

Figure 13-7 shows three ways in which a solid might change its dimensions when forces act on it. In Fig. 13-7*a*, a cylinder is stretched. In Fig. 13-7*b*, a cylinder is deformed by a force perpendicular to its axis, much as we might deform a pack of cards or a book. In Fig. 13-7*c*, a solid object, placed in a fluid under high pressure, is compressed uniformly on all sides. What the three deformation types have in common is that a **stress,** or deforming force per unit area, produces a **strain,** or unit deformation. In Fig. 13-7, *tensile stress* (associated with stretching) is illustrated in (*a*), *shearing stress* in (*b*), and *hydraulic stress* in (*c*).

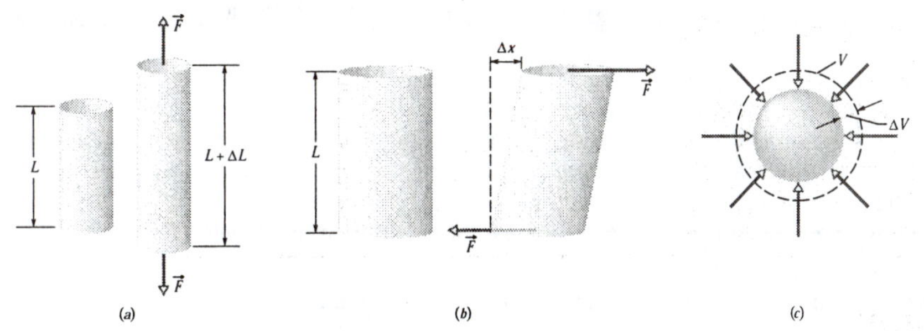

Fig. 13-7: (*a*) A cylinder subject to *tensile stress* stretches by an amount ΔL. (*b*) A cylinder subject to *shearing stress* deforms by an amount Δx, somewhat like a pack of playing cards would. (*c*) A solid sphere subject to uniform *hydraulic stress* from a fluid shrinks in volume by an amount ΔV. All the deformations shown are greatly exaggerated.

The stresses and the strains take different forms in the three situations of Fig. 13-7, but—over the range of engineering usefulness—stress and strain are proportional to each other. The constant of proportionality is called a **modulus of elasticity,** so that

$$\text{stress} = \text{modulus} \times \text{strain}. \tag{13-17}$$

In a standard test of tensile properties, the tensile stress on a test cylinder (like that in Fig. 13-8) is slowly increased from zero to the point at which the cylinder fractures, and the strain is carefully measured and plotted. The result is a graph of stress versus strain like that in Fig. 13-9. For a substantial range of applied stresses, the stress-strain relation is linear, and the specimen recovers its original dimensions when the stress is removed; it is here that Eq. 13-17 applies. If the stress is increased beyond the **yield strength** S_y of the specimen, the specimen becomes permanently deformed. If the stress continues to increase, the specimen eventually ruptures, at a stress called the **ultimate strength** S_u.

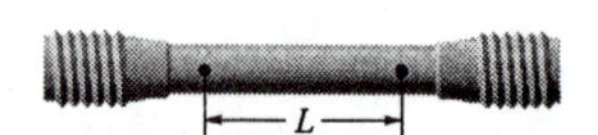

Fig. 13-8: A test specimen, used to determine a stress-strain curve such as that of Fig. 13-9. The change ΔL that occurs in a certain length L is measured in a tensile stress-strain test.

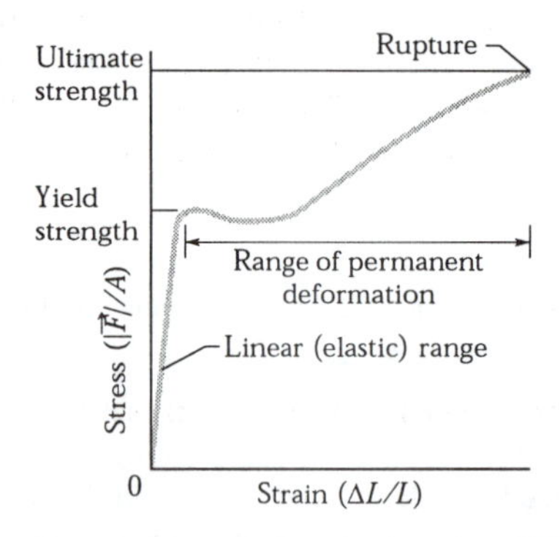

Fig. 13-9: A stress-strain curve for a steel test specimen such as that of Fig. 13-8. The specimen deforms permanently when the stress is equal to the *yield strength* of the material. It ruptures when the stress is equal to the *ultimate strength* of the material.

Tension and Compression

For simple tension or compression, the stress on an object is defined as $|\vec{F}|/A$, where $|\vec{F}|$ is the magnitude of the force applied perpendicularly to area A on the object. The strain, or unit deformation, is then the dimensionless quantity $\Delta L/L$, the fractional (or sometimes percentage) change in the length of the specimen. If the specimen is a long rod and the stress does not exceed the yield strength, then not only the entire rod but also every section of it experiences the same strain when a given stress is applied. Because the strain is dimensionless, the modulus in Eq. 13-17 has the same dimensions as the stress, namely, force per unit area.

The modulus for tensile and compressive stresses is called the **Young's modulus** and is represented in engineering practice by the symbol E. Equation 13-17 becomes

$$\frac{|\vec{F}|}{A} = E\frac{\Delta L}{L}. \tag{13-18}$$

The strain $\Delta L/L$ in a specimen can often be measured conveniently with a *strain gauge* (Fig. 13-10). This simple and useful device, which can be attached directly to operating machinery with an adhesive, is based on the principle that its electrical properties are dependent on the strain it undergoes.

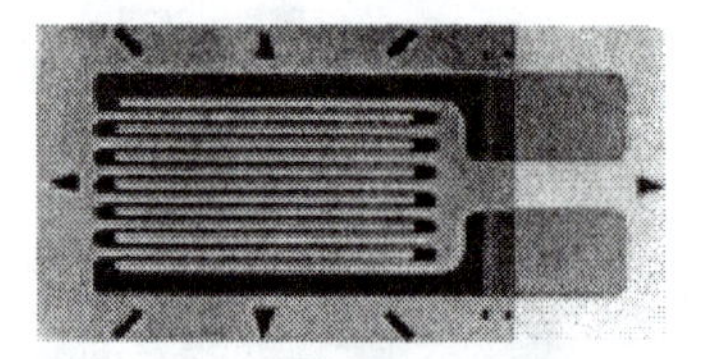

Fig. 13-10: A strain gauge of overall dimensions 9.8 mm by 4.6 mm. The gauge is fastened with adhesive to the object whose strain is to be measured; it experiences the same strain as the object. The electrical resistance of the gauge varies with the strain, permitting strains up to about 3 mm to be measured.

Although the Young's modulus for an object may be almost the same for tension and compression, the object's ultimate strength may well be different for the two types of stress. Concrete, for example, is very strong in compression but is so weak in tension that it is almost never used in that manner. Table 13-1 shows the Young's modulus and other elastic properties for some materials of engineering interest.

TABLE 13-1: Some Elastic Properties Of Selected Materials Of Engineering Interest

Material	Density ρ (kg/m^3)	Young's Modulus E ($10^9\ N/m^2$)	Ultimate Strength S_u ($10^6\ N/m^2$)	Yield Strength S_y ($10^6\ N/m^2$)
Steel[a]	7860	200	400	250
Aluminum	2710	70	110	95
Glass	2190	65	50[b]	—
Concrete[c]	2320	30	40[b]	—
Wood[d]	525	13	50[b]	—
Bone	1900	9[b]	170[b]	—
Polystyrene	1050	3	48	—

[a]Structural steel (ASTM-A36). [b]In compression. [c]High strength. [d]Douglas fir.

Shearing

In the case of shearing, the stress is also a force per unit area, but the force vector lies in the plane of the area rather than perpendicular to it. The strain is the dimensionless ratio $\Delta x/L$, with the quantities defined as shown in Fig. 13-7*b*. The corresponding modulus, which is given the symbol G in engineering practice, is called the **shear modulus.** For shearing, Eq. 13-17 is written as

$$\frac{|\vec{F}_\perp|}{A} = G\frac{\Delta x}{L}. \tag{13-19}$$

Shearing stresses play a critical role in the buckling of shafts that rotate under load and in bone fractures caused by bending.

Hydraulic Stress

In Fig. 13-7*c*, the stress is the fluid pressure P on the object. As you will see in Eq. 15-1, if the forces that act on an area A are uniform, then pressure is the ratio of the force perpendicular to an area and the area. The strain is $\Delta V/V$, where V is the original volume of the specimen and ΔV is the absolute value of the change in volume. The corresponding modulus, with symbol B, is called the **bulk modulus** of the material. The object is said to be under *hydraulic compression,* and the pressure can be called the *hydraulic stress.* For this situation, we write Eq. 13-17 as

$$P = B\frac{\Delta V}{V}. \tag{13-20}$$

The bulk modulus is $2.2 \times 10^9\ N/m^2$ for water, and is $16\times10^{10}\ N/m^2$ for steel. The pressure at the bottom of the Pacific Ocean, at its average depth of about $4000\ m$, is $4.0\times10^7\ N/m^2$. The fractional compression $\Delta V/V$ of a volume of water due to this pressure is 1.8%; that for a steel object is only about 0.025%. In general, solids—with their rigid atomic lattices—are less compressible than liquids, in which the atoms or molecules are less tightly coupled to their neighbors.

Touchstone Example 13-5-1, at the end of this chapter, illustrates how to use what you learned in this section.

TE

Touchstone Example 13-3-1

In Fig. TE13-1*a*, a uniform beam, of length L and mass m = 1.8 kg, is at rest with its ends on two scales. A uniform block, with mass M = 2.7 kg, is at rest on the beam, with its center a distance $L/4$ from the beam's left end. What do the scales read?

SOLUTION: The first steps in the solution of *any* problem about static equilibrium are these: Clearly define the system to be analyzed and then draw a free-body diagram of it, indicating all the forces on the system. Here, let us choose the system as the beam and block taken together. Then the forces on the system are shown in the free-body diagram of Fig. TE13-1*b*. (Choosing the system takes experience and often there can be more than one good choice.

The normal forces on the beam from the scales are $\vec{F}_l$ on the left and $\vec{F}_r$ on the right. The scale readings that we want are equal to the magnitudes of those forces. The gravitational force $\vec{F}_{g,beam}$ on the beam acts at the beam's center of mass and is equal to $m\vec{g}$. Similarly, the gravitational force $\vec{F}_{g,block}$ on the block acts at the block's center of mass and is equal to $M\vec{g}$. However, to simplify Fig. TE13-1*b*, the block is represented by a dot within the boundary of the beam and the vector $\vec{F}_{g,block}$ is drawn with its tail on that dot. (This downward shift of vector $\vec{F}_{g,block}$ along its line of action does not alter the torque due to $\vec{F}_{g,block}$ about any axis perpendicular to the figure.)

The **Key Idea** here is that, because the system is in static equilibrium, we can apply the balance of forces equations (Eqs. 13-7 and 13-8) and the balance of torques equation (13-9) to it. The forces have no x components, so Eq. 13-7 ($F_{\text{net},x} = 0$) provides no information. For the y components, Eq. 13-8 ($F_{\text{net},y} = 0$) gives us

$$F_l + F_r - Mg - mg = 0. \qquad \text{(TE13-1)}$$

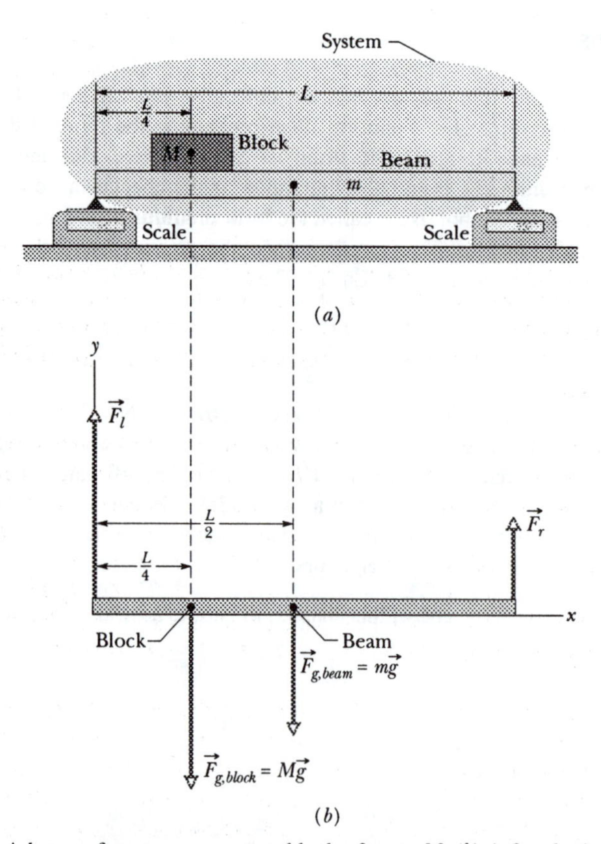

Fig. TE13-1 (*a*) A beam of mass m supports a block of mass M. (*b*) A free-body diagram, showing the forces that act on the system *beam + block*.

This equation contains two unknowns, the forces $\vec{F}_l$ and $\vec{F}_r$, so we also need to use Eq. 13-9, the balance of torques equation. We can apply it to *any* rotation axis perpendicular to the plane of Fig. TE13-1. Let us choose a rotation axis through the left end of the beam. We shall also use our general rule for assigning signs to torques: If a torque would cause an initially stationary body to rotate clockwise about the rotation axis, the z-componet of the torque is negative. If the rotation would be counterclockwise, the z-componet of the torque is positive. Finally, we shall write the torques in the form $r_\perp|\vec{F}|$, where the moment arm $r_\perp$ is 0 for $\vec{F}_l$, $L/4$ for $M\vec{g}$, $L/2$ for $m\vec{g}$, and L for $\vec{F}_r$.

We now can write the balancing equation ($\tau_{\text{net},z} = 0$) as

$$(0)(|\vec{F}_l|) - (L/4)(Mg) - (L/2)(mg) + (L)(|\vec{F}_r|) = 0,$$

which gives us

$$\begin{aligned} |\vec{F}_r| &= \tfrac{1}{4}Mg + \tfrac{1}{2}mg \\ &= \tfrac{1}{4}(2.7 \text{ kg})(9.8 \text{ m/s}^2) + \tfrac{1}{2}(1.8 \text{ kg})(9.8 \text{ m/s}^2) \\ &= 15.44 \text{ N} \approx 15 \text{ N}. \qquad \text{(Answer)} \end{aligned}$$

Now, solving Eq. TE13-1 for F_l and substituting this result, we find

$$\begin{aligned} |\vec{F}_l| &= (M + m)g - |\vec{F}_r| \\ &= (2.7 \text{ kg} + 1.8 \text{ kg})(9.8 \text{ m/s}^2) - 15.44 \text{ N} \\ &= 28.66 \text{ N} \approx 29 \text{ N}. \qquad \text{(Answer)} \end{aligned}$$

Notice the strategy in the solution: When we wrote an equation for the balance of force components, we got stuck with two unknowns. If we had written an equation for the balance of torques around some *arbitrary* axis, we would have again gotten stuck with those two unknowns. However, because we chose the axis to pass through the point of application of one of the unknown forces, here $\vec{F}_l$, we did not get stuck. Our choice neatly eliminated that force from the torque equation, allowing us to solve for the other unknown force magnitude F_r. Then we returned to the equation for the balance of force components to find the remaining unknown force magnitude.

Touchstone Example 13-3-2

In Fig. TE13-2*a*, a ladder of length $L = 12$ m and mass $m = 45$ kg leans against a slick (frictionless) wall. Its upper end is at height $h = 9.3$ m above the pavement on which the lower end rests (the pavement is not frictionless). The ladder's center of mass is $L/3$ from the lower end. A firefighter of mass $M = 72$ kg climbs the ladder until her center of mass is $L/2$ from the lower end. What then are the magnitudes of the forces on the ladder from the wall and the pavement?

SOLUTION: First, we choose our system as being the firefighter and ladder, together, and then we draw the free-body diagram of Fig. TE13-2*b*. The firefighter is represented with a dot within the boundary of the ladder. The gravitational force on her is represented with its equivalent $M\vec{g}$, and that vector has been shifted along its line of action, so that its tail is on the dot. (The shift does not alter a torque due to $M\vec{g}$ about any axis perpendicular to the figure.)

The only force on the ladder from the wall is the horizontal force $\vec{F}_w$ (there cannot be a frictional force along a frictionless wall). The force $\vec{F}_p$ on the ladder from the pavement has a horizontal component F_{px} that is a static frictional force and a vertical component F_{py} that is a normal force.

A **Key Idea** here is that the system is in static equilibrium, so the balancing equations (Eqs. 13-7 through 13-9) apply to it. Let us start with Eq. 13-9 ($\tau_{\text{net},z} = 0$). To choose an axis about which to calculate the torques, note that we have unknown forces ($\vec{F}_w$ and $\vec{F}_p$) at the two ends of the ladder. To eliminate, say, $\vec{F}_p$ from the calculation, we place the axis at point O, perpendicular to the figure. We also place the origin of an xy coordinate system at O. We can find torques about O with any of Eqs. 11-27 through 11-31, but Eq. 11-31 ($|\vec{\tau}| = r_\perp|\vec{F}|$) is easiest to use here.

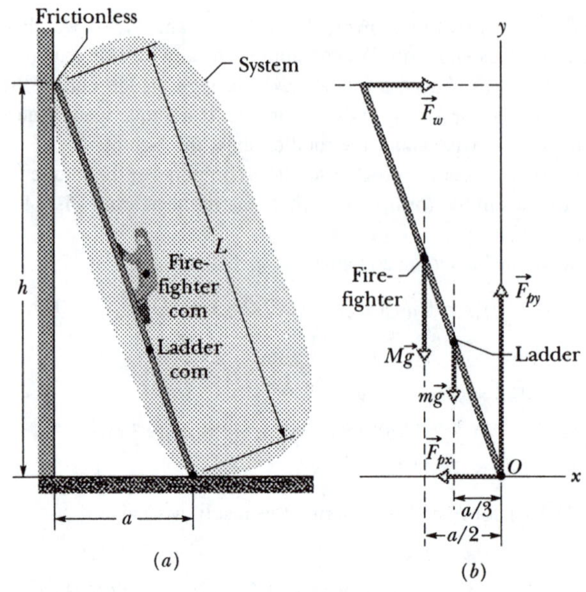

Fig. TE13-2 (*a*) A firefighter climbs halfway up a ladder that is leaning against a frictionless wall. The pavement beneath the ladder is not frictionless. (*b*) A free-body diagram, showing the forces that act on the firefighter–ladder system. The origin O of a coordinate system is placed at the point of application of the unknown force $\vec{F}_p$ (whose vector components $\vec{F}_{px}$ and $\vec{F}_{py}$ are shown).

To find the moment arm $r_\perp$ of $\vec{F}_w$, we draw a line of action through that vector (Fig. TE13-2*b*). Then $r_\perp$ is the perpendicular distance between O and the line of action. In Fig. TE13-2*b*, it extends along the y axis and is equal to the height h. We similarly draw lines of action for $M\vec{g}$ and $m\vec{g}$ and see that their moment arms extend along the x axis. For the distance a shown in Fig. TE13-2*a*, the moment arms are $a/2$ (the firefighter is halfway up the ladder) and $a/3$ (the ladder's center of mass is one-third of the way up the ladder), respectively. The moment arms for $\vec{F}_{px}$ and $\vec{F}_{py}$ are zero.

Now, the torques can be written in the form $r_\perp F$. The balancing equation $\tau_{\text{net},z} = 0$ becomes

$$-(h)(|\vec{F}_w|) + (a/2)(Mg) + (a/3)(mg) + (0)(|F_{px}|) + (0)(|F_{py}|) = 0. \quad \text{(TE13-2)}$$

(Recall our rule: A positive torque corresponds to counterclockwise rotation and a negative torque corresponds to clockwise rotation.)

Using the Pythagorean theorem, we find that

$$a = \sqrt{L^2 - h^2} = 7.58 \text{ m}.$$

Then Eq. TE13-2 gives us

$$|\vec{F}_w| = \frac{ga(M/2) + m/3)}{h}$$

$$= \frac{(9.8 \text{ m/s}^2)(7.58 \text{ m})(72/2 \text{ kg} + 45/3 \text{ kg})}{9.3 \text{ m}}$$

$$= 407 \text{ N} \approx 410 \text{ N}. \quad \text{(Answer)}$$

But since $\vec{F}_w$ points to the right, its component F_w is also $+410$ N.

Now we need to use the force balancing equations. The equation $F_{\text{net},x} = 0$ gives us

$$F_w + F_{px} = 0, \quad \text{(TE13-3)}$$

so

$$F_{px} = -F_w = -410 \text{ N}, \quad \text{(Answer)}$$

where the minus sign tells us F_{px} points to the left.

Since gravitational forces are negative while the local gravitational strength g and the force component F_{py} are positive, the equation $F_{\text{net},y} = 0$ gives us

$$F_{py} - Mg - mg = 0, \qquad \text{(TE13-4)}$$

so

$$F_{py} = (M + m)g = (72 \text{ kg} + 45 \text{ kg})(9.8 \text{ m/s}^2)$$
$$= 1146.6 \text{ N} \approx 1100 \text{ N}. \qquad \text{(Answer)}$$

Touchstone Example 13-3-3

Figure TE13-3*a* shows a safe, of mass $M = 430$ kg, hanging by a rope from a boom with dimensions $a = 1.9$ m and $b = 2.5$ m. The boom consists of a hinged beam and a horizontal cable that connects the beam to a wall. The uniform beam has a mass m of 85 kg; the mass of the cable and rope are negligible.

(a) What is the tension T_c in the cable? In other words, what is the magnitude of the force $\vec{T}_c$ on the beam from the cable?

SOLUTION: The system here is the beam alone, and the forces on it are shown in the free-body diagram of Fig. TE13-3*b*. The force from the cable is $\vec{T}_c$. The gravitational force on the beam acts at the beam's center of mass (at the beam's center) and is represented by its equivalent $m\vec{g}$. The vertical component of the force on the beam from the hinge is F_v, and the horizontal component of the force from the hinge is F_h. The force componet from the rope supporting the safe is $\vec{T}_r$. Because beam, rope, and safe are stationary, the magnitude of $\vec{T}_r$ is equal to the weight of the safe: $|\vec{T}_r| = Mg$. We place the origin O of an xy coordinate system at the hinge.

One **Key Idea** here is that our system is in static equilibrium, so the balancing equations apply to it. Let us start with Eq. 13-9 ($\tau_{\text{net},z} = 0$). Note that we are asked for the magnitude of force $\vec{T}_c$ and not of forces $\vec{F}_h$ and $\vec{F}_v$ acting at the hinge, at point O. Thus, a second **Key Idea** is that, to eliminate $\vec{F}_h$ and $\vec{F}_v$ from the torque calculation, we should calculate torques about an axis that is perpendicular to the figure at point O. Then $\vec{F}_h$ and $\vec{F}_v$ will have moment arms of zero. The lines of action for $\vec{T}_c$, $\vec{T}_r$, and $m\vec{g}$ are dashed in Fig. TE13-3*b*. The corresponding moment arms are a, b, and $b/2$.

Writing torques in the form of $r_\perp|\vec{F}|$ and using our rule about signs for torques, the balancing equation $\tau_{\text{net},z} = 0$ becomes

$$(a)(|\vec{T}_c|) - (b)(|\vec{T}_r|) - (\tfrac{1}{2}b)(|mg|) = 0.$$

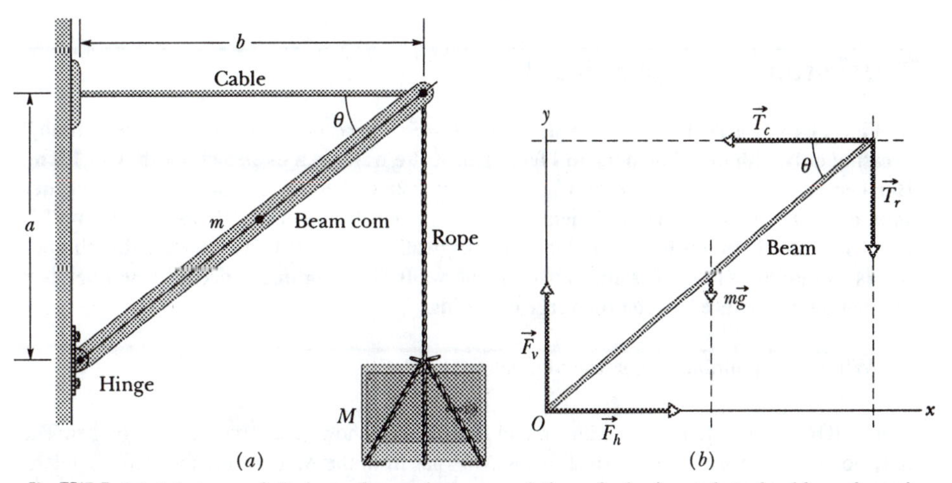

Fig. TE13-3 (*a*) A heavy safe is hung from a boom consisting of a horizontal steel cable and a uniform beam. (*b*) A free-body diagram for the beam.

Substituting Mg for $|\vec{T}_r|$ and solving for $|\vec{T}_c|$, we find that

$$|\vec{T}_c| = \frac{gb(M + \frac{1}{2}m)}{a}$$

$$= \frac{(9.8 \text{ m/s}^2)(2.5 \text{ m})(430 \text{ kg} + 85/2 \text{ kg})}{1.9 \text{ m}}$$

$$= 6093 \text{ N} \approx 6100 \text{ N}.$$

Since $\vec{T}_c$ points along the negative x-axis, its component T_c is negative, so $T_c = |\vec{T}_c| - 6100$ N.

(Answer)

(b) Find the magnitude $|\vec{F}|$ of the net force on the beam from the hinge.

SOLUTION: Now we want to know the values of the force components F_h and F_v so we can combine them to get $|\vec{F}|$. Because we know T_c, our **Key Idea** here is to apply the force balancing equations to the beam. For the horizontal balance, we write $F_{\text{net},x} = 0$ as

$$F_h + T_c = 0,$$

and so

$$F_h = -T_c = +6093 \text{ N}.$$

For the vertical balance, we write $F_{\text{net},y} = 0$ as

$$F_v - mg - T_r = 0.$$

Substituting $-Mg$ for T_r and solving for F_v, we find that

$$F_v = (m + M)g = (85 \text{ kg} + 430 \text{ kg})(9.8 \text{ m/s}^2)$$

$$= +5047 \text{ N}.$$

The net force vector is then

$$\vec{F} = F_h\hat{\text{i}} + F_v\hat{\text{j}}$$

$$= (6093 \text{ N})\hat{\text{i}} + (5047 \text{ N})\hat{\text{j}}$$

Also, from the Pythagorean theorem, we now have

$$|\vec{F}| = \sqrt{F_h^2 + F_v^2}$$

$$= \sqrt{(6093 \text{ N})^2 + (5047 \text{ N})^2} \approx 7900 \text{ N}. \qquad \text{(Answer)}$$

Note that $|\vec{F}|$ is substantially greater than either the combined weights of the safe and the beam, 5000 N, or the tension in the horizontal wire, 6100 N.

Touchstone Example 13-3-4

In Fig. TE13-4, a rock climber with mass m = 55 kg rests during a "chimney climb," pressing only with her shoulders and feet against the walls of a fissure of width w = 1.0 m. Her center of mass is a horizontal distance d = 0.20 m from the wall against which her shoulders are pressed. The coefficient of static friction between her shoes and the wall is μ_1 = 1.1, and between her shoulders and the wall it is μ_2 = 0.70. To rest, the climber wants to minimize her horizontal push on the walls. The minimum occurs when her feet and her shoulders are both on the verge of sliding.

(a) What is that minimum horizontal push on the walls?

SOLUTION: Our system is the climber, and Fig. TE13-4 shows the forces that act on her. The only horizontal forces are the normal forces $\vec{N}$ on her from the walls, at her feet and shoulders. The static frictional forces on her are $\vec{f}_1$ and $\vec{f}_2$, directed upward. The gravitational force $\vec{F}_g = m\vec{g}$ acts at her center of mass.

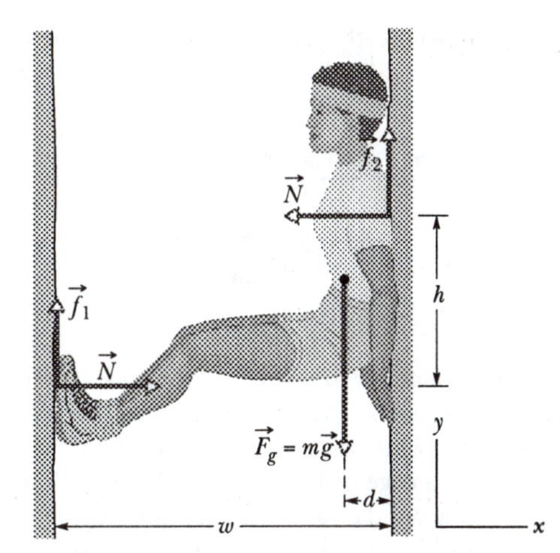

Fig. TE13-4 The forces on a climber resting in a rock chimney. The push of the climber on the chimney walls results in a rise to the normal forces $\vec{N}$ and the static frictional forces $\vec{f}_1$ and $\vec{f}_2$.

A **Key Idea** is that, because the system is in static equilibrium, we can apply the force balancing equations (Eqs. 13-7 and 13-8) to it. The equation $F_{net,x} = 0$ tells us that the two normal forces on her must be equal in magnitude and opposite in direction. We seek the magnitude $|\vec{N}|$ of these two forces, which is also the magnitude of her push against either wall.

The balancing equation $F_{net,y} = 0$ gives us

$$f_1 + f_2 + F_g = 0, \text{ where } F_g = -mg \qquad \text{(TE13-5)}$$

We want the climber to be on the verge of sliding at both her feet and her shoulders. That means we want the static frictional forces there to be at their maximum values. Those maximum values are, from Eq. 6-7 ($|\vec{f}_{s,\max}| = \mu_s |\vec{N}|$), since $\vec{f}_1$ and $\vec{f}_2$ act upwards in the positive y-direction, their magnitudes are

$$|\vec{f}_1| = +\mu_1 |\vec{N}| \quad \text{and} \quad |\vec{f}_2| = +\mu_2 |\vec{N}|. \qquad \text{(TE13-6)}$$

Substituting these expressions into Eq. TE13-5 and solving for the magnitude of $|\vec{N}|$ gives to us

$$|\vec{N}| = \frac{mg}{\mu_1 + \mu_2} = \frac{(55 \text{ kg})(9.8 \text{ m/s}^2)}{1.1 + 0.70} = 299 \text{ N} \approx 300 \text{ N}.$$

Thus, her minimum horizontal push must be about 300 N.

(b) For that push, what must be the vertical distance *h* between her feet and her shoulders if she is to be stable?

SOLUTION: A **Key Idea** here is that the climber will be stable if the torque balancing equation giving the z-component of torque ($\tau_{net,z} = 0$) applies to her. This means that the forces on her must not produce a net torque about any rotation axis. Another **Key Idea** is that we are free to choose a rotation axis that helps simplify the calculation. We shall write the torques in the form $r_\perp |\vec{F}|$, where $r_\perp$ is the moment arm of force $\vec{F}$. In Fig. TE13-4, we choose a rotation axis at her shoulders, perpendicular to the figure's plane. Then the moment arms of the forces acting there ($\vec{N}$ and $\vec{f}_2$) are zero. Frictional force $\vec{f}_1$, the normal force $\vec{N}$ at her feet, and the gravitational force $\vec{F}_g = m\vec{g}$ have the corresponding moment arms *w*, *h*, and *d*.

Recalling our rule about the signs of torques and the corresponding directions, we can now write the torque component $\tau_{net,z} = 0$ as

$$-(w)(|\vec{f}_1|) + (h)(|\vec{N}|) + (d)(mg) + (0)(|\vec{f}_2|) + (0)(|\vec{N}|) = 0. \qquad \text{(TE13-7)}$$

(Note how the choice of rotation axis neatly eliminates $|\vec{f}_2|$ from the calculation.) Next, solving Eq. TE13-7 for h, setting $|\vec{f}_1| = \mu_1|\vec{N}|$, and substituting $|\vec{N}| = 299$ N and other known values, we find that

$$h = \frac{|\vec{f}_1|w - mgd}{|\vec{N}|} = \frac{\mu_1|\vec{N}|w - mgd}{|\vec{N}|} = \mu_1 w - \frac{mgd}{|\vec{N}|}$$

$$= (1.1)(1.0\text{ m}) - \frac{(55\text{ kg})(9.8\text{ m/s}^2)(0.20\text{ m})}{299\text{ N}}$$

$$= 0.739\text{ m} \approx 0.74\text{ m}. \qquad \text{(Answer)}$$

We would find the same required value of h if we wrote the torques about any other rotation axis perpendicular to the page, such as one at her feet.

If h is more than *or* less than 0.74 m, she must exert a force greater than 299 N on the walls to be stable. Here, then, is the advantage of knowing the physics before you climb a chimney. When you need to rest, you will avoid the (dire) error of novice climbers who place their feet too high or too low. Instead, you will know that there is a "best" distance between shoulders and feet, requiring the least push, and giving you a good chance to rest.

Touchstone Example 13-5-1

A structural steel rod has a radius R of 9.5 mm and a length L of 81 cm. A 62 kn force $\vec{F}$ stretches it along its length. What are the stress on the rod and the elongation and strain of the rod?

SOLUTION: The first **Key Idea** here has to do with what is meant by the second sentence in the problem statement. We assume the rod is held stationary by, say, a clamp or vise at one end. Then force $\vec{F}$ is applied at the other end, parallel to the length of the rod and thus perpendicular to the end face there. Therefore, the situation is like that in Fig. 13-7*a*.

The next **Key Idea** is that we assume the force is applied uniformly across the end face and thus over an area $A = \pi R^2$. Then the stress on the rod is given by the left side of Eq. 13-18:

$$\text{stress} = \frac{|\vec{F}|}{A} = \frac{|\vec{F}|}{\pi R^2} = \frac{6.2 \times 10^4\text{ N}}{(\pi)(9.5 \times 10^{-3}\text{ m})^2}$$

$$= 2.2 \times 10^8\text{ N/m}^2. \qquad \text{(Answer)}$$

The yield strength for structural steel is 2.5×10^8 N/m^2, so this rod is dangerously close to its yield strength.

Another **Key Idea** is that the elongation of the rod depends on the stress, the original length L, and the type of material in the rod. The last determines which value we use for Young's modulus E (from Table 13-1). Using the value for steel, Eq. 13-18 gives us

$$\Delta L = \frac{(|\vec{F}|/A)L}{E} = \frac{(2.2 \times 10^8\text{ N/m}^2)(0.81\text{ m})}{2.0 \times 10^{11}\text{ N/m}^2}$$

$$= 8.9 \times 10^{-4}\text{ m} = 0.89\text{ mm}. \qquad \text{(Answer)}$$

The last **Key Idea** we need here is that strain is the ratio of the change in length to the original length, so we have

$$\frac{\Delta L}{L} = \frac{8.9 \times 10^{-4}\text{ m}}{0.81\text{ m}}$$

$$= 1.1 \times 10^{-3} = 0.11\%. \qquad \text{(Answer)}$$

14 Gravitation

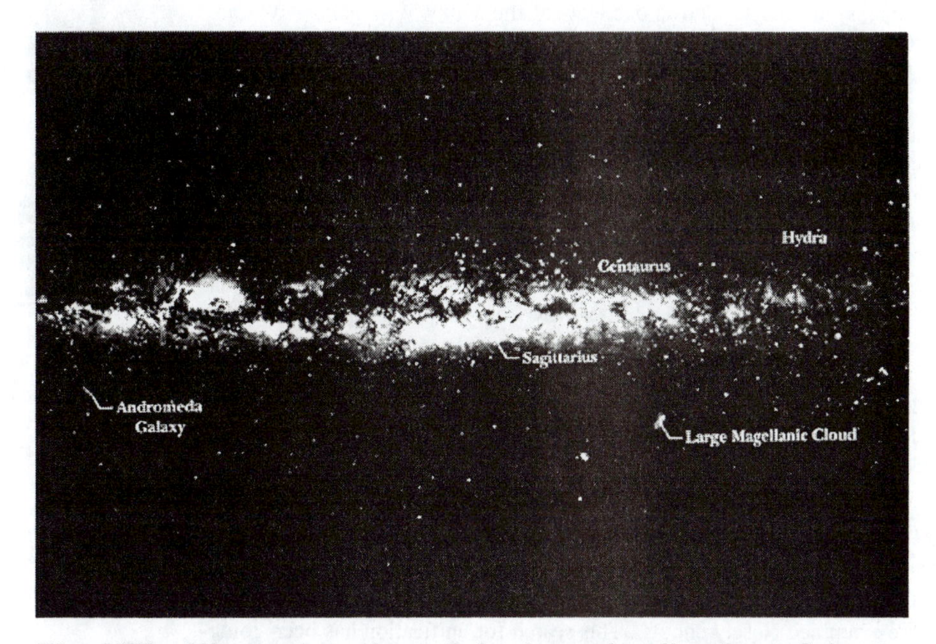

The Milky Way galaxy is a disk-shaped collection of dust, planets, and billions of stars, including our Sun and solar system. The force that binds it or any other galaxy together is the same force that holds Earth's moon in orbit and you on Earth—the gravitational force. That force is also responsible for one of nature's strangest objects, the black hole, a star that has completely collapsed onto itself. The gravitational force near a black hole is so strong that not even light can escape it.

If that is the case, how can a black hole be detected?

The answer is in this chapter.

14-1 The World and the Gravitational Force

The drawing that opens this chapter shows our view of the Milky Way galaxy. We are near the edge of the disk of the galaxy, about 26 000 light-years (2.5×10^{20} m) from its center, which in the drawing lies in the star collection known as Sagittarius. Our galaxy is a member of the Local Group of galaxies, which includes the Andromeda galaxy (Fig. 14-1) at a distance of 2.3×10^6 light-years, and several closer dwarf galaxies, such as the Large Magellanic Cloud shown in the opening drawing.

The Local Group is part of the Local Supercluster of galaxies. Measurements taken during and since the 1980s suggest that the Local Supercluster and the supercluster consisting of the clusters Hydra and Centaurus are all moving toward an exceptionally massive region called the Great Attractor. This region appears to be about 300 million light-years away, on the opposite side of the Milky Way from us, past the clusters Hydra and Centaurus.

The force that binds together these progressively larger structures, from star to galaxy to supercluster, and may be drawing them all toward the Great Attractor, is the gravitational force. That force not only holds you on Earth but also reaches out across intergalactic space.

Fig. 14-1: The Andromeda galaxy. Located 2.3×10^6 light-years from us, and faintly visible to the naked eye, it is very similar to our home galaxy, the Milky Way.

14-2 Newton's Law of Gravitation

Physicists like to study seemingly unrelated phenomena to show that a relationship can be found if they are examined closely enough. This search for unification has been going on for centuries. In 1665, the 23-year-old Isaac Newton made a basic contribution to physics when he showed that the force that holds the Moon in its orbit is the same force that makes an apple fall. We take this so much for granted now that it is not easy for us to comprehend the ancient belief that the motions of earthbound bodies and heavenly bodies were different in kind and were governed by different laws.

Newton concluded that not only does Earth attract an apple and the Moon but every body in the universe attracts every other body; this tendency of bodies to move toward each other is called **gravitation.** Newton's conclusion takes a little getting used to, because the familiar attraction of Earth for earthbound bodies is so great that it overwhelms the attraction that earthbound bodies have for each other. For example, Earth attracts an apple with a force magnitude of about 0.8 N. You also attract a nearby apple (and it attracts you), but the force of attraction has less magnitude than the weight of a speck of dust.

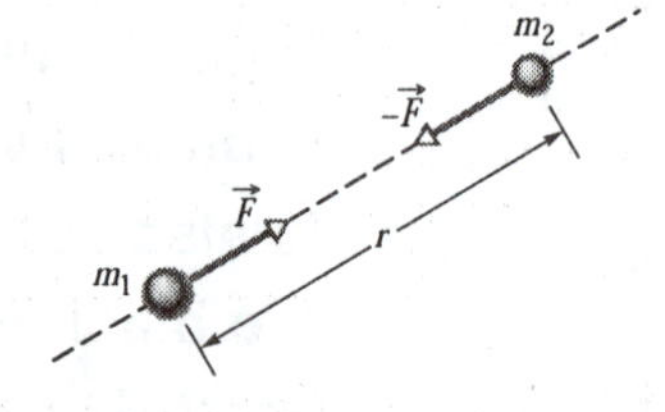

Fig. 14-2: Two particles, of masses m_1 and m_2 and with separation r, attract each other according to Newton's law of gravitation, Eq. 14-1. The forces of attraction, F and $-F$, are equal in magnitude and in opposite directions.

Quantitatively, Newton proposed a *force law* that we call **Newton's law of gravitation:** Every particle attracts any other particle with a **gravitational force** whose magnitude is given by

$$|\vec{F}| = |\vec{F}_{2\to 1}| = |\vec{F}_{1\to 2}| = G\frac{m_1 m_2}{r^2} \quad \text{(Newton's law of gravitation).} \tag{14-1}$$

Here m_1 and m_2 are the masses of the particles, r is the distance between them, and G is the **gravitational constant,** whose value is now known to be

$$\begin{aligned} G &= 6.67 \times 10^{-11}\ \text{N}\cdot\text{m}^2/\text{kg}^2 \\ &= 6.67 \times 10^{-11}\ \text{m}^3/\text{kg}\cdot\text{s}^2. \end{aligned} \tag{14-2}$$

As Fig. 14-2 shows, a particle m_2 attracts a particle m_1 with a gravitational force $\vec{F}$ that is directed toward particle m_2, and particle m_1 attracts particle m_2 with a gravitational force $-\vec{F}$ that is directed toward m_1. The forces $\vec{F}$ and $-\vec{F}$ form a third-law force pair; they are opposite in direction but equal in magnitude. They depend on the separation of the two particles, but not on their location: the particles could be in a deep cave or in deep space. Also forces $\vec{F}$ and $-\vec{F}$ are not altered by the presence of other bodies, even if those bodies lie between the two particles we are considering.

The strength of the gravitational force—that is, how strongly two particles with given masses at a given separation attract each other—depends on the value of the gravitational constant G. If G—by some miracle—were suddenly multiplied by a factor of 10, you would be crushed to the floor by Earth's attraction. If G were divided by this factor, Earth's attraction would be weak enough that you could jump over a building.

Although Newton's law of gravitation applies strictly to particles, we can also apply it to real objects as long as the sizes of the objects are small compared to the distance between them. The Moon and Earth are far enough apart so that, to a good approximation, we can treat them both as particles—but what about an apple and Earth? From the point of view of the apple, the broad and level Earth, stretching out to the horizon beneath the apple, certainly does not look like a particle.

Newton solved the apple-Earth problem by proving an important theorem called the *shell theorem:*

➤A uniform spherical shell of matter attracts a particle that is outside the shell as if all the shell's mass were concentrated at its center.

Earth can be thought of as a nest of such shells, one within another, and each attracting a particle outside Earth's surface as if the mass of that shell were at the center of the shell. Thus, from the apple's point of view, Earth *does* behave like a particle, one that is located at the center of Earth and has a mass equal to that of Earth.

Suppose, as in Fig. 14-3, that Earth pulls down on an apple with a force of magnitude 0.80 N. The apple must then pull up on Earth with a force of magnitude 0.80 N, which we take to act at the center of Earth. Although the forces are matched in magnitude, they produce different accelerations when the apple is released. For the apple, the acceleration is about 9.8 m/s^2, the familiar acceleration of a falling body near Earth's surface. For Earth, the acceleration measured in a reference frame attached to the center of mass of the apple-Earth system is only about $1 \times 10^{-25} \text{ m/s}^2$.

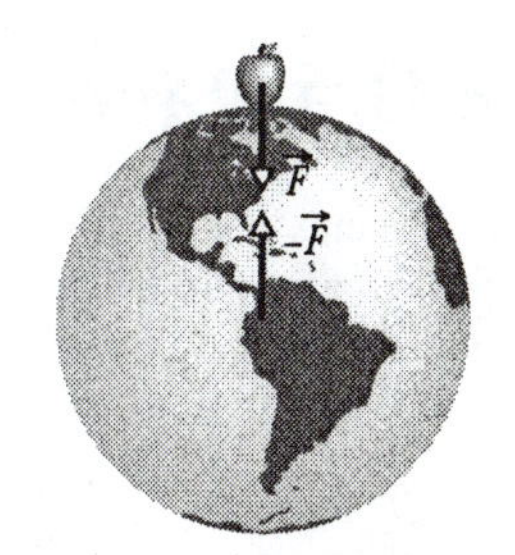

Fig. 14-3: The apple pulls up on Earth just as hard as Earth pulls down on the apple.

READING EXERCISE 14-1: A particle is to be placed, in turn, outside four objects, each of mass m: (1) a large uniform solid sphere, (2) a large uniform spherical shell, (3) a small uniform solid sphere, and (4) a small uniform shell. In each situation, the distance between the particle and the center of the object is d. Rank the objects according to the magnitude of the gravitational force they exert on the particle, greatest first.

14-3 Gravitation and the Principle of Superposition

Given a group of particles, we find the net (or resultant) gravitational force on any one of them from the others by using the **principle of superposition.** This is a general principle that says a net effect is the sum of the individual effects. Here, the principle means that we first compute the gravitational force that acts on our selected particle due to each of the other particles, in turn. We then find the net force by adding these forces vectorially, as usual.

For n interacting particles, we can write the principle of superposition for gravitational forces as

$$\vec{F}_{1,\text{net}} = \vec{F}_{2\to1} + \vec{F}_{3\to1} + \vec{F}_{4\to1} + \vec{F}_{5\to1} + \cdots + \vec{F}_{n\to1}. \quad (14\text{-}3)$$

Here $\vec{F}_{1,\text{net}}$ is the net force on particle 1 and, for example, $\vec{F}_{3\to1}$ is the force on particle 1 from particle 3. We can express this equation more compactly as a vector sum:

$$\vec{F}_{1,\text{net}} = \sum_{i=2}^{n} \vec{F}_{i\to1}. \quad (14\text{-}4)$$

What about the gravitational force on a particle from a real extended object? The force is found by dividing the object into parts small enough to treat as particles and then using Eq. 14-4 to find the vector sum of the forces on the particle from all the parts. In the limiting case, we can divide the extended object into differential parts of mass dm, each of which produces only a differential force $d\vec{F}$ on the particle. In this limit, the sum of Eq. 14-4 becomes an integral and we have

$$\vec{F}_1 = \int d\vec{F}, \qquad (14\text{-}5)$$

in which the integral is taken over the entire extended object and we drop the subscript "net." If the object is a uniform sphere or a spherical shell, we can avoid the integration of Eq. 14-5 by assuming that the object's mass is concentrated at the object's center and using Eq. 14-1.

Touchstone Example 14-3-1, at the end of this chapter, illustrates how to use what you learned in this section.

TE

14-4 Gravitation Near Earth's Surface

Let us assume that Earth is a uniform sphere of mass M. The magnitude of the gravitational force from Earth on a particle of mass m, located outside Earth a distance r from Earth's center, is then given by Eq. 14-1 as

$$|\vec{F}| = G\frac{Mm}{r^2} \qquad (14\text{-}6)$$

If the particle is released, it will fall toward the center of Earth, as a result of the gravitational force $\vec{F}$, with a **gravitational acceleration** $\vec{a}_g$. Newton's second law tells us that magnitudes $\vec{F}$ and $\vec{a}_g$ are related by

$$|\vec{F}| = m|\vec{a}_g| \qquad (14\text{-}7)$$

Now, substituting $|\vec{F}|$ from Eq. 14-6 into Eq. 14-7 and solving for $|\vec{a}_g|$, we find

$$|\vec{a}_g| = \frac{GM}{r^2} \qquad (14\text{-}8)$$

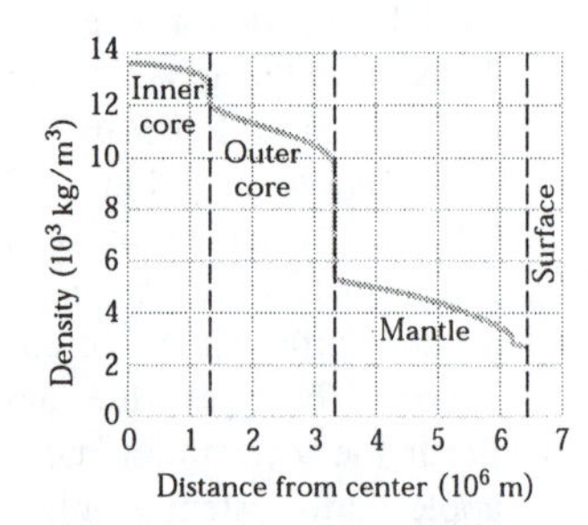

Fig. 14-4: The density of Earth as a function of distance from the center. The limits of the solid inner core, the largely liquid outer core, and the solid mantle are shown, but the crust of Earth is too thin to show clearly on this plot.

Table 14-1 shows values of $|\vec{a}_g|$ computed for various altitudes above Earth's surface.

Since Section 6-3, we have assumed that Earth is an inertial frame by neglecting its actual rotation. This simplification has allowed us to assume that the local gravitational strength g of a particle is the same as the magnitude of the gravitational acceleration (which we now call $|\vec{a}_g|$). Furthermore, we assumed that g has the constant value of 9.8 m/s^2 over Earth's surface. However, the g we would measure differs from the $|\vec{a}_g|$ we would calculate with Eq. 14-8 for three reasons: (1) Earth is not uniform, (2) it is not a perfect sphere, and (3) it rotates. Moreover, because g differs from $|\vec{a}_g|$, the measured weight mg of the particle differs from the magnitude of the gravitational force on the particle as given by Eq. 14-6 for the same three reasons. Let us now examine those reasons.

1. ***Earth is not uniform.*** The density (mass per unit volume) of Earth varies radially as shown in Fig. 14-4, and the density of the crust (or outer section) of Earth varies from region to region over Earth's surface. Thus, g varies from region to region over the surface.

TABLE 14-1: Variation of $|\vec{a}_g|$ with Altitude

Altitude (km)	$\|\vec{a}_g\|$ (m/s^2)	Altitude Example
0	9.83	Mean Earth surface
8.8	9.80	Mt. Everest
36.6	9.71	Highest manned balloon
400	8.70	Space shuttle orbit
35 700	0.225	Communications satellite

2. ***Earth is not a sphere.*** Earth is approximately an ellipsoid, flattened at the poles and bulging at the equator. Its equatorial radius is greater than its polar radius by 21 km. Thus, a point at the poles is closer to the dense core of Earth than is a point on the equator. This is one reason the free-fall acceleration g increases as one proceeds, at sea level, from the equator toward either pole.

3. ***Earth is rotating.*** The rotation axis runs through the north and south poles of Earth. An object located on Earth's surface anywhere except at those poles must rotate in a circle about the rotation axis and thus must have a centripetal acceleration directed toward the center of the circle. This centripetal acceleration requires a centripetal net force that is also directed toward that center.

To see how Earth's rotation causes the local gravitational strength at the Earth's surface, g, to differ from $|\vec{a}_g|$, let us analyze a simple situation in which a crate of mass m is on a scale at the equator. Figure 14-5*a* shows this situation as viewed from a point in space above the north pole.

Figure 14-5*b*, a free-body diagram for the crate, shows the two forces on the crate, both acting along a radial axis r that extends from Earth's center. The normal force $\vec{N}$ on the crate from the scale is directed outward, in the positive direction of axis r. The gravitational force, represented with its equivalent $m\vec{a}_g$, is directed inward. Because the crate travels in a circle about the center of Earth as Earth turns, it has a centripetal acceleration $\vec{a}$ directed inward. From Eq. 11-20, we know the magnitude of this acceleration is equal to $\omega^2 R$, where ω is Earth's angular speed and R is the circle's radius (approximately Earth's radius). Thus, we can write Newton's Second Law in component form for the r axis ($F_{\text{net},r} = ma_r$) as

$$\left|\vec{N} + m\vec{a}_g\right| = m(-\omega^2 R). \quad (14\text{-}9)$$

The magnitude $|\vec{N}|$ of the normal force is equal to the weight mg read on the scale. With mg substituted for $|\vec{N}|$ and with the fact that $\vec{N}$ and $\vec{a}_g$ point in opposite directions, Eq. 14-9 gives us

$$mg - m|\vec{a}_g| = -m(\omega^2 R), \quad (14\text{-}10)$$

or

$$mg = m|\vec{a}_g| - m(\omega^2 R),$$

which says

(measured weight) = (magnitude of gravitational force) – (mass times centripetal acceleration).

Thus, the measured weight is actually less than the magnitude of the gravitational force on the crate, because of Earth's rotation.

To find a corresponding expression for g and $|\vec{a}_g|$, we cancel m from Eq. 14-10 to write

$$g = |\vec{a}_g| - \omega^2 R, \quad (14\text{-}11)$$

which says

(free-fall acceleration) = (local gravitational strength) – (centripetal acceleration).

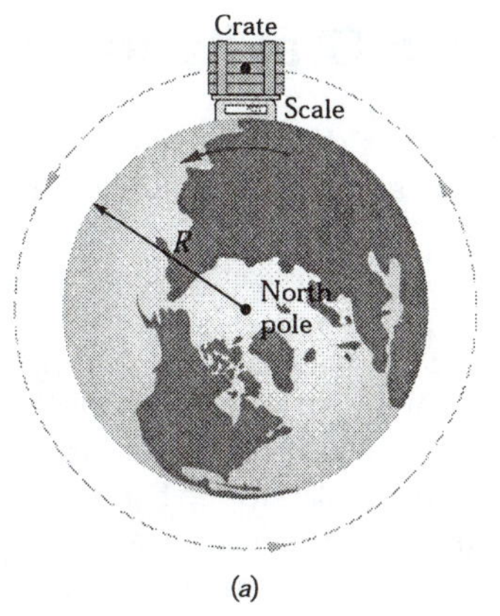

(*a*)

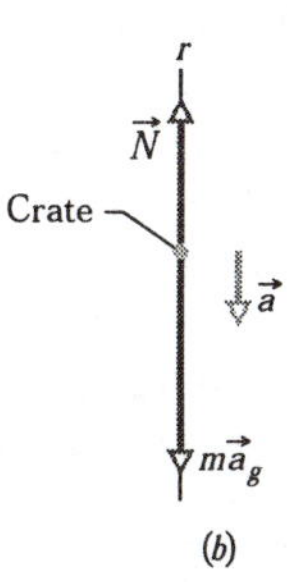

(*b*)

Fig. 14-5: (*a*) A crate lies on a scale at Earth's equator, as seen along Earth's rotation axis from above the north pole. (*b*) A free-body diagram for the crate, with a radially outward r axis. The gravitational force on the crate is represented with its equivalent $m\vec{a}_g$. The normal force on the crate from the scale is $\vec{N}$. Because of Earth's rotation, the crate has a centripetal acceleration $\vec{a}$ that is directed toward Earth's center.

Thus, the measured free-fall acceleration is actually less than the local gravitational strength, because of Earth's rotation.

The difference between the local gravitational strength g and the magnitude of the gravitational acceleration $|\vec{a}_g|$ is equal to $\omega^2 R$ and is greatest on the equator (for one reason, the radius of the circle traveled by the crate is greatest there). To find the difference, we can use Eq. 11-5 ($\omega = \Delta\theta/\Delta t$) and Earth's radius $R = 6.37 \times 10^6$ m. For one rotation of Earth, θ is 2π rad and the time period Δt is about 24 h. Using these values (and converting hours to seconds), we find that g is less than $|\vec{a}_g|$ by only about 0.034 m/s² (compared to 9.8 m/s²). Therefore, neglecting the difference in g and $|\vec{a}_g|$ is often justified. Similarly, neglecting the difference between weight and the magnitude of the gravitational acceleration is also often justified.

14-5 Gravitation Inside Earth

Newton's shell theorem can also be applied to a situation in which a particle is located *inside* a uniform shell, to show the following:

➤A uniform shell of matter exerts no *net* gravitational force on a particle located inside it.

Caution: This statement does *not* mean that the gravitational forces on the particle from the various elements of the shell magically disappear. Rather, it means that the *sum* of the force vectors on the particle from all the elements is zero.

If the density of Earth were uniform, the gravitational force acting on a particle would be a maximum at Earth's surface and would decrease as the particle moved outward. If the particle were to move inward, perhaps down a deep mine shaft, the gravitational force would change for two reasons. (1) It would tend to increase because the particle would be moving closer to the center of Earth. (2) It would tend to decrease because the thickening shell of material lying outside the particle's radial position would not exert any net force on the particle.

For a uniform Earth, the second influence would prevail and the force on the particle would steadily decrease to zero as the particle approached the center of Earth. However, for the real (nonuniform) Earth, the force on the particle actually increases as the particle begins to descend. The force reaches a maximum at a certain depth; only then does it begin to decrease as the particle descends farther.

Touchstone Example 14-5-1, at the end of this chapter, illustrates how to use what you learned in this section.

TE

14-6 Gravitational Potential Energy

In Section 10-3, we discussed the gravitational potential energy of a particle-Earth system. We were careful to keep the particle near Earth's surface, so that we could regard the gravitational force as constant. We then chose some reference configuration of the system as having a gravitational potential energy of zero. Often, in this configuration the particle was on Earth's surface. For particles not on Earth's surface, the gravitational potential energy decreased when the separation between the particle and Earth decreased.

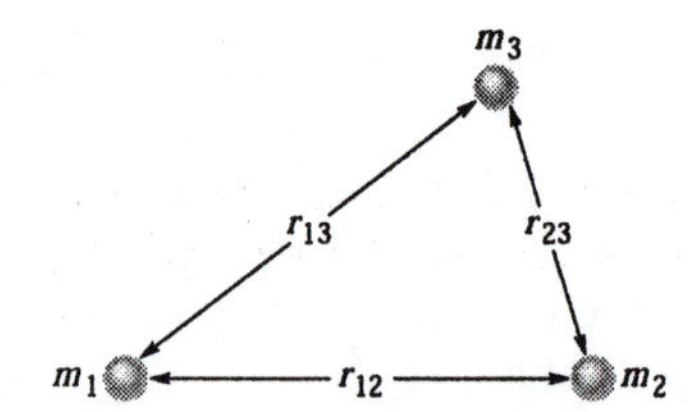

Fig. 14-6: Three particles form a system. (The separation for each pair of particles is labeled with a double subscript to indicate the particles.) The gravitational potential energy *of the system* is the sum of the gravitational potential energies of all three pairs of particles.

Here, we broaden our view and consider the gravitational potential energy U of two particles, of masses m and M, separated by a distance r. We again choose a reference configuration with U equal to zero. However, to simplify the equations, the separation distance r in the reference configuration is now large enough to be approximated as *infinite*. As before, the gravitational potential energy decreases when the separation decreases. Since $U = 0$ for $r = \infty$, the potential energy is negative for any finite separation and becomes progressively more negative as the particles move closer together.

With these facts in mind and as we shall justify next, we take the gravitational potential energy of the two-particle system to be

$$U = -\frac{GMm}{r} \quad \text{(gravitational potential energy).} \tag{14-12}$$

Note that $U(r)$ approaches zero as r approaches infinity and that for any finite value of r, the value of $U(r)$ is negative.

The potential energy given by Eq. 14-12 is a property of the system of two particles rather than of either particle alone. There is no way to divide this energy and say that so much belongs to one particle and so much to the other. However, if $M >> m$, as is true for Earth (mass M) and a baseball (mass m), we often speak of "the potential energy of the baseball." We can get away with this because, when a baseball moves in the vicinity of Earth, changes in the potential energy of the baseball-Earth system appear almost entirely as changes in the kinetic energy of the baseball, since changes in the kinetic energy of Earth are too small to be measured. Similarly, in Section 14-8 we shall speak of "the potential energy of an artificial satellite" orbiting Earth, because the satellite's mass is so much smaller than Earth's mass. When we speak of the potential energy of bodies of comparable mass, however, we have to be careful to treat them as a system.

If our system contains more than two particles, we consider each pair of particles in turn, calculate the gravitational potential energy of that pair with Eq. 14-12 as if the other particles were not there, and then algebraically sum the results. Applying Eq. 14-12 to each of the three pairs of Fig. 14-6, for example, gives the potential energy of the system as

$$U = -\left(\frac{Gm_1m_2}{r_{12}} + \frac{Gm_1m_3}{r_{13}} + \frac{Gm_2m_3}{r_{23}}\right). \tag{14-13}$$

Proof of Equation 14-12

Let us shoot a baseball directly away from Earth along the path shown in Fig. 14-7. We want to find an expression for the gravitational potential energy U of the ball at point P along its path, at radial distance R from Earth's center. To do so, we first find the work W done on the ball by the gravitational force as the ball travels from point P to a great (infinite) distance from Earth. Because the gravitational force $\vec{F}(r)$ is a variable force (its magnitude depends on r), we must use the techniques of Section 9-5 to find the work. In vector notation, we can write

$$W = \int_R^\infty \vec{F}(r) \cdot d\vec{r}. \tag{14-14}$$

The integral contains the scalar (or dot) product of the force $\vec{F}(r)$ and the differential displacement vector $d\vec{r}$ along the ball's path. We can expand that product as

$$\vec{F}(r) \cdot d\vec{r} = F(r)\, dr \cos\phi, \tag{14-15}$$

where ϕ is the angle between the directions of $\vec{F}(r)$ and $d\vec{r}$. When we substitute 180° for ϕ and Eq. 14-1 for $F(r)$, Eq. 14-15 becomes

$$\vec{F}(r) \cdot d\vec{r} = -\frac{GMm}{r^2} dr$$

where M is Earth's mass and m is the mass of the ball.

Substituting this into Eq. 14-14 and integrating gives us

$$\begin{aligned} W &= -GMm\int_R^\infty \frac{1}{r^2} dr = \left[\frac{GMm}{r}\right]_R^\infty \\ &= 0 - \frac{GMm}{R} = -\frac{GMm}{R}. \end{aligned} \tag{14-16}$$

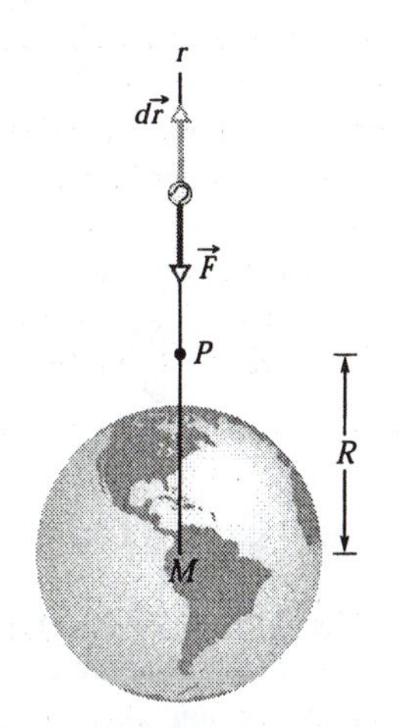

Fig. 14-7: A baseball is shot directly away from Earth, through point P at radial distance R from Earth's center. The gravitational force $\vec{F}$ and a differential displacement vector $d\vec{r}$ are shown, both directed along a radial r axis.

W in Eq. 14-16 is the work required to move the ball from point P (at distance R) to infinity. Equation 10-4 ($\Delta U = -W$) tells us that we can also write that work in terms of potential energies as

$$U_\infty - U = -W.$$

The potential energy U_∞ at infinity is zero, and U is the potential energy at P. Thus, with Eq. 14-16 substituted for W, the previous equation becomes

$$U = W = -\frac{GMm}{R}$$

Switching R to r gives us Eq. 14-12, which we set out to prove.

Path Independence

In Fig. 14-8, we move a baseball from point A to point G along a path consisting of three radial lengths and three circular arcs (centered on Earth). We are interested in the total work W done by Earth's gravitational force $\vec{F}$ on the ball as it moves from A to G. The work done along each circular arc is zero, because the direction of $\vec{F}$ is perpendicular to the arc at every point. Thus, the only works done by $\vec{F}$ are along the three radial lengths, and the total work W is the sum of those works.

Now, suppose we mentally shrink the arcs to zero. We would then be moving the ball directly from A to G along a single radial length. Does that change W? No. Because no work was done along the arcs, eliminating them does not change the work. The path taken from A to G now is clearly different, but the work done by $\vec{F}$ is the same.

We discussed such a result in a general way in Section 10-2. Here is the point: The gravitational force is a conservative force. Thus, the work done by the gravitational force on a particle moving from an initial point i to a final point f is independent of the actual path taken between the points. From Eq. 10-4, the change ΔU in the gravitational potential energy from point i to point f is given by

$$\Delta U = U_f - U_i = -W. \qquad (14\text{-}17)$$

Since the work W done by a conservative force is independent of the actual path taken, the change ΔU in gravitational potential energy is *also independent* of the actual path taken.

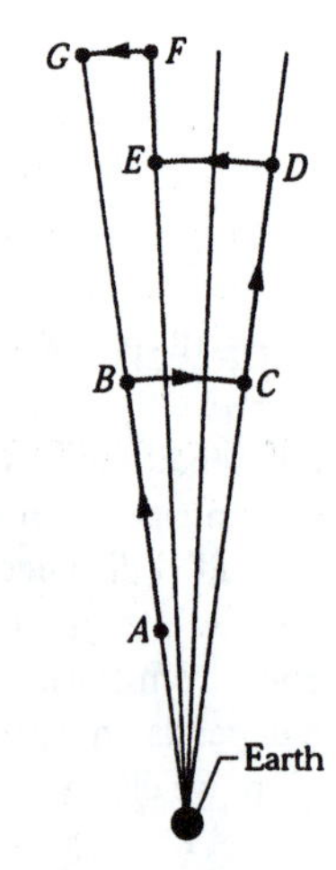

Fig. 14-8: Near Earth, a baseball is moved from point A to point G along a path consisting of radial lengths and circular arcs.

Potential Energy and Force

In the proof of Eq. 14-12, we derived the potential energy function $U(r)$ from the force function $\vec{F}(r)$. We should be able to go the other way—that is, to start from the potential energy function and derive the force function. Guided by Eq. 10-22, we can write the radial force component F_r as

$$\begin{aligned} F_r &= -\frac{dU}{dr} = -\frac{d}{dr}\left(-\frac{GMm}{r}\right) \\ &= -\frac{GMm}{r^2}. \end{aligned} \qquad (14\text{-}18)$$

This is Newton's law of gravitation (Eq. 14-1). The minus sign indicates that the force on mass m points radially inward, toward mass M.

Escape Speed

If you fire a projectile upward, usually it will slow, stop momentarily, and return to Earth. There is, however, a certain minimum initial speed that will cause it to move upward forever, theoretically coming to rest only at infinity. This initial speed is called the (Earth) **escape speed.**

Consider a projectile of mass m, leaving the surface of a planet (or some other astronomical body or system) with escape speed v. It has a kinetic energy K given by $\frac{1}{2}mv^2$ and a potential energy U given by Eq. 14-12:

$$U = -\frac{GMm}{R}$$

in which M is the mass of the planet, and R is its radius.

When the projectile reaches infinity, it stops and thus has no kinetic energy. It also has no potential energy because this is our zero-potential-energy configuration. Its total energy at infinity is therefore zero. From the principle of conservation of energy, its total energy at the planet's surface must also have been zero, so

$$K + U = \tfrac{1}{2}mv^2 + \left(-\frac{GMm}{R}\right) = 0$$

This yields

$$v = \sqrt{\frac{2GM}{R}}. \qquad (14\text{-}19)$$

The escape speed v does not depend on the direction in which a projectile is fired from a planet. However, attaining that speed is easier if the projectile is fired in the direction the launch site is moving as the planet rotates about its axis. For example, rockets are launched eastward at Cape Canaveral to take advantage of the Cape's eastward speed of 1500 km/h due to Earth's rotation.

Equation 14-19 can be applied to find the escape speed of a projectile from any astronomical body, provided we substitute the mass of the body for M and the radius of the body for R. Table 14-2 shows escape speeds from some astronomical bodies.

TABLE 14-2: Some Escape Speeds

Body	Mass (kg)	Radius (m)	Escape Speed (km/s)
Ceres[a]	1.17×10^{21}	3.8×10^{5}	0.64
Earth's moon	7.36×10^{22}	1.74×10^{6}	2.38
Earth	5.98×10^{24}	6.37×10^{6}	11.2
Jupiter	1.90×10^{27}	7.15×10^{7}	59.5
Sun	1.99×10^{30}	6.96×10^{8}	618
Sirius B[b]	2×10^{30}	1×10^{7}	5200
Neutron star[c]	2×10^{30}	1×10^{4}	2×10^{5}

[a]The most massive of the asteroids.
[b]A *white dwarf* (a star in a final stage of evolution) that is a companion of the bright star Sirius.
[c]The collapsed core of a star that remains after that star has exploded in a *supernova* event.

READING EXERCISE 14-2: You move a ball of mass m away from a sphere of mass M. (a) Does the gravitational potential energy of the ball-sphere system increase or decrease? (b) Is positive or negative work done by the gravitational force between the ball and the sphere?

Touchstone Example 14-6-1, at the end of this chapter, illustrates how to use what you learned in this section.

TE

14-7 Planets and Satellites: Kepler's Laws

The motions of the planets, as they seemingly wander against the background of the stars, have been a puzzle since the dawn of history. The "loop-the-loop" motion of Mars, shown in Fig. 14-9, was particularly baffling. Johannes Kepler (1571-1630), after a lifetime of study, worked out the empirical laws that govern these motions. Tycho Brahe (1546-1601), the last of the great astronomers to make observations without the help of a telescope, compiled the extensive data from which Kepler was able to derive the three laws of planetary motion that now bear his name. Later, Newton (1642-1727) showed that his law of gravitation leads to Kepler's laws.

In this section we discuss each of Kepler's laws in turn. Although here we apply the laws to planets orbiting the Sun, they hold equally well for satellites, either natural or artificial, orbiting Earth or any other massive central body.

Fig. 14-9: The path of the planet Mars as it moved against a background of the constellation Capricorn during 1971. Its position on four selected days is marked. Both Mars and Earth are moving in orbits around the Sun so that we see the position of Mars relative to us; this sometimes results in an apparent loop in the path of Mars.

➤**1. THE LAW OF ORBITS:** All planets move in elliptical orbits, with the Sun at one focus.

Figure 14-10 shows a planet of mass m moving in such an orbit around the Sun, whose mass is M. We assume that $M >> m$, so that the center of mass of the planet-Sun system is approximately at the center of the Sun.

The orbit in Fig. 14-10 is described by giving its **semimajor axis** a and its **eccentricity** e, the latter defined so that ea is the distance from the center of the ellipse to either focus f or f'. *An eccentricity of zero corresponds to a circle,* in which the two foci merge to a single central point. The eccentricities of the planetary orbits are not large, so—sketched on paper—the orbits look circular. The eccentricity of the ellipse of Fig. 14-10, which has been exaggerated for clarity, is 0.74. The eccentricity of Earth's orbit is only 0.0167.

➤**2. THE LAW OF AREAS:** A line that connects a planet to the Sun sweeps out equal areas in the plane of the planet's orbit in equal times; that is, the rate dA/dt at which it sweeps out area A is constant.

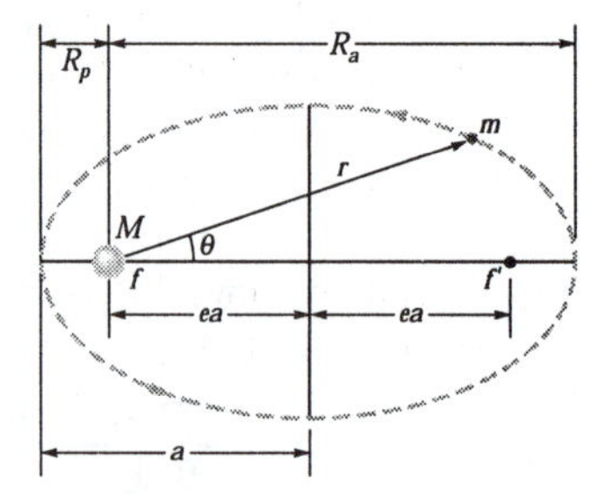

Fig. 14-10: A planet of mass m moving in an elliptical orbit around the Sun. The Sun, of mass M, is at one focus F of the ellipse. The other focus is F', which is located in empty space. Each focus is a distance ea from the ellipse's center, with e being the eccentricity of the ellipse. The semimajor axis a of the ellipse, the perihelion (nearest the Sun) distance R_p, and the aphelion (farthest from the Sun) distance R_a are also shown.

Qualitatively, this second law tells us that the planet will move most slowly when it is farthest from the Sun and most rapidly when it is nearest to the Sun. As it turns out, Kepler's second law is totally equivalent to the law of conservation of angular momentum. Let us prove it.

The area of the shaded wedge in Fig. 14-11*a* closely approximates the area swept out in time Δt by a line connecting the Sun and the planet, which are separated by a distance r. The area ΔA of the wedge is approximately the area of a triangle with base $r\,\Delta\theta$ and height r. Since the area of a triangle is one-half of the base times the height, $\Delta A >> \frac{1}{2}r^2\,\Delta\theta$. This expression for ΔA becomes more exact as Δt (hence $\Delta\theta$) approaches zero. The instantaneous rate at which area is being swept out is then

$$\frac{dA}{dt} = \frac{1}{2}r^2\frac{d\theta}{dt} = \frac{1}{2}r^2\omega, \tag{14-20}$$

in which ω is the angular speed of the rotating line connecting Sun and planet.

Figure 14-11*b* shows the linear momentum $\vec{p}$ of the planet, along with its radial and perpendicular components. From Eq. 12-23 ($|\vec{L}| = r|\vec{p}_\perp|$), the magnitude of the angular momentum $\vec{L}$ of the planet about the Sun is given by the product of r and $p_\perp$, the component of $\vec{p}$ perpendicular to r. Here, for a planet of mass m,

$$\begin{aligned}|\vec{L}| &= r|\vec{p}_\perp| = (r)(m|\vec{v}_\perp|) = (r)(m\omega r)\\ &= mr^2\omega,\end{aligned} \tag{14-21}$$

where we have replaced $v_\perp$ with its equivalent ωr (Eq. 11-15). Eliminating $r^2\omega$ between Eqs. 14-20 and 14-21 leads to

$$\frac{dA}{dt} = \frac{|\vec{L}|}{2m} \tag{14-22}$$

If dA/dt is constant, as Kepler said it is, then Eq. 14-22 means that $|\vec{L}|$ must also be constant—angular momentum is conserved. Kepler's second law is indeed equivalent to the law of conservation of angular momentum.

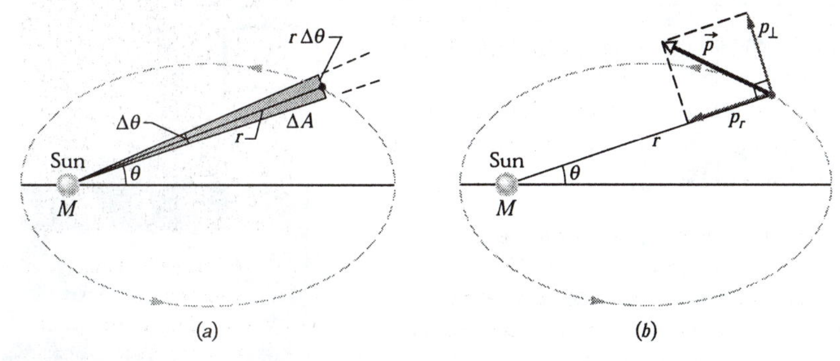

Fig. 14-11: (*a*) In time t, the line r connecting the planet to the Sun (of mass M) sweeps through an angle θ, sweeping out an area A (shaded). (*b*) The linear momentum $\vec{p}$ of the planet and its components.

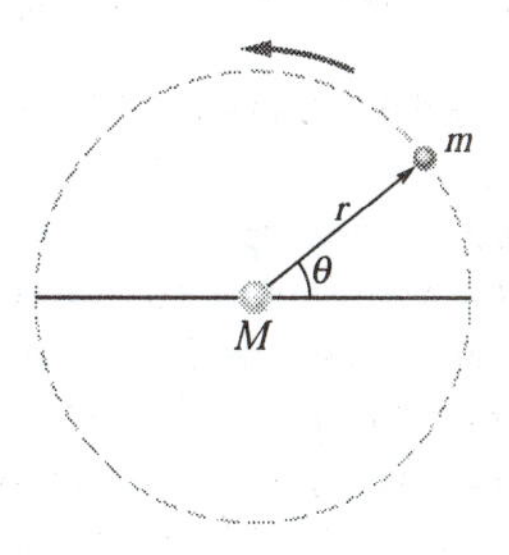

Fig. 14-12: A planet of mass m moving around the Sun in a circular orbit of radius r.

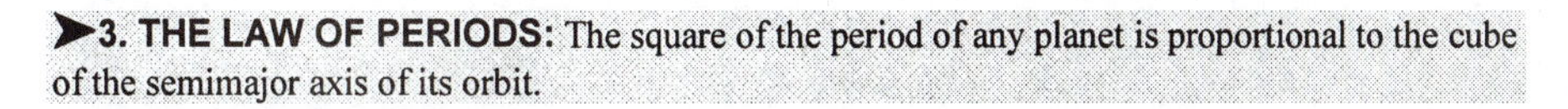

➤3. THE LAW OF PERIODS: The square of the period of any planet is proportional to the cube of the semimajor axis of its orbit.

To see this, consider the circular orbit of Fig. 14-12, with radius r (the radius of a circle is equivalent to the semimajor axis of an ellipse). Applying Newton's Second Law, $(|\vec{F}| = m|\vec{a}|)$, to the orbiting planet in Fig. 14-12 yields

$$\frac{GMm}{r^2} = (m)(\omega^2 r). \tag{14-23}$$

Here we have substituted from Eq.14-1 for the force magnitude $|\vec{F}|$ and used Eq. 11-20 to substitute $\omega^2 r$ for the centripetal acceleration. If we use Eq. 11-17 to replace ω with $2\pi/T$, where T is the period of the motion, we obtain Kepler's third law:

$$T^2 = \left(\frac{4\pi^2}{GM}\right) r^3 \quad \text{(law of periods).} \tag{14-24}$$

The quantity in parentheses is a constant that depends only on the mass M of the central body about which the planet orbits.

Equation 14-24 holds also for elliptical orbits, provided we replace r with a, the semimajor axis of the ellipse. This law predicts that the ratio T^2/a^3 has essentially the same value for every planetary orbit around a given massive body. Table 14-3 shows how well it holds for the orbits of the planets of the solar system.

TABLE 14-3:
Kepler's Law of Periods for the Solar System

Planet	Semimajor Axis a (10^{10} m)	Period T(y)	T^2/a^3 (10^{-34} y^2/m^3)
Mercury	5.79	0.241	2.99
Venus	10.8	0.615	3.00
Earth	15.0	1.00	2.96
Mars	22.8	1.88	2.98
Jupiter	77.8	11.9	3.01
Saturn	143	29.5	2.98
Uranus	287	84.0	2.98
Neptune	450	165	2.99
Pluto	590	248	2.99

READING EXERCISE 14-3: Satellite 1 is in a certain circular orbit about a planet, while satellite 2 is in a larger circular orbit. Which satellite has (a) the longer period and (b) the greater speed?

Touchstone Examples 14-7-1 and 14-7-2, at the end of this chapter, illustrate how to use what you learned in this section.

TE

14-8 Satellites: Orbits and Energy

On February 7, 1984, at a height of 102 km above Hawaii and with a speed of about 29,000 km/h, Bruce McCandless stepped (untethered) into space from a space shuttle and became the first human satellite.

As a satellite orbits Earth on its elliptical path, both its speed, which fixes its kinetic energy K, and its distance from the center of Earth, which fixes its gravitational potential energy U, fluctuate with fixed periods. However, the mechanical energy E of the satellite remains constant. (Since the satellite's mass is so much smaller than Earth's mass, we assign U and E for the Earth-satellite system to the satellite alone.)

The potential energy of the system is given by Eq. 14-12 and is

$$U = -\frac{GMm}{r}$$

(with $U = 0$ for infinite separation). Here r is the radius of the orbit, assumed for the time being to be circular, and M and m are the masses of Earth and the satellite, respectively.

To find the kinetic energy of a satellite in a circular orbit, we write Newton's second law, in terms of vector magnitudes as $(|\vec{F}| = m|\vec{a}|)$, as

$$\frac{GMm}{r^2} = m\frac{v^2}{r} \tag{14-25}$$

where v^2/r is the centripetal acceleration of the satellite. Then, from Eq. 14-25, the kinetic energy is

$$K = \tfrac{1}{2}mv^2 = \frac{GMm}{2r} \tag{14-26}$$

which shows us that for a satellite in a circular orbit,

$$K = -\frac{U}{2} \quad \text{(circular orbit).} \tag{14-27}$$

The total mechanical energy of the orbiting satellite is

$$E = K + U = \frac{GMm}{2r} - \frac{GMm}{r}$$

or

$$E = -\frac{GMm}{2r} \quad \text{(circular orbit).} \tag{14-28}$$

This tells us that for a satellite in a circular orbit, the total energy E is the negative of the kinetic energy K:

$$E = -K \quad \text{(circular orbit).} \tag{14-29}$$

For a satellite in an elliptical orbit of semimajor axis a, we can substitute a for r in Eq. 14-28 to find the mechanical energy as

$$E = -\frac{GMm}{2a} \quad \text{(elliptical orbit).} \tag{14-30}$$

Equation 14-30 tells us that the total energy of an orbiting satellite depends only on the semimajor axis of its orbit and not on its eccentricity e. For example, four orbits with the same semimajor axis are shown in Fig. 14-13; the same satellite would have the same total mechanical energy E in all four orbits. Figure 14-14 shows the variation of K, U, and E with r for a satellite moving in a circular orbit about a massive central body.

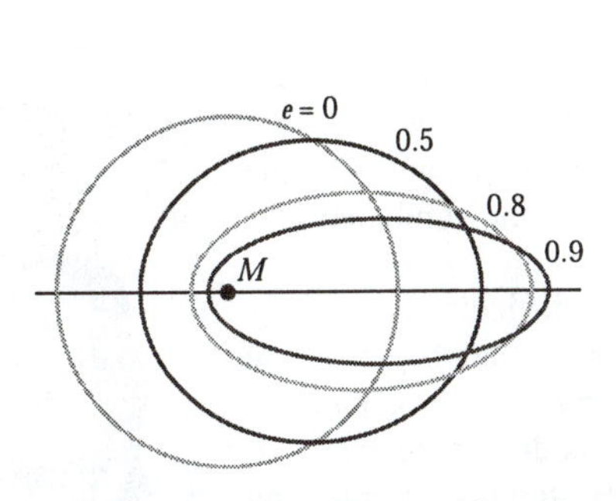

Fig. 14-13: Four orbits about an object of mass M. All four orbits have the same semimajor axis a and thus correspond to the same total mechanical energy E. Their eccentricities e are marked.

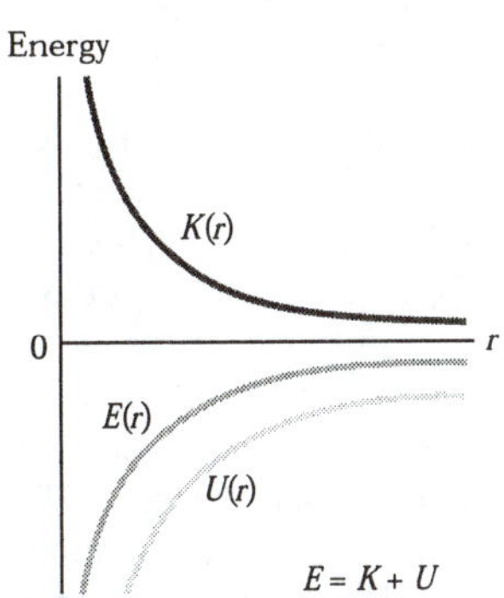

Fig. 14-14: The variation of kinetic energy K, potential energy U, and total energy E with radius r for a satellite in a circular orbit. For any value of r, the values of U and E are negative, the value of K is positive, and $E = -K$. As r approaches infinity, all three energy curves approach a value of zero.

READING EXERCISE 14-4: In the figure, a space shuttle is initially in a circular orbit of radius r about Earth. At point P, the pilot briefly fires a forward-pointing thruster to decrease the shuttle's kinetic energy K and mechanical energy E. (a) Which of the dashed elliptical orbits shown in the figure will the shuttle then take? (b) Is the orbital period T of the shuttle (the time to return to P) then greater than, less than, or the same as in the circular orbit?

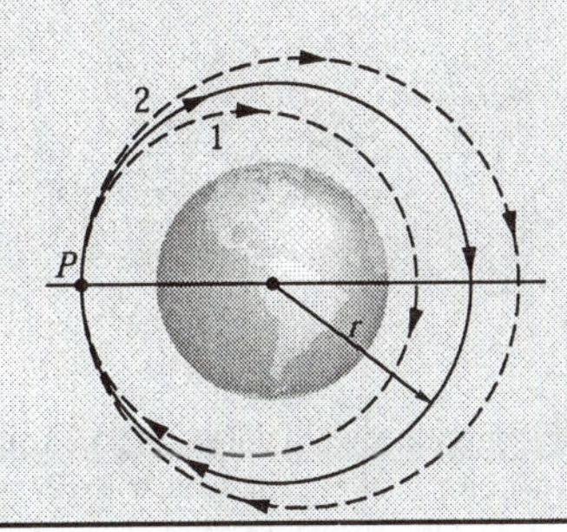

14-9 Einstein and Gravitation

Principle of Equivalence

Albert Einstein once said: "I was…in the patent office at Bern when all of a sudden a thought occurred to me: 'If a person falls freely, he will not feel his own weight.' I was startled. This simple thought made a deep impression on me. It impelled me toward a theory of gravitation."

Thus Einstein tells us how he began to form his **general theory of relativity.** The fundamental postulate of this theory about gravitation (the gravitating of objects toward each other) is called the **principle of equivalence,** which says that gravitation and acceleration are equivalent. If a physicist were locked up in a small box as in Fig. 14-15, he would not be able to tell whether the box was at rest on Earth (and subject only to Earth's gravitational force), as in Fig. 14-15*a*, or accelerating through interstellar space at

9.8 m/s² (and subject only to the force producing that acceleration), as in Fig. 14-15*b*. In both situations he would feel the same and would read the same value for his weight on a scale. Moreover, if he watched an object fall past him, the object would have the same acceleration relative to him in both situations.

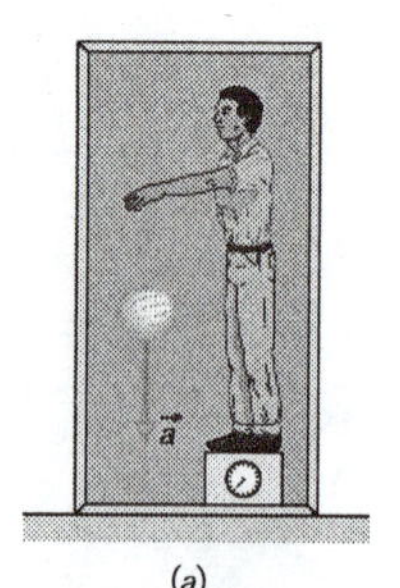

(*a*)

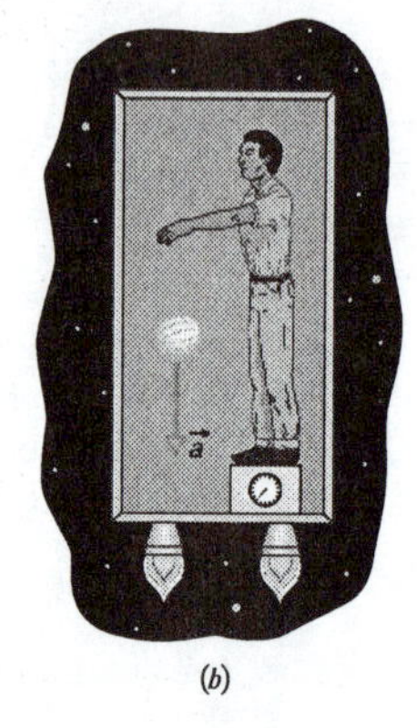

(*b*)

Fig. 14-15: (*a*) A physicist in a box resting on Earth sees a cantaloupe falling with acceleration $a = 9.8\ \mathrm{m/s^2}$. (*b*) If he and the box accelerate in deep space at $9.8\ \mathrm{m/s^2}$, the cantaloupe has the same acceleration relative to him. It is not possible, by doing experiments within the box, for the physicist to tell which situation he is in. For example, the platform scale on which he stands reads the same weight in both situations.

Curvature of Space

We have thus far explained gravitation as due to a force between masses. Einstein showed that, instead, gravitation is due to a curvature (or shape) of space that is caused by the masses. (As is discussed later in this book, space and time are entangled so the curvature of which Einstein spoke is really a curvature of *spacetime,* the combined four dimensions of our universe.)

Picturing how space (such as vacuum) can have curvature is difficult. An analogy might help: Suppose that from orbit we watch a race in which two boats begin on the equator with a separation of 20 km and head due south (Fig. 14-16*a*). To the sailors, the boats travel along flat, parallel paths. However, with time the boats draw together until, nearer the south pole, they touch. The sailors in the boats can interpret this drawing together in terms of a force acting on the boats. However, we can see that the boats draw together simply because of the curvature of Earth's surface. We can see this because we are viewing the race from "outside" that surface.

Figure 14-16*b* shows a similar race: Two horizontally separated apples are dropped from the same height above Earth. Although the apples may appear to travel along parallel paths, they actually move toward each other because they both fall toward Earth's center. We can interpret the motion of the apples in terms of the gravitational force on the apples from Earth. We can also interpret the motion in terms of a curvature of the space near Earth, due to the presence of Earth's mass. This time we cannot see the curvature because we cannot get "outside" the curved space, as we got "outside" the curved Earth in the boat example. However, we can depict the curvature with a drawing like Fig. 14-16*c*; there the apples would move along a surface that curves toward Earth because of Earth's mass.

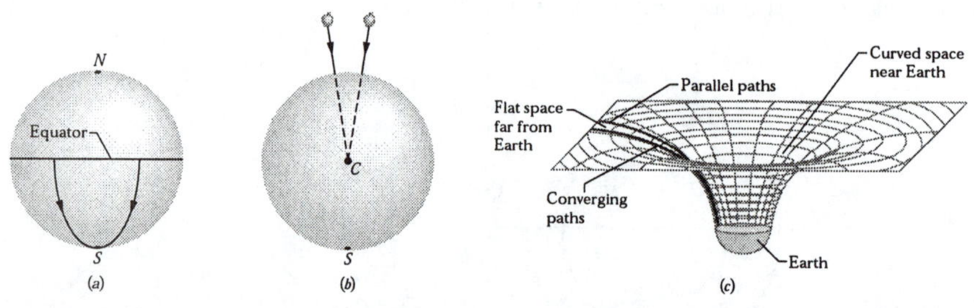

Fig. 14-16: (*a*) Two objects moving along lines of longitude toward the south pole converge because of the curvature of Earth's surface. (*b*) Two objects falling freely near Earth move along lines that converge toward the center of Earth because of the curvature of space near Earth. (*c*) Far from Earth (and other masses), space is flat and parallel paths remain parallel. Close to Earth, the parallel paths begin to converge because space is curved by Earth's mass.

When light passes near Earth, its path bends slightly because of the curvature of space there, an effect called *gravitational lensing.* When it passes a more massive structure, like a galaxy or a black hole having large mass, its path can be bent more. If such a massive structure is between us and a quasar (an extremely bright, extremely distant source of light), the light from the quasar can bend around the massive structure and toward us (Fig. 14-17*a*). Then, because the light seems to be coming to us from a number of slightly different directions in the sky, we see the same quasar in all those different directions. In some situations, the quasars we see blend together to form a giant luminous arc, which is called an *Einstein ring* (Fig. 14-17*b*).

Should we attribute gravitation to the curvature of spacetime due to the presence of masses or to a force between masses? Or should we attribute it to the actions of a type of fundamental particle called a *graviton,* as conjectured in some modern physics theories? We do not know.

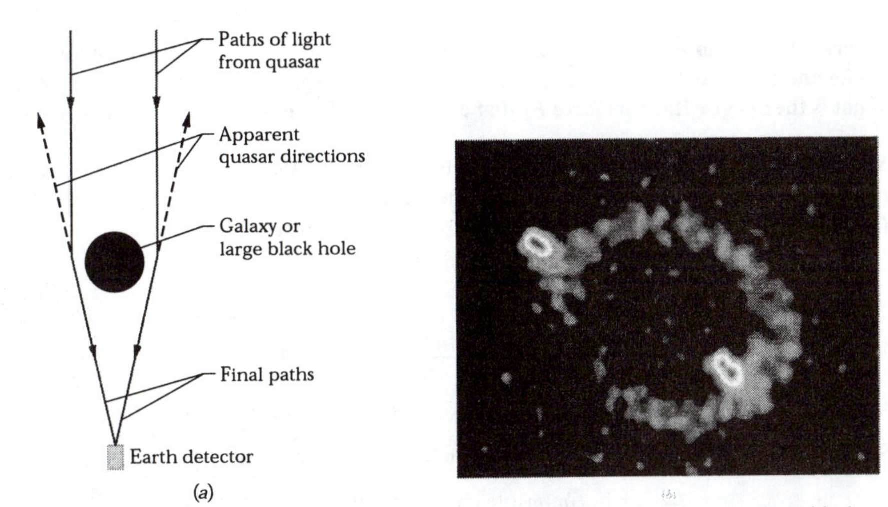

Fig. 14-17: (*a*) Light from a distant quasar follows curved paths around a galaxy or a large black hole because the mass of the galaxy or black hole has curved the adjacent space. If the light is detected, it appears to have originated along the backward extensions of the final paths (dashed lines). (*b*) The Einstein ring known as MG1131+0456 on the computer screen of a telescope. The source of the light (actually, radio waves, which are a form of invisible light) is far behind the large, unseen galaxy that produces the ring; a portion of the source appears as the two bright spots seen along the ring.

Touchstone Example 14-3-1

Figure TE14-1 shows an arrangement of three particles, particle 1 having mass $m_1 =$ 6.0 kg and particles 2 and 3 having mass $m_2 = m_3 = 4.0$ kg, and with distance $a = 2.0$ cm. What is the net gravitational force $\vec{F}_1$ that acts on particle 1 due to the other particles?

SOLUTION: One **Key Idea** here is that, because we have particles, the magnitude of the gravitational force on particle 1 due to either of the other particles is given by Eq. 14-1 ($|\vec{F}| = Gm_1m_2/r^2$). Thus, the magnitude of the force $\vec{F}_{2\to1}$ on particle 1 from particle 2 is

$$|\vec{F}_{2\to1}| = \frac{Gm_1m_2}{a^2}$$
$$= \frac{(6.67\times10^{-11}\ \mathrm{m^3/kg\cdot s^2})(6.0\ \mathrm{kg})(4.0\ \mathrm{kg})}{(0.020\ \mathrm{m})^2}$$
$$= 4.00\times10^{-6}\ \mathrm{N}.$$

Similarly, the magnitude of force $\vec{F}_{3\to1}$ on particle 1 from particle 3 is

$$|\vec{F}_{3\to1}| = \frac{Gm_1m_3}{(2a)^2}$$
$$= \frac{(6.67\times10^{-11}\ \mathrm{m^3/kg\cdot s^2})(6.0\ \mathrm{kg})(4.0\ \mathrm{kg})}{(0.040\ \mathrm{m})^2}$$
$$= 1.00\times10^{-6}\ \mathrm{N}.$$

To determine the directions of $\vec{F}_{2\to1}$ and $\vec{F}_{3\to1}$, we use this **Key Idea:** Each force on particle 1 is directed toward the particle responsible for that force. Thus, $\vec{F}_{2\to1}$ is directed in the positive direction of y (Fig. TE14-1*b*) and has only the y component $F_{2\to1}$. Similarly, $\vec{F}_{3\to1}$ is directed in the negative direction of x and has only the x component $-F_{3\to1}$.

To find the net force $\vec{F}_{1,\mathrm{net}}$ on particle 1, we first use this very important **Key Idea:** Because the forces are not directed along the same line, we *cannot* simply add or subtract their magnitudes or their components to get their net force. Instead, we must add them as vectors.

We can do so on a vector-capable calculator. However, here we note that $-F_{3\to1}$ and $F_{2\to1}$ are actually the x and y components of $\vec{F}_{1,\mathrm{net}}$. Therefore, we shall follow the guide of Eq. 4-6 to find first the magnitude and then the direction of $\vec{F}_{1,\mathrm{net}}$. The magnitude is

$$F_{1,\mathrm{net}} = \sqrt{(F_{2\to1})^2 + (-F_{3\to1})^2}$$
$$= \sqrt{(4.00\times10^{-6}\ \mathrm{N})^2 + (-1.00\times10^{-6}\ \mathrm{N})^2}$$
$$= 4.1\times10^{-6}\ \mathrm{N}. \qquad \text{(Answer)}$$

Relative to the positive direction of the x axis, Eq. 4-6 gives the direction of $\vec{F}_{1,\mathrm{net}}$ as

$$\theta = \tan^{-1}\frac{F_{2\to1}}{-F_{3\to1}} = \tan^{-1}\frac{4.00\times10^{-6}\ \mathrm{N}}{-1.00\times10^{-6}\ \mathrm{N}} = -76°.$$

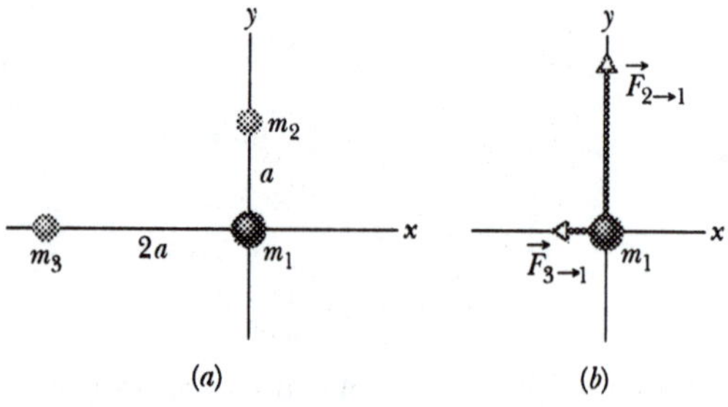

Fig. TE14-1 (*a*) An arrangement of three particles. (*b*) The forces acting on the particle of mass m_1 due to the other particles.

Is this a reasonable direction? No, the direction of $\vec{F}_{1,\text{net}}$ must be between the directions of $\vec{F}_{2\rightarrow 1}$ and $\vec{F}_{3\rightarrow 1}$. A calculator displays only one of the two possible answers to a $\tan^{-1}$ function. We find the other answer by adding 180°. That gives us

$$-76° + 180° = 104°, \qquad \text{(Answer)}$$

which *is* a reasonable direction for $\vec{F}_{1,\text{net}}$.

Touchstone Example 14-5-1

In *Pole to Pole,* an early science fiction story by George Griffith, three explorers attempt to travel by capsule through a naturally formed (and, of course, fictional) tunnel directly from the south pole to the north pole (Fig. TE14-2). According to the story, as the capsule approaches Earth's center, the gravitational force on the explorers becomes alarmingly large and then, exactly at the center, it suddenly but only momentarily disappears. Then the capsule travels through the second half of the tunnel, to the north pole.

Check Griffith's description by finding the gravitational force on the capsule of mass *m* when it reaches a distance *r* from Earth's center. Assume that Earth is a sphere of uniform density *r* (mass per unit volume).

SOLUTION: Newton's shell theorem gives us three **Key Ideas** here:

1. When the capsule is at a radius r from Earth's center, the portion of Earth that lies outside a sphere of radius r does *not* produce a net gravitational force on the capsule.
2. The portion that lies inside that sphere *does* produce a net gravitational force on the capsule.
3. We can treat the mass M_{ins} of that inside portion of Earth as being the mass of a particle located at Earth's center.

All three ideas tell us that we can write Eq. 14-1, for the magnitude of the gravitational force on the capsule, as

$$|\vec{F}| = \frac{GmM_{\text{ins}}}{r^2}. \qquad \text{(TE14-1)}$$

To write the mass M_{ins} in terms of the radius r, we note that the volume V_{ins} containing this mass is $\frac{4}{3}\pi r^3$. Also, its density is Earth's density ρ. Thus, we have

$$M_{\text{ins}} = \rho V_{\text{ins}} = \rho\frac{4\pi r^3}{3}. \qquad \text{(TE14-2)}$$

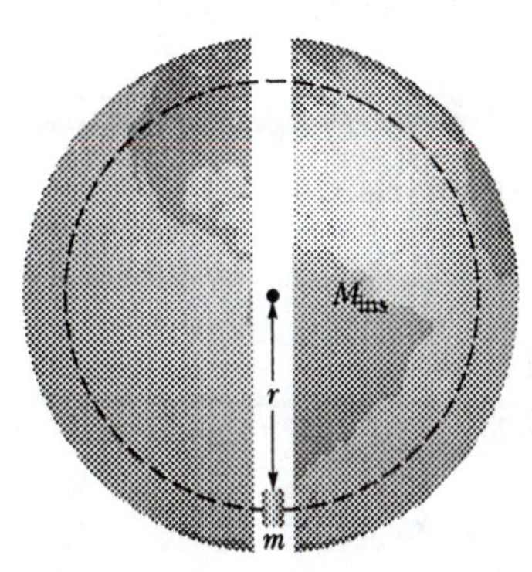

Fig. TE14-2 A capsule of mass m falls from rest through a tunnel that connects Earth's south and north poles. When the capsule is at distance r from Earth's center, the portion of Earth's mass that is contained in a sphere of that radius is M_{ins}.

Then, after substituting this expression into Eq. TE14-1 and canceling, we have

$$|\vec{F}| = \frac{4\pi Gm\rho}{3} r. \qquad \text{(Answer)} \quad \text{(TE14-3)}$$

This equation tells us that the force magnitude $|\vec{F}|$ depends linearly on the capsule's distance r from Earth's center. Thus, as r decreases, $|\vec{F}|$ also decreases (opposite of Griffith's description), until it is zero at Earth's center. At least Griffith got zero-at-the-center correct.

Equation TE14-3 can also be written in terms of the force vector $\vec{F}$ and the capsule's position vector $\vec{r}$ along a radial axis extending from Earth's center. Let K represent the collection of constants $4\pi Gm\rho/3$. Then, Eq. TE14-3 becomes

$$\vec{F} = -K\vec{r}, \qquad \text{(TE14-4)}$$

in which we have inserted a minus sign to indicate that $\vec{F}$ and $\vec{r}$ have opposite directions. Equation TE14-4 has the form of Hooke's law (Eq. 9-18). Thus, under the idealized conditions of the story, the capsule would oscillate like a block on a spring, with the center of the oscillation at Earth's center. After the capsule had fallen from the south pole to Earth's center, it would travel from the center to the north pole (as Griffith said) and then back again.

Touchstone Example 14-6-1

An asteroid, headed directly toward Earth, has a speed of 12 km/s relative to the planet when it is at a distance of 10 Earth radii from Earth's center. Neglecting the effects of Earth's atmosphere on the asteroid, find the asteroid's speed v_f when it reaches Earth's surface.

SOLUTION: One **Key Idea** is that, because we are to neglect the effects of the atmosphere on the asteroid, the mechanical energy of the asteroid–Earth system is conserved during the fall. Thus, the final mechanical energy (when the asteroid reaches Earth's surface) is equal to the initial mechanical energy. We can write this as

$$K_f + U_f = K_i + U_i, \qquad \text{(TE14-5)}$$

where K is kinetic energy and U is gravitational potential energy.

A second **Key Idea** is that, if we assume the system is isolated, the system's linear momentum must be conserved during the fall. Therefore, the momentum change of the asteroid and that of Earth must be equal in magnitude and opposite in sign. However, because Earth's mass is so great relative to the asteroid's mass, the change in Earth's speed is negligible relative to the change in the asteroid's speed. So, the change in Earth's kinetic energy is also negligible. Thus, we can assume that the kinetic energies in Eq. TE14-5 are those of the asteroid alone.

Let m represent the asteroid's mass and M represent Earth's mass (5.98×10^{24} kg). The asteroid is initially at the distance $10R_E$ and finally at the distance R_E, where R_E is Earth's radius (6.37×10^6 m). Substituting Eq. 14-12 for U and $\frac{1}{2}mv^2$ for K, we rewrite Eq. TE14-5 as

$$\tfrac{1}{2}mv_f^2 - \frac{GMm}{R_E} = \tfrac{1}{2}mv_i^2 - \frac{GMm}{10R_E}.$$

Rearranging and substituting known values, we find

$$\begin{aligned} v_f^2 &= v_i^2 + \frac{2GM}{R_E}\left(1 - \frac{1}{10}\right) \\ &= (12 \times 10^3 \text{ m/s})^2 \\ &\quad + \frac{2(6.67 \times 10^{-11} \text{ m}^3/\text{kg}\cdot\text{s}^2)(5.98 \times 10^{24} \text{ kg})}{6.37 \times 10^6 \text{ m}} 0.9 \\ &= 2.567 \times 10^8 \text{ m}^2/\text{s}^2, \end{aligned}$$

and thus

$$|\vec{v}_f| = 1.60 \times 10^4 \text{ m/s} = 16 \text{ km/s}. \qquad \text{(Answer)}$$

At this speed, the asteroid would not have to be particularly large to do considerable damage at impact. As an example, if it were only 5 m across, the impact could release about as much energy as the nuclear explosion at Hiroshima. Alarmingly, about 500 million asteroids of this size are near Earth's orbit, and in 1994 one of them apparently penetrated Earth's atmosphere and exploded at an altitude of 20 km near a remote South Pacific island (setting off nuclear-explosion warnings on six military satellites). The impact of an asteroid 500 m across (there may be a million of them near Earth's orbit) could end modern civilization and almost eliminate humans worldwide.

Touchstone Example 14-7-1

Comet Halley orbits about the Sun with a period of 76 years and, in 1986, had a distance of closest approach to the Sun, its *perihelion distance* R_p, of 8.9×10^{10} m. Table 14-3 shows that this is between the orbits of Mercury and Venus.

(a) What is the comet's farthest distance from the Sun, its *aphelion distance* R_a?

SOLUTION: One **Key Idea** comes from Fig. 14-10, in which we see that $R_a + R_p = 2a$, where a is the semimajor axis of the orbit of comet Halley. Thus, we can find R_a if we first find a. A second **Key Idea** is that we can relate a to the given period via the law of periods (Eq. 14-24) if we simply substitute the semimajor axis a for r. Doing so and then solving for a, we have

$$a = \left(\frac{GMT^2}{4\pi^2}\right)^{1/3}. \tag{TE14-6}$$

If we substitute the mass M of the Sun, 1.99×10^{30} kg, and the period T of the comet, 76 years or 2.4×10^9 s, into Eq. TE14-6, we find that $a = 2.7 \times 10^{12}$ m. Now we have

$$\begin{aligned} R_a &= 2a - R_p \\ &= (2)(2.7 \times 10^{12}\text{ m}) - 8.9 \times 10^{10}\text{ m} \\ &= 5.3 \times 10^{12}\text{ m}. \end{aligned} \quad \text{(Answer)}$$

Table 14-3 shows that this is a little less than the semimajor axis of the orbit of Pluto. Thus, the comet does not get farther from the Sun than Pluto.

(b) What is the eccentricity e of the orbit of comet Halley?

SOLUTION: The **Key Idea** here is that we can relate e, a, and R_p via Fig. 14-10. We see there that $ea = a - R_p$, or

$$\begin{aligned} e &= \frac{a - R_p}{a} = 1 - \frac{R_p}{a} \\ &= 1 - \frac{8.9 \times 10^{10}\text{ m}}{2.7 \times 10^{12}\text{ m}} = 0.97. \end{aligned} \quad \text{(Answer)}$$

This cometary orbit, with an eccentricity approaching unity, is a long thin ellipse.

Touchstone Example 14-7-2

***Hunting a black hole.* Observations of the light from a certain star indicate that it is part of a binary (two-star) system. This visible star has orbital speed $|\vec{v}| = 270$ km/s, orbital period $T = 1.70$ days, and approximate mass $m_1 = 6M_s$, where M_s is the Sun's mass,**

1.99×10^{30} kg. Assuming that the visible star and its companion star, which is dark and unseen, are both in circular orbits (see Fig. TE14-3), determine the approximate mass m_2 of the dark star.

SOLUTION: Some of the **Key Ideas** in this challenging problem are as follows:

1. The two stars are in circular orbits, not about each other, but about the center of mass of this two-star system.
2. As with two-particle systems of Section 8-2, the center of mass of the two-star system must lie along a line connecting the centers of the stars—that is, at point O in Fig. TE14-3. The visible star orbits at radius r_1, the dark star at radius r_2.
3. The center of mass of the system is not even approximately at the center of a central, massive object (like the Sun). Therefore, Kepler's law of periods, Eq. 14-24, does *not* apply here and we cannot easily find mass m_2 with it.
4. The centripetal force causing each star to move in a circle is the gravitational force due to the other star. The magnitude of the force is Gm_1m_2/r^2, where r is the distance between the centers of the stars.
5. From Eq. 5-19, the magnitude of centripetal acceleration $|\vec{a}_c|$ of the visible star is v^2/r_1.

These ideas lead us to write Newton's second law ($|\vec{F}| = m|\vec{a}|$) for the visible star as

$$\frac{Gm_1m_2}{r^2} = m_1\frac{v^2}{r_1}. \qquad \text{(TE14-7)}$$

This equation contains the required mass m_2, but to find it we need expressions for r and r_1. (Note that m_1 cancels out.)

We start by locating the center of mass relative to the visible star, using Eq. 8-8. That star is at distance zero relative to itself, the center of mass is at distance r_1, and the dark star is at distance r. Equation 8-8 then becomes

$$r_1 = \frac{m_1(0) + m_2r}{m_1 + m_2}, \qquad \text{(TE14-8)}$$

which yields

$$r = r_1\frac{m_1 + m_2}{m_2}. \qquad \text{(TE14-9)}$$

To find an expression for r_1, we note that the visible star is moving in a circle of radius r_1, at speed $|\vec{v}|$, and with period T. Thus, from Eq. 5-20, we have $|\vec{v}| = 2\pi r_1/T$ or

$$r_1 = \frac{|\vec{v}|T}{2\pi}. \qquad \text{(TE14-10)}$$

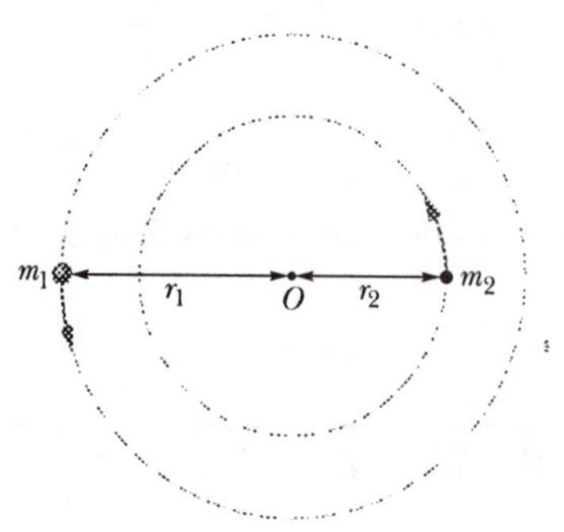

Fig. TE14-3 A visible star with mass m_1 and a dark, unseen star with mass m_2 orbit around the center of mass of the two-star system at O.

Substituting this for r_1 in Eq. TE14-9 results in

$$r = \frac{|\vec{v}|T}{2\pi} \frac{m_1 + m_2}{m_2}. \tag{TE14-11}$$

Now we return to Eq. TE14-7 and substitute for r with Eq. TE14-11, for r_1 with Eq. TE14-10, and for m_1 with the given $6M_s$. Rearranging the result and substituting known data then give us

$$\begin{aligned} \frac{m_2^3}{(6M_s + m_2)^2} &= \frac{|\vec{v}^{\,3}|T}{2\pi G} \\ &= \frac{(2.7 \times 10^5 \text{ m/s})^3 (1.70 \text{ days})(86\,400 \text{ s/day})}{(2\pi)(6.67 \times 10^{-11} \text{ N}\cdot\text{m}^2/\text{kg}^2)} \\ &= 6.90 \times 10^{30} \text{ kg}, \end{aligned}$$

or

$$\frac{m_2^3}{(6M_s + m_2)^2} = 3.47 M_s. \tag{TE14-12}$$

We can solve this cubic equation for m_2 with a polynomial solver on a calculator. Instead, since we are working with approximate masses anyway, we can substitute integer multiples of M_s for m_2 until we find one that makes Eq. TE14-12 nearly true. This occurs for

$$m_2 \approx 9M_s. \qquad \text{(Answer)}$$

The data here approximate those for the binary system LMC X-3 in the Large Magellanic Cloud (shown in the figure that begins this chapter). From other data, the dark object is known to be especially compact: It may be a star that collapsed under its own gravitational pull to become a neutron star or a black hole. Since a neutron star cannot have a mass larger than about $2M_s$, the result $m_2 \approx 9M_s$ strongly suggests that the dark object is a black hole.

Thus, we can detect the presence of a black hole provided it is part of a binary system with a visible star whose mass, orbital speed, and orbital period can be measured.

15 Fluids

The force exerted by water on the body of a descending diver increases noticeably, even for a relatively shallow descent to the bottom of a swimming pool. However, in 1975, using scuba gear with a special gas mixture for breathing, William Rhodes emerged from a chamber that had been lowered 300 m into the Gulf of Mexico, and he then swam to a record depth of 350 m. Strangely, a novice scuba diver practicing in a swimming pool might be in more danger from the force exerted by the water than was Rhodes. Occasionally, novice scuba divers die because they have neglected that danger.

What is this potentially lethal risk?

The answer is in this chapter.

15-1 Fluids And The World Around Us

Fluids—which include both liquids and gases—play a central role in our daily lives. We breathe and drink them, and a rather vital fluid circulates in the human cardiovascular system. The Earth's oceans and atmosphere consist of fluids.

Cars and jet planes need many different fluids to operate including fluids in their tires, fuel tanks, and engine combustion chambers. They also need fluids for their air conditioning, lubrication and hydraulic systems. Windmills transform the kinetic energy in air to electrical energy while hydroelectric plants convert the gravitational potential energy in water to electrical energy. Over long time periods air and water carve out and reshape the Earth's landscape.

In our study of fluids, we will start by examining simple physical situations that we encounter in everyday experiences. First, we will study the forces that act on fluids that are in static equilibrium and consider the buoyant forces on objects in fluids. Then we will examine how a hydraulic system can be used as a lever. Later in the chapter, we will study the motions of ideal fluids as they flow through pipes and around objects.

15-2 What is a Fluid?

In order to understand what we mean by the term "fluid," let us consider our everyday experiences at the Earth's surface to compare solids, liquids and gases. A solid vertical column that rests on a table can retain its shape without external support. Since a gravitational force is acting on each of the columns shown in Fig. 15-1, each exerts a downward normal force on the table that is equal in magnitude to its weight. What happens if we try to make a column out of a liquid? Without external support the gravitational forces on the liquid will cause it to collapse and flow into a puddle. (In more formal terms, liquids cannot withstand shear stresses, but solids can.)

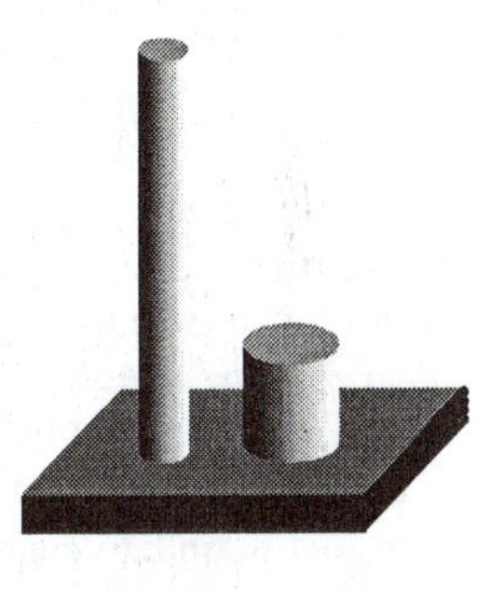

Fig. 15-1: Two columns resting on a table have the same mass, so they each exert the same downward normal force on the table.

However, we can maintain a vertical column of liquid if we provide it with solid walls. In this case, the liquid presses sideways against the walls and the walls press back against the liquid. Thus, the vertical columns of the liquid and solid differ in that the column of liquid needs external forces acting on it to maintain its shape while the solid does not. However, both a solid column and a container full of liquid will exert normal forces on a table.

When external forces are present, a **fluid**, unlike a solid, can flow until it conforms to the boundaries of any container that we put it in. Obviously gases, such as the air that surrounds us, are also fluids as they can conform to the shape of a container quite rapidly. Some gooey materials, such as pitch, heavy syrup, and silly putty take a longer time to conform to the boundaries of a container. But since they can do so eventually we also classify them as fluids.

You may wonder why we classify both liquids and gases as fluids. After all, the liquid form of water is quite different from both steam and ice. However, ice cannot conform itself to the boundaries of a container. Ice and other solids have their constituent atoms or molecules organized in an orderly and fairly rigid three-dimensional array.

15-3 Pressure and Density

Defining Pressure for Uniform Forces

Let us consider the properties of the two solid columns shown in Fig. 15-1. Since they both have the same weight, they exert the same downward forces on the table. However, if you placed your hand under each of the columns, you would *feel* a difference. Why? Because the forces are spread out over different size areas. The difference in sensory perceptions when one places one's hand under a narrow versus wide column of equal mass is a detected difference in *pressure*. If a force is evenly distributed over every point of an area, (as is the case for the normal forces exerted by the cylinders), we say that it is *uniform* over the area. For a force that is both uniformly distributed over an area and perpendicular or normal to it, the **pressure** P on a surface is defined as the magnitude of the net force acting on the surface divided by its area. Thus, it can be expressed by the equation

$$P \equiv \frac{|\vec{F}_{\perp}|}{A} \qquad \text{(uniform forces normal to area } A\text{),} \qquad (15\text{-}1)$$

where, as usual, the symbol ≡ is used to signify that the equation holds "by definition." Using this definition it is clear that the column on the left in Fig. 15-1 has one fourth the area that the right hand column does, so it exerts four times more pressure on the table. Later on in this section we will refine our discussion of pressure to handle situations in which the normal forces acting on a surface are not uniform.

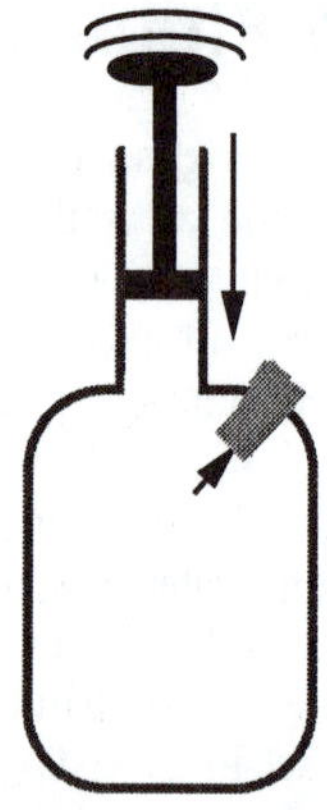

Fig. 15-2: Suppose a hole is drilled at some random place on a bottle and plugged with a cork. If an airtight plunger is thrust down the bottle's neck, the increased pressure in the bottle can cause the cork to pop out no matter what direction it faces. This can happen whenever the bottle contains either a gas (such as air) or a liquid (such as water).

Is Pressure a Vector or Scalar?

In order to think about the idea of pressure exerted by a fluid, let us do a thought experiment. Consider a bottle full of a fluid that has a piston on top. Although there is a hole in the bottle, it is plugged with a cork as shown in Fig. 15-2. If we press on the piston, the cork will pop out. This indicates that the fluid exerts a perpendicular force on the face of the cork that is sticking into the bottle. What's remarkable is that no matter where the hole and cork are located on the bottle, the cork would still pop out! Somehow, the downward force we apply with the piston to one part of the fluid is translated into "internal forces" that act in all directions. Thus, *the fluid pressure acting at the surface of a container appears to have no preferred direction.*

Let us consider a fluid that is not moving so that we can define it as being in a state of static equilibrium. What is the pressure like inside the fluid? Does it still have "internal forces" that act normal to elements of surface areas we insert in the fluid? Does pressure have a preferred direction inside the fluid? We can consider these questions both experimentally and theoretically.

Experimental Results: If we want to measure the pressure exerted by a fluid at a point inside a container of fluid we can design a very simple little pressure sensor like that shown in Fig. 15-3*a*. The sensor consists of a piston having a small area ΔA that fits snugly in a cylinder. The cylinder is evacuated so it contains no matter other than a coiled spring that is lodged behind the piston. By measuring the spring compression we can determine magnitude of the normal force the fluid exerts on the piston and then use Eq. 15-1 to calculate the fluid pressure from our measurements. If we place the center of the piston at the point of interest in a fluid to measure its pressure, we find that the magnitude of force on the piston is the same no matter how we orient the sensor. An array of tiny pressure sensors having different orientations is shown in Fig. 15-3*b*. Hence, if we do get the reported results for our experiment, we must conclude that:

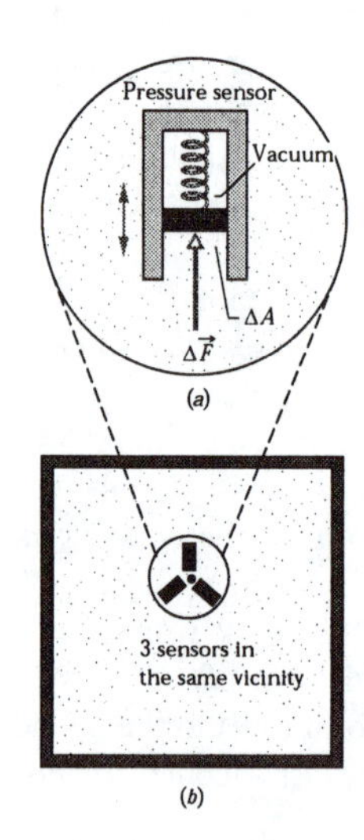

Fig. 15-3: (*a*) A tiny pressure sensor that uses spring compression to measure the net force normal to the area, ΔA, of a piston. (*b*) When an array of pressure sensors pointing in different directions are placed in the vicinity of a single point in a fluid their pressure measurements are identical.

➤Experiments reveal that at a given point in a fluid that is in static equilibrium, the pressure P has the same value in all directions. In other words, pressure is a scalar, having no directional properties.

Agreement between Experiment and Theory: We can use the fact that we have chosen to examine a fluid that is in static equilibrium to see why we should indeed expect the pressure near a point in the fluid to be non-directional. Let us start by transporting a container of fluid to a location in outer space where there are essentially no gravitational forces present. Next we can draw an imaginary cubical boundary around a tiny parcel of fluid (Fig. 15-4) centered on some point in the container. Since the parcel of fluid is in equilibrium it cannot be accelerating and we must conclude that the net force on its boundaries is zero as shown in Fig. 15-4. Since the force vectors on opposite faces of the cubical parcel must be equal in magnitude and opposite in direction, the pressure on opposite faces must be the same. Furthermore, there are no gravitational forces acting in our special case and thus there is no preferred direction. These facts allow us to conclude that if no part of the fluid is accelerating, the pressure must be the same in all directions throughout the entire container. So far we have considered a very special shape for our parcel in the absence of gravitational forces. In Section 15-4 we consider what happens to the pressure in fluids in static equilibrium close to the Earth's surface or at other locations where gravitational forces must be taken into account.

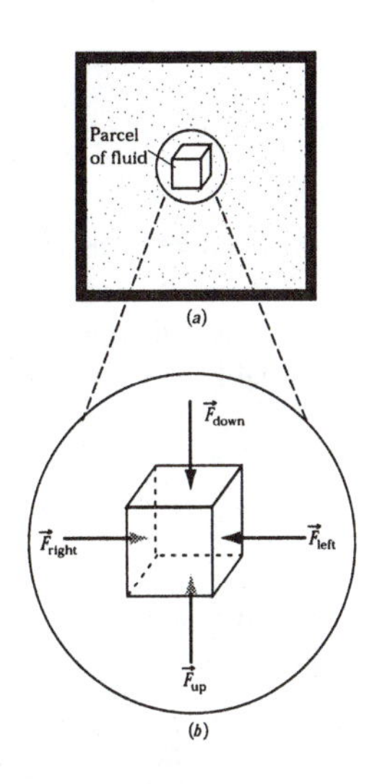

Fig. 15-4: (*a*) A tiny parcel of fluid in in static equilibrium in a container in outer space. (*b*) Since the parcel does not accelerate, the net force on it due to the surrounding fluid must be zero, so the pressure on the parcel is the same in all directions. For clarity, force vectors on the front and back parcel faces are not shown.

Defining Pressure for Non-Uniform Forces and Surfaces

If the forces on an area are not uniform or if our area is curved, we can still use our basic definition of pressure by breaking our area A into segments. Our area segments must be small enough so that the normal forces acting on each segment are uniform and each area segment is essentially flat (Fig. 15-5). If we do this, the pressure at the location of the ith segment of area can be defined as

$$P_i \equiv \lim_{\Delta A_i \to 0} \frac{|\vec{F}_i|}{\Delta A_i} \quad \text{(pressure at a point, non-uniform forces).} \quad (15\text{-}2)$$

where $|\vec{F}_i|$ is the magnitude of the net force normal to the ith area. That is, the pressure at any point is the limit of this ratio as the area ΔA_i centered on that point, is made smaller and smaller. Obviously, the net force acting on smaller areas will be smaller so the ratio is still physically meaningful.

The SI unit of pressure is the newton per square meter, which is given a special name, the **pascal** (Pa). In metric countries, tire pressure gauges are calibrated in kilopascals. The pascal is related to some other common (non-SI) pressure units as follows:

$$1 \text{ atm} = 1.01 \times 10^5 \text{ Pa} = 760 \text{ torr} = 14.7 \text{ lb/in}^2.$$

The *atmosphere* (atm) is, as the name suggests, the approximate average pressure of the atmosphere at sea level. The *torr* (named for Evangelista Torricelli, who invented the mercury barometer in 1674) was formerly called the *millimeter of mercury* (mm Hg). The pound per square inch is often abbreviated psi. Table 15-1 shows some pressures.

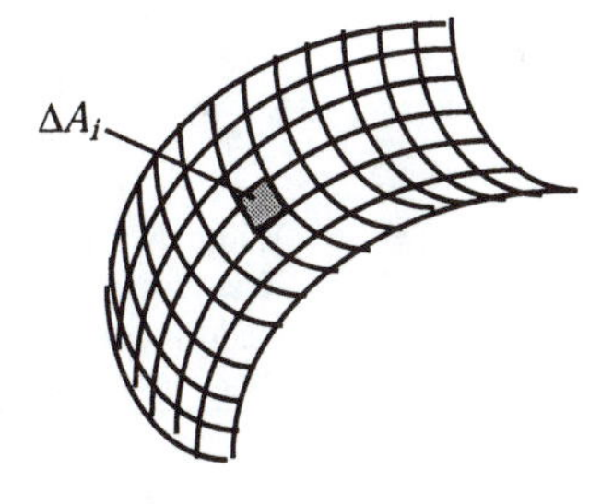

Fig. 15-5: In order to calculate the pressure exerted by fluids on curved surfaces or surfaces that have non-uniform forces acting on them, a surface area must be divided into a large number of small area elements, $\Delta A_1, \Delta A_2$, and so on. The ith area is shown in the diagram above.

Density

Let us return one more time to the column of solid we discussed above. Clearly, the weight of a column of a given size and height depends on what the column is made of. If a certain column was constructed of a material like Styrofoam® it would be much lighter than if it were made of lead. Hence, it would be convenient to have a way to predict the weight of an object that has a certain size and shape.

For this purpose, we invent a new quantity that is a measure of the mass of one cubic centimeter of a material. To determine this value for a given substance, we measure the total mass M in a measured volume V of the material and calculate M/V. This quantity is called *density*. In general, density is a measure of mass per unit of volume. The standard symbol for density is the Greek letter rho ρ.

Table 15-2 shows the densities of several substances and the average densities of some objects. Note that the density of a gas (see Air in the table) varies considerably with pressure, but the density of a liquid (see Water) does not; that is, gases are readily *compressible* but liquids are not.

The density of a fluid is not always uniform. For example, the density of the gas molecules and other particles that make up the Earth's atmosphere is much greater close to the surface of the Earth than in the stratosphere. As is the case for pressure, we can find the density ρ of any fluid at point i, if we isolate a small volume element ΔV_i around that point and measure the mass Δm_i of the fluid contained within that element. The **density** is then

$$\rho_i = \frac{\Delta m_i}{\Delta V_i} \quad \text{(density in the vicinity of a point } i\text{).} \qquad (15\text{-}3)$$

In theory, the density at any point in a fluid is the limit of this ratio as the volume element ΔV at that point is made smaller and smaller. In practice, many fluid samples are large compared to atomic dimensions and are thus "smooth" rather than "lumpy" with atoms. If it is reasonable to assume further that the sample has a uniform density, we can simplify Eq. 15-3 to

$$\rho = \frac{m}{V} \quad \text{(uniform density),} \qquad (15\text{-}4)$$

where m and V are the mass and volume of the sample. Density is a scalar property; its SI unit is the kilogram per cubic meter.

TABLE 15-1: Some Pressures

	Pressures (Pa)
Center of the Sun	2×10^{16}
Center of Earth	4×10^{11}
Highest sustained laboratory pressure	1.5×10^{10}
Deepest ocean trench (bottom)	1.1×10^{8}
Spike heels on a dance floor	1×10^{6}
Automobile tire[a]	2×10^{5}
Atmosphere at sea level	1.0×10^{5}
Normal blood pressure[a,b]	1.6×10^{4}
Best laboratory vacuum	10^{-12}

[a]Pressure in excess of atmospheric

[b]The systolic pressure (120 torr on a physician's pressure gauge).

TABLE 15-2: Some Densities

Material or Object	Density (kg/m³)
Stray atoms in interstellar space	10^{-20}
Air remaining in the best laboratory vacuum	10^{-17}
Air: 20°C and 1 atm pressure	1.21
20°C and 50 atm pressure	60.5
Styrofoam	1×10^{2}
Ice	0.917×10^{3}
Water: 20°C and 1 atm	0.998×10^{3}
20°C and 50 atm	1.000×10^{3}
Seawater: 20°C and 1 atm	1.024×10^{3}
Whole blood	1.060×10^{3}
Iron	7.9×10^{3}
Mercury (the metal)	13.6×10^{3}
Earth: average	5.5×10^{3}
core	9.5×10^{3}
crust	2.8×10^{3}
Sun: average	1.4×10^{3}
core	1.6×10^{5}
White dwarf star (core)	10^{10}
Uranium nucleus	3×10^{17}
Neutron star (core)	10^{18}
Black hole (1 solar mass)	10^{19}

Density and pressure are fundamental concepts in regard to fluids. When we discuss solids, we are concerned with particular lumps of matter, such as wooden blocks, baseballs, or metal rods. Physical quantities that we find useful, and in whose terms we express Newton's laws, are *mass* and *force*. We might speak, for example, of a 3.6 kg block acted on by a 25 N force. With fluids, we are more interested in the extended substance, and in properties that can vary from point to point in that substance. In these cases, it is more useful to speak of *density* and *pressure* than of mass and force.

READING EXERCISE 15-1: Estimate the pressure in pascals exerted on a dance floor by just the spike heels worn by a 125 lb woman who is standing on both feet. Assume that half of her weight is on the spike heels and half is on the front soles of her shoes. How does your estimate compare with the number given in Table 15-1? Discuss why it is possible for the spike heels to exert more pressure on the floor than the pressure exerted on the road by a single tire holding up its share of a 2500 lb automobile.

READING EXERCISE 15-2: Examine the densities listed in Table 15-2. Use the fact that air and water have significantly different densities to develop a plausible explanation for the fact that air is a much more compressible fluid than water is.

READING EXERCISE 15-3: Consider a book of dimensions 8 inches by 10 inches. Show that the downward force on the book by the atmosphere is about 1200 pounds. The downward force on a smooth thick rubber mat of the same dimensions is also 1200 pounds. When the rubber mat is placed on a smooth Formica table you cannot lift it. However, you can easily lift the book. Can you explain this phenomenon?

Touchstone Example 15-3-1, at the end of this chapter, illustrates how to use what you learned in this section.

TE

15-4 Gravitational Forces and Fluids at Rest

In the last section we considered the forces on a parcel of fluid that was "at rest" in outer space. We found that the pressure in any fluid in a container was the same in every direction at every location in the container. This is not so in fluids which experience gravitational forces. In this section we will consider how the presence of gravitational forces leads to pressure differences at different levels in a container of fluid. As every diver knows, the pressure *increases* with depth below the air-water interface. The diver's depth gauge, in fact, is a pressure sensor much like that of Fig. 15-3*b*. As every mountaineer knows, the pressure *decreases* with altitude as one ascends into the atmosphere. The pressures encountered by the diver and the mountaineer are usually called *hydrostatic pressures,* because they are due to fluids that are static (at rest). Here we want to find an expression for hydrostatic pressure as a function of depth or altitude.

Let us consider a fluid such as air or water near the surface of the Earth. What happens to its pressure when one changes from an initial level y_1 to a final level y_2? Since a fluid is made up of lots of molecules, we can pick any subset of them as our "object" and apply Newton's laws to that parcel of fluid. So let us imagine a parcel of the fluid consisting of all the molecules contained in a cylindrical column of cross sectional area A that extends between the two levels y_1 and y_2 as shown in Fig. 15-6*a*.

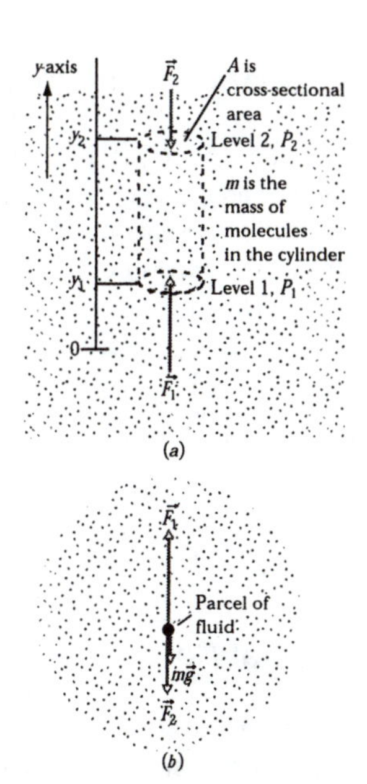

Fig. 15-6: (*a*) A parcel of fluid such as air or water is contained in an imaginary cylinder of cross-sectional area A. Forces $\vec{F}_1$ and $\vec{F}_2$ act, respectively, on the bottom and top of the cylinder. The gravitational force of the parcel of fluid is $\vec{F}_g = m\vec{g}$. (*b*) A free-body diagram of the forces that act on the parcel of fluid in the cylinder.

If the parcel of fluid is at rest, or *static equilibrium*, the horizontal forces on it from the sides must add up to zero. Similarly, static equilibrium requires that the vector sum of the vertical forces on the parcel of fluid must be zero too. There are three vertical forces that act on the parcel of fluid in the cylinder. Figure 15-6*b* shows a free-body diagram of these forces. Force $\vec{F}_1$ acts at the bottom surface of the cylinder and is due to the water below the cylinder. Similarly, force $\vec{F}_2$ acts at the top surface of the cylinder and is due to the fluid above the cylinder. The gravitational force on the water in the cylinder can be represented by $\vec{F}_g = m\vec{g}$. The net force comprised of these forces must be zero, so that

$$\vec{F}_{\text{net}} = \vec{F}_1 + \vec{F}_2 + \vec{F}_g = 0.$$

Since we know that all the forces act in the vertical direction we can rewrite this equation in terms of the y-components of the vectors as

$$F_{1y} + F_{2y} + F_g = F_{1y} + F_{2y} - mg = 0 \qquad (15\text{-}5)$$

We know that force $\vec{F}_1$ acts in an upward direction and is inherently positive while force $\vec{F}_2$ acts in a downward direction and is inherently negative. So we can replace the force *components* with the corresponding pressures and areas, with

$$F_{1y} = +P_1 A \text{ and } F_{2y} = -P_2 A. \qquad (15\text{-}6)$$

We use the explicit minus sign in front of the inherently positive $P_2 A$ term to signify the fact that the force component F_{2y} must be negative. Note that the mass m of the fluid in the cylinder is $m = \rho V$, where ρ represents the density of the fluid and where V

represents the volume of the cylinder. Note that since the volume is the product of its face area A and its height $\Delta y = y_2 - y_1$, the mass m is equal to $\rho A(y_2 - y_1)$. We also find that using conventional coordinates, the components of gravitational force $F_g = -mg$ where g is an inherently positive factor representing the local gravitational strength. Using these facts and substituting Eq. 15-6 into Eq. 15-5, we get

$$P_2 = P_1 - \rho g(y_2 - y_1) = P_1 - \rho g\Delta y \qquad \text{(only if } \rho \text{ is uniform)} \qquad (15\text{-}7a)$$

or

$$\Delta P = -\rho g\Delta y \qquad (15\text{-}7b)$$

where the units of pressure must be pascals. Equation 15-7*a* can be used to find pressure changes in either a liquid (as a function of depth) or in the atmosphere (as a function of altitude or height).

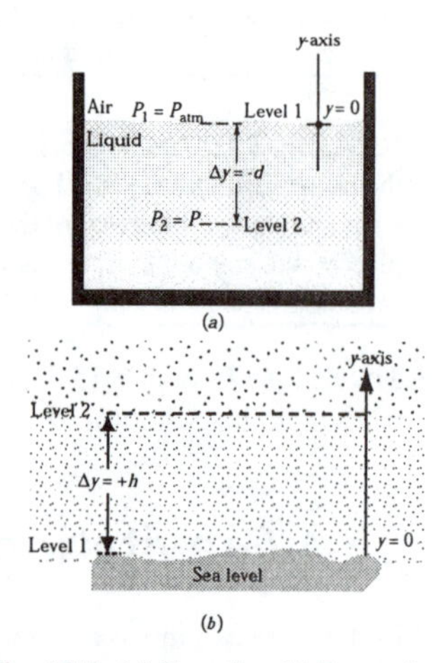

Fig. 15-7: (*a*) Equation 15-7*a* can be used to determine the pressure underwater or (*b*) the atmospheric pressure above the surface of the Earth.

Special Case 1: Pressure in the Earth's Atmosphere

Equation 15-7*a* can be used to determine the pressure of the atmosphere at a given distance above sea level in terms of the atmospheric pressure P_{atm} at sea level (assuming that the atmospheric density is uniform over that distance). For example, to find the atmospheric pressure at a distance *h* above the ground level 1 as shown in Fig. 15-7*b*, we substitute

$$y_1 = 0,\ P_1 = P_{atm} \text{ at sea level and } y_2 = +h,\ P_2 = P$$

into Eq. 15-7*a* with $\rho = \rho_{air}$. Thus, we obtain an expression for pressure which decreases with altitude,

$$P = P_{atm} - \rho_{air} gh. \qquad (15\text{-}8a)$$

Special Case 2: Underwater Pressure Below Sea Level

Suppose we want to calculate the pressure *P* at a depth *d* below the surface of a body of water with its surface located at sea level. Then we choose level 1 to be the surface, level 2 to be a distance *d* below it (as in Fig. 15-7*a*), and P_{atm} to represent the atmospheric pressure at sea level. Since we typically refer to depth as being a positive number such as 13 meters or 40,000 leagues, we must substitute the following into Eq. 15-7*a*,

$$y_1 = 0,\ P_1 = P_{atm} \text{ and } y_2 = -d,\ P_2 = P,$$

so that Eq. 15-7*a* reduces to

$$P = P_{atm} + \rho_{water} gd \qquad \text{(at depth } d \text{ below sea level in water).} \qquad (15\text{-}8b)$$

Note that the pressure in a liquid increases with depth but does not depend on the horizontal location of the parcel of liquid. If we have a liquid other than water, obviously we need to use the density of that liquid in place of the density of water.

➤When gravitational forces are present, the pressure at a point in a fluid in static equilibrium depends on the depth of that point but not on any horizontal dimension of the fluid or the shape of its container.

Gauge Pressure in Liquids

In Equation 15-8*b*, *P* is said to be the total pressure or **absolute pressure** at level 2. To see why, note in Fig. 15-7 that the pressure *P* at level 2 consists of two contributions: (1) P_{atm}, the pressure due to the atmosphere and (2) $\rho_{liquid} gd$, the additional pressure due to the mass of the liquid above level 2. In general, the difference between an absolute pressure and an atmospheric pressure is called the **gauge pressure.** (The name comes from the use of a gauge to measure this difference in pressures.) For the situation of Fig. 15-7, the gauge pressure for water or an other liquid is $\rho_{\text{liquid}} g d$.

READING EXERCISE 15-4: Scuba divers know that if they descend to a depth of 10 m the pressure they experience doubles. However, alpine mountain climbers must ascend to about 5.5 km to cut the atmospheric pressure in half. What factor in Eq. 15-7*a* accounts for the fact that the pressure change with distance is so much greater in air than in water?

READING EXERCISE 15-5: The figure shows four containers of olive oil. Rank them according to the pressure at depth *d*, greatest first.

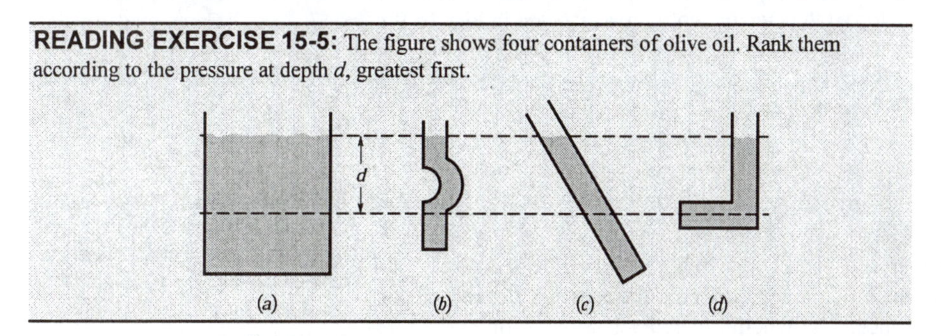

Touchstone Examples 15-4-1 and 15-4-2, at the end of this chapter, illustrate how to use what you learned in this section.

15-5 Measuring Pressure

Electronic Pressure Sensors

One of the most popular methods for measuring absolute pressure in a gas is to use an electronic sensor. Many electronic sensors work in much the same way as our test sensor shown in Fig. 15-3. A common electronic pressure sensor has a flexible membrane with a vacuum chamber on one side of it. The flexing of the membrane under pressure is sensed electronically. These devices are used in recording barometers found in weather stations and in physics laboratories.

Typically an electronic gas sensor will be damaged when immersed in a liquid. However, when tubing connected to a gas sensor is immersed in a liquid, the pressure at various depths in the liquid can also be measured.

(*a*)

(*b*)

Fig. 15-8: Two popular gas pressure sensors used in contemporary physics laboratories can record between 0 and about 7 atmospheres of pressure.
(*a*) Vernier pressure sensor.
(*b*) PASCO pressure sensor.
Photos used with permission of Vernier Software and Technology and PASCO scientific.

The Mercury Barometer

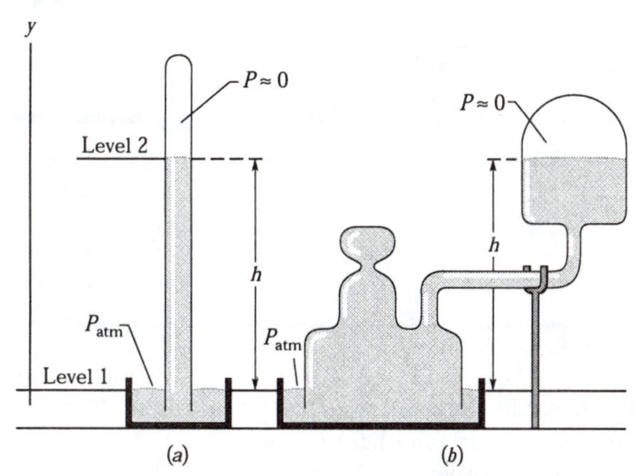

Fig. 15-9: (*a*) A mercury barometer. (*b*) Another mercury barometer. The distance *h* is the same in both cases.

For historical and practical reasons, the *mercury barometer* and the open tube *manometer* are still popular methods for measuring atmospheric pressure and pressures near atmospheric pressure.

Figure 15-9*a* shows a very basic *mercury barometer,* a device used to measure the pressure of the atmosphere. The long glass tube is filled with mercury and inverted with its open end in a dish of mercury, as the figure shows. The space above the mercury column contains only mercury vapor, whose pressure is so small at ordinary temperatures that it can be neglected.

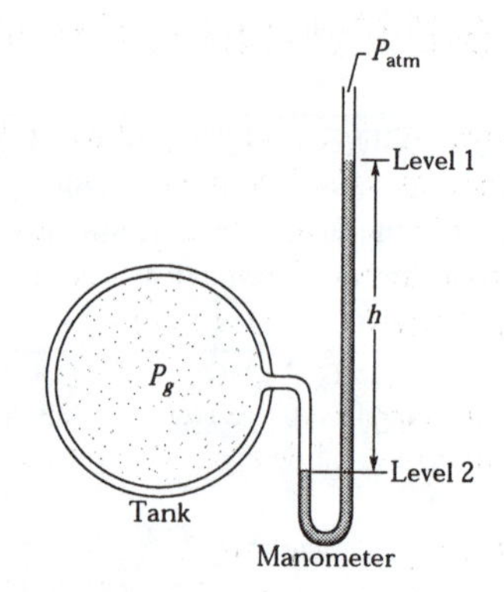

Fig. 15-10: An open-tube manometer, connected to measure the gauge pressure of the gas in the tank on the left. The right arm of the U-tube is open to the atmosphere.

We can use Eq. 15-7*b* to find the atmospheric pressure P_{atm} in terms of the height h of the mercury column. We choose level 1 of Fig. 15-9*a* to be that of the air-mercury interface and level 2 to be that of the top of the mercury column, as labeled in Fig. 15-9*a*. We then substitute

$$y_1 = 0,\ P_1 = P_{atm} \text{ and } y_2 = h,\ P_2 = 0$$

into Eq. 15-7*b*, finding that

$$P_{atm} = \rho g h \tag{15-9}$$

where ρ is the density of the mercury.

For a given pressure, the height h of the mercury column does not depend on the cross-sectional area of the vertical tube. The fanciful mercury barometer of Fig. 15-9*b* gives the same reading as that of Fig. 15-9*a*; all that counts is the vertical distance h between the mercury levels.

Equation 15-9 shows that, for a given pressure, the height of the column of mercury depends on the value of g at the location of the barometer and on the density of mercury, which varies only slightly with temperature. The column height (in millimeters) is numerically equal to the pressure (in torr) *only* if the barometer is at a place where g has its accepted standard value of $+9.80665\ \text{m/s}^2$ *and* the temperature of the mercury is $0°\text{C}$. If these conditions do not prevail (and they rarely do), small corrections must be made before the height of the mercury column can be transformed into a pressure.

The Open-Tube Manometer

An *open-tube manometer* (Fig. 15-10) measures the gauge pressure P_g of a gas. It consists of a U-tube containing a liquid, with one end of the tube connected to the vessel whose gauge pressure we wish to measure and the other end open to the atmosphere. Looking at the figure, we see that the "U" of fluid below the line marked "Level 2" is being pushed on the left by the force from the pressure of the gas in the tank and is being pushed on the right by the force arising from the pressure built up by everything above it, including the weight of the column between levels 1 and 2 and the pressure of the atmosphere. When the column is in equilibrium (no rising or falling), these forces must balance. Keeping this in mind, we can use Eq. 15-7*b* to find the gauge pressure in terms of the height h shown in Fig. 15-10. Let us choose levels 1 and 2 as shown in Fig. 15-10. We then substitute

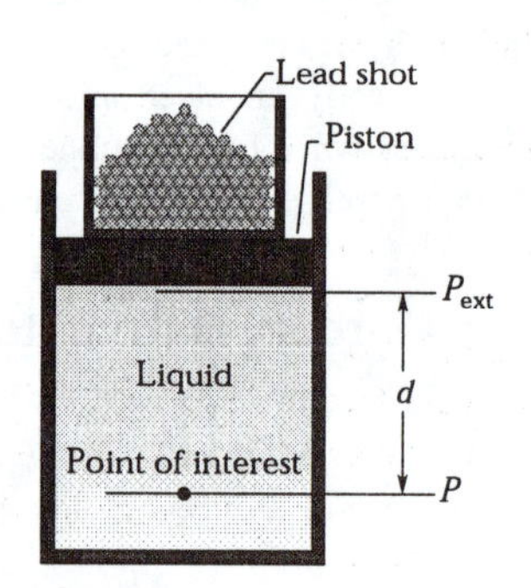

Fig. 15-11: Lead shot (small balls of lead) loaded onto the piston create a pressure P_{ext} at the top of the enclosed (incompressible) liquid. If P_{ext} is increased, by adding more lead shot, the pressure increases by the same amount at all points within the liquid.

$$y_1 = 0,\ P_1 = P_{atm} \text{ and } y_2 = -h,\ P_2 = P$$

into Eq. 15-7, finding that

$$P_g = P - P_{atm} = \rho g h, \tag{15-10}$$

where ρ is the density of the liquid in the tube. The gauge pressure P_g is directly proportional to h.

The gauge pressure can be positive or negative, depending on whether $P > P_{atm}$ or $P < P_{atm}$. In inflated tires or the human circulatory system, the (absolute) pressure is greater than atmospheric pressure, so the gauge pressure is a positive quantity, sometimes called the *overpressure.* If you suck on a straw to pull fluid up the straw, the (absolute) pressure in your lungs is actually less than atmospheric pressure. The gauge pressure in your lungs is then a negative quantity.

15-6 Pascal's Principle

When you squeeze one end of a tube to get toothpaste out the other end, you are watching **Pascal's principle** in action. This principle is also the basis for the Heimlich maneuver, in which a sharp pressure increase properly applied to the abdomen is transmitted to the throat, forcefully ejecting food lodged there. The principle was first stated clearly in 1652 by Blaise Pascal (for whom the unit of pressure is named):

➤A change in the pressure applied to an enclosed fluid is transmitted undiminished to every portion of the fluid and to the walls of its container.

This is just want we found at the start of our discussion of pressure when we applied a force to the piston in Fig. 15-2 and found that all the corks in the bottle popped out.

Demonstrating Pascal's Principle

Consider the case in which the incompressible fluid is a liquid contained in a tall cylinder, as in Fig. 15-11. The cylinder is fitted with a piston on which a container of lead shot rests. The atmosphere, container, and shot put pressure P_{ext} on the piston and thus on the liquid. The pressure P at any point in the liquid a distance d below the piston is then

$$P = P_{\text{ext}} + \rho g d. \tag{15-11}$$

Let us add a little more lead shot to the container to increase P_{ext} by an amount ΔP_{ext}. The quantities ρ, g, and d in Eq. 15-11 are unchanged, so the pressure change at any point is

$$\Delta P = \Delta P_{\text{ext}}. \tag{15-12}$$

This pressure change is independent of d, so it must hold for all points within the liquid, as Pascal's principle states.

Pascal's Principle and the Hydraulic Lever

Figure 15-12 shows how Pascal's principle can be made the basis of a hydraulic lever. In operation, let an external force of magnitude F_i be directed downward on the left-hand (or input) piston, whose area is A_i. An incompressible liquid in the device then produces an upward force of magnitude F_o on the right-hand (or output) piston, whose area is A_o. To keep the system in equilibrium, there must be a downward force of magnitude F_o on the output piston from an external load (not shown). The force $\vec{F}_i$ applied on the left and the downward force $\vec{F}_o$ from the load on the right produce a change ΔP in the pressure of the liquid that is given by

$$\Delta P = \frac{|\vec{F}_i|}{A_i} = \frac{|\vec{F}_o|}{A_o},$$

so

$$|\vec{F}_o| = |\vec{F}_i| \frac{A_o}{A_i}. \tag{15-13}$$

Equation 15-13 shows that the output force $\vec{F}_o$ on the load must be greater than the input force $\vec{F}_i$ if $A_o > A_i$, as is the case in Fig. 15-12.

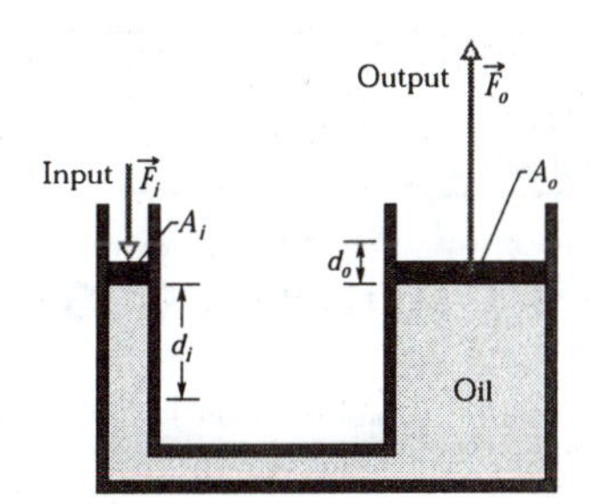

Fig. 15-12: A hydraulic arrangement that can be used to magnify a force F_i. The work done is, however, not magnified and is the same for both the input and output forces.

If we move the input piston downward a distance d_i, the output piston moves upward a distance d_o, such that the same volume V of the incompressible liquid is displaced at both pistons. Then

$$V = A_i d_i = A_o d_o,$$

which we can write as

$$d_o = d_i \frac{A_i}{A_o}. \tag{15-14}$$

This shows that, if $A_o > A_i$ (as in Fig. 15-12), the output piston moves a smaller distance than the input piston moves.

From Eqs. 15-13 and 15-14 we can write the output work as

$$W = F_o d_o = \left(F_i \frac{A_o}{A_i}\right)\left(d_i \frac{A_i}{A_o}\right) = F_i d_i, \tag{15-15}$$

which shows that the work W done *on* the input piston by the applied force is equal to the work W done *by* the output piston in lifting the load placed on it.

The advantage of a hydraulic lever is this:

➤With a hydraulic lever, a given force applied over a given distance can be transformed to a greater force applied over a smaller distance.

The product of force and distance remains unchanged so that the same work is done. However, there is often tremendous advantage in being able to exert the larger force. Most of us, for example, cannot lift an automobile directly but can with a hydraulic jack, even though we have to pump the handle farther than the automobile rises. In this device, the displacement d_i is accomplished not in a single stroke but over a series of small strokes.

READING EXERCISE 15-6: Consider the cylinder of a real hydraulic jack which is filled with oil. The oil is slightly compressible. Discuss how the relations presented in this section are affected by the compressibility of the oil. For example, is the work put in still equal to the work out? If not, which is larger? What (if any) are the energy transformations that take place?

READING EXERCISE 15-7: The pressure on the bottom of a container with sloping walls is determined by the height of the central column. Relate this observation with the concepts addressed in this section.

15-7 Archimedes' Principle

Let us think about what happens when we immerse an object in a fluid. Your first guess might be that the weight of the water above the object would push it to the bottom. But think about what happens when we try to push a beach ball down under the surface in a swimming pool. It will be hard to push it down. If we do get it under the surface and release it, it will go shooting up into the air. What's going on here? It turns out that the critical issue is that pressure is a scalar. You can get forces in all directions—up as well as down. Let us see how this works.

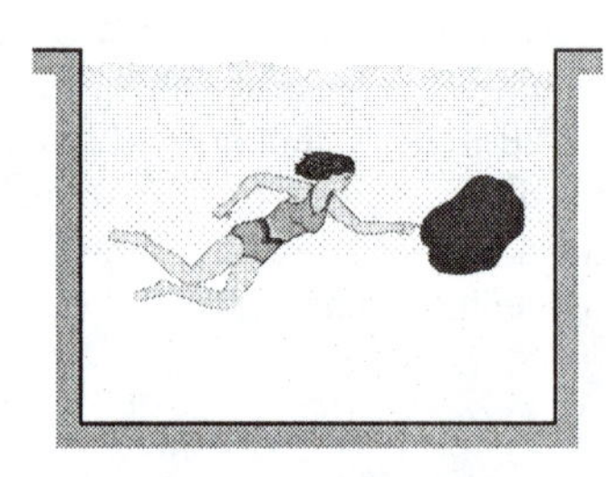

Fig. 15-13: A thin-walled plastic sack of water is in static equilibrium in the pool. Its weight must be balanced by a net upward force on the water in the sack from the surrounding water.

Figure 15-13 shows a student in a swimming pool, manipulating a very thin plastic sack (of negligible mass) that is filled with water. She finds that the sack and its contained water are in static equilibrium, tending neither to rise nor to sink. The downward gravitational force $\vec{F}_g$ on the contained water must be balanced by a net upward force from the water surrounding the sack.

This net upward force on an object in a fluid is called a **buoyant force** $\vec{F}_b$. It exists because the pressure in the surrounding water increases with depth below the surface.

Thus, the pressure near the bottom of the sack is greater than the pressure near the top. Then the forces on the sack due to this pressure are greater in magnitude near the bottom of the sack than near the top. Some of the forces are represented in Fig. 15-14*a*, where the space occupied by the sack has been left empty. Note that the force vectors drawn near the bottom of that space (with upward components) have longer lengths than those drawn near the top of the sack (with downward components). If we vectorially add all the forces on the sack from the water, the horizontal components cancel and the vertical components add to yield the upward buoyant force $\vec{F}_b$ on the sack. (Force $\vec{F}_b$ is shown to the right of the pool in Fig. 15-14*a*.)

Because the sack of water is in static equilibrium, the magnitude $|\vec{F}_b|$ is equal to the magnitude $m_f g$ of the gravitational force $\vec{F}_g$ on the sack of water: $|\vec{F}_b| = m_f g$. (Subscript *f* refers to *fluid*, here the water.) In words, the magnitude of the buoyant force is equal to the weight of the water in the sack.

In Fig. 15-14*b*, we have replaced the sack of water with a stone that exactly fills the hole in Fig. 15-14*a*. The stone is said to *displace* the water, meaning that it occupies space that would otherwise be occupied by water. The object pushes the water out of the way. We have changed nothing about the shape of the hole, so the forces at the hole's surface must be the same as when the water-filled sack was in place. Thus, the same upward buoyant force that acted on the water-filled sack now acts on the stone; that is, the magnitude $|\vec{F}_b|$ of the buoyant force is equal to $m_f g$, the weight of the water displaced by the stone.

Unlike the water-filled sack, the stone is not in static equilibrium. The magnitude of the downward gravitational force $|\vec{F}_g|$ on the stone is greater in magnitude than that of the upward buoyant force, as is shown in the free-body diagram to the right of the pool in Fig. 15-14*b*. The stone thus accelerates downward, sinking to the bottom of the pool.

Let us next exactly fill the hole in Fig. 15-14*a* with a block of light-weight wood, as in Fig. 15-14*c*. Again, nothing has changed about the forces at the hole's surface, so the magnitude $|\vec{F}_b|$ of the buoyant force is still equal to $m_f g$, the weight of the displaced water. Like the stone, the block is not in static equilibrium. However, this time the magnitude of the gravitational force $|\vec{F}_g|$ is less than the buoyant force (as shown to the right of the pool), and so the block accelerates upward, rising to the top surface of the water.

Our results with the sack, stone, and block apply to all fluids and are summarized in **Archimedes' principle:**

➤When a body is fully or partially submerged in a fluid, a buoyant force $\vec{F}_g$ from the surrounding fluid acts on the body. The force is directed upward and has a magnitude equal to the weight $m_f g$ of the fluid that has been displaced by the body.

The buoyant force on a body in a fluid has the magnitude

$$|\vec{F}_b| = m_f g \qquad \text{(buoyant force magnitude),} \qquad (15\text{-}16)$$

where m_f is the mass of the fluid that is displaced by the body.

In the late evening of August 21, 1986, something (possibly a volcanic tremor) disturbed Cameroon's Lake Nyos, which has a high concentration of dissolved carbon dioxide. The disturbance caused that gas to form bubbles. Being lighter than the surrounding fluid (the water), those bubbles were buoyed to the surface, where they released the carbon dioxide. The gas, being heavier than the surrounding fluid (now the air), rushed down the mountainside like a river, asphyxiating 1700 persons and the scores of animals seen here.

Floating

When we release a block of light-weight wood just above the water in a pool, it moves into the water because the gravitational force on it pulls it downward. As the block displaces more and more water, the magnitude $|\vec{F}_b|$ of the upward buoyant force acting on it increases. Eventually, $|\vec{F}_b|$ is large enough to equal the magnitude $|\vec{F}_g|$ of the downward gravitational force on the block, and the block comes to rest. The block is then in static equilibrium and is said to be *floating* in the water. In general,

➤When a body floats in a fluid, the magnitude $|F_b|$ of the buoyant force on the body is equal to the magnitude $|F_g|$ of the gravitational force on the body.

We can write this statement as

$$|\vec{F}_b| = |\vec{F}_g| \quad \text{(floating).} \tag{15-17}$$

From Eq. 15-16, we know that $|\vec{F}_b| = m_f g$. Thus,

➤When a body floats in a fluid, the magnitude $|\vec{F}_g|$ of the gravitational force on the body is equal to the weight $m_f g$ of the fluid that has been displaced by the body.

We can write this statement as

$$|\vec{F}_g| = m_f g \quad \text{(floating).} \tag{15-18}$$

In thinking about whether an object will sink or float, density is the key consideration, *not* the total mass of the object. If the density is less than that of water, the object can displace a mass of water equal to its own weight by being only partially submerged. If the density is greater than that of water, even completely immersing the object produces a buoyant force that is less than the object's weight. In that case, part of the object's weight will be unbalanced; there will be a net downward force and the object will sink to the bottom.

Apparent Weight in a Fluid

If we place a stone on a scale that is calibrated to measure weight, then the reading on the scale is the stone's weight. However, if we do this underwater, the upward buoyant force on the stone from the water decreases the reading. That reading is then an apparent weight. In general, an apparent weight is related to the actual weight of a body and the buoyant force on the body by

(apparent weight) = (actual weight) − (magnitude of buoyant force),

which we can write as

$$\text{weight}_{\text{app}} = \text{weight} - \left|\vec{F}_b\right| \quad \text{(apparent weight).} \quad (15\text{-}19)$$

If, in some strange test of strength, you had to lift a heavy stone, you could do it more easily with the stone underwater. Then your applied force would need to exceed only the stone's apparent weight, not its larger actual weight, because the upward buoyant force would help you lift the stone.

The magnitude of the buoyant force on a floating body is equal to the body's weight. Equation 15-19 thus tells us that a floating body has an apparent weight of zero—the body would produce a reading of zero on a scale. (When astronauts prepare to perform a complex task in space, they practice the task floating underwater, where their apparent weight is zero as it is in space.)

READING EXERCISE 15-8: A penguin floats first in a fluid of density ρ_0, then in a fluid of density $0.95\rho_0$, and then in a fluid of density $1.10\rho_0$. (a) Rank the densities according to the magnitude of the buoyant force on the penguin, greatest first. (b) Rank the densities according to the amount of fluid displaced by the penguin, greatest first.

Touchstone Example 15-7-1, at the end of this chapter, illustrates how to use what you learned in this section.

TE

15-8 Ideal Fluids in Motion

The motion of *real fluids* is very complicated and not yet fully understood. Instead, we shall discuss the motion of an **ideal fluid,** which is simpler to handle mathematically and yet provides useful results. Here are four assumptions that we make about our ideal fluid; they all are concerned with flow:

1. ***Steady flow*** In *steady* (or *laminar*) *flow,* the velocity of the moving fluid at any fixed point does not change with time, either in magnitude or in direction. The gentle flow of water near the center of a quiet stream is steady; that in a chain of rapids is not. Figure 15-15 shows a transition from steady flow to *nonsteady* (or *turbulent*) flow for a rising stream of smoke. The speed of the smoke particles increases as they rise and, at a certain critical speed, the flow changes from steady to nonsteady (that is, from laminar to *nonlaminar* flow).

Fig. 15-15: At a certain point, the rising flow of smoke and heated gas changes from steady to turbulent.

2. ***Incompressible flow*** We assume, as we have already done for fluids at rest, that our ideal fluid is incompressible; that is, its density has a constant, uniform value.
3. ***Nonviscous flow*** Roughly speaking, the viscosity of a fluid is a measure of how resistive the fluid is to flow. For example, thick honey is more resistive to flow than water, and so honey is said to be more viscous than water. Viscosity is the fluid analog of friction between solids; both are mechanisms by which the kinetic energy of moving objects can be transferred to thermal energy. In the absence of friction, a block could glide at constant speed along a horizontal surface. In the same way, an object moving through a nonviscous fluid would experience no *viscous drag force*—that is, no resistive force due to viscosity; it could move at constant speed through the fluid. The British scientist Lord Rayleigh noted that in an ideal fluid a ship's propeller would not work but, on the other hand, a ship (once set into motion) would not need a propeller!

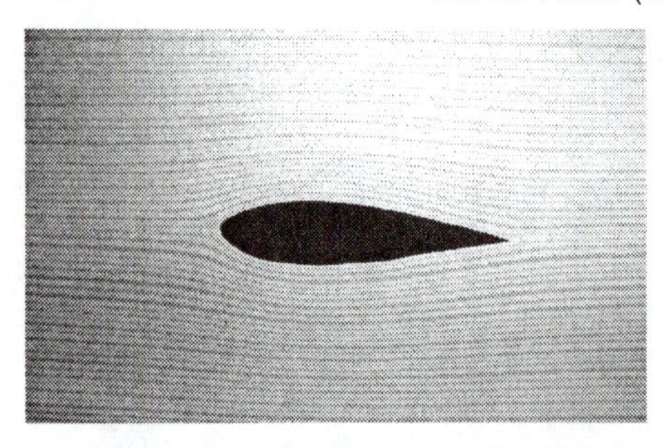

Fig. 15-16: Streamlined flow around an airfoil.

Fig. 15-17: Smoke reveals streamlines in airflow past a car in a wind-tunnel test.

We can make the flow of a fluid visible by adding a *tracer.* This might be a dye injected into many points across a liquid stream (Fig. 15-16) or smoke particles added to a gas flow (Figs. 15-15 and 15-17). Each bit of a tracer follows a *streamline,* which is the path that a tiny element of the fluid would take as the fluid flows. Recall from Chapter 2 that the velocity of a particle is always tangent to the path taken by the particle. Here the particle is the fluid element, and its velocity $\vec{v}$ is always tangent to a streamline (Fig. 15-18). For this reason, two streamlines cannot intersect with a finite fluid velocity, since a fluid element cannot flow in two directions simultaneously. However, at a point of zero velocity, a stagnation point, there is no direction defined and two or more streamlines may intersect at a stagnation point.

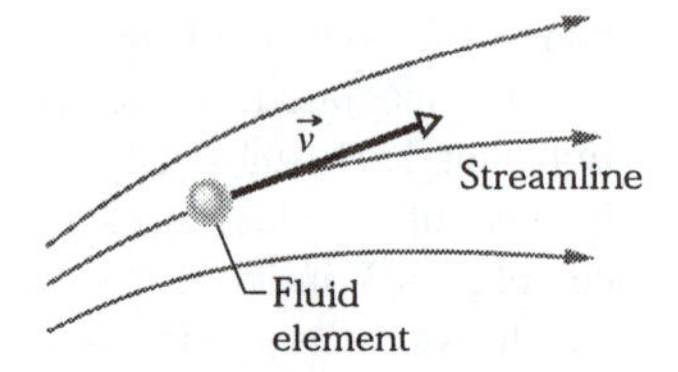

Fig. 15-18: A fluid element P traces out a streamline as it moves. The velocity vector of the element is tangent to the streamline at every point.

15-9 The Equation of Continuity

You may have noticed that you can increase the speed of the water emerging from a garden hose by partially closing the hose opening with your thumb. Apparently the speed $|\vec{v}|$ of the water depends on the cross-sectional area A through which the water flows. This makes sense when we realize that the faucet is putting water out at a certain rate. If that much water has to get through a smaller hole, it has to go faster.

Here we wish to derive an expression that relates $|\vec{v}|$ and A for the steady flow of an ideal fluid through a tube with varying cross section, like that in Fig. 15-19. The flow there is toward the right, and the tube segment shown (part of a longer tube) has length L. The fluid has speeds $|\vec{v}_1|$ at the left end of the segment and $|\vec{v}_2|$ at the right end. The tube has cross-sectional areas A_1 at the left end and A_2 at the right end. Suppose that in a time interval Δt a volume ΔV of fluid enters the tube segment at its left end (that volume is colored dark grey in Fig. 15-19*a*). Then, because the fluid is incompressible, an identical volume ΔV must emerge from the right end of the segment (it is colored light grey in Fig. 15-19*b*).

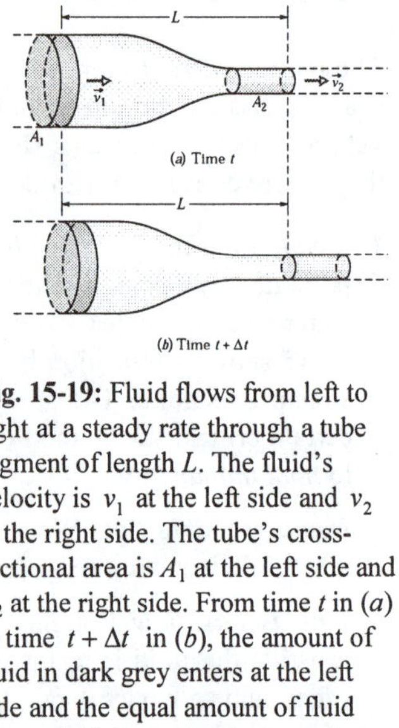

Fig. 15-19: Fluid flows from left to right at a steady rate through a tube segment of length L. The fluid's velocity is v_1 at the left side and v_2 at the right side. The tube's cross-sectional area is A_1 at the left side and A_2 at the right side. From time t in (*a*) to time $t + \Delta t$ in (*b*), the amount of fluid in dark grey enters at the left side and the equal amount of fluid shown in light grey emerges at the right side.

We can use this common volume ΔV to relate the speeds and areas. To do so, we first consider Fig. 15-20, which shows a side view of a tube of *uniform* cross-sectional area A. In Fig. 15-20*a*, a fluid element e is about to pass through the dashed line drawn across the tube width. The element's speed is $|\vec{v}|$, so during a time interval Δt, the element moves along the tube a distance $\Delta x = |\vec{v}|\Delta t$. The volume ΔV of fluid that has passed through the dashed line in that time interval Δt is

$$\Delta V = A\,\Delta x = A|\vec{v}|\,\Delta t. \qquad (15\text{-}20)$$

Applying Eq. 15-20 to both the left and right ends of the tube segment in Fig. 15-19, we have

$$\Delta V = A_1|\vec{v}_1|\Delta t = A_2|\vec{v}_2|\Delta t$$

or

$$A_1|\vec{v}_1| = A_2|\vec{v}_2| \qquad \text{(equation of continuity).} \qquad (15\text{-}21)$$

This relation between speed and cross-sectional area is called the **equation of continuity** for the flow of an ideal fluid. It tells us that the flow speed increases when we decrease the cross-sectional area through which the fluid flows (as when we partially close off a garden hose with a thumb).

Equation 15-21 applies not only to an actual tube but also to any so-called *tube of flow,* or imaginary tube whose boundary consists of streamlines. Such a tube acts like a real tube because no fluid element can cross a streamline; thus, all the fluid within a tube of flow must remain within its boundary. Figure 15-21 shows a tube of flow in which the cross-sectional area increases from area A_1 to area A_2 along the flow direction. From Eq. 15-21 we know that, with the increase in area, the speed must decrease, as is indicated by the greater spacing between streamlines at the right in Fig. 15-21. Similarly, you can see that in Fig. 15-16 the speed of the flow is greatest just above and just below the cylinder.

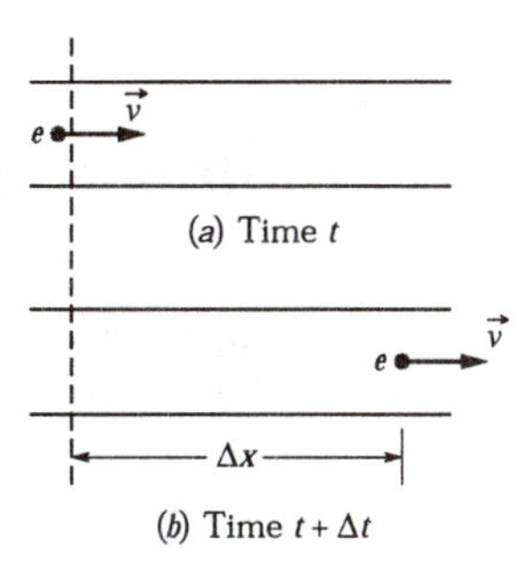

Fig. 15-20: Fluid flows at a constant speed $|\vec{v}|$ through a tube. (*a*) At time *t*, fluid element *e* is about to pass the dashed line. (*b*) At time $t+\Delta t$, element *e* is a distance $\Delta x = v_x \Delta t$ from the dashed line where v_x is the *x*-component of v .

We can rewrite Eq. 15-21 as

$$R_V = A|\vec{v}| = \text{a constant} \qquad \text{(volume flow rate, equation of continuity)}, \qquad (15\text{-}22)$$

in which R_V is the **volume flow rate** of the fluid (volume per unit time). Its SI unit is the cubic meter per second (m^3/s). If the density ρ of the fluid is uniform, we can multiply Eq. 15-22 by that density to get the **mass flow rate** R_m (mass per unit time):

$$R_m = \rho R_V = \rho A v = \text{a constant} \qquad \text{(mass flow rate)}. \qquad (15\text{-}23)$$

The SI unit of mass flow rate is the kilogram per second (kg/s). Equation 15-23 says that the mass that flows into the tube segment of Fig. 15-19 each second must be equal to the mass that flows out of that segment each second.

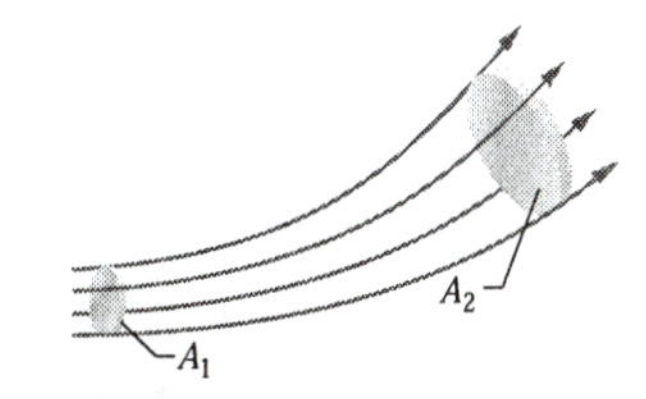

Fig. 15-21: A tube of flow is defined by the streamlines that form the boundary. The volume flow rate must be the same for all cross sections of the tube of flow.

READING EXERCISE 15-9: The figure shows a pipe and gives the volume flow rate (in cm/s) and the direction of flow for all but one section. What are the volume flow rate and the direction of flow for that section?

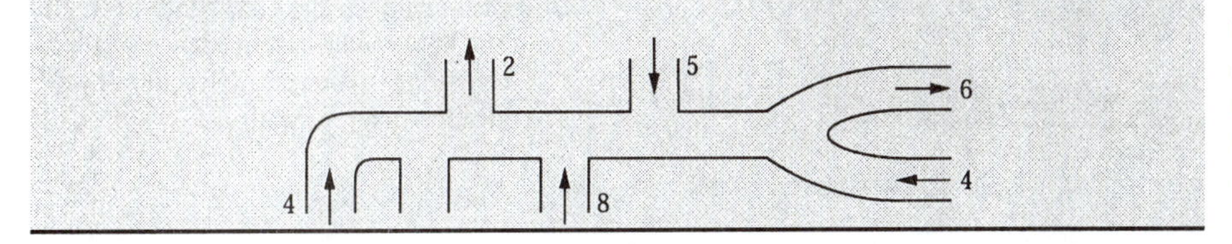

Engineers can use velocity measurements and the equation of continuity as a tool for determining volume flow rates of incompressible fluids such as oil flowing in a pipe of variable cross section or water flowing in a stream of variable cross section. But Equation 15-23 can only be used under certain conditions. In deriving the equation of continuity we assumed that fluid isn't being added to or subtracted from the system. For example, we assume that our oil pipe doesn't leak or that new water not is being added to a stream by a tributary as it flows along. In cases where fluid is added for subtracted from a tube of flow we would find that $|\vec{v}_1|A_1 \neq |\vec{v}_2|A_2$.

Even when the continuity condition holds, whenever we use Eq. 15-23 we assume that we have a uniform flow of fluid over a cross sectional area. In other words, we assume that the velocity of each of the elements of the fluid is moving at the same speed in a direction that is perpendicular to a cross-sectional area. In real pipes fluid flows more slowly near the surface of the pipe than in the middle and it is not obvious how to calculate the product $|\vec{v}|A$ at a given location along our tube.

In the next section we will introduce the mathematical concept of *flux* to help us deal with more realistic situations in which the simplified application of the equation of continuity cannot be used either because we do not have true continuity or because we do not have a nice uniform flow of fluid over a cross sectional area.

Touchstone Examples 15-9-1 and 15-9-2, at the end of this chapter, illustrate how to use what you learned in this section.

TE

15-10 Volume Flux

The term *volume flux* is a synonym for volume flow rate, R_V. In this section we consider a more careful mathematical definition of volume flux that will allow us to apply the equation of continuity to more realistic situations. The word "flux" comes from the Latin word meaning "to flow." Often the word "flux" is used in science and engineering to describe the rate of flow or penetration of matter or energy through a surface. However, later in this text we will introduce definitions of electric and magnetic flux. Even though the mathematical definitions of electric and magnetic flux are analogous to that of volume flow rate, it is surprising to find that nothing at all is flowing.

Volume Flux in a Stream

In this section we will develop the concept of flux to calculate the rate at which water flows though a cross sectional area of a stream. Suppose we have a stream of water that behaves like an ideal fluid undergoing steady laminar flow (as described in Sections 15-8 and 15-9). Assume all the water is traveling in the same direction through a wide shallow channel as shown in Fig. 15-22 but that the elements of water closer to the stream bottom and sides are moving more slowly than the water in the top central part of the stream. We cannot simply find a single term speed, $|\vec{v}|$, at all points along our cross sectional area.

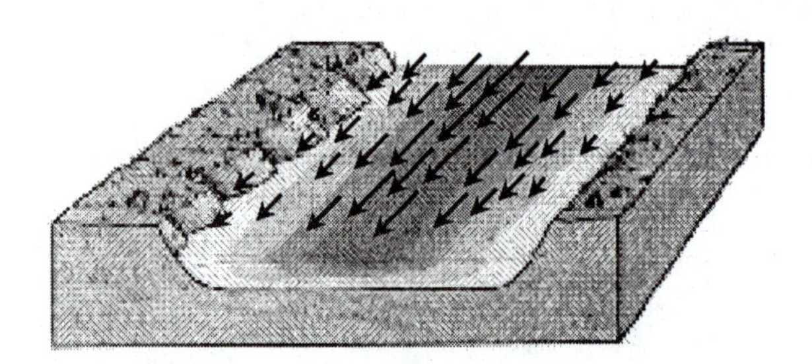

Fig. 15-22: Water moving in laminar flow through a wide shallow channel. The darker regions in the diagram with longer velocity vectors indicate areas farther from the sides and bottom of the stream where the water encounters less drag and hence flows faster. The flow can be described by a vector field consisting of the velocity vectors at each location in the stream.

Let us make our definition of flux more precise by considering stream water passing through many small elements of cross-sectional area. We can denote the ith area element as ΔA_i, where we use the notation Δ to signify that the area is part of a larger area. In fact, we typically choose the area elements ΔA_i to be small enough that the velocity of the part of the stream flowing through it is essentially constant. Suppose that, as in Fig. 15-23, the volume flux, Φ_i, represents the rate at which water flows through the ith element of area.

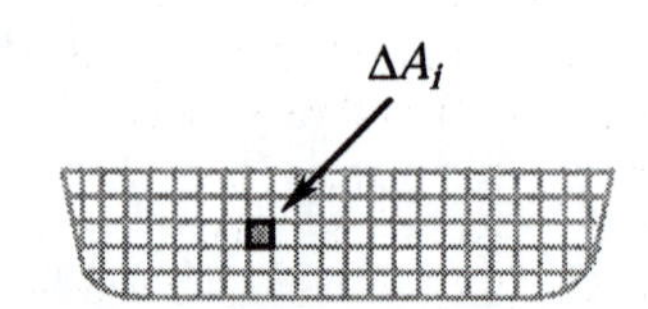

Fig. 15-23: The ith of many, many small area elements that makes up the cross-sectional area of a stream channel. The size of the elements is chosen to be so small that the stream velocity vector is constant over an element of area.

➤ **Volume flux** is defined as the rate at which something passes through an area.

Obviously if the water velocity $\vec{v}$ is parallel to the plane of the area element ΔA_i, water passes by the area without going through it. In that case the flux element, Φ_i, is zero (Fig. 15-24*a*). On the other hand, if $\vec{v}$ is perpendicular to the plane of the area, the flux element, Φ_i, is a maximum (Fig. 15-24*b*). If the area is oriented so it is between perpendicular and parallel the volume flux is in between zero and its maximum value (Fig. 15-24*c*). Thus, the volume flux depends on the *angle* between the velocity vector $\vec{v}$ representing the flow of water and the orientation of the areas shown in Fig. 15-24.

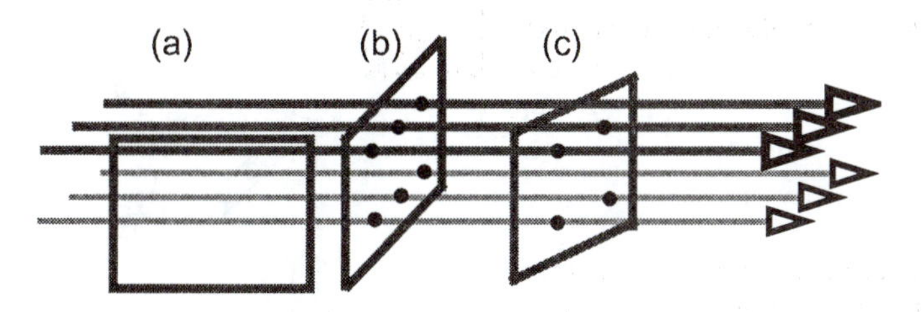

Fig. 15-24: The velocity vector field of a stream is represented here as imaginary streamlines. The amount of water that passes through a small imaginary area depends on its orientation. Three orientations are shown for the same area element ΔA_i. (*a*) No water passes through the area when it is parallel to the velocity vector, (*b*) the largest volume of water passes through an area that is perpendicular to the velocity vector, and (*c*) less water passes through when the orientation of the plane of the area is between perpendicular and parallel.

As shown in Equation 15-22, the volume of water, ΔV, that passes in a time Δt through a small area element, ΔA_i, perpendicular to the stream's direction of flow is given by $\Delta V = |\Delta x|\Delta A = (|v_x|\Delta t)\Delta A$ (Fig. 15-24). Thus, when an area is perpendicular to the stream velocity v_x, the flux element, $\Phi_\perp$, is given by the volume rate of flow through the area

$$\Phi_\perp = \frac{\Delta V}{\Delta t} = \frac{\Delta x \Delta A}{\Delta t} = |v_x|\Delta A. \qquad (15\text{-}24)$$

Defining the Normal Vector for an Area

In order to find the flux at intermediate angles it is convenient to represent the orientation of our small area mathematically. An area uses two of the three dimensions of space, but a direction perpendicular to the plane of the area lies along a single line. Thus it is easier to define the orientation of an area mathematically by a vector having a magnitude of ΔA that points in a direction *perpendicular* to the plane of the area. Since the term normal is a synonym for perpendicular, a vector used to describe the direction and magnitude of an area is known as a **normal vector** and can be denoted as $\Delta\vec{A}$. But what direction should our normal vector point? If the element of area is part of an imaginary closed container, it is conventional to represent our normal vector for a flat surface as an arrow that points *out* of a surface rather than into a surface. The box in Fig. 15-25 has an inside and outside even if it has no front, nonetheless it has an obvious inside and outside. Thus, we can point the normal vectors representing each element of area outward. If we put a front on the box, we would define it as a *closed surface* because if it were a real box, we could trap or enclose something inside of it.

How does the angle between the normal to an area and a stream velocity vector affect the flux? Suppose water in a straight stream channel is moving from left to right as it flows through a small imaginary box shown in Fig. 15-26. What will the flux be through each face of the box as a function of the angle θ between the normal vector of a box face and the stream velocity vector $\vec{v}_i$? Since the flux is a measure of the volume of water that flows through a surface per unit time, it will depend on the component of $\vec{v}_i$ normal to the plane of the area element and the magnitude of the area $|\Delta\vec{A}_i|$ (Fig. 15-25). But since the component of $\vec{v}_i$ along the direction of the normal vector to an area is $|\vec{v}|\cos\theta$, the flux element Φ_i is given by

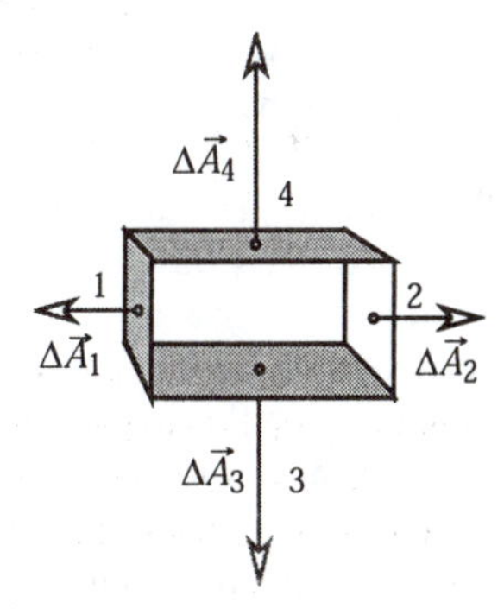

Fig. 15-25: The directions of the normal vectors are shown for four of the six faces that make up a container. Each normal vector points out of the container. Since area elements 3 and 4 have twice the area as 1 and 2, their normal vectors have twice the length.

$$\Phi_i = (|\vec{v}_i|\cos\theta)|\Delta\vec{A}_i| = \vec{v}_i \cdot \Delta\vec{A}_i$$

where $|\Delta\vec{A}_i|$ represents the magnitude of the ith area element. In order to make our definition of flux more generally useful, we will develop an equation for flux in terms the velocity and area vectors $\vec{v}$ and $\Delta\vec{A}$. We define the flux of water through the ith element of area mathematically using the scalar product (introduced in Section 9-4) as

$$\Phi_i \equiv \vec{v}_i \cdot \Delta\vec{A}_i \qquad \text{(volume flux definition for a small area element).} \qquad (15\text{-}25)$$

Let us consider how to calculate the net flow of water through an imaginary surface placed in a stream such as the box shown in Fig. 15-26. We can assign a velocity vector to each point in the stream of water. If our area elements are small enough, the velocity vector is the same everywhere on the surface of a given area element. Thus, if we know the area and orientation of each face and the magnitude and direction of the stream velocity vectors, we can use Eq. 15-25 to calculate the net flux through the cross sectional area of the stream by adding up the products of the normal velocity vector components and the area elements through which water flows. Mathematically this is given by

$$\Phi_{net} = \Phi_1 + \Phi_2 + \ldots \Phi_N = \vec{v}_1 \cdot \Delta\vec{A}_1 + \vec{v}_2 \cdot \Delta\vec{A}_2 + \ldots \vec{v}_N \cdot \Delta\vec{A}_N = \sum_{n=1}^{N} \vec{v}_n \cdot \Delta\vec{A}_n \qquad (15\text{-}26)$$

where $\vec{v}_1, \vec{v}_2, \vec{v}_3$, and so on represent the velocity vectors at the location of each of the N area elements.

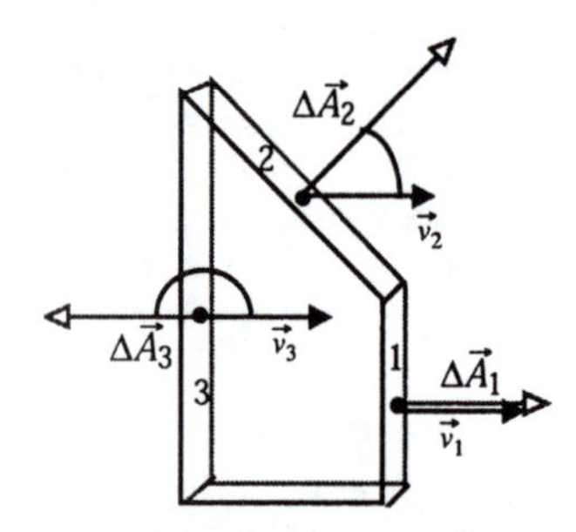

Fig. 15-26: A small box with six faces (or surfaces) is placed in an area of the stream where the velocity of the water is uniform over all the faces of the box. The angle between the velocity vector and the normal vectors for three of the six faces of the box are shown.

Net Fluid Volume Flux through a Closed Container

Let us consider how fluid flows into face 3 and out of faces 1 and 2 of a small imaginary box we have placed in our stream (Fig. 15-26). If our fluid, in this case water, is incompressible and is not created or destroyed inside our imaginary closed surface, we expect that the rate at which it flows into the box should be the same as the rate at which it flows out of the box. The total flux through all the faces of a closed container is known as the **net flux**. Thus, we expect the net flux of an incompressible fluid through the box to be zero. Does the mathematics of our method for finding net flux tell us this?

We have chosen our area elements to be small enough that the flow velocity vector at the location of any particular area element is uniform. However, that doesn't mean that magnitudes of the velocity vectors and relative angle between the velocity vectors and the normal vectors are necessarily the same at the location of each flat surface area. Thus, in general the net flux through a surface is the sum of the flux through each surface area and is still given by the application of Eq. 15-26 to our new situation. Note that our definition of net flux allows us to deal with a general case for which we have different velocity vectors, and normal vector orientations at each face. This might be the case if the stream is very turbulent.

When the velocity and area vectors point in the same general direction as they do in faces 1 and 2, the scalar product of $\vec{v}$ and $\Delta\vec{A}$ is positive. This tells us that water *flowing out of a surface* is defined as a *positive* flux. But the flux at face 3 is different. The velocity and area vectors are in opposite directions and their scalar product is negative. Thus, when water flows *into a surface* the flux is *negative*. Another feature of our box is that the areas associated with the bottom face and the front and back faces all have normal vectors that are perpendicular to the stream velocity vectors. The scalar product rules give us no flux or volume flow through these additional faces. It can be shown that the sum of the negative flux of water into face 3, the positive flux of water out of face 1, and the positive flux out of face 2 adds up to zero.

If we refine the equation of continuity to take the direction of flow of fluid into a closed surface as negative volume flux and the flow out of a closed surface as positive volume flux, we indeed expect the net flux through our imaginary box to be zero. This makes sense physically so long as water is incompressible and so long as we can't spontaneously create new water inside the box or remove water from the box. Later when we define electric and magnetic flux, we will see that it may be possible to have a net flux at the boundaries of a closed surface.

READING EXERCISE 15-10: Consider the imaginary box in Fig. 15-26 and assume that it has a width of 4.0 cm, a depth of 1.0 cm, a height on the left side of 8.0 cm and a height on the right side of 4.0 cm. (a) Find the magnitude of the area of faces 1, 2, and 3 respectively. Report your answer in square meters, and (b) What is the total surface area of the box? (Be sure to include the areas of all six area elements (faces) in your calculation.)

READING EXERCISE 15-11: Consider the imaginary box in Fig. 15-26 and assume that it is placed in a stream that moves from left to right through the box with a uniform velocity. In other words we make the simplifying assumption that the velocity vector is the same at every point and on the surface of the small box. Suppose the box has a width of 4.0 cm, a depth of 1.0 cm, a height on the left side of 8.0 cm and a height on the right side of 4.0 cm. If the magnitude of the stream velocity at the location of the box is 0.50 m/s (a) Find the flux through faces 1, 2, and 3 respectively in m^3/s, (b) explain why only faces 1, 2, and 3 have non-zero flux, and (c) show that the net flux through the closed surface defined by the box is zero.

15-11 Bernoulli's Equation

Figure 15-27 represents a tube through which an ideal fluid is flowing at a steady rate. In a time interval Δt, suppose that a volume of fluid ΔV, colored dark purple in Fig. 15-27*a*, enters the tube at the left (or input) end and an identical volume, colored light green in Fig. 15-27*b*, emerges at the right (or output) end. The emerging volume must be the same as the entering volume because the fluid is incompressible, with an assumed constant density ρ.

Let y_1, $|\vec{v}_1|$, and P_1 be the elevation, speed, and pressure of the fluid entering at the left, and y_2, $|\vec{v}_2|$, and P_2 be the corresponding quantities for the fluid emerging at the right. By applying the principle of conservation of energy to the fluid, we shall show that these quantities are related by

$$P_1 + \tfrac{1}{2}\rho v_1^2 + \rho g y_1 = P_2 + \tfrac{1}{2}\rho v_2^2 + \rho g y_2. \qquad (15\text{-}27)$$

We can also write this equation as

$$P + \tfrac{1}{2}\rho v^2 + \rho g y = \text{a constant} \quad \text{(Bernoulli's equation).} \qquad (15\text{-}28)$$

Equations 15-27 and 15-28 are equivalent forms of **Bernoulli's equation,** named after Daniel Bernoulli, who studied fluid flow in the 1700s.* Like the equation of continuity (Eq. 15-22), Bernoulli's equation is not a new principle but simply the reformulation of a familiar principle in a form more suitable to fluid mechanics. As a check, let us apply Bernoulli's equation to fluids at rest, by putting $|\vec{v}_1| = |\vec{v}_2| = 0$ in Eq. 15-27. The result is

$$P_2 = P_1 + \rho g(y_1 - y_2),$$

which is Eq. 15-7 with a slight change in notation.

A major prediction of Bernoulli's equation emerges if we take y to be a constant ($y = 0$, say) so that the fluid does not change elevation as it flows. Equation 15-27 then becomes

$$P_1 + \tfrac{1}{2}\rho v_1^2 = P_2 + \tfrac{1}{2}\rho v_2^2, \qquad (15\text{-}29)$$

which tells us that:

➤If the speed of a fluid element increases as it travels along a horizontal streamline, the pressure of the fluid must decrease, and conversely.

Put another way, where the streamlines are relatively close together (that is, where the velocity is relatively great), the pressure is relatively low, and conversely.

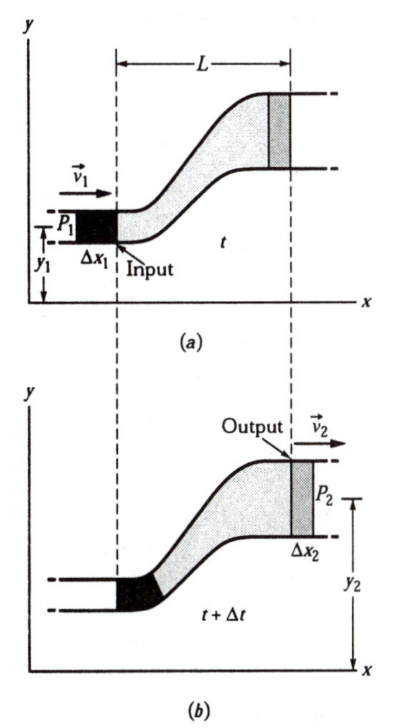

Fig. 15-27: Fluid flows at a steady rate through a length L of a tube, from the input end at the left to the output end at the right. From time t in (*a*) to time $t+\Delta t$ in (*b*), the parcel of fluid of mass m shown in dark purple enters the input end and the equal amount shown in light green emerges from the output end.

---

*For irrotational flow (which we assume), the constant in Eq. 15-28 has the same value for all points within the tube of flow; the points do not have to lie along the same streamline. Similarly, the points 1 and 2 in Eq. 15-27 can lie anywhere within the tube of flow.

The link between a change in speed and a change in pressure makes sense if you consider a fluid element. When the element nears a narrow region, the higher pressure behind it accelerates it so that it then has a greater speed in the narrow region. When it nears a wide region, the higher pressure ahead of it decelerates it so that it then has a lesser speed in the wide region.

Bernoulli's equation is strictly valid only to the extent that the fluid is ideal. If viscous forces are present, thermal energy will be involved. We take no account of this in the derivation that follows.

Proof of Bernoulli's Equation

Let us take as our system all of the (ideal) fluid shown in Fig. 15-27. We shall apply the principle of conservation of energy to the entire volume of the system as it moves from its initial state (Fig. 15-27*a*) to its final state (Fig. 15-27*b*). We assume that the fluid lying between the two vertical planes separated by a distance L in Fig. 15-27 does not change its properties during this process; thus we need be concerned only with changes that take place at the input and output ends.

The work-kinetic energy theorem tells us that if energy is conserved

$$W_{net} = \Delta K. \tag{15-30}$$

In other words, the change in the kinetic energy of our system must equal the net work done on it. But the change in kinetic energy results from the change in speed of the parcels of fluid at the ends of the tube. Thus,

$$\begin{aligned} \Delta K &= \tfrac{1}{2}mv_2^2 - \tfrac{1}{2}mv_1^2 \\ &= \tfrac{1}{2}\rho V(v_2^2 - v_1^2), \end{aligned} \tag{15-31}$$

in which $m = \rho V$ is the mass of the parcel of fluid of volume V that enters at the input end and the parcel that leaves at the output end during a small time interval Δt .

The work done on the system arises from two sources. One is the work W_g done by the gravitational force ($m\vec{g}$) on the parcel of fluid of mass m during the vertical lift of the mass from the input level to the output level is

$$\begin{aligned} W_g &= -mg(y_2 - y_1) \\ &= -\rho g V(y_2 - y_1). \end{aligned} \tag{15-32}$$

This work is negative because the upward displacement and the downward gravitational force have opposite directions.

The other source of work is the net work done on the entire system of moving fluid as a result of the pressure difference between its input and output ends. We start by finding the equation for the work needed to move a parcel of fluid through a distance Δx in a tube of cross section A. If the parcel of fluid is under pressure P, it is given by

$$W = |\vec{F}|\Delta x = (PA)\Delta x = P(\mathrm{A}\,\Delta x) = P\,\Delta V.$$

We now consider the system consisting of the entire volume of fluid shown in Fig. 15-27 (not just the volume V of a parcel of the fluid). Since there is a pressure difference between the left end of the system and the right end of the system the net work done on the system due to forces at the two ends is then given by

$$\begin{aligned} W_s &= P_1 V - P_2 V \\ &= -(P_2 - P_1)V. \end{aligned} \tag{15-33}$$

The work-kinetic energy theorem of Eq. 15-30 now becomes

$$W = W_g + W_s = \Delta K.$$

Substituting from Eqs. 15-31, 15-32, and 15-33 yields

$$-\rho g V(y_2 - y_1) - V(P_2 - P_1) = \tfrac{1}{2}\rho V(v_2^2 - v_1^2).$$

This, after a slight rearrangement, matches Eq. 15-27, which we set out to prove.

READING EXERCISE 15-12: Water flows smoothly through the pipe shown in the figure, descending in the process. Rank the four numbered sections of pipe according to (a) the volume flow rate R_V through them, (b) the flow speed $|\vec{v}|$ through them, and (c) the water pressure P within them, greatest first.

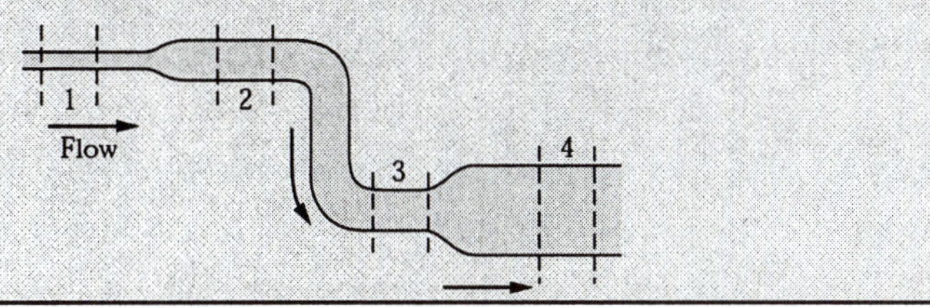

Touchstone Example 15-11-1, at the end of this chapter, illustrates how to use what you learned in this section.

TE

Touchstone Example 15-3-1

A living room has floor dimensions of 3.5 m and 4.2 m and a height of 2.4 m.

(a) What does the air in the room weigh when the air pressure is 1.0 atm?

SOLUTION: The **Key Ideas** here are these: (1) The air's weight is equal to mg, where m is its mass. (2) Mass m is related to the air density ρ and the air's volume V by Eq. 15-4 ($\rho = m/V$). Putting these two ideas together and taking the density of air at 1.0 atm from Table 15-2, we find

$$\begin{aligned} mg &= (\rho V)g \\ &= (1.21\ \text{kg/m}^3)(3.5\ \text{m} \times 4.2\ \text{m} \times 2.4\ \text{m})(9.8\ \text{m/s}^2) \\ &= 418\ \text{N} \approx 420\ \text{N}. \end{aligned} \quad \text{(Answer)}$$

This is the weight of about 110 cans of Pepsi.

(b) What is the magnitude of the atmosphere's force on the floor of the room?

SOLUTION: The **Key Idea** here is that the atmosphere pushes up on the ceiling with a force of magnitude $|\vec{F}|$ that is uniform over the ceiling. Thus, it produces a pressure that is related to $|\vec{F}|$ and the flat area A of the ceiling by Eq. 15-1 ($P = |\vec{F}|/A$), which gives us

$$\begin{aligned} |\vec{F}| &= PA = (1.0\ \text{atm})\left(\frac{1.01 \times 10^5\ \text{N/m}^2}{1.0\ \text{atm}}\right)(3.5\ \text{m})(4.2\ \text{m}) \\ &= 1.5 \times 10^6\ \text{N}. \end{aligned} \quad \text{(Answer)}$$

Touchstone Example 15-4-1

A novice scuba diver practicing in a swimming pool takes enough air from his tank to fully expand his lungs before abandoning the tank at depth L and swimming to the surface. He ignores instructions and fails to exhale during his ascent. When he reaches the surface, the difference between the external pressure on him and the air pressure in his lungs is 9.3 kPa. From what depth does he start? What potentially lethal danger does he face?

SOLUTION: The **Key Idea** here is that when the diver fills his lungs at depth d, the external pressure on him (and thus the air pressure within the lungs) is greater than normal and given by Eq. 15-8b as

$$P = P_{\text{atm}} + \rho_{\text{water}} g d,$$

where P_{atm} is atmospheric pressure and ρ_{water} is the water's density (998 kg/m^3, from Table 15-2). As he ascends, the external pressure on him decreases, until it is atmospheric pressure P_{atm} at the surface. His blood pressure also decreases, until it is normal. However, because he does not exhale, the air pressure in his lungs remains at the value it had at depth d. At the surface, the pressure difference between the higher pressure in his lungs and the lower pressure on his chest is

$$\Delta P = P - P_{\text{atm}} = \rho_{\text{water}} g d,$$

from which we find

$$\begin{aligned} d &= \frac{\Delta P}{\rho g} = \frac{9300\ \text{Pa}}{(998\ \text{kg/m}^3)(9.8\ \text{m/s}^2)} \\ &= 0.95\ \text{m}. \end{aligned} \quad \text{(Answer)}$$

This is not deep! Yet, the pressure difference of 9.3 kPa (about 9% of atmospheric pressure) is sufficient to rupture the diver's lungs and force air from them into the depressurized blood, which then carries the air to the heart, killing the diver. If the diver follows instructions and gradually exhales as he ascends, he allows the pressure in his lungs to equalize with the external pressure, and then there is no danger.

Touchstone Example 15-4-2

The U-tube in Fig. TE15-1 contains two liquids in static equilibrium: Water of density ρ_w (= 998 kg/m^3) is in the right arm, and oil of unknown density ρ_x is in the left. Measurement gives l = 135 mm and d = 12.3 mm. What is the density of the oil?

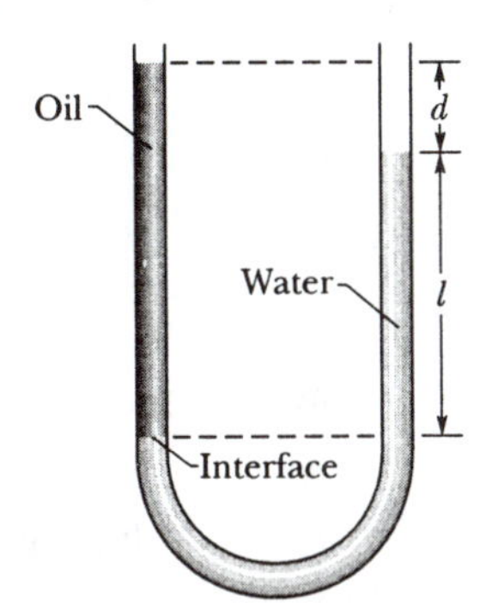

Fig. TE15-1 The oil in the left arm stands higher than the water in the right arm because the oil is less dense than the water. Both fluid columns produce the same pressure P_{int} at the level of the interface.

SOLUTION: One **Key Idea** here is that the pressure P_{int} at the oil–water interface in the left arm depends on the density ρ_x and height of the oil above the interface. A second **Key Idea** is that the water in the right arm *at the same level* must be at the same pressure P_{int}. The reason is that, because the water is in static equilibrium, pressures at points in the water at the same level must be the same even if the points are separated horizontally.

In the right arm, the interface is a distance l below the free surface of the *water* and we have, from Eq. 15-8*b*,

$$P_{int} = P_{atm} + \rho_w g l \qquad \text{(right arm).}$$

In the left arm, the interface is a distance $l + d$ below the free surface of the *oil* and we have, again from Eq. 15-8*b*,

$$P_{int} = P_{atm} + \rho_x g(l + d) \qquad \text{(left arm).}$$

Equating these two expressions and solving for the unknown density yield

$$\rho_x = \rho_w \frac{l}{l + d} = (998\ \text{kg/m}^3)\frac{135\ \text{mm}}{135\ \text{mm} + 12.3\ \text{mm}}$$
$$= 915\ \text{kg/m}^3. \qquad \text{(Answer)}$$

Note that the answer does not depend on the atmospheric pressure P_{atm} or the free-fall acceleration g.

Touchstone Example 15-7-1

What fraction of the volume of an iceberg floating in seawater is visible?

SOLUTION: Let V_i be the total volume of the iceberg. The nonvisible portion is below water and thus is equal to the volume V_f of the fluid (the seawater) displaced by the iceberg. We seek the fraction (call it frac)

$$\text{frac} = \frac{V_i - V_f}{V_i} = 1 - \frac{V_f}{V_i}, \qquad \text{(TE15-1)}$$

but we know neither volume. A **Key Idea** here is that, because the iceberg is floating, Eq. 15-18 ($|\vec{F}_g| = m_f g$) applies. We can write that equation as

$$m_i g = m_f g,$$

from which we see that $m_i = m_f$. Thus, the mass of the iceberg is equal to the mass of the displaced fluid (seawater). Although we know neither mass, we can relate them to the densities of ice and seawater given in Table 15-2 by using Eq. 15-4 ($\rho = m/V$). Because $m_i = m_f$, we can write

$$\rho_i V_i = \rho_f V_f$$

or

$$\frac{V_f}{V_i} = \frac{\rho_i}{\rho_f}.$$

Substituting this into Eq. TE15-1 and then using the known densities, we find

$$\text{frac} = 1 - \frac{\rho_i}{\rho_f} = 1 - \frac{917 \text{ kg/m}^3}{1024 \text{ kg/m}^3}$$

$$= 0.10 \text{ or } 10\%. \qquad \text{(Answer)}$$

Touchstone Example 15-9-1

The cross-sectional area A_0 of the aorta (the major blood vessel emerging from the heart) of a normal resting person is 3 cm², and the speed $|v_0|$ of the blood through it is 30 cm/s. A typical capillary (diameter ≈ 6 μm) has a cross-sectional area A of 3×10^{-7} cm² and a flow speed $|\vec{v}|$ of 0.05 cm/s. How many capillaries does such a person have?

SOLUTION: The **Key Idea** here is that all the blood that passes through the capillaries must have passed through the aorta. Therefore, the volume flow rate through the aorta must equal the total volume flow rate through the capillaries. Let us assume that the capillaries are identical, with the given cross-sectional area A and flow speed $|\vec{v}|$. Then, from Eq. 15-21 we have

$$A_0|\vec{v}_0| = nA|\vec{v}|$$

where n is the number of capillaries. Solving for n yields

$$n = \frac{A_0|\vec{v}_0|}{A|\vec{v}|} = \frac{(3 \text{ cm}^2)(30 \text{ cm/s})}{(3 \times 10^{-7} \text{ cm}^2)(0.05 \text{ cm/s})}$$

$$= 6 \times 10^9 \text{ or 6 billion.} \qquad \text{(Answer)}$$

You can easily show that the combined cross-sectional area of the capillaries is about 600 times the cross-sectional area of the aorta.

Touchstone Example 15-9-2

Figure TE15-2 shows how the stream of water emerging from a faucet "necks down" as it falls. The indicated cross-sectional areas are $A_0 = 1.2$ cm² and $A = 0.35$ cm². The two levels are separated by a vertical distance $h = 45$ mm. What is the volume flow rate from the tap?

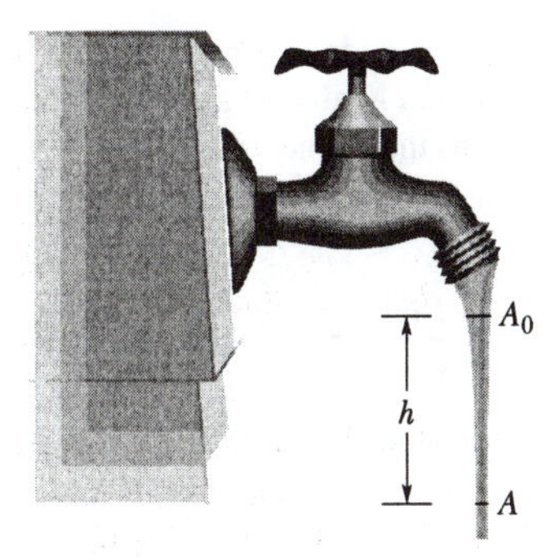

Fig. TE15-2 As water falls from a tap, its speed increases. Because the flow rate must be the same at all cross sections, the stream must "neck down."

SOLUTION: The **Key Idea** here is simply that the volume flow rate through the higher cross section must be the same as that through the lower cross section. Thus, from Eq. 15-21, we have

$$A_0|\vec{v}_0| = A|\vec{v}|, \qquad \text{(TE15-2)}$$

where $|\vec{v}_0|$ and $|\vec{v}|$ are the water speeds at the levels corresponding to A_0 and A. From Eq. 2-16 we can also write, because the water is falling freely with acceleration g,

$$v^2 = v_0^2 + 2gh. \qquad \text{(TE15-3)}$$

Eliminating $|\vec{v}|$ between Eqs. TE15-2 and TE15-3 and solving for $|\vec{v}_0|$, we obtain,

$$|\vec{v}_0| = \sqrt{\frac{2ghA^2}{A_0^2 - A^2}}$$

$$= \sqrt{\frac{(2)(9.8\ \text{m/s}^2)(0.045\ \text{m})(0.35\ \text{cm}^2)^2}{(1.2\ \text{cm}^2)^2 - (0.35\ \text{cm}^2)^2}}$$

$$= 0.286\ \text{m/s} = 28.6\ \text{cm/s}.$$

From Eq. 15-22, the volume flow rate R_V is then

$$R_V = A_0|\vec{v}_0| = (1.2\ \text{cm}^2)(28.6\ \text{cm/s})$$

$$= 34\ \text{cm}^3/\text{s}. \qquad \text{(Answer)}$$

Touchstone Example 15-11-1

Ethanol of density $\rho = 791$ kg/m^3 flows smoothly through a horizontal pipe that tapers in cross-sectional area from $A_1 = 1.20 \times 10^{-3}$ m^2 to $A_2 = A_1/2$. The pressure difference between the wide and narrow sections of pipe is 4120 Pa. What is the volume flow rate R_V of the ethanol?

SOLUTION: One **Key Idea** here is that, because the fluid flowing through the wide section of pipe must entirely pass through the narrow section, the volume flow rate R_V must be the same in the two sections. Thus, from Eq. 15-22,

$$R_V = |\vec{v}_1|A_1 = |\vec{v}_2|A_2. \qquad \text{(TE15-4)}$$

However, with two unknown speeds, we cannot evaluate this equation for R_V.

A second **Key Idea** is that, because the flow is smooth, we can apply Bernoulli's equation. From Eq. 15-27, we can write

$$P_1 + \tfrac{1}{2}\rho v_1^2 + \rho g y = P_2 + \tfrac{1}{2}\rho v_2^2 + \rho g y, \qquad \text{(TE15-5)}$$

where subscripts 1 and 2 refer to the wide and narrow sections of pipe, respectively, and y is their common elevation. This equation hardly seems to help because it does not contain the desired volume flow R_V and it contains the unknown speeds $|\vec{v}_1|$ and $|\vec{v}_2|$.

However, there is a neat way to make it work for us: First, we can use Eq. TE15-4 and the fact that $A_2 = A_1/2$ to write

$$|\vec{v}_1| = \frac{R_V}{A_1} \quad \text{and} \quad |\vec{v}_2| = \frac{R_V}{A_2} = \frac{2R_V}{A_1}. \qquad \text{(TE15-6)}$$

Then we can substitute these expressions into Eq. TE15-5 to eliminate the unknown speeds and introduce the desired volume flow rate. Doing this and solving for R_V yield

$$R_V = A_1\sqrt{\frac{2(P_1 - P_2)}{3\rho}}. \qquad \text{(TE15-7)}$$

We still have a decision to make: We know that the pressure difference between the two sections is 4120 Pa, but does that mean that $P_1 - P_2$ is 4120 Pa or -4120 Pa? We could guess the former is true, or otherwise the square root in Eq. TE15-7 would give us an imaginary number. Instead of guessing, however, let's try some reasoning. From Eq. TE15-4 we see that speed $|\vec{v}_2|$ in the narrow section (small A_2) must be greater than speed $|\vec{v}_1|$ in the wider section (larger A_2). Recall that if the speed of a fluid increases as it travels along a horizontal path (as here), the pressure of the fluid must decrease. Thus, P_1 is greater than P_2, and $P_1 - P_2 = 4120$ Pa. Inserting this and known data into Eq. TE15-7 gives

$$R_V = 1.20 \times 10^{-3}\ \text{m}^2\sqrt{\frac{(2)(4120\ \text{Pa})}{(3)(791\ \text{kg/m}^3)}}$$

$$= 2.24 \times 10^{-3}\ \text{m}^3/\text{s}. \qquad \text{(Answer)}$$

16 Oscillations

On September 19, 1985, seismic waves from an earthquake that originated along the west coast of Mexico caused terrible and widespread damage in Mexico City, about 400 km from the origin.

Why did the seismic waves cause such extensive damage in Mexico City but almost none on the way there?

The answer is in this chapter.

16-1 Periodic Motion: An Overview

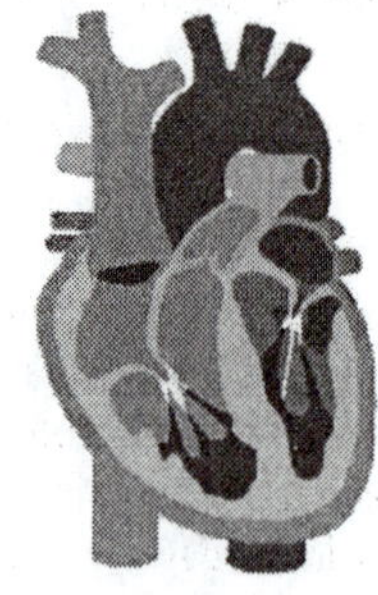

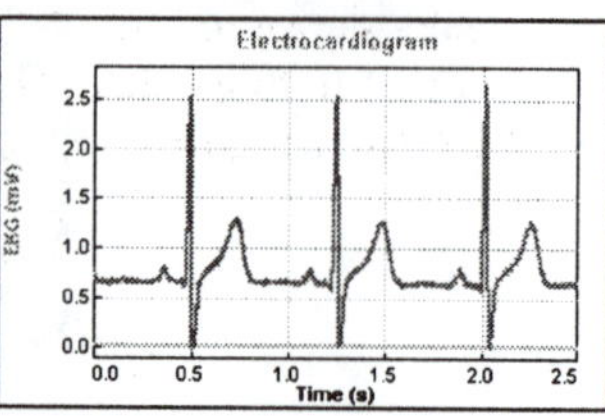

Fig. 16-1: An electrocardiogram showing the periodic pattern of electrical signals that drive human heart beats. Data recorded with an MBL system EKG sensor. (Courtesy of Vernier Software and Technology.)

Any measurable quantity that repeats itself at regular time intervals is defined as undergoing **periodic** behavior. We are surrounded by systems with quantities that vary periodically. The systems with periodic behavior that are most familiar involve obvious mechanical oscillations or motions. There are swinging chandeliers, boats bobbing at anchor, and the surging pistons in the engines of cars. There are oscillating guitar strings, drum heads, bells, diaphragms in telephones and speaker systems, and quartz crystals in wristwatches.

The motions associated with some periodic behavior are not obvious. For example, we cannot see the oscillations of the air molecules that transmit the sensation of sound, the oscillations of the atoms in a solid that convey the sensation of temperature, and the oscillations of the electrons in the antennas of radio and TV transmitters that convey information. Some examples of periodic changes are shown in Figs. 16-1 and 16-2. They include electrical signals associated with human heartbeats and the air pressure changes that occur when musical instruments are played.

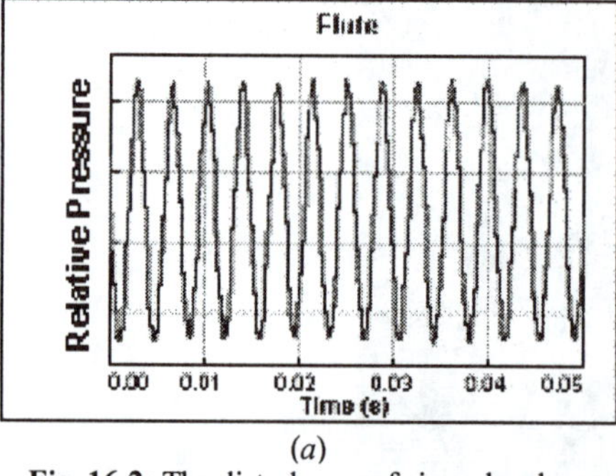

(*a*)

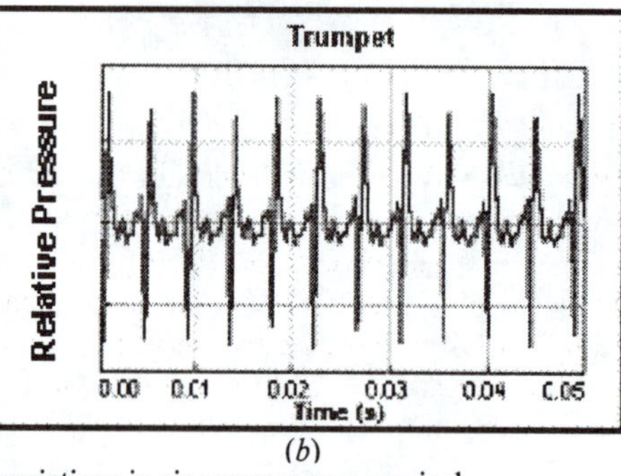

(*b*)

Fig. 16-2: The disturbance of air molecules causes variations in air pressure near musical instruments. The pattern of these pressure variations repeats itself at regular time intervals so that the sound is periodic. The pressure variations which are proportional to the voltage output of a small microphone can be recorded with an MBL system. (*a*) A sustained note from a trumpet and (*b*) from a flute. (Data courtesy of Vernier Software and Technology.)

In this chapter you will study the motion of mechanical systems which undergo very regular sinusoidal oscillations known as **simple harmonic motion** (or **SHM**). The air pressure variations of the flute shown in Fig. 16-2*a* over time looks like a sine or cosine function, and this tells us that the pressure variations are due to air molecules undergoing simple harmonic motion. Although the motions of the electrons in the body which cause the EKG pattern in Fig. 16-1 and the motions of the air molecules near the bell of a trumpet in Fig. 16-2*b* are periodic, they are not simple harmonic motions. However, a mathematical equation for periodic motion can be constructed from a linear combination of simple harmonic motions.

Mastering the mathematical description of simple harmonic motion is critical to acquiring a full understanding of periodic physical systems. It is vital to obtaining a full appreciation of the rotational motions already introduced in Chapters 5, 11, and 12. In addition, the material introduced in this chapter is needed in the study of mechanical and sound waves treated in Chapters 17 and 18. SHM concepts will help you in the study of magnetism, light and other electromagnetic waves encountered in Chapters 30-37. Finally, an understanding of mechanical oscillations will help you understand modern physics including relativity and the wave behavior of photons and matter that you will encounter in Chapters 38-45.

16-2 Simple Harmonic Motion

Simple Harmonic Motion and Uniform Circular Motion

In 1610, Galileo, using his newly constructed telescope, discovered the four principal moons of Jupiter. Over weeks of observation, each moon seemed to him to be moving back and forth relative to the planet in what today we would call simple harmonic motion; the disk of the planet was the midpoint of the motion. The record of Galileo's observations, written in his own hand, is still available. A. P. French of MIT used Galileo's data to work out the position of the moon Callisto relative to Jupiter. In the results shown in Fig. 16-3, the circles are based on Galileo's observations and the curve is a best fit to the data. The curve looks sinusoidal with a full oscillation that takes about 16.8 days as can be seen on the plot.

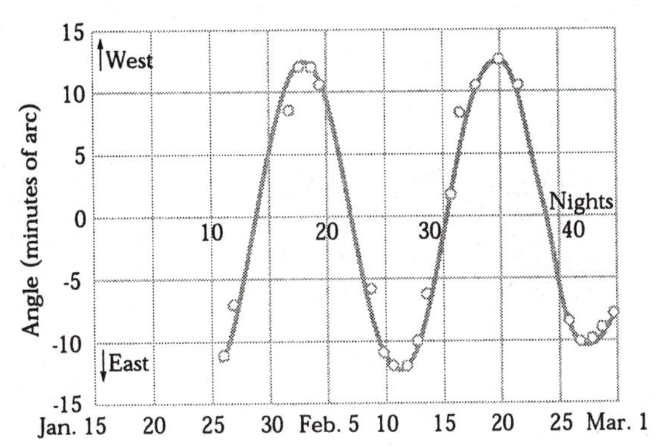

Fig. 16-3: The angle between Jupiter and its moon Callisto as seen from Earth. The circles are based on Galileo's 1610 measurements. The curve is a best fit, strongly suggesting simple harmonic motion. At Jupiter's mean distance, 10 minutes of arc corresponds to about 2×10^6 km. (Adapted from A. P. French, *Newtonian Mechanics*, W. W. Norton & Company, New York, 1971, p. 288.)

Actually, Callisto moves with essentially constant speed in an essentially circular orbit around Jupiter. Its true motion—far from being simple harmonic—is uniform circular motion. What Galileo saw—and what you can see with a good pair of binoculars and a little patience—is the projection of this uniform circular motion on a line in the plane of the motion. We are led by Galileo's remarkable observations to the conclusion that simple harmonic motion is uniform circular motion viewed edge-on. In more formal language:

➤Simple harmonic motion is the projection of uniform circular motion on a diameter of the circle in which the latter motion occurs.

We can explore the relationship between the sinusoidal oscillations associated with simple harmonic motion and uniform circular motion that we studied in Chapter 5 more carefully using an everyday object instead of a moon that orbits around a distant planet. Consider a spot on a disk that is rotating about an axis at a constant rotational velocity. Sample frames of a movie depicting this motion are shown in Fig. 16-4. For this case we have chosen to point the x-axis up and the y-axis to the right. If we take a series of side views of the disk (so you only see one of the dimensions it is moving in), the projection of the spot on the x-axis as a function of time gives us a sinusoidal graph as shown in Fig. 16-5.

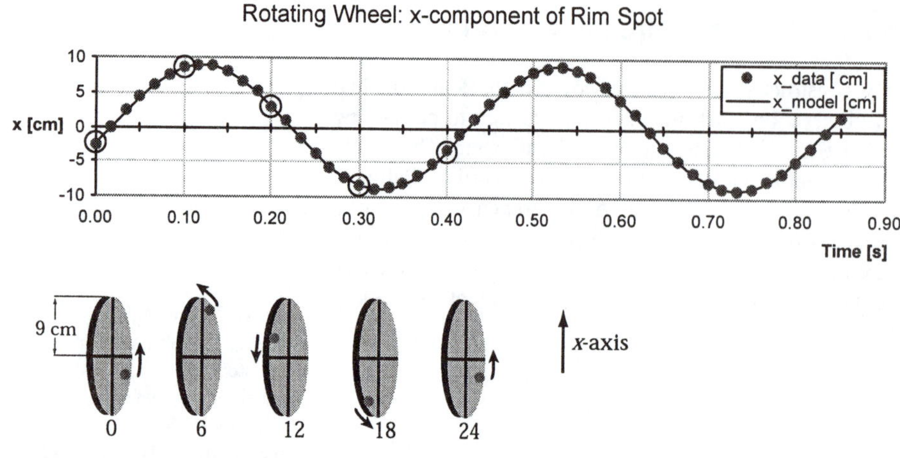

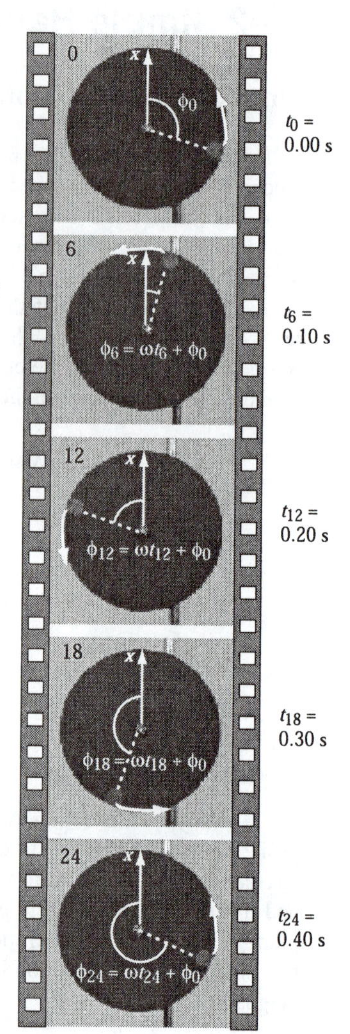

Fig. 16-4: This selection of frames shows the positions of a spot on a rotating disk every 1/10th of a second. They represent every 6th frame of a video sequence recorded at 60 frames/second. The angle the spot makes with respect to the chosen *x*-axis increases at a constant rate.

Fig. 16-5: The motion of a disk with a constant rotational velocity was recorded at 60 frames a second. A plot of the *x*-component of a spot on the disk has a sinusoidal shape. Thus it oscillates with SHM. The location of the spot on our sideways depictions of the disk is shown below the graph for every 6th frame.

Period and Frequency

As part of our study of uniform circular motion in Section 5-7 we introduced the idea of a period as the time it takes an object rotating about an axis at a regular rate to complete a single revolution. For example, the video frames of a rotating disk in Fig. 16-4 show that it has a period of 0.40 s. In Figure 16-5, we see that the projection of the spot on our chosen *x* axis, appears to oscillate up and down with a period of T = 0.40 s. Thus, the time for one oscillation of the spot's *x*-component is the same as the period of the rotation of the disk.

While the period tells us the time for one rotation or oscillation, **frequency** is a related quantity that tells us how many oscillations or cycles there are in a given time. For example, we can see from the Fig. 16-5 graph that the spot has a frequency of 2.5 s^{-1} because that's the number of complete oscillations that would occur if we had taken data for one second rather than for only 0.85 s. The symbol for frequency is *f*, and its SI unit is the **hertz** (abbreviated Hz), where

$$1 \text{ hertz} = 1 \text{ Hz} = 1 \text{ oscillation per second} = 1 \text{ s}^{-1}. \qquad (16\text{-}1)$$

Clearly when the period of oscillation is very short there are many more oscillations in a second so the frequency goes up. The converse is true also, when the period is long the frequency goes down. In fact, we see that the period and frequency of the oscillations shown in Fig. 16-5 are inversely related to each other. This inverse relationship holds in general so that

$$T = \frac{1}{f}. \qquad (16\text{-}2)$$

The Equation Describing Simple Harmonic Motion

We have rather glibly described the graphs shown in Figs. 16-2*a* and 16-5 as representing sinusoidal functions. We do this because the graphs look like that of a sine or cosine as a function of angle. Recall that the cosine of an angle, ϕ, is defined as the ratio of the distance of a point of interest from the *y*-axis, denoted as *x*, and the magnitude of the distance of the point from the origin denoted as r as shown in Fig. 16-6. The sine function is similarly defined in terms of a ratio involving the distance from the x-axis:

$$\cos\phi \equiv \frac{x}{r}, \qquad (16\text{-}3)$$

$$\sin\phi \equiv \frac{y}{r}. \qquad (16\text{-}4)$$

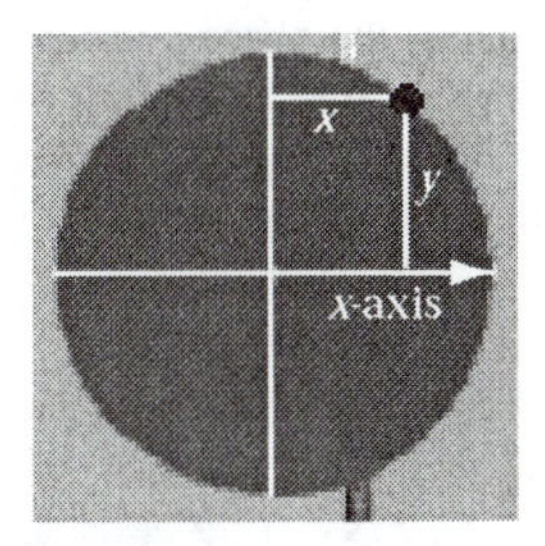

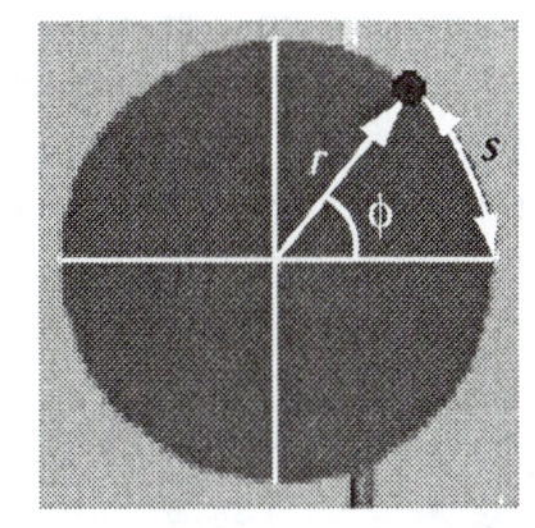

Fig. 16-6: The location of a spot on a disk can be described in either Cartesian or Polar coordinates. It is conventional to measure the rotational position ϕ with respect to the x-axis where a counterclockwise arc from the axis is defined as a positive ϕ.

In considering rotational positions or phases for oscillations, we continue the convention of describing angles in radians (or rads) used in Chapters 11 and 12. Recall that in Eq. 11-1 the magnitude of the radian is defined as the magnitude of the ratio of the arc length s of a rotating object and the object's distance from its axis of rotation ($|\phi| = s/r$). As the object moves through a complete cycle it sweeps out an arc length of $s = 2\pi r$ and thus an angle of $2\pi r/r = 2\pi$ radians. Although it would be possible to use degrees, this choice would not be as convenient for reasons that will become more apparent later.

How do our definitions of the sinusoidal functions in Eq. 16-3 and Eq. 16-4 lead to the graph shapes shown in Figs. 16-2*a* and 16-5? Let's use the cosine function as an example. In Fig. 16-4, a spot on a disk is turning with a constant rotational velocity ω. If we denote the initial angular position of the disk spot when $t = 0$ s as ϕ_0, then the angular position increases according to the equation $\theta = \omega t + \phi_0$. Let us think about the first cycle. In the time period where θ is near zero the value of x does not change very rapidly but after a quarter-turn near $\theta = \pi/2$ the value of x is changing very rapidly. The rate of change of x slows down again near $\theta = \pi$. Between π and 2π, x is negative but its rate of change speeds up and slows down again. This clearly leads to a wave-like sinusoidal graph shape that goes on and on as the disk spot turns round and round.

As we already stated, simple harmonic motion is defined as the type of periodic sinusoidal motion described by the x-axis projection of our rotating disk spot shown in Fig. 16-5. In general for such a sinusoidal motion, the displacement x of the particle from the origin as a function of time *can be described by either a sine or cosine function.* Using the cosine function we find the SHM can be represented adequately by the equation

$$x(t) = x_m \cos(\omega t + \phi_0) \quad \text{(SHM displacement)}, \quad (16\text{-}5)$$

where x_m, ω, and ϕ_0 are constants. The quantities that determine the shape of the graph are displayed in Fig. 16-7 with their names. We now shall define those quantities.

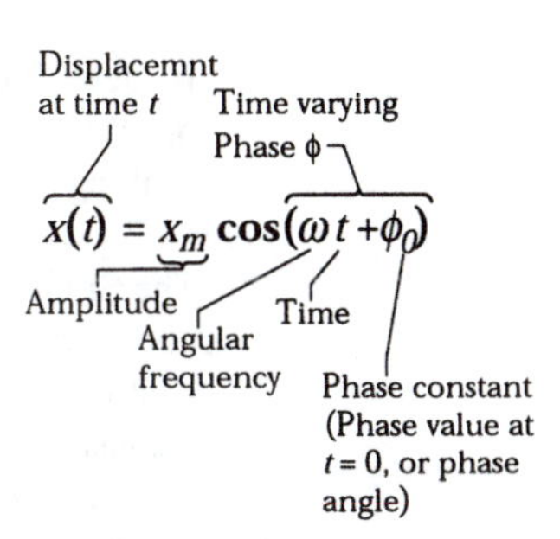

Fig. 16-7: A handy reference to the quantities in Eq. 16-5 for simple harmonic motion.

The quantity x_m, called the **amplitude** of the motion, is a positive constant whose value depends on how the motion was started. The subscript m stands for *maximum* because the amplitude is the magnitude of the maximum displacement of the particle in either direction. The cosine function in Eq. 16-5 varies between the limits ± 1, so the displacement $x(t)$ varies between the limits $\pm x_m$. For example, in Fig. 16-5 the amplitude of the cosine curve is obviously 9 cm. This is also the distance from the axis of rotation of the disk to the spot near its rim.

The time-varying quantity $\phi = (\omega t + \phi_0)$ shown in Eq. 16-5 is called the **time varying phase** of the motion. The constant ϕ_0 is called the **initial phase or phase constant** which is also sometimes referred to as the phase angle. The value of the initial ϕ_0 depends on the displacement and velocity of an oscillating particle at time $t = 0$.

Some of the time-varying phases for the rotating disk are shown in Fig. 16-5. Each "frame" shown consists of every 6th frame of a more complete set of video images. The phases in each of the frames in the figure are denoted as ϕ_0, $\phi_6 = \omega t_6 + \phi_0$, $\phi_{12} = \omega t_{12} + \phi_0$ and so on where the corresponding times are $t_0 = 0.00$ s, $t_6 = 0.10$ s, $t_{12} = 0.30$ s. Note that the initial phase (or phase constant), allows us to determine that value of the displacement x at time $t = 0.00$ s. Thus, the phase constant plays the same role as the x_0 or y_0 terms in the kinematic equations because it determines the initial value of the function.

For simplicity, in the $x(t)$ plots of Fig. 16-8*a*, the phase constant ϕ_0 has been set to zero.

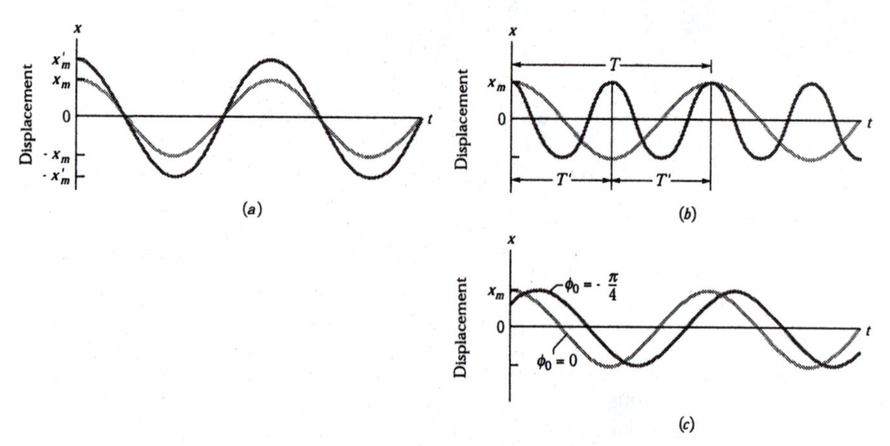

Fig. 16-8: In each case, the light grey curve is obtained from Eq. 16-5 with $\phi_0 = 0$. (*a*) The darker grey curve differs from the light grey curve *only* in that its amplitude x'_m is greater (the dark grey-curve extremes of displacement are higher and lower). (*b*) The darker grey curve differs from the light grey curve *only* in that its period is $T' = T/2$ (the darker grey curve is compressed horizontally). (*c*) The darker grey curve differs from the light grey curve *only* in that $\phi_0 = -\pi/4$ rad rather than zero (the negative value of ϕ_0 shifts the dark grey curve to the right).

In describing SHM, the constant ω is known as the **angular frequency** of the motion. When the SHM equation represents the projection of a spot on a steadily rotating object along an axis, *the rotational velocity of the object and the angular frequency of the projection are identical in value*. This is certainly the case in Fig. 16-5.

The SI unit of angular frequency is the radian per second. (To be consistent, then, ϕ_0 must be in radians.) Figure 16-8 compares $x(t)$ for two simple harmonic motions that differ either in amplitude, in period (and thus in frequency and angular frequency), or in phase constant.

We can use the fact that the rotational velocity of a spinning object and the angular frequency of a corresponding oscillation are identical to derive an important relationship. In particular, we can relate the angular frequency, ω, of an oscillating object to its oscillation frequency *f*. The derivation goes as follows: A rotating object undergoing uniform circular motion sweeps through an angle of 2π radians in a single period *T* so that its rotational velocity is given by $\omega = 2\pi/T$. But since the frequency is inversely proportional to the period (that is, since $f = 1/T$), it is obvious that

$$\omega = \frac{2\pi}{T} = 2\pi f. \tag{16-6}$$

READING EXERCISE 16-1: Although it is not conventional to do so, the equation $x(t) = x'_m \sin(\omega' t + \phi'_0)$ can also be used to describe simple harmonic motion. Suppose the same motion has been described by both the cosine function in Eq. 16-5 and by the sine function shown here. Consider the amplitude, angular frequency and phase constant associated with the motion. Which factors stay the same? Which will change? Explain.
Hint: You may want to think about the spot on a disk as it undergoes uniform circular motion.

READING EXERCISE 16-2: A particle undergoing simple harmonic oscillation of period T (like that in Fig. 16-1) is at $-x_m$ at time $t = 0$. Is it at $-x_m$, at $+x_m$, at 0, between $-x_m$ and 0, or between 0 and $+x_m$ when (a) $t = 2.00T$, (b) $t = 3.50T$, and (c) $t = 5.25T$?

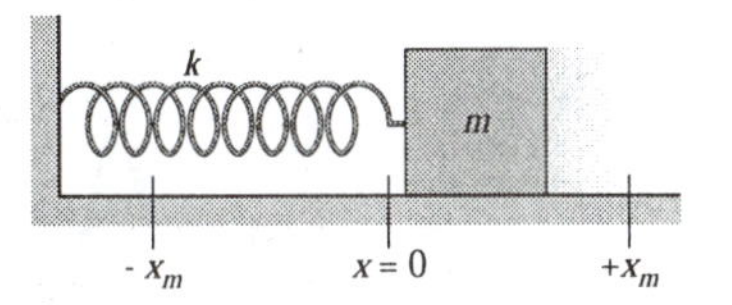

Fig. 16-9: We can imagine a *horizontal oscillator* that consists of a mass-spring system. In theory the mass can oscillate back and forth on a frictionless surface when it is displaced from equilibrium. In practice, this system is very difficult to set up.

16-3 The Mass-Spring System and the SHM Force Law

So far we have defined how to describe simple harmonic motion mathematically as shown in Eq. 16-5. In this section we are interested in using Newton's Second Law to deduce what kind of *forces* are required to cause a system to undergo simple harmonic motion. We will use a mass-spring system as a model.

In physics we usually analyze a simplified model system before considering more complex "real-world" systems. Unfortunately, in picking a model mass-spring system we are presented with a dilemma. The simplest system to model mathematically is a *horizontal oscillator* with a partially extended spring attached to a block that slides on a perfectly frictionless surface as shown in Fig. 16-9. However, it is not feasible to set up an experiment using such a system. The simplest system to set up experimentally is a *vertical oscillator* in which a mass hangs down from a spring. However, it is easy to explain the net forces a mass experiences are the **same** for a horizontal and vertical oscillator. Thus, for simplicity, we will consider our model system to be a mass hanging down vertically from a spring.

We start by presenting the results of measurements made on our vertical mass-spring system. Next we show how our experimental knowledge of forces on the system can be used in conjunction with Newton's Second Law to derive its position vs. time equation theoretically. At the same time we can determine theoretically how factors such as mass and spring stiffness influence the period of oscillation. In a later section we will show how the forces experienced by a pendulum bob have the same mathematical characteristics as those that drive a mass on a spring.

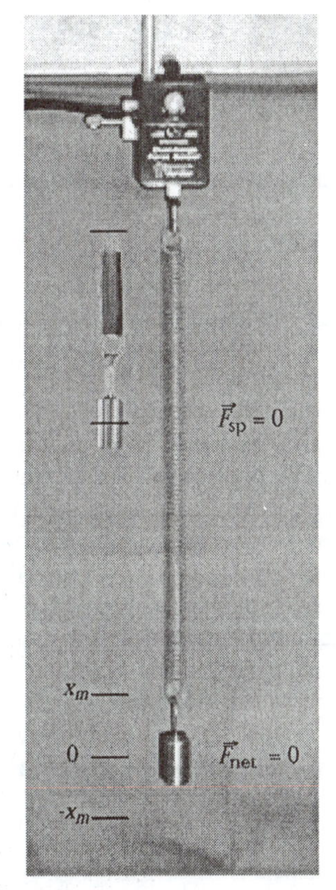

Fig. 16-10: A common way to measure forces and displacements for a vertical mass-spring oscillator is to hang a light spring with an attached mass from an electronic force sensor. An ultrasonic motion detector can be placed underneath the mass to measure displacements. If we displace the mass vertically from its equilibrium point, it will oscillate.

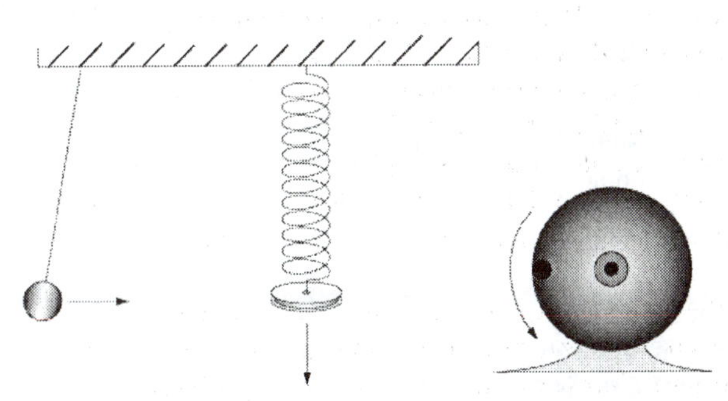

Fig. 16-11: We know that the pendulum, the mass on the spring and the Cartesian components of a spot on a disk undergo SHM. What are the mathematical characteristics of forces needed to induce simple harmonic motion in a mass-spring system? In a pendulum? Are they the same?

Experimental Findings: Displacement vs. Time

A 100 g mass is attached to a light 10.1 g spring and suspended from an electronic force sensor. An ultrasonic motion detector is placed underneath it to record its displacement from equilibrium. The mass is pulled down to a maximum displacement of about 4.0 cm and released. Shortly thereafter a microcomputer-based laboratory or MBL system (discussed in Section 1-8) is used to measure the forces exerted by the spring on the mass. At the same time, the displacement of the mass is tracked with the motion detector. The experimental results for the displacement from equilibrium as a function of time are shown in Fig. 16-12.

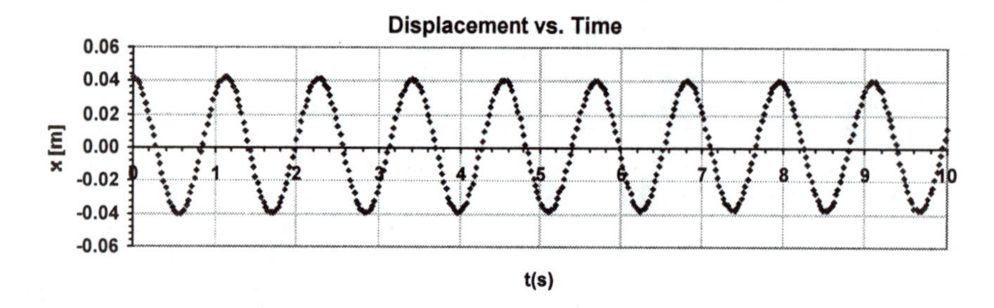

Fig. 16-12: Experimental data shows that the net force exerted on a 100 g mass by a spring causes the mass to oscillate with what appears to be SHM.

Careful examination of the graph in Fig. 16-12 gives a period of oscillation for our 100 g mass is $T_{exp} = 1.14\text{ s}$. A model of the plotted data shows that the equation that describes the displacement vs. time has exactly the same form as Eq. 16-5. In particular,

$$x(t) = x_m \cos(\omega t + \phi_0) = (0.040\text{ m})\cos\{(5.5\,\text{rad/s})t + 0.0\text{ rad}\}. \qquad (16\text{-}7)$$

The value of the maximum displacement or amplitude x_m is 0.040 m, or only 4.0 cm, the angular frequency, ω, is 5.5 rad/s, and the initial phase (or phase constant), ϕ_0, is 0.0 rad. Equation 16-7 provides experimental verification that our vertical spring-mass system oscillates with *simple harmonic motion.*

Experimental Results: F_{net} vs. Displacement from Equilibrium

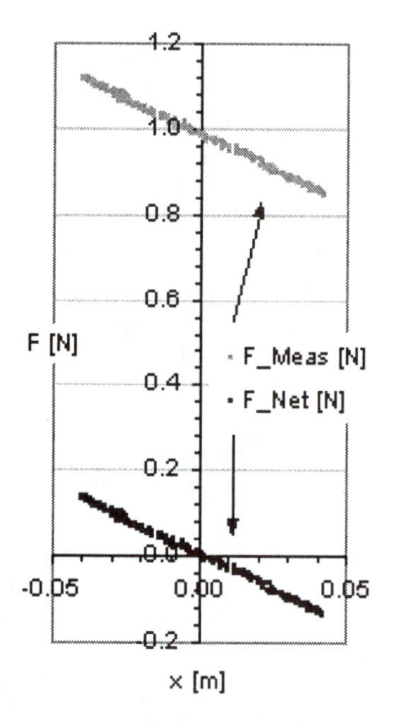

Fig. 16-13: The net force on a mass hung vertically from a spring is the vector sum of the downward component of gravitational force on the mass, $-mg$, and the measured force due to the total spring stretch.

In order to use Newton's Second Law to understand how the hanging mass should move, we are interested in finding the *net force* on it. The net force is clearly a combination of the spring force acting upward and the gravitational force acting downward. But, we know from Section 6-4 that the tension forces at the ends of a massless stretched spring are equal and opposite so that the magnitude of the downward force the spring exerts on the force sensor has the same magnitude as the upward force exerted on the hanging mass. Thus, if we can assume that the spring mass is negligible, the force sensor gives us a direct measure of only the spring force acting on the mass. For a massless spring we can calculate the net force along the vertical axis as the vector sum of the vertical spring and gravitational forces. Thus, the force components can be combined in the equation

$$F_{\text{net},x} = F_{\text{spring},x} + F_{\text{grav},x} = F_{\text{spring},x} - mg \qquad \text{(massless spring).}$$

The top line in Fig. 16-13 shows the x-component of the measured spring force as a function of the displacement of the mass from equilibrium. The bottom line shows the net force when the downward component of the gravitational force is "added" vectorially to it.

A least squares analysis of the negative slope of the net force vs. displacement data shown in the bottom graph in Fig. 16-13 shows that the net force component which lies on the $F_{net} = -kx$ where $k = 3.23$ [N/m]. This result—a restoring force component in the x direction that is proportional to the displacement component but opposite in sign—is familiar. It is Hooke's law as described earlier in Eq. 9-19,

$$F_{net} = -kx.$$

We can use the fact that Eq. 9-19 describes the spring forces that cause our mass to oscillate sinusoidally to develop an alternative definition of simple harmonic motion. It says:

➤Simple harmonic motion is the motion executed by a particle of mass m subject to a force that is proportional to the displacement of the particle but opposite in sign.

Theoretical Considerations

We can combine Newton's Second Law with our experimental verification of Eq. 9-19 for a spring-mass system that undergoes SHM to get

$$F_{net} = ma_x = -kx, \tag{16-8}$$

where F_{net}, a_x, and x are the respective x-components of force, acceleration and displacement. This equation can be used to explain why Eq. 16-5 does indeed describe the motion of our mass-spring system. To do this, we need to express the acceleration of the mass as the second derivative of the displacement. By doing this we can re-write Eq. 16-8 as

$$F_{net} = ma_x = m\frac{d^2x}{dt^2} = -kx \text{ so that } \frac{d^2x}{dt^2} = -\frac{k}{m}x \tag{16-9}$$

What happens when we actually take the first and then the second derivative of Eq. 16-5? Do we get a minus sign and a positive constant that can be associated with the ratio k/m? The first derivative is

$$\frac{dx}{dt} = \frac{d(x_m \cos(\omega t + \phi_0))}{dt} = -\omega x_m \sin(\omega t + \phi_0),$$

while the second derivative is

$$\frac{d^2x}{dt^2} = \frac{d(-\omega x_m \sin(\omega t + \phi_0))}{dt} = -\omega^2 x_m \cos(\omega t + \phi_0) = -\omega^2 x. \tag{16-10}$$

We do indeed get a negative sign times a positive constant in front of the x term, but in order for the theoretical equation derived from Newton's Second Law to match our experimentally-determined equation for displacement vs. time, we must have the angular frequency be equal to

$$\omega = \sqrt{\frac{k}{m}} \quad \text{(angular frequency).} \tag{16-11}$$

By combining Eqs. 16-6 and 16-11, we can write, for the **period** of the linear oscillator shown in Fig. 16-10 as

$$T = 2\pi\sqrt{\frac{m}{k}} \quad \text{(period).} \tag{16-12}$$

Equations 16-11 and 16-12 tell us that a large angular frequency (and thus a small period) goes with a stiff spring (large k) and a low mass object (small m). This seems like a very reasonable pair of relationships. Let us verify whether our data for the mass spring system satisfies Eq. 16-12. In order to simplify our theoretical model *we have chosen to ignore the relatively small 10.1 g mass of the spring* and assume that the mass of the system can be adequately represented by only the hanging mass. Using this assumption our theoretical model tells us that a spring with k of 3.23 N/m and hanging mass, m, of 0.100 kg will have a period of

$$T_{\text{theory}} = 2\pi\sqrt{\frac{m}{k}} = 2\pi\sqrt{\frac{0.100 \text{ kg}}{3.23 \text{ N/m}}} = 1.11 \text{ s}$$

This predicted period is quite close to our experimental value of $T_{\text{exp}} = 1.14$ s! Actually, the fact that the experimental period is slightly larger than the theoretically determined period means that we can't completely ignore the relatively small mass of our spring.

Describing How Spring Forces Cause Oscillations

We can now see why an object that experiences a spring force oscillates. If we hold the object out at some displacement and release it, the net force acts opposite to the displacement so the object will start to accelerate towards the undisplaced position. It moves faster and faster towards the undisplaced position but as it gets there, the force is becoming smaller, so when it gets to its undisplaced position moving with some velocity, there is nothing to stop it. By Newton's First Law it keeps going and overshoots the equilibrium position. The force now acts in the opposite direction to slow it down, but by the time the force has brought the object to a stop the object has another displacement, this time on the other side. The process repeats, and if there is no friction or damping, the oscillations will go on forever.

Every oscillating system such as a diving board or a violin string has an element of "springiness" and an element of "inertia" or mass, and thus resembles the linear oscillator of Fig. 16-10. In the linear oscillator of Fig. 16-10, these elements are located in separate parts of the system: The springiness is in the spring, which we assume to be massless, and the inertia is entirely in the mass, which we assume to be rigid. In a violin string, however, the two elements are both within the string itself, as you will see in Section 17-6.

An Angular Simple Harmonic Oscillator

Figure 16-14 shows an angular version of a simple harmonic oscillator; the element of springiness or elasticity is associated with the twisting of a suspension wire rather than the extension and compression of a spring as we previously had. The device is called a **torsion oscillator** (or torsion pendulum), with *torsion* referring to the twisting.

If we rotate the disk in Fig. 16-14 by some angular displacement θ from its rest position (where the reference line is at $\theta = 0$) and release it, it will oscillate about that position in **angular simple harmonic motion.** Rotating the disk through an angle θ in either direction introduces a restoring torque given by

$$\tau = -\kappa\theta. \qquad (16\text{-}13)$$

Here κ (Greek *kappa*) is a constant, called the **torsion constant,** that depends on the length, diameter, and material of the suspension wire.

Comparison of Eq. 16-13 with Eq. 16-8 leads us to suspect that Eq. 16-13 is the angular form of Hooke's law, and that we can transform Eq. 16-12, which gives the period of linear SHM, into an equation for the period of angular SHM: we replace the spring constant k in Eq. 16-12 with its equivalent, the constant κ of Eq. 16-13, and we replace the mass m in Eq. 16-12 with *its* equivalent, the rotational inertia I of the oscillating disk. These replacements lead to

$$T = 2\pi\sqrt{\frac{I}{\kappa}} \quad \text{(torsion oscillator),} \qquad (16\text{-}14)$$

which is the correct equation for the period of an angular simple harmonic oscillator, or torsion pendulum.

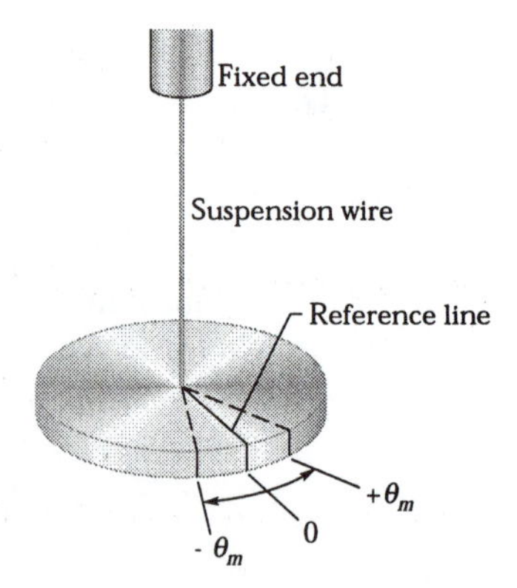

Fig. 16-14: An angular simple harmonic oscillator, or torsion oscillator, is an angular version of the linear simple harmonic oscillator of Fig. 16-10. The disk oscillates in a horizontal plane; the reference line oscillates with angular amplitude θ_m. The twist in the suspension wire stores potential energy as a spring does and provides the restoring torque.

READING EXERCISE 16-3: The experimental period of the 100 g vertical mass oscillating on the spring shown in Fig. 16-10 is about 3% larger than the period calculated using a simplified theoretical model in which the spring is assumed to be massless, but we know the spring actually has a mass of 10.1 g. Consider the nature of Eq. 16-12 and explain why the measured period should be a bit larger than the theoretical value we reported. No calculation is needed.

READING EXERCISE 16-4: Which of the following relationships between the force $\vec{F}$ on a particle and the particle's position x implies simple harmonic oscillation: (a) $\vec{F} = (-5\ \text{N/m})x$, (b) $\vec{F} = (-400\ \text{N/m}^2)x^2$, (c) $\vec{F} = (+10\ \text{N/m})x$, (d) $\vec{F} = (3\ \text{N/m}^2)x^2$?

16-4 Velocity and Acceleration for SHM

The Velocity for Simple Harmonic Motion

Let us imagine how the velocity of the mass on the spring in Fig. 16-10 changes as it moves through a complete oscillation cycle. Is the magnitude of the velocity a maximum, a minimum or zero when the magnitude of its displacement is the greatest? Obviously the mass has a velocity of zero when it is turning around. But the mass turns around when the magnitude of the displacement is a maximum. This means that the velocity of the mass is out of phase with its displacement in the same way that the cosine and sine functions are out of phase with each other. By differentiating Eq. 16-5, we find that the expression for the velocity of a particle moving with simple harmonic motion is indeed a sine function whenever the displacement is a cosine function; that is,

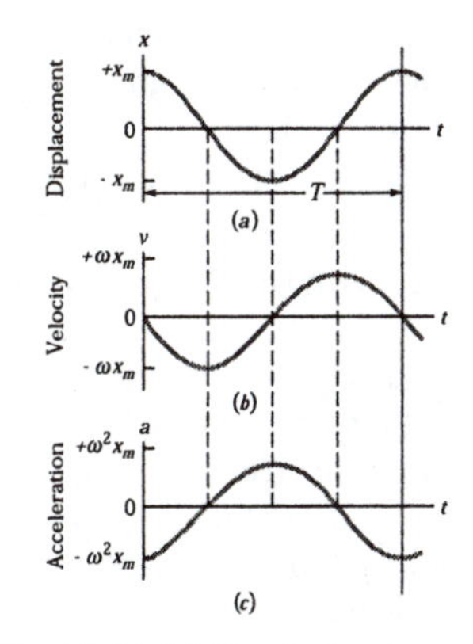

Fig. 16-15: (*a*) The displacement $x(t)$ of a particle oscillating in SHM with phase constant ϕ_0 equal to zero. The period T marks one complete oscillation. (*b*) The velocity $v(t)$ of the particle. (*c*) The acceleration $a(t)$ of the particle.

$$v(t) = \frac{dx(t)}{dt} = \frac{d}{dt}[x_m \cos(\omega t + \phi_0)]$$

or

$$v(t) = -\omega x_m \sin(\omega t + \phi_0) \quad \text{(velocity)}. \qquad (16\text{-}15)$$

Figure 16-15*a* is a plot of Eq. 16-5 with $\phi_0 = 0$ rad. Fig. 16-15*b* shows Eq. 16-15, also with $\phi_0 = 0$ rad. Analogous to the amplitude x_m in Eq. 16-5, the positive quantity ωx_m in Eq. 16-15 is called the **velocity amplitude** v_m. As you can see in Fig. 16-15*b*, the velocity of the oscillating particle varies between the limits $\pm v_m = \pm\omega x_m$. Note also in that figure that the curve of $v(t)$ is *shifted* (to the left) from the curve of $x(t)$ by one-quarter period; when the magnitude of the displacement is greatest (that is, $x(t) = x_m$), the magnitude of the velocity is least (that is, $v(t) = 0$ m/s). When the magnitude of the displacement is least (that is, zero), the magnitude of the velocity is greatest (that is, $v_m = \omega x_m$).

The Acceleration of SHM

Knowing the velocity $v(t)$ for simple harmonic motion, we can find an expression for the acceleration of the oscillating particle by differentiating once more. Thus, we have, from Eq. 16-15,

$$a(t) = \frac{dx(t)}{dt} = \frac{d}{dt}[-\omega x_m \sin(\omega t + \phi_0)],$$

or

$$a(t) = -\omega^2 x_m \cos(\omega t + \phi_0) \quad \text{(acceleration)}. \qquad (16\text{-}16)$$

Figure 16-15*c* is a plot of Eq. 16-16 for the case $\phi_0 = 0$ rad. The positive quantity $\omega^2 x_m$ in Eq. 16-16 is called the **acceleration amplitude** a_m; that is, the acceleration of the particle varies between the limits $\pm a_m = \pm\omega^2 x_m$, as Fig. 16-15*c* shows. Note also that the curve of $a(t)$ is shifted (to the left) by one-quarter period relative to the curve of $v(t)$.

We can combine Eqs. 16-5 and 16-16 to yield

$$a(t) = -\omega^2\, x(t), \qquad (16\text{-}17)$$

which is the hallmark of simple harmonic motion:

➤In SHM, the acceleration is proportional to the displacement but opposite in sign, and the two quantities are related by the square of the angular frequency.

Thus, as Fig. 16-15 shows, when the displacement has its greatest positive value, the acceleration has its greatest negative value, and conversely. When the displacement is zero, the acceleration is also zero.

16-5 Gravitational Pendulums

We turn now to a class of simple harmonic oscillators in which the springiness is associated with the gravitational force rather than with the elastic properties of a twisted wire or a compressed or stretched spring. Since oscillators that depend on gravitational forces for their springiness hang, they are all considered to be types of pendulums.

The Simple Pendulum Oscillating at a Small Angle

If you hang a small particle of mass m (called a *bob*) from the end of a long unstretchable massless thread and fix the thread at its upper end, you have constructed a **simple pendulum.** An example of a simple pendulum is shown in Fig. 16-16. What happens when you set the bob swinging back and forth a small horizontal distance compared to the length of the string? You easily see that the bob's motion is periodic. Is it, in fact, simple harmonic motion? If so, what is the period T?

We can use MBL data to measure the horizontal components of displacement, velocity, and acceleration as a function of time for the pendulum shown in Fig. 16-16. The results are presented graphically in Fig. 16-17.

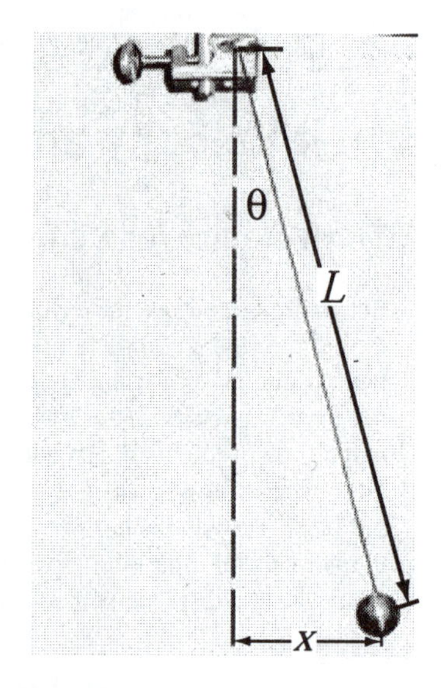

Fig. 16-16: A pendulum bob swings back and forth at a small angle. Its horizontal displacement x from its vertical equilibrium is measured using an ultrasonic motion detector attached to an MBL system. The length L of this pendulum measured from the pivot to the center of the bob is 32 cm.

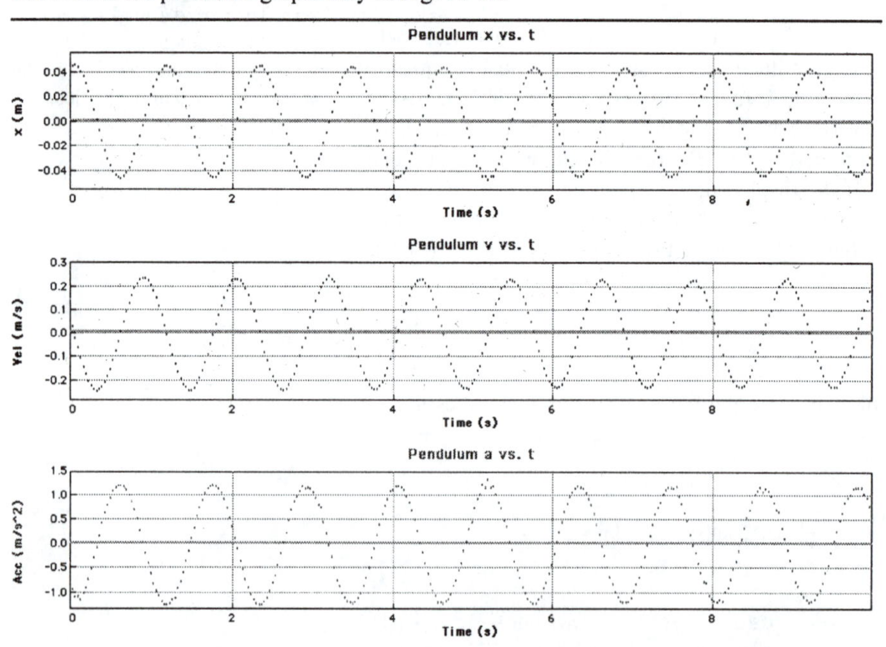

Fig. 16-17: Measurements for the horizontal displacement of the simple pendulum shown in Fig. 16-16 were obtained using an MBL system outfitted with a motion detector. Graphs of v_x, and a_x are also constructed from measurements of the horizontal displacement x. The length of the pendulum L was set at 32 cm.

Careful examination of the x vs. t graph in Fig. 16-17 gives a period of oscillation for our pendulum bob of $T_{\exp} = 1.14$ s. A model of the plotted data shows that the equation that describes the horizontal displacement vs. time has exactly the *same form* as Eq. 16-5. In particular,

$$x(t) = x_m \cos(\omega t + \phi_0) = (0.045 \text{ m})\cos\{(5.5 \text{ rad/s})t + 0.0 \text{ rad}\}. \quad (16\text{-}7)$$

The value of the maximum displacement or amplitude x_m is 0.045 m, or only 4.5 cm, the angular frequency, ω, is 5.5 rad/s, and the initial phase (or phase constant), ϕ_0, is 0.0 rad. Since

$$\sin\theta = \frac{x}{L}, \text{ then } \theta = \sin^{-1}\left(\frac{x}{L}\right) = \sin^{-1}\left(\frac{4.5}{32}\right) = 0.14 \text{ rad or } 8.1°,$$

the maximum angle of displacement is only about 8°. Equation 16-7 provides experimental verification that our simple pendulum oscillates with *simple harmonic motion* when it is pulled back and released at a small angle. This experimental verification strongly suggests that the net horizontal force on the pendulum bob is proportional to its horizontal displacement—at least when angles are small.

Let us see if our experimental results could have been predicted theoretically.

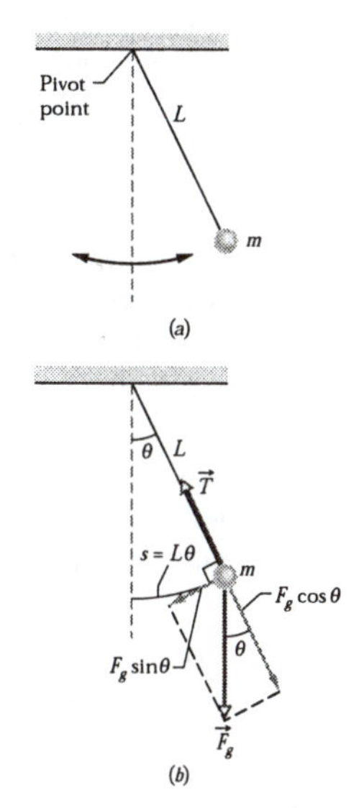

Fig. 16-18: (*a*) A simple pendulum. (*b*) The forces acting on the bob are the gravitational force $\vec{F}_g$ and the force $\vec{T}$ from the string. The tangential component $F_g \sin\theta$ of the gravitational force is a restoring force that tends to bring the pendulum back to its central position.

Theoretical Derivation of Simple Pendulum Forces

To find the net horizontal forces on the pendulum, we can set·up a free body diagram using methods introduced in Chapter 6. The forces acting on the bob are the force $\vec{T}$ from the string and the gravitational force $\vec{F}_g$, as shown in Fig. 16-18*b* where the string makes an angle θ with the vertical. We resolve $\vec{F}_g$ into a radial component $F_g \cos\theta$ and a component $F_g \sin\theta$ that is tangent to the path taken by the bob. This tangential component produces a restoring torque about the pendulum's pivot point, because it always acts opposite the displacement of the bob so as to bring the bob back toward its central location. That location is called the *equilibrium position* $(\theta = 0)$, because the pendulum would be at rest there were it not swinging.

From Eq. 11-30 $(|\vec{\tau}| = r_\perp |\vec{F}|)$, we can write this restoring torque as

$$\tau = -L(|\vec{F}_g|\sin\theta), \quad (16\text{-}18)$$

where the minus sign indicates that the torque acts to reduce θ, and L is the moment arm of the force component $F_g \sin\theta$ about the pivot point. Substituting Eq. 16-18 into Eq. 11-32 $(\tau = I\alpha)$ and then substituting mg as the magnitude of F_g, we obtain

$$-L(mg\sin\theta) = I\alpha, \quad (16\text{-}19)$$

where I is the pendulum's rotational inertia about the pivot point and α is its angular acceleration about that point.

We want to focus on the nature of the motion when the maximum (and minimum) angle of displacement is small. Note that whenever an angle θ is small, then the arc length *s that the pendulum bob sweeps through* (with respect to its equilibrium) and the value of x, have essentially the same magnitude as shown in Fig. 16-19, so that

$$\sin\theta = \frac{x}{L} \approx \frac{s}{L} = \theta \quad \text{(approximation for small } \theta\text{)}. \quad (16\text{-}20)$$

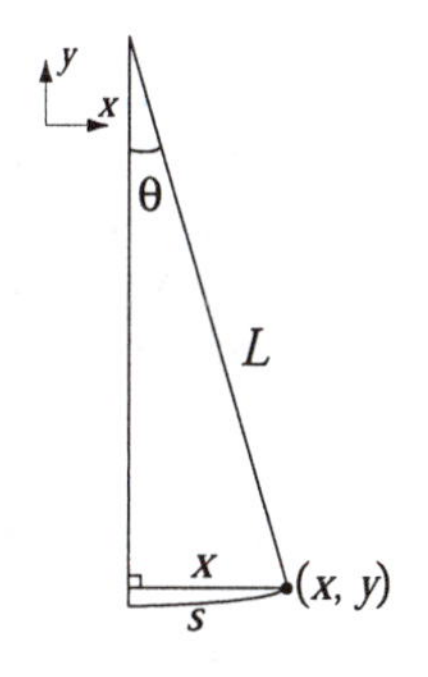

Fig. 16-19: This diagram illustrates that the arc length s and the value of y are approximately the same when the angle θ is small.

Thus when the angle θ is small, then we can replace sin θ with θ (expressed in radian measure). (As an example, if $\theta = 8.00° = 0.140$ rad, then $\sin\theta = 0.139$, a difference of only about 0.7%.) Using this approximation and rearranging terms, we then have

$$\alpha \approx -\frac{mgL}{I}\theta. \quad (16\text{-}21)$$

This equation is the angular equivalent of Eq. 16-5, the hallmark of SHM. It tells us that the angular acceleration α of the pendulum is proportional to the angular displacement $F_{\text{net},x} = ma_x = -(mg/L)x$ but opposite in sign. We can also express this same relationship in terms of the horizontal displacement x. For Figure 16-19, we can see that at small angles the arc length, s, swept out by the pendulum and the horizontal displacement x, are approximately the same or $x \approx s$, since $s = L\theta$ and for the pendulum $I = mL^2$ and $\alpha = a/L$. Given all this, we can now write

$$F_{\text{net},x} = ma_x = -\frac{mg}{L}x. \qquad (16\text{-}22)$$

This form of Eq. 16-8 tells us that the linear horizontal acceleraton is proportional to the horizontal displacement which we suspected must be the case as a result of our simple pendulum experiment. Only here, for a given pendulum the mg/L assumes the role of k for the mass-spring system.

Thus, as the pendulum bob moves to, say, the right as in Fig. 16-18*a*, its acceleration *to the left* increases until it stops and begins moving to the left. Then, when it is on the left, its acceleration to the right tends to return it to the right, and so on, as it swings back and forth in SHM. More precisely, we have verified both experimentally and theoretically that the motion of a *simple pendulum swinging through small angles* is approximately SHM. We can state this restriction to small angles another way: To be correct on predicting the period of a motion to within about 0.7%, the **angular amplitude** θ_m of the motion (the maximum angle of swing) must be about 8° or less.

Comparing Eqs. 16-22 and Eq. 16-17, we see that the angular frequency of the pendulum is $\omega = \sqrt{mgL/I}$. Next, if we substitute this expression for ω into Eq. 16-6 $(\omega = 2\pi/T)$, we see that the period of the pendulum may be written as

$$T = 2\pi\sqrt{\frac{I}{mgL}}. \qquad (16\text{-}23)$$

All the mass of a simple pendulum is concentrated in the mass m of the particle-like bob, which is at radius L from the pivot point. Thus, we can use Eq. 11-23 $(I = mr^2)$ to write $I = mL^2$ for the rotational inertia of the pendulum. Substituting this into Eq. 16-23 and simplifying yields

$$T = 2\pi\sqrt{\frac{L}{g}} \quad \text{(simple pendulum, small amplitude)}, \qquad (16\text{-}24)$$

as a simpler expression for the period of a simple pendulum swinging through only small angles. (We also assume small-angle swinging in the problems of this chapter.) Of course we can use Eq. 16-24 to predict the period of the pendulum of length $L = 32$ cm described in Figs. 16-16 and 16-17. We get

$$T_{\text{theory}} = 2\pi\sqrt{\frac{L}{g}} = 2\pi\sqrt{\frac{0.32\text{m}}{9.80\ \text{N/kg}}} = 1.14\ \text{s}$$

This result matches our experimental result to three significant figures. This period is also identical to that of our hanging mass. This is because just for fun we chose the pendulum length L so that we would get the same period for the two systems!

A very surprising outcome of this theoretical derivation is *that the period of the simple pendulum does not depend on its bob mass*. If you reflect on this, you should be able to see that the motion of the simple pendulum bob is mass-independent for the same reason that the motion of a falling object does not depend on its mass.

Measuring *g* with a Simple Pendulum

Geologists often use a pendulum to determine the local gravitational strength, *g*, at a particular location on Earth's surface. If a simple pendulum is used, we can solve Eq. 16-24, for *g* to get

$$g = \frac{4\pi^2 L}{T^2}. \qquad (16\text{-}25)$$

Thus, by measuring *L* and the period *T*, we can find the value of *g*. In order to make more precise measurements, a number of refinements are needed. Geophysicists often use a physical pendulum consisting of a solid rod in conjunction with a more sophisticated equation than that shown in Eq. 16-25. They can also place the pendulum in an evacuated chamber.

The Physical Pendulum

A "real" pendulum that isn't just a point mass suspended from a massless string, is usually called a **physical pendulum,** and it can have a complicated distribution of mass, much different from that of a simple pendulum. Does a physical pendulum also undergo SHM? If so, what is its period?

Figure 16-20 shows an arbitrary physical pendulum displaced to one side by angle θ. The gravitational force $\vec{F}_g$ acts at its center of mass *C*, at a distance h from the pivot point *O*. In spite of their shapes, comparison of Figs. 16-20 and 16-18*b* reveals only one important difference between an arbitrary physical pendulum and a simple pendulum. For a physical pendulum, the restoring component $F_g \sin\theta$ of the gravitational force has a moment arm of distance *h* about the pivot point, rather than of string length *L*. In all other respects, an analysis of the physical pendulum would duplicate our analysis of the simple pendulum up through Eq. 16-23. Again, (for small θ_m) we would find that the motion is approximately SHM.

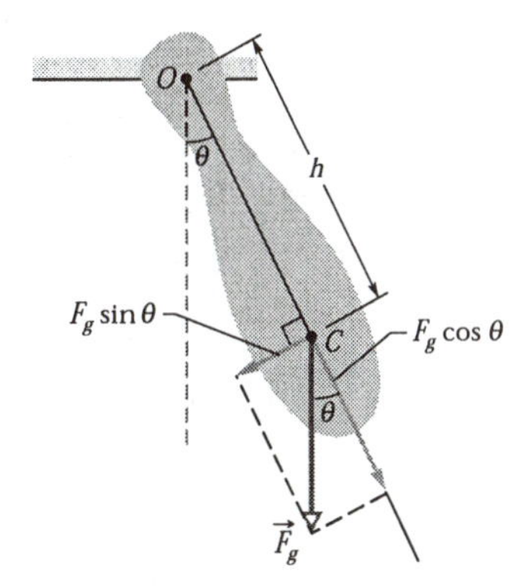

Fig. 16-20: A physical pendulum. The restoring torque is $hF_g \sin\theta$. When $\theta = 0$, the center of mass *C* hangs directly below pivot point *O*.

If we replace *L* with h in Eq. 16-23, we can write the period of a physical pendulum as

$$T = 2\pi\sqrt{\frac{I}{mgh}} \quad \text{(physical pendulum, small amplitude).} \qquad (16\text{-}26)$$

As with the simple pendulum, *I* is the rotational inertia of the pendulum about *O*. However, now *I* is not simply mL^2 (it depends on the shape of the physical pendulum), but it is still proportional to *m*.

A physical pendulum will not swing if it pivots at its center of mass. Formally, this corresponds to putting $h = 0$ in Eq. 16-26. That equation then predicts $T \to \infty$, which implies that such a pendulum will never complete one swing.

Corresponding to any physical pendulum that oscillates about a given pivot point *O* with period *T* is a simple pendulum of length L_0 with the same period *T*. We can find L_0 with Eq. 16-24. The point along the physical pendulum at distance L_0 from point *O* is called the *center of oscillation* of the physical pendulum for the given suspension point.

READING EXERCISE 16-5: The vertical acceleration of a falling object is independent of its mass. Likewise the period of a simple pendulum oscillating at small angles is independent of bob mass. Can you explain why in each case? What is similar about the two situations?

READING EXERCISE 16-6: Three physical pendulums, of masses m_0, $2m_0$, and $3m_0$, have the same shape and size and are suspended at the same point. Rank the masses according to the periods of the pendulum, greatest period first.

16-6 Energy in Simple Harmonic Motion

In Chapter 10 we saw that the energy of a simple pendulum swinging at a small angle transfers energy back and forth between kinetic energy and potential energy, while the sum of the two—the mechanical energy E of the oscillating pendulum—remains constant. What about the linear oscillator made up of a mass-spring system that was considered in Section 16-3? Does it trade energy back and forth as it oscillates? Let us try to answer this question for the linear oscillator using theoretical considerations.

The potential energy of a linear oscillator like that of Fig. 16-9 is associated entirely with the spring. Its value depends on how much the spring is stretched or compressed, that is, on $x(t)$. We can use Eqs. 10-14 and 16-5 to find

$$U(t) = \tfrac{1}{2}kx^2 = \tfrac{1}{2}kx_m^2\cos^2(\omega t + \phi). \tag{16-27}$$

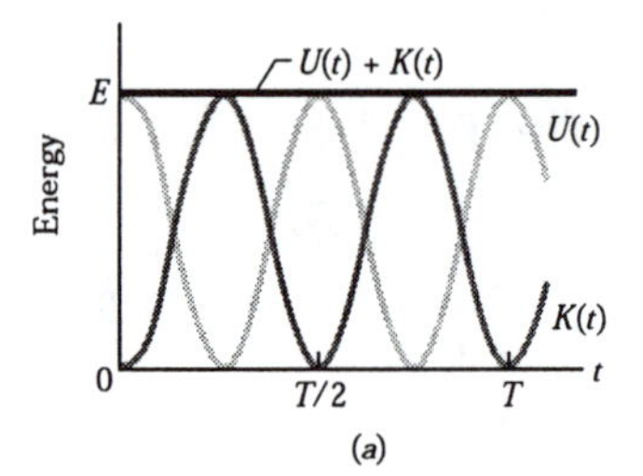

Note carefully that a function written in the form $\cos^2 A$ (as here) means $(\cos A)^2$ and is *not* the same as one written $\cos A^2$, which means $\cos(A^2)$.

The kinetic energy of the system of Fig. 16-9 is associated entirely with the block. Its value depends on how fast the block is moving—that is, on $v(t)$. We can use Eq. 16-15 to find

$$K(t) = \tfrac{1}{2}mv^2 = \tfrac{1}{2}m\omega x_m^2\sin^2(\omega t + \phi). \tag{16-28}$$

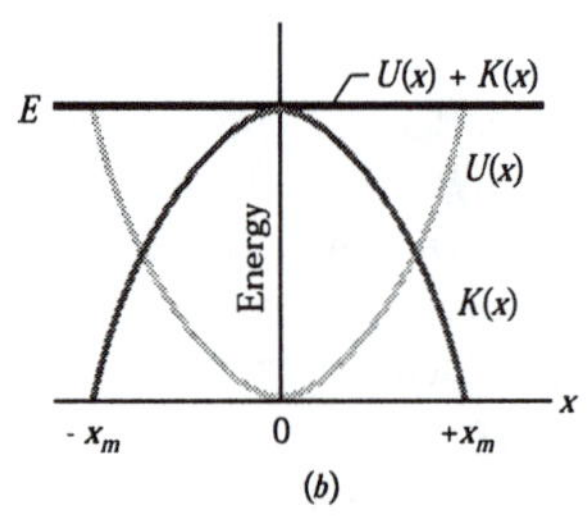

Fig. 16-21: (*a*) Potential energy $U(t)$, kinetic energy $K(t)$, and mechanical energy E as functions of time t for a linear harmonic oscillator. Note that all energies are positive and that the potential energy and the kinetic energy peak twice during every period.
(*b*) Potential energy $U(t)$, kinetic energy $K(x)$, and mechanical energy E as functions of position x for a linear harmonic oscillator with amplitude x_m. For $x = 0$ the energy is all kinetic, and for $x = \pm x_m$ it is all potential.

If we use Eq. 16-11 to substitute k/m for ω^2, we can write Eq. 16-28 as

$$K(t) = \tfrac{1}{2}mv^2 = \tfrac{1}{2}kx_m^2\sin^2(\omega t + \phi). \tag{16-29}$$

The mechanical energy follows from Eqs. 16-27 and 16-29 and is

$$\begin{aligned} E &= U + K \\ &= \tfrac{1}{2}kx_m^2\cos^2(\omega t + \phi) + \tfrac{1}{2}kx_m^2\sin^2(\omega t + \phi) \\ &= \tfrac{1}{2}kx_m^2[\cos^2(\omega t + \phi) + \sin^2(\omega t + \phi)]. \end{aligned}$$

For any angle α,

$$\cos^2\alpha + \sin^2\alpha = 1.$$

Thus, the quantity in the square brackets above is unity and we have

$$E = U + K = \tfrac{1}{2}kx_m^2. \tag{16-30}$$

The mechanical energy of a linear oscillator is indeed constant and independent of time. The potential energy and kinetic energy of a linear oscillator are shown as functions of time t in Fig. 16-21*a*, and as functions of displacement x in Fig. 16-21*b*.

You might now understand why an oscillating system normally contains an element of springiness and an element of inertia: The former stores its potential energy and the latter stores its kinetic energy.

READING EXERCISE 16-7: In Fig. 16-9, the block has a kinetic energy of 3 J and the spring has an elastic potential energy of 2 J when the block is at $x = +20$ cm. (a) What is the kinetic energy when the block is at $x = 0$? What are the elastic potential energies when the block is at (b) $x = -2.0$ cm and (c) $x = -x_m$?

16-7 Damped Simple Harmonic Motion

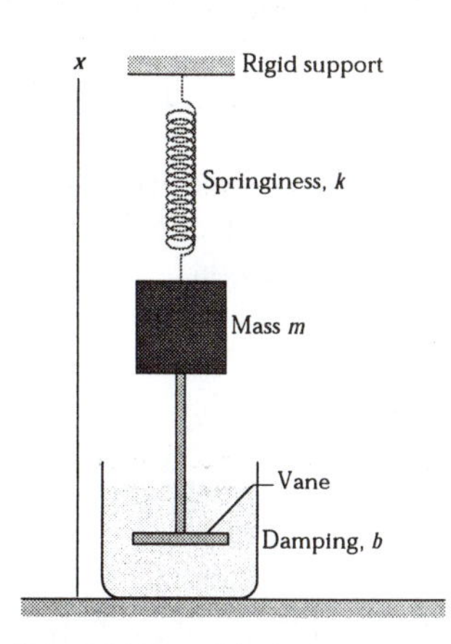

Fig. 16-22: An idealized damped simple harmonic oscillator. A vane immersed in a liquid exerts a damping force on the block as the block oscillates parallel to the x axis.

One thing you might have noticed about the harmonic motion we have described is that it goes on forever. Whatever starting value of t you put into Eq. 16-5, it will oscillate from then on with the same amplitude it started with. Oscillations in the real world usually die out gradually, transferring mechanical energy to thermal energy by the action of frictional forces. A pendulum will swing only briefly under water, because the water exerts a drag force on the pendulum that quickly eliminates the motion. A pendulum swinging in air does better, but still the motion dies out eventually, because the air exerts a drag force on the pendulum and friction acts at its support. These forces reduce the energy of the pendulum's motion.

When the motion of an oscillator is reduced by external friction or drag forces, the oscillator and its motion are said to be **damped.** An idealized example of a damped oscillator is shown in Fig. 16-22, where a block with mass m oscillates vertically on a spring with spring constant k. From the block, a rod extends to a vane (both assumed to have negligible mass) that is submerged in a liquid. As the vane moves up and down, the liquid exerts an inhibiting drag force on it and thus on the entire oscillating system. With time, the mechanical energy of the block-spring system decreases, as energy is transferred to thermal energy of the liquid and vane.

Let us assume the liquid exerts a **damping force** $\vec{F}_d$ that is proportional in magnitude to the velocity $\vec{v}$ of the vane and block (an assumption that is accurate if the vane moves slowly). Then, for components along the x-axis in Fig. 16-22, we have

$$\vec{F}_d = -b\vec{v}, \tag{16-31}$$

where b is a **damping constant** that depends on the characteristics of both the vane and the liquid and has the SI unit of kilogram per second. The minus sign indicates that $\vec{F}_d$ opposes the motion.

The force on the block from the spring is $\vec{F}_s = -kx$. Let us assume that the gravitational force on the block is negligible compared to $\vec{F}_d$ and $\vec{F}_s$. Then we can write Newton's Second Law for components along the x axis ($F_{\text{net},x} = ma_x$) as

$$-b\vec{v} - kx = m\vec{a}. \tag{16-32}$$

Substituting dx/dt for $\vec{v}$ and d^2x/dt^2 for $\vec{a}$ and rearranging give us the differential equation

$$m\frac{d^2x}{dt^2} + b\frac{dx}{dt} + kx = 0. \tag{16-33}$$

The solution of this equation is

$$x(t) = x_m e^{-bt/2m}\cos(\omega' t + \phi), \tag{16-34}$$

where x_m is the amplitude and ω' is the angular frequency of the damped oscillator. This angular frequency is given by

$$\omega' = \sqrt{\frac{k}{m} - \frac{b^2}{4m^2}}. \tag{16-35}$$

If $b = 0$ (there is no damping), then Eq. 16-35 reduces to Eq. 16-11 ($\omega = \sqrt{k/m}$) for the angular frequency of an undamped oscillator, and Eq. 16-34 reduces to Eq. 16-5 for the displacement of an undamped oscillator. If the damping constant is small but not zero (so that $b << \sqrt{km}$), then $\omega' \approx \omega$.

We can regard Eq. 16-34 as a cosine function whose amplitude, which is $x_m e^{-bt/2m}$, gradually decreases with time, as Fig. 16-23 suggests. For an undamped oscillator, the mechanical energy is constant and is given by Eq. 16-30 ($E = \frac{1}{2}kx_m^2$). If the oscillator is damped, the mechanical energy is not constant but decreases with time. If the damping is small, we can find $E(t)$ by replacing x_m in Eq. 16-30 with $x_m e^{-bt/2m}$, the amplitude of the damped oscillations. By doing so, we find that

$$E(t) \approx \tfrac{1}{2}kx_m^2 e^{-bt/m}, \qquad (16\text{-}36)$$

which tells us that, like the amplitude, the mechanical energy decreases exponentially with time.

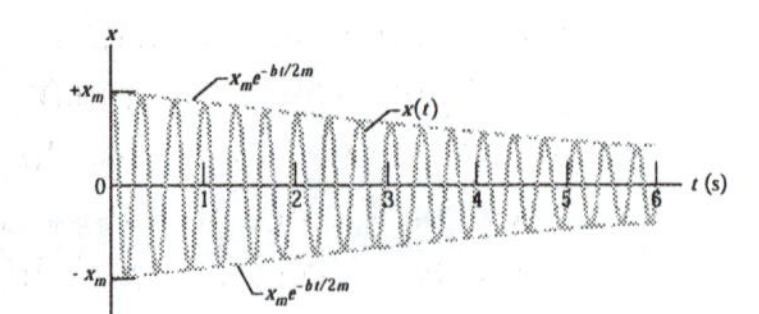

Fig. 16-23: The displacement function $x(t)$ for the damped oscillator of Fig. 16-22, with $m = 250\,g$, $k = 85\ \text{N/m}$, and $b = 70\ \text{g/s}$. The amplitude, which is $x_m e^{-bt/2m}$, decreases exponentially with time.

READING EXERCISE 16-8: Here are three sets of values for the spring constant, damping constant, and mass for the damped oscillator of Fig. 16-22. Rank the sets according to the time required for the mechanical energy to decrease to one-fourth of its initial value, greatest first.

Set 1	$2k_0$	b_0	m_0
Set 2	k_0	$6b_0$	$4m_0$
Set 3	$3k_0$	$3b_0$	m_0

16-8 Forced Oscillations and Resonance

Sometimes we would like to maintain oscillations longer than they would naturally continue because of the damping forces. For example, if you were on a swing that was given only one big push, you would go up and back a few times before the mechanical energy was completely lost and you came to a stop. Although we cannot totally eliminate such loss of mechanical energy, we can replenish the energy from some source. As an example, you know that by swinging your legs or torso you can "pump" a swing to maintain or enhance the oscillations. In doing this, you transfer biochemical energy to mechanical energy of the oscillating system.

A person swinging in a swing without anyone pushing it is an example of *free oscillation.* However, if someone pushes the swing periodically, the swing has *forced,* or *driven, oscillations. Two* angular frequencies are associated with a system undergoing driven oscillations: (1) the *natural* angular frequency ω of the system, which is the angular frequency at which it would oscillate if it were suddenly disturbed and then left to oscillate freely, and (2) the angular frequency ω_d of the external driving force causing the driven oscillations.

We can use Fig. 16-22 to represent an idealized forced simple harmonic oscillator if we cause the structure marked "rigid support" to move up and down at an angular frequency ω_d that we can adjust. This forced oscillator will settle down to oscillate at our chosen angular frequency ω_d of the driving force, and its displacement $x(t)$ is given by

$$x(t) = x_m \cos(\omega_d t + \phi), \qquad (16\text{-}37)$$

where x_m is the amplitude of the oscillations.

How large the displacement amplitude x_m is depends on a complicated function of ω_d and ω. The velocity amplitude v_m of the oscillations is easier to describe: it is greatest when

$$\omega_d = \omega \qquad \text{(resonance)}, \qquad (16\text{-}38)$$

a condition called **resonance.** Equation 16-38 is also *approximately* the condition at which the displacement amplitude x_m of the oscillations is greatest. Thus, if you push a swing at its natural angular frequency, the displacement and velocity amplitudes will increase to large values, a fact that children learn quickly by trial and error. If you push at other angular frequencies, either higher or lower, the displacement and velocity amplitudes will be smaller.

Figure 16-24 shows how the displacement amplitude of an oscillator depends on the angular frequency ω_d of the driving force, for three values of the damping coefficient b. Note that for all three the amplitude is approximately greatest when $\omega_d/\omega = 1$—that is, when the resonance condition of Eq. 16-38 is satisfied. The curves of Fig. 16-24 show that less damping gives a taller and narrower *resonance peak.*

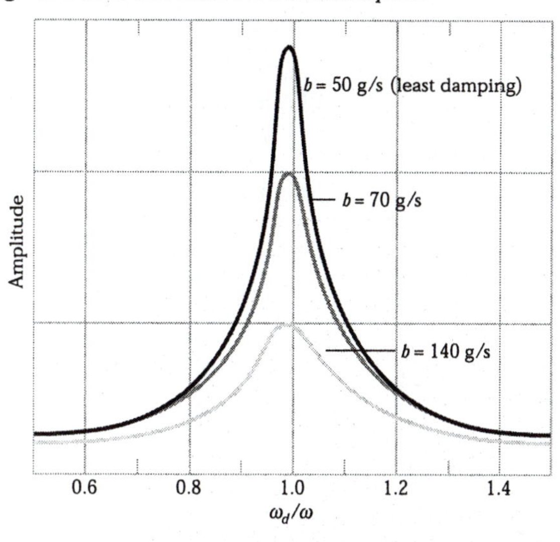

Fig. 16-24: The displacement amplitude x_m of a forced oscillator varies as the angular frequency ω_d of the driving force is varied. The amplitude is greatest *approximately* at $\omega_d/\omega = 1$, the resonance condition. The curves here correspond to three values of the damping constant b.

All mechanical structures have one or more natural angular frequencies, and if a structure is subjected to a strong external driving force that matches one of these angular frequencies, the resulting oscillations of the structure may rupture it. Thus, for example, aircraft designers must make sure that none of the natural angular frequencies at which a wing can oscillate matches the angular frequency of the engines in flight. A wing that flaps violently at certain engine speeds would obviously be dangerous.

Mexico's earthquake in September 1985 was a major earthquake (8.1 on the Richter scale), but the seismic waves from it should have been too weak to cause extensive damage when they reached Mexico City about 400 km away. However, Mexico City is largely built on an ancient lake bed, where the soil is still soft with water. Although the amplitude of the seismic waves was weak in the firmer ground en route to Mexico City, their amplitude substantially increased in the loose soil of the city. Acceleration amplitudes of the waves were as much as $0.20g$, and the angular frequency was (surprisingly) concentrated around 3 rad/s. Not only did the ground begin oscillating with large amplitudes, but many of the buildings with intermediate height had resonant angular frequencies of about 3 rad/s. Most of those buildings collapsed during the violent shaking, while shorter buildings (higher resonant angular frequencies) and taller buildings (with lower resonant angular frequencies) remained standing.

Touchstone Example 16-4-1

A block whose mass m is 680 g is fastened to a spring whose spring constant k is 65 N/m. The block is pulled a distance $x = 11$ cm from its equilibrium position at $x = 0$ and released from rest at $t = 0$.

(a) What are the angular frequency, the frequency, and the period of the resulting motion?

SOLUTION: The **Key Idea** here is that the block–spring system forms a linear simple harmonic oscillator, with the block undergoing SHM. Then the angular frequency is given by Eq. 16-11:

$$\omega = \sqrt{\frac{k}{m}} = \sqrt{\frac{65 \text{ N/m}}{0.68 \text{ kg}}} = 9.78 \text{ rad/s}$$

$$\approx 9.8 \text{ rad/s.} \qquad \text{(Answer)}$$

The frequency follows from Eq. 16-6, which yields

$$f = \frac{\omega}{2\pi} = \frac{9.78 \text{ rad/s}}{2\pi \text{ rad}} = 1.56 \text{ Hz} \approx 1.6 \text{ Hz.} \qquad \text{(Answer)}$$

The period follows from Eq. 16-2, which yields

$$T = \frac{1}{f} = \frac{1}{1.56 \text{ Hz}} = 0.64 \text{ s} = 640 \text{ ms.} \qquad \text{(Answer)}$$

(b) What is the amplitude of the oscillation?

SOLUTION: The **Key Idea** here is that, the mechanical energy of the spring–block system is conserved. The block is released from rest 11 cm from its equilibrium position, with zero kinetic energy and the elastic potential energy of the system at a maximum. Thus, the block will have zero kinetic energy whenever it is again 11 cm from its equilibrium position, which means it will never be farther than 11 cm from that position. The magnitude of its maximum displacement is 11 cm:

$$x_m = 11 \text{ cm.} \qquad \text{(Answer)}$$

(c) What is the maximum speed v_m of the oscillating block, and where is the block when it occurs?

SOLUTION: The **Key Idea** here is that the maximum speed v_m is defined as the velocity amplitude ωx_m in Eq. 16-15, since x_m is always defined as being positive, v_m is also positive,

$$v_m = \omega x_m = (9.78 \text{ rad/s})(0.11 \text{ m})$$

$$= 1.1 \text{ m/s.} \qquad \text{(Answer)}$$

This maximum speed occurs when the oscillating block is rushing through the origin; compare Figs. 16-15*a* and 16-15*b*, where you can see that the speed is a maximum whenever $x = 0$.

(d) What is the magnitude $|a_m|$ of the maximum acceleration of the block?

SOLUTION: The **Key Idea** this time is that the magnitude a_m of the maximum acceleration is the acceleration amplitude $\omega^2 x_m$ in Eq. 16-16; that is,

$$a_m = \omega^2 x_m = (9.78 \text{ rad/s})^2(0.11 \text{ m})$$

$$= 11 \text{ m/s}^2. \qquad \text{(Answer)}$$

This maximum magnitude of acceleration occurs when the block is at the ends of its path. At those points, the force acting on the block has its maximum magnitude; compare Figs. 16-15*a* and 16-15*c*, where you can see that the magnitudes of the displacement and acceleration are maximum at the same times.

(e) What is the phase constant ϕ for the motion?

SOLUTION: Here the **Key Idea** is that Eq. 16-5 gives the displacement of the block as a function of time. We know that at time $t = 0$, the block is located at $x = x_m$. Substituting these *initial conditions,* as they are called, into Eq. 16-5 and canceling x_m give us

$$1 = \cos \phi \qquad \text{(TE16-1)}$$

Taking the inverse cosine then yields

$$\phi = 0 \text{ rad.} \qquad \text{(Answer)}$$

(Any angle that is an integer multiple of 2π rad also satisfies Eq. TE16-1; we chose the smallest angle.)

(f) What is the displacement function $x(t)$ for the spring–block system?

SOLUTION: The **Key Idea** here is that $x(t)$ is given in general form by Eq. 16-5. Substituting known quantities into that equation gives us

$$\begin{aligned} x(t) &= x_m \cos(\omega t + \phi) \\ &= (0.11 \text{ m}) \cos[(9.8 \text{ rad/s})t + 0] \qquad \text{(Answer)} \end{aligned}$$

Touchstone Example 16-4-2

In Fig. TE16-1, a penguin (obviously skilled in aquatic sports) dives from a uniform board that is hinged at the left and attached to a spring at the right. The board has length $L =$ 2.0 m and mass $m =$ 12 kg; the spring constant k is 1300 N/m. When the penguin dives, it leaves the board and spring oscillating with a small amplitude. Assume that the board is stiff enough not to bend, and find the period T of the oscillations.

SOLUTION: Since a spring is involved, we might guess that the oscillations are SHM, but we shall not assume that. Instead, we use the following **Key Idea:** If the board is in SHM, then the acceleration and displacement of the oscillating end of the board must be related by an expression in the form of Eq. 16-17 ($a = -\omega^2 x$). If so, we shall be able to find ω and then the desired T from the expression. Let us check by finding the relation between the acceleration and displacement of the board's right end.

Because the board rotates about the hinge as one end oscillates, we are concerned with a torque $\vec{\tau}$ on the board about the hinge. That torque is due to the force $\vec{F}$ on the board from the spring. Because $\vec{F}$ varies with time, $\vec{\tau}$ must also. However, at any given instant we can relate the magnitudes of $\vec{\tau}$ and $\vec{F}$ with Eq. 11-29 ($|\vec{\tau}| = |\vec{r}||\vec{F}| \sin \phi$). Here we have

$$|\vec{\tau}| = L|\vec{F}| \sin 90°, \qquad \text{(TE16-2)}$$

Fig. TE16-1 Touchstone Example 16-4-2. The dive by the penguin causes the board and spring to oscillate; the board pivots about the hinge at the left.

where L is the moment arm of force $\vec{F}$ and 90° is the angle between the moment arm and the force's line of action. Combining Eq. TE16-2 with Eq. 11-32 ($\vec{\tau} = I\vec{\alpha}$) gives us

$$L|\vec{F}| = I|\vec{\alpha}| \tag{TE16-3}$$

where I is the board's rotational inertia about the hinge, and α is its angular acceleration about that point. We may treat the board as a thin rod pivoted about one end so the board's rotational inertia I is $\frac{1}{3}mL^2$.

Now let us mentally erect a vertical x-axis through the oscillating right end of the board, with the positive direction upward. Then the force on the right end of the board from the spring is $F = -kx$, where x is the vertical displacement of the right end.

Substituting these expressions for I and $|\vec{F}|$ into Eq. TE16-3 gives us

$$Lk|x| = \frac{mL^2|\vec{\alpha}|}{3}. \tag{TE16-4}$$

We now have a mixture of the magnitude of the vertical linear displacement component $|x|$ and magnitude of rotational acceleration $|\vec{\alpha}|$ (about the hinge). We can replace $|\vec{\alpha}|$ in Eq. TE16-4 with the componet of (linear) acceleration $|\alpha_x|$ along the x axis by substituting according to Eq. 11-19 ($|\vec{a}_x| = |\vec{\alpha}|r$) for tangential acceleration. Here the tangential acceleration is $|a_x|$ and the radius of rotation r is L, so $|\vec{\alpha}| = |\alpha_x|/L$. With that substitution, Eq. TE16-4 becomes

$$Lk|x| = \frac{mL^2|a_x|}{3L},$$

which yields

$$|a_x| = \frac{3k}{m}|x|.$$

But, we can see from the diagram that when x is negative, F and $|a_x|$ are positive and vice versa. So we can rewrite the equation in terms of the components $|a_x|$ and $|x|$ as

$$|a_x| = -\frac{3k}{m}|x|. \tag{TE16-5}$$

Equation TE16-5 is, in fact, of the same form as Eq. 16-17 ($a = -\omega^2 x$). Therefore, the board does indeed undergo SHM, and comparison of Eqs. TE16-5 and 16-17 shows that

$$\omega^2 = \frac{3k}{m},$$

which gives $\omega = \sqrt{3k/m}$. Using Eq. 16-6 ($\omega = 2\pi/T$) to find T then gives us

$$T = 2\pi\sqrt{\frac{m}{3k}} = 2\pi\sqrt{\frac{12\text{ kg}}{3(1300\text{ N/m})}}$$

$$= 0.35\text{ s.} \qquad \text{(Answer)}$$

Perhaps surprisingly, the period is independent of the board's length L.

Touchstone Example 16-7-1

For the damped oscillator of Fig. 16-22, m = 250 g, k = 85 N/m, and b = 70 g/s.

(a) What is the period of the motion?

SOLUTION: The **Key Idea** here is that because $b \ll \sqrt{km} = 4.6$ kg/s, the period is approximately that of the undamped oscillator. From Eq. 16-1, we then have

$$T = 2\pi\sqrt{\frac{m}{k}} = 2\pi\sqrt{\frac{0.25\text{ kg}}{85\text{ N/m}}} = 0.34\text{ s.} \qquad \text{(Answer)}$$

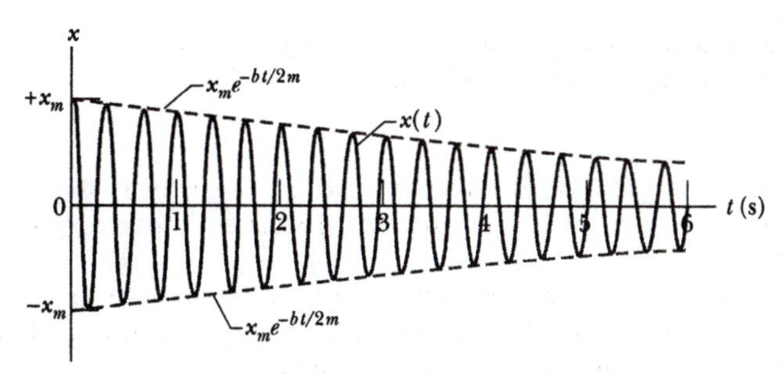

Fig. TE16-2 Touchstone Example 16-7-1. The displacement function $x(t)$ for the damped oscillator of Fig. 16-22, with values given in Touchstone Example 16-7-1. The amplitude, which is $x_m\, e^{-bt/2m}$, decreases exponentially with time.

(b) How long does it take for the amplitude of the damped oscillations to drop to half its initial value?

SOLUTION: Now the **Key Idea** is that the amplitude at time t is displayed in Eq. 16-34 as $x_m\, e^{-bt/2m}$. It has the value x_m at $t = 0$. Thus, we must find the value of t for which

$$x_m\, e^{-bt/2m} = \tfrac{1}{2}x_m.$$

Canceling x_m and taking the natural logarithm of the equation that remains, we have $\ln \frac{1}{2}$ on the right side and

$$\ln(e^{-bt/2m}) = -bt/2m$$

on the left side. Thus,

$$t = \frac{-2m \ln \frac{1}{2}}{b} = \frac{-(2)(0.25 \text{ kg})(\ln \frac{1}{2})}{0.070 \text{ kg/s}}$$

$$= 5.0 \text{ s.} \qquad \text{(Answer)}$$

Because $T = 0.34$ s, this is about 15 periods of oscillation.

(c) How long does it take for the mechanical energy to drop to one-half its initial value?

SOLUTION: Here the **Key Idea** is that, from Eq. 16-36, the mechanical energy at time t is $\frac{1}{2}kx_m^2\, e^{-bt/m}$. It has the value $\frac{1}{2}kx_m^2$ at $t = 0$. Thus, we must find the value of t for which

$$\tfrac{1}{2}kx_m^2\, e^{-bt/m} = \tfrac{1}{2}(\tfrac{1}{2}kx_m^2).$$

If we divide both sides of this equation by $\frac{1}{2}kx_m^2$ and solve for t as we did above, we find

$$t = \frac{-m \ln \frac{1}{2}}{b} = \frac{-(0.25 \text{ kg})(\ln \frac{1}{2})}{0.070 \text{ kg/s}} = 2.5 \text{ s.} \qquad \text{(Answer)}$$

This is exactly half the time we calculated in (b), or about 7.5 periods of oscillation. Figure TE16-2 was drawn to illustrate this Touchstone Example.

17 Waves—I

When a beetle moves along the sand within a few tens of centimeters of this sand scorpion, the scorpion immediately turns toward the beetle and dashes to it (for lunch). The scorpion can do this without seeing (it is nocturnal) or hearing the beetle.

How can the scorpion so precisely locate its prey?

The answer is in this chapter.

17-1 Waves and Particles

Two ways to get in touch with a friend in a distant city are to write a letter and to use the telephone.

The first choice (the letter) involves the concept of "particles": a material object moves from one point to another, carrying with it information and energy. Most of the preceding chapters deal with particles or with systems of particles.

The second choice (the telephone) involves the concept of "waves," the subject of this chapter and the next. In your telephone call, a sound wave carries your message from your vocal cords to the telephone. There, an electromagnetic wave takes over, passing along a copper wire or an optical fiber or through the atmosphere, possibly by way of a communications satellite. At the receiving end there is another sound wave, from a telephone to your friend's ear. Although the message is passed, nothing that you have touched reaches your friend. For example, if you have the flu, she can't catch it by talking to you on the phone since no matter (and therefore no virus) is passed between the two of you. In a wave, information and energy move from one point to another but no material object makes that journey.

Leonardo da Vinci understood waves when he wrote of water waves: "It often happens that the wave flees the place of its creation, while the water does not; like the waves made in a field of grain by the wind, where we see the waves running across the field while the grain remains in place."

Particle and *wave* are the two great concepts in classical physics, in the sense that we seem able to associate almost every branch of the subject with one or the other. The two concepts are quite different. The word *particle* suggests a tiny concentration of matter capable of transmitting energy. The word *wave* suggests just the opposite—namely, a broad distribution of energy, filling the space through which it passes. The job at hand is to put aside particles for a while and to learn something about waves.

17-2 Types of Waves

Waves are of three main types:

1. ***Mechanical waves.*** These waves are most familiar because we encounter them almost constantly; common examples include water waves, sound waves, and seismic waves. All these waves have certain central features: They are governed by Newton's laws, and they can exist only within a material medium, such as water, air, and rock.

2. ***Electromagnetic waves.*** These waves are less familiar, but you use them constantly; common examples include visible and ultraviolet light, radio and television waves, microwaves, x rays, and radar waves. These waves require no material medium to exist. Light waves from stars, for example, travel through the vacuum of space to reach us. All electromagnetic waves travel through a vacuum at the same speed c, given by

$$c = 299\ 792\ 458\ \text{m/s} \qquad \text{(speed of light)}. \qquad (17\text{-}1)$$

3. ***Matter waves.*** Although these waves are commonly used in modern technology, their type is probably very unfamiliar to you. These waves are associated with electrons, protons, and other fundamental particles, and even atoms and molecules. Because we commonly think of these things as constituting matter, such waves are called matter waves.

Much of what we discuss in this chapter applies to waves of all kinds. However, for specific examples we shall refer to mechanical waves because they are the most familiar to us.

17-3 Pulses and Waves

Consider Figure 17-1 which shows a "Slinky wave demonstrator" consisting of a series of connected Slinkies that hang from long, evenly-spaced strings. If we give a quick pull outward (to the left) on the left end of the Slinky and then give it a quick push back inward (to the right), a compression-expansion disturbance like that shown in Fig. 17-1*a* will propagate along the length of the slinky. (You need to look carefully at Fig. 17-1*a*, not 17-1*b*, to see the effect). Since the back-and-forth oscillations of the slinky coils are parallel to the direction that the disturbance travels, the motion is said to be **longitudinal.**

On the other hand, if we give the end of the Slinky a quick jerk back and forth at right angles to the line of Slinky, a disturbance like that shown in Fig. 17-1*b* results and moves down the length of the Slinky. In this case, the displacement of the coils in the Slinky is *perpendicular* to the direction that the disturbance travels (i.e., along the length of the Slinky). Such a disturbance is called **transverse**.

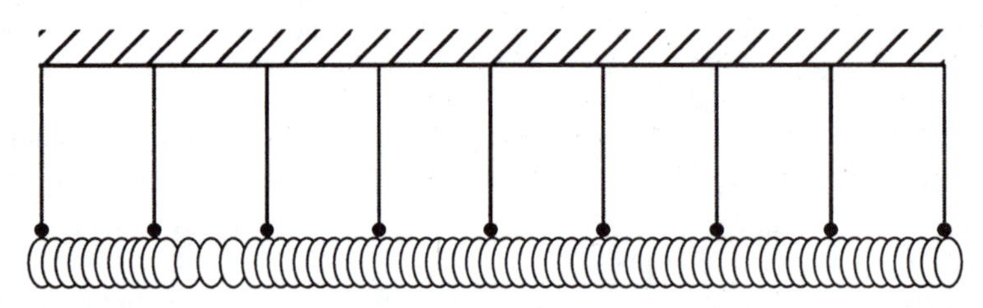

Fig. 17-1(*a*): A large Slinky demonstrator consisting of a series of connected Slinkies hanging from long, evenly-spaced strings. Note the compression-expansion disturbances shown around the second and third strings from the left.

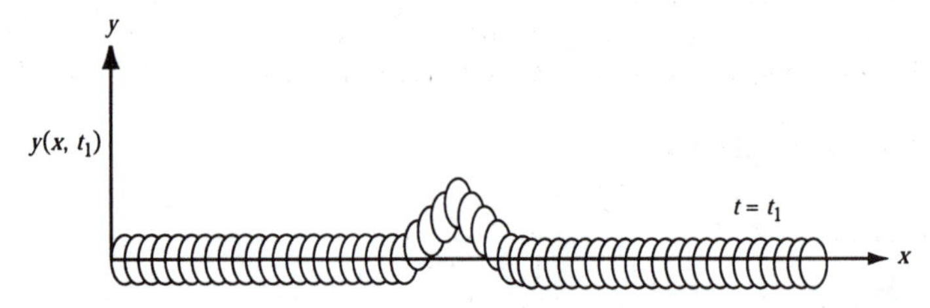

Fig. 17-1(*b*): A transverse or sideways disturbance or pulse moving from left to right along a Slinky wave demonstrator. In this case the viewer is lying on the floor underneath the Slinky looking up at time $t = t_1$.

In both cases discussed above, the singular disturbance is referred to as a **pulse.** If we repeat the motion that causes the disturbance (either the up-down jerk or the push-pull motion) at regular time intervals, the result is a traveling and repeating disturbance which is referred to as a **wave**. Just as with single pulses, waves can be longitudinal or transverse.

For example, if you give one end of a stretched string a single up-and-down jerk, a single *pulse* travels along the string as in Fig. 17-2*a*. This pulse and its motion can occur because the string is under tension. During the upward part of the jerk, when you pull your end of the string upward, it begins to pull upward on the adjacent section of the string via tension between the two sections. As the adjacent section moves upward, it begins to pull the next section upward, and so on. Meanwhile, suppose that you have now pulled down on your end of the string, completing the up-down stroke. As each section is moving upward in turn, it begins to be pulled back downward by neighboring sections that are already on the way down. The net result is that a distortion in the string's shape (the pulse) moves along the string at some velocity $\vec{v}$. Note that although the disturbance moves down the string to the right, the parts of the string itself just move up and down. This is an extremely important, but subtle, point regarding pulses and waves. The wave or pulse travels, but the particles that make up the medium (string or otherwise) do not.

Transverse and Longitudinal Waves

If, instead of a single up-down stroke, you move your hand up and down in continuous simple harmonic motion as in Fig. 17-2*b*, a continuous *wave* travels along the string at velocity $\vec{v}$. The vertical displacement of the string is perpendicular to the direction that the wave propagates (along the string), and so this wave is called a **transverse wave.** The larger the vertical displacement of your hand during the up-down stroke, the higher the peak and the lower the valley of the wave will be. This characteristic, measured as the magnitude of the maximum displacement of a small bit of string from its equilibrium position as the wave passes through it, is called the **amplitude** y_m of the wave. This definition of amplitude is very similar to the one developed in the last chapter in regard to oscillations.

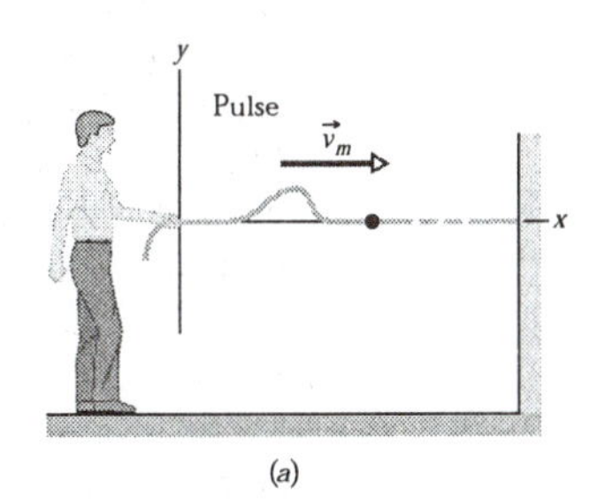

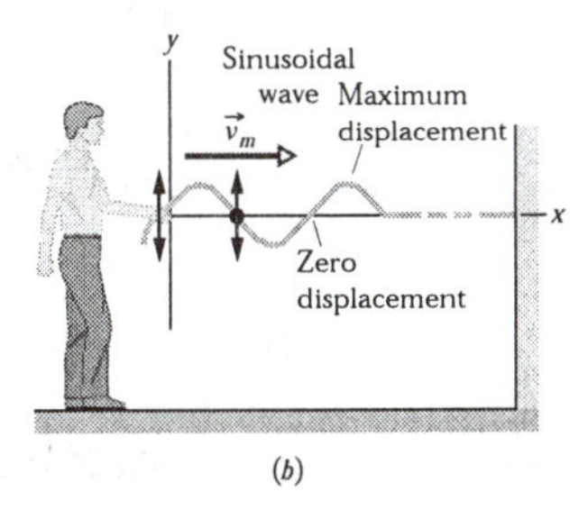

Fig. 17-2: (*a*) A single pulse is sent along a stretched string. A typical string segment (marked with a dot) moves up once and then down as the pulse passes. The segment's motion is perpendicular to the wave's direction of travel, so the pulse is a *transverse wave.* (*b*) A sinusoidal wave is sent along the string. A typical string segment moves up and down continuously as the wave passes. This too is a transverse wave.

If we take a photograph of a transverse wave (perhaps the wave in the string of Fig. 17-2*b*) at some time $t = t_1$, we can see the wave shape and try to find a mathematical function that describes that shape or **wave form.** For example, if the motion of your hand is a sinusoidal function of time, the wave has a sinusoidal shape at any given instant, as in Fig. 17-2*b*. That is, the wave has the shape of a continuously repeating sine curve or cosine curve. (We consider here only an "ideal" string, in which no friction like forces within the string cause the wave to die out as it travels along the string. In addition, we assume that the string is so long that we need not consider a wave rebounding from the far end.)

The wave in Fig. 17-2*b* travels along a single line, so we call it a one-dimensional wave. (Sound waves, for example, travel in more than one dimension.) For the wave in Fig. 17-2*b*, or any other one-dimensional wave or pulse, the convention is to define the line of travel as the x-axis. As we can see from Fig. 17-2*b*, the vertical displacement of a bit of string at a given time depends on how far down along the string we make the measurement. Hence, we say that the vertical displacement of the string is a function of the horizontal position, x. The distance (in this example the horizontal distance) over which the wave undergoes one cycle, for example, the peak to peak or valley to valley distance, is called the **wavelength** λ.

Now consider Figure 17-3. Like the longitudinal pulse set up in the Slinky of Fig. 17-1*a*, if you suddenly move the piston in Fig. 17-3 leftward and then rightward, you will send a compression-expansion pulse along the pipe. The leftward motion of the piston drags the air next to it leftward, changing the air pressure there. The decreased air pressure then pulls leftward on the molecules of air somewhat farther along the pipe. This process proceeds as a "chain reaction" along the length of the tube. Moving the piston rightward increases the air pressure next to it which causes adjacent air molecules to move rightward. That process also propagates down the tube. That is, after having moved leftward, the molecules nearest to the piston, and then molecules farther away, move back rightward. Thus, the motion of the air and the change in air pressure travel rightward along the pipe as a pulse. Pulses of sound and sound waves are expansion-compression disturbances like these.

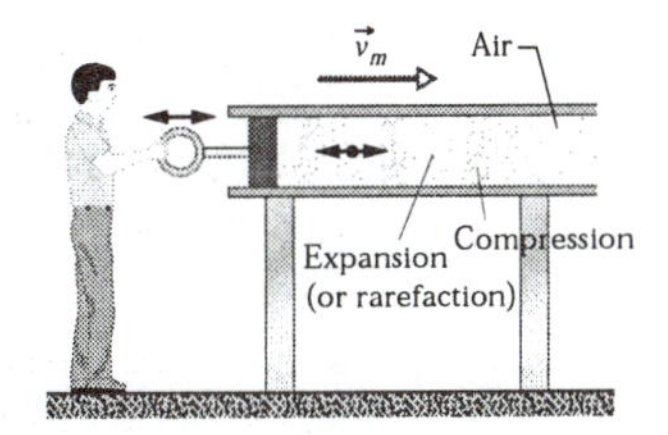

Fig. 17-3: A sound wave is set up in an air-filled pipe by moving a piston back and forth. Because the oscillations of an segment of the air (represented by the black dot) are parallel to the direction in which the wave travels, the wave is a *longitudinal wave.*

If you push and pull on the piston in Fig. 17-3 in simple harmonic motion, a sinusoidal wave travels along the pipe. Because the back-and-forth oscillations are parallel to the direction the disturbance travels, the wave is said to be a **longitudinal wave.** In this chapter we concentrate on transverse waves, and string waves in particular; in Chapter 18 we shall concentrate on longitudinal waves, and sound waves in particular.

Both a transverse wave and a longitudinal wave are said to be **traveling waves** because they both travel from one point to another, as from one end of the string to the other end in Fig. 17-2 or from one end of the pipe to the other end in Fig. 17-3. Remember though that it is the wave that moves from end to end, not the material (string or air) through which the wave moves.

The sand scorpion shown in the photograph opening this chapter uses waves of both transverse and longitudinal motion to locate its prey. When a beetle even slightly disturbs the sand, it sends pulses along the sand's surface (Fig. 17-4). One set of pulses is longitudinal, traveling with speed $|\vec{v}_L| = 150 \text{ m/s}$. A second set is transverse, traveling with speed $|\vec{v}_T| = 50 \text{ m/s}$.

The scorpion, with its eight legs spread roughly in a circle about 5 cm in diameter, intercepts the faster longitudinal pulses first and learns the direction of the beetle; it is in the direction of whichever leg is disturbed earliest by the pulses. The scorpion then senses the time interval Δt between that first interception and the interception of the slower transverse waves and uses it to determine the distance d to the beetle. This distance is given by

$$\Delta t = \frac{d}{|\vec{v}_T|} - \frac{d}{|\vec{v}_L|},$$

and it turns out to be

$$d = (75 \text{ m/s})\,\Delta t.$$

For example, if $\Delta t = 4.0$ ms, then $d = 30$ cm, which gives the scorpion a perfect fix on the beetle.

Of course, the scorpion no more does these calculations in its head as it hunts a beetle than you do a conscious recalculation of the location of your center of mass as you decide in what way to throw out your arms when you slip. Instead, instinct and habits based on successful personal experiences rule the scorpion (and you) in such situations. But there is physics behind such " instincts." They are not arbitrarily successful.

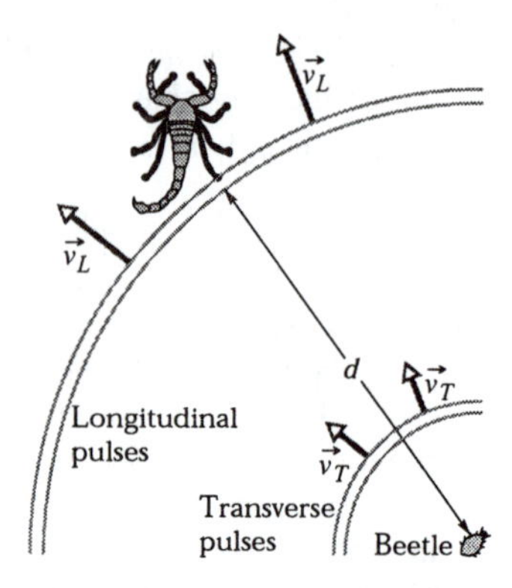

Fig. 17-4: A beetle's motion sends fast longitudinal pulses and slower transverse pulses along the sand's surface. The sand scorpion first intercepts the longitudinal pulses; here, it is the rear-most right leg that senses the pulses earliest.

Waves and Oscillations

As we discussed above, if we take a picture of a wave, we can freeze the wave in time and investigate the spatial characteristics of the wave, like the wave's amplitude and its length. Alternatively, we could pick a special *place,* x_1, along the wave and peek at it through a slit, as in Fig. 17-5. As we peek thought the slit, we would see the bits of string or Slinky rise and fall as the wave travels through space. If we make measurements while peeking, we could plot the displacement as a function of time for this location $x = x_1$. Doing so allows us to focus in on the characteristics of the wave that are associated with the passage of time. When we do this, we see several aspects of a wave that remind us of the oscillations that we learned about in the last chapter.

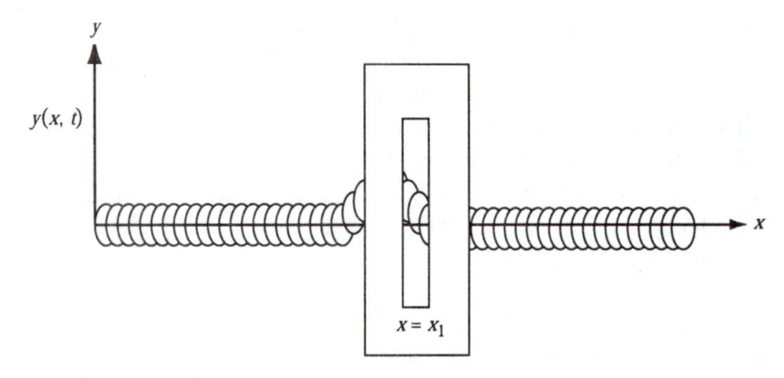

Fig. 17-5: A transverse wave travelling along a hanging Slinky. The viewer is lying on the floor, looking through a slit at location $x = x_1$. The viewer peeks through a slit to determine the displacement as a function of time at a fixed position x_1.

For example, let us examine the small piece of string boxed in Fig. 17-6. This figure shows five "snapshots" of a sinusoidal wave traveling in the positive direction of an x axis. The movement of the wave is indicated by the rightward progress of the short arrow pointing to a high point of the wave. From snapshot to snapshot, the short arrow moves to the right with the wave shape, but the pieces of the string move *only* parallel to the y axis. Let us follow the motion of the boxed string segment at $x = 0$. In the first snapshot (Fig. 17-6*a*), it is at displacement $y = 0$. In the next snapshot, it is at its extreme downward displacement because a *valley* (or extreme low point) of the wave is passing through it. It then moves back up through $y = 0$. In the fourth snapshot, it is at its extreme upward displacement because a *peak* (or extreme high point) of the wave is passing through it. In the fifth snapshot, it is again at $y = 0$, having completed one full oscillation. Notice that the wave in Fig. 17-6 moves to the right by $\frac{1}{4}\lambda$ from one snapshot to the next. Thus, by the fifth snapshot, it has moved to the right by 1λ.

A graph of the displacement of this segment of string as a function of time is shown in Figure 17-7. The time required for the cycle rising and falling to repeat is called the **period** T of the wave and is shown explicitly in Fig. 17-7.

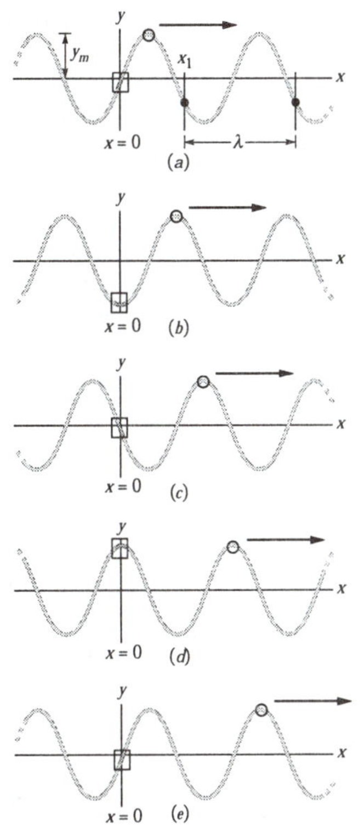

Fig. 17-6: Five "snapshots" of a traveling string wave. A particular bit of string (the piece around $x = 0$) has been boxed so you can see how the string segment moves up and down vertically as the wave disturbance travels horizontally to the right. The amplitude y_m is indicated. A typical wavelength λ, measured from an arbitrary position x_1, is also indicated. The circle in each snapshot shows the location of the wave crest as it moves from left to right.

READING EXERCISE 17-1: A snapshot of an asymmetric pulse traveling along a string from left to right is shown in the figure below. For this pulse, which graph best represents the shape of the displacement y versus position for a *fixed time*?

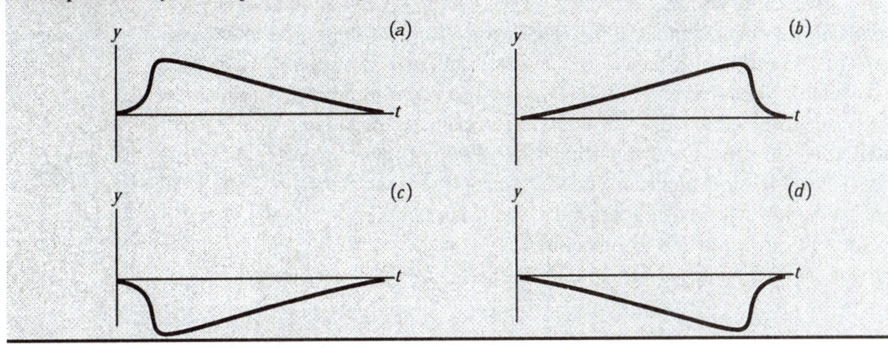

17-4 The Mathematical Expression for a Sinusoidal Wave

We found that the disturbance (whether pulse or wave, transverse or longitudinal) depends on both position, x and time, t. If we call the displacement y, we can write $y = f(x,t)$ or $y(x,t)$ to represent this functional dependence on time and position. In the example of the transverse pulse traveling along a Slinky as pictured in Fig. 17-1*b*, $y(x,t)$ represents the transverse (vertical) displacement of the Slinky rings from their equilibrium position at given position x and time t. (Alternatively, in the longitudinal wave on the Slinky shown in Fig. 17-1*a*, $y(x,t)$ could represent the number of Slinky coils per centimeter at a given x and t.)

We can completely describe any wave or pulse that does not change shape over time and travels at a constant velocity using the relation $y = f(x,t)$, in which y is the displacement as a function f of the time t and the position x. In general, a wave can have any shape so long as it is not too sharp. The trick then is to find the correct expression for the function, $f(x,t)$.

Fortunately, it turns out that any shape pulse or wave can be constructed by adding up different sinusoidal oscillations. This makes sinusoidal waves especially interesting and important to us. So, for the rest of this section we'll discuss the properties and descriptions of sine waves. Recall that a sinusoidal shape like the wave in Fig. 17-2*b* can be described with either a sine function or a cosine function; both give the same general shape for the wave. Although we could use the cosine function, in this chapter we choose to use the sine function.

Imagine a sinusoidal wave like that of Fig. 17-2*b*. As the wave sweeps through succeeding segments (that is, very short sections) of the string, the segments move up and then down. Again, we will call the direction that the wave travels the *x*-axis. Notice that here x does not represent a displacement like it did in the previous chapter. It is a label which tells which bit of the string we are talking about. At time t, the displacement y of the segment located at position x is given by

$$y(x,t) = A\sin(\phi). \qquad (17\text{-}2)$$

Amplitude and Phase

The argument ϕ of the sine function is called the phase of the wave. Since the displacement, y, is a function of both time t and position x, the phase ϕ must also be a function of both time and position. So, ϕ is really a function $\phi(x,t)$ which we can represent in a general way as $\phi(x,t) = kx \pm \omega t$ where k and ω are constants that we can determine for a particular wave.

Although the phase ϕ is a function of both time and position, it is *neither* a time *nor* a position. Rather, the phase ϕ *must* be an angle because we can only take the sine of angles.

The *phase* of the wave is the argument $kx \pm \omega t$ of the sine in the equation above. As the wave sweeps through a string segment at a particular position x, the phase changes linearly with time t. This means that the sine also changes, oscillating between +1 and −1. Its extreme positive value (+1) corresponds to a peak of the wave moving through the segment; then, the value of y at position x is y_m. Its extreme negative value (−1) corresponds to a valley of the wave moving through the segment; then, the value of y at position x is $-y_m$. Thus, the sine function and the time-dependent phase of a wave correspond to the oscillation of a string segment, and the amplitude of the wave determines the extremes of the segment's displacement.

Taking these ideas into account, we can rewrite the general expression for our sine wave as

$$y(x,t) = y_m \sin(kx \pm \omega t), \qquad (17\text{-}3)$$

where y_m is the amplitude of the wave and k and ω are constants.

Wavelength and Wave Number

In order to come up with a more useful mathematical expression for the wave, we must more carefully investigate the nature of the phase of the wave. We know that the sine function repeats itself every 2π radians. We also know that the wavelength λ of a wave is the distance (parallel to the direction of the wave's travel) between repetitions of the shape of the wave (or *wave form*). A typical wavelength is marked in Fig. 17-6*a*, which is a snapshot of the wave at time $t = 0$. At that time, $y(x,t) = y_m \sin(kx \pm \omega t)$ becomes

$$y(x,0) = y_m \sin kx. \qquad (17\text{-}4)$$

By definition, the displacement y is the same at both ends of one wavelength—that is, y is the same at $x = x_1$ and at $x = x_1 + \lambda$. (Again, x does not represent a displacement here. It is a label which tells which bit of the string we are talking about.) By the equation above,

$$\begin{aligned} y_m \sin kx_1 &= y_m \sin k(x_1 + \lambda) \\ &= y_m \sin(kx_1 + k\lambda). \end{aligned} \qquad (17\text{-}5)$$

A sine function begins to repeat itself when its angle (or argument) is increased by $2\pi\,\text{rad}$, so in the equation above, we must have $k\lambda = 2\pi$, or

$$k = \frac{2\pi}{\lambda} \quad \text{(wave number)}. \qquad (17\text{-}6)$$

We call k the **wave number**. The wave number is inversely proportional to the wavelength. Its SI unit is the radian per meter, or the inverse meter. (Note that the symbol k here does *not* represent a spring constant as previously.)

Period, Angular Frequency, and Frequency

In contrast to the graphs in Fig. 17-6 which represent pictures of the string at a particular instant, we are now going to focus on a particular bit of string and consider how it moves as a function of time. As we noted earlier, if you were to peek at the string through a slit, you would see that the single segment of the string at that position moves up and down in simple harmonic motion. Figure 17-7 shows a graph of the displacement y of a bit of string versus time t at a certain position along the string, taken to be $x = 0$. This motion is described by $y(x,t) = y_m \sin(kx \pm \omega t)$. Setting $x = 0$ and arbitrarily choosing the time dependent term to be positive we get

$$y(0,t) = y_m \sin(\omega t). \tag{17-7}$$

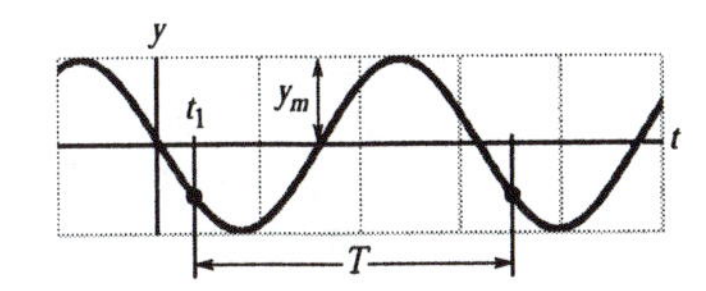

Fig. 17-7: A graph of the displacement of the string segment at $x = 0$ as a function of time, as the sinusoidal wave of Fig. 17-6 passes through it. The amplitude y_m is indicated. A typical period T, measured from an arbitrary time t_1, is also indicated.

Figure 17-7 is a graph of this equation. Be careful though, this figure *does not* show the shape of the wave. It shows a graph of the time-dependent variation in the displacement of a small bit of the string.

Since we have defined the period of oscillation T of a wave to be the time any string segment takes to move through one full oscillation, we can apply this equation to both ends of this time interval. Equating the results yield

$$\begin{aligned} y_m \sin \omega t_1 &= y_m \sin \omega(t_1 + T) \\ y_m \sin \omega t_1 &= y_m \sin(\omega t_1 + \omega T) \end{aligned} \tag{17-8}$$

which can be true only if $\omega T = 2\pi$. In other words,

$$\omega = \frac{2\pi}{T} \quad \text{(angular frequency)}. \tag{17-9}$$

We call ω the **angular frequency** of the wave; its SI unit is the radian per second.

You should recall from Chapter 16, the **frequency** f of an oscillation is defined as $1/T$. Hence, its relationship to angular frequency is

$$f = \frac{1}{T} = \frac{\omega}{2\pi} \quad \text{(frequency)}. \tag{17-10}$$

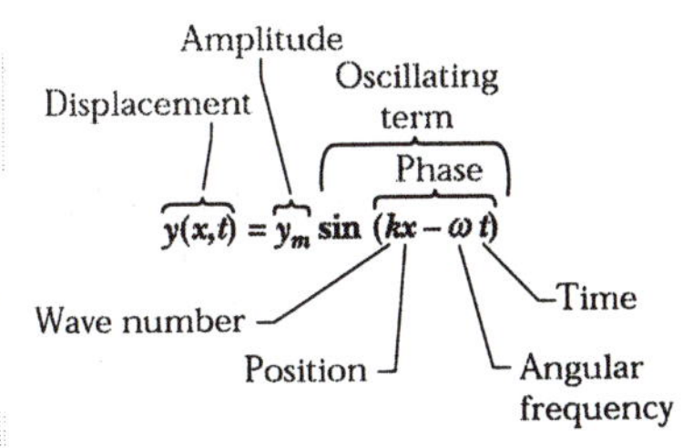

Fig. 17-8: The names of the quantities in Eq. 17-3, for a transverse sinusoidal wave.

Like the frequency of simple harmonic motion in Chapter 16, this frequency f is a number of oscillations per unit time—here, the number made by a string segment as the wave moves through it. As in Chapter 16, f is usually measured in hertz or its multiples, such as kilohertz.

To summarize this discussion, the names of the quantities in $y(x,t) = y_m \sin(kx \pm \omega t)$ (Eq. 17-3) are displayed in Fig. 17-8 for your reference.

READING EXERCISE 17-2: The figure is a composite of three snapshots, each of a wave traveling along a particular string. The phases for the waves are given by (a) $(2\,\text{rad/m})x - (4\,\text{rad/s})t$, (b) $(4\,\text{rad/m})x - (8\,\text{rad/s})t$, and (c) $(8\,\text{rad/m})x - (16\,\text{rad/s})t$. Which phase corresponds to which wave in the figure?

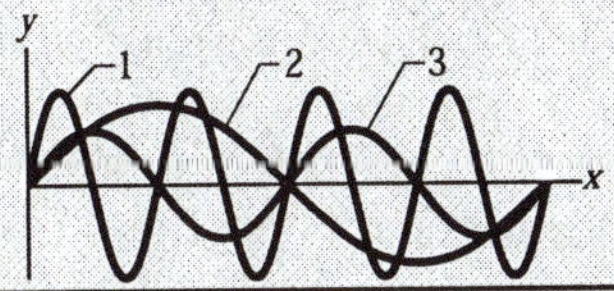

17-5 The "Wave" Speed of a Traveling Distortion

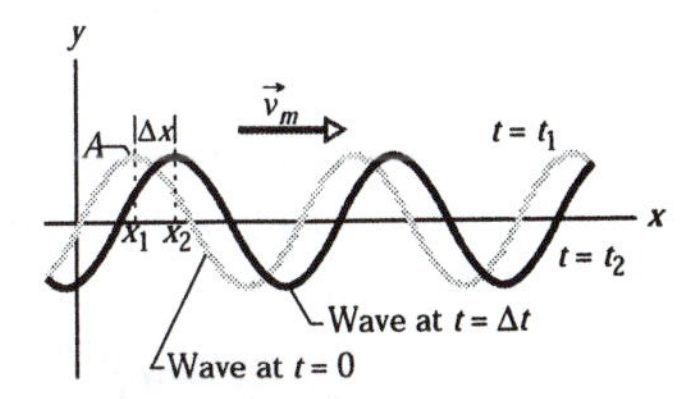

Fig. 17-9: One end of a string that is under tension is moved up and down continuously in a sinusoidal fashion as shown in Fig. 17-2*b*. Two snapshots of the sinusoidal wave that results are shown, at time $t = t_1$ and then at time t_2. As the wave moves to the right with an x-component of velocity v_w, the entire sine curve shifts by Δx during Δt. The string segment at x_1 moves down during $\Delta t = t_2 - t_1$ while the string element at x_2 moves up as the local wave crest moves from left to right.

Figure 17-9 shows two snapshots of a sinusoidal wave taken a small time interval Δt apart. The wave is traveling in the positive x-direction (to the right in Fig. 17-9*b*), the entire wave pattern is moving a distance Δx in that direction during the interval Δt. The ratio $\Delta x/\Delta t$ (or, in the differential limit, dx/dt) is the **wave speed** component along the x axis v_w. How can we find the magnitude $|\vec{v}_w|$ of the wave speed?

As the wave in Fig. 17-9 moves, each point of the moving wave form, such as point A marked on a peak, retains its displacement y. (Points on the string do not retain their displacement, but points on the wave *form* do.) That is, time passes and the location of point A changes (so x and t are both changing), but the value of $y(x, t)$ associated with point A does not change; it remains the maximum value. If point A retains its displacement as it moves, the phase in $y(x,t) = y_m \sin(kx \pm \omega t)$ must remain a constant. So,

$$kx \pm \omega t = \text{ a constant.} \qquad (17\text{-}11)$$

Note that although this argument is constant, in this case, both x and t are increasing. (The time t always increases and x increases because the wave is moving in the (rightward) positive direction). In order for the phase $kx \pm \omega t$ to remain constant with both t and x increasing, the negative sign must be chosen. In other words,

➤The equation for a sinusoidal wave traveling right is $y(x,t) = y_m \sin(kx - \omega t)$ and the equation for a sinusoidal wave moving left is $y(x,t) = y_m \sin(kx + \omega t)$.

By assuming that the wave moves at a constant speed $|\vec{v}_w|$, we can express its speed in terms of how far an imaginary point on the waveform has moved in a short time interval, or

$$|\vec{v}_w| = \frac{|\Delta \vec{x}|}{\Delta t}.$$

If we take the time interval under consideration to be the period of the wave T, then we know that the wave travels one wavelength λ. So,

$$|\vec{v}_w| = \frac{|\Delta \vec{x}|}{\Delta t},$$

leads to the following relationship:

$$|\vec{v}_w| = \frac{\lambda}{T} = \lambda f = \frac{\omega}{k} \qquad \text{(wave speed).} \qquad (17\text{-}12)$$

Are all waves traveling waves? No. Consider a wave of arbitrary shape, given by

$$y(x,t) = h(kx \pm \omega t), \qquad (17\text{-}13)$$

where h represents *any* function, the sine function being one possibility. Since the variables x and t enter in the combination $kx \pm \omega t$, this wave is a traveling wave. All traveling waves *must* be of the form of Eq. 17-13 directly above. Thus, $y(x,t) = \sqrt{ax + b^2 t}$ represents a possible traveling wave. The function $y(x,t) = \sin(ax^2 - bt)$, on the other hand, does *not* represent a traveling wave (because the x is squared).

Particle Velocities in Mechanical Waves

In discussing the velocity of a wave above, it is important we remember that the magnitude of this velocity is the speed at which the wave moves, *not* the speed of the bits of string through which the wave travels. To repeat the quote of Leonardo da Vinci: consider "...waves made in a field of grain by the wind, where we see the waves running across the field while the grain remains in place."

We have already begun to develop an understanding of the velocity of the wave itself. We now consider the velocity of the particles that are displaced as a transverse wave moves through. Let v_y be the transverse (for example vertical) velocity component an oscillating string segment. To find the velocity component v_y, we simply need to differentiate the expression for the displacement of the string segment, $y(x,t) = y_m \sin(kx \pm \omega t)$, with respect to time while holding x constant:

$$v_y = \frac{\partial y}{\partial t} = \pm \omega y_m \cos(kx \pm \omega t). \qquad (17\text{-}14)$$

If we differentiate this expression again, we have an expression for the acceleration of the segment of string:

$$a_y = \frac{\partial v_y}{\partial t} = \frac{\partial^2 y}{\partial t^2} = -\omega^2 y_m \sin(kx \pm \omega t). \qquad (17\text{-}15)$$

Touchstone Examples 17-5-1 and 17-5-2, at the end of this chapter, illustrate how to use what you learned in this section.

TE

17-6 Wave Speed on a Stretched String

If we distort a string that is under tension, we observe that this distortion (or wave) will travel rapidly along the string (Fig. 17-1). This is true whether the distortion is a single pulse (Figs. 17-1*b* or 17-2*a*) or continuous sinusoidal pattern (Fig. 17-2*b*). In Section 17-3 we asserted that a net force generated by the tension forces acting in different directions on each string segment should cause the distortion to propagate along the string as a wave.

In general, the speed of a wave is determined by the properties of the medium through which it travels. For instance, how do the properties of the stretched string affect the speed of a traveling wave? Intuitively we expect that the wave will travel faster when the string is stretched to a higher tension and that the wave will travel more slowly on a thicker string where each segment of string has a greater mass. In this section we use the impulse-momentum theorem—a form of Newton's Second Law—to derive an equation that relates the speed of the traveling wave to the tension of the string and its massiveness.

Consider a horizontal stretched string as shown in Fig. 17-10. Suppose the string is under tension—a non-vector quantity we denote as T. What happens if one end of the string is jerked abruptly upward and then returned to its equilibrium position? The pulse that results propagates along the string as a wave like the one shown in Fig. 17-10. Although the amplitude in Fig. 17-10 is exaggerated for clarity, we assume the wave amplitude is small enough that: (1) the tension in the string does not change significantly and (2) the angle θ between any displaced string element and the horizontal is so small that $\sin\theta \approx \tan\theta \approx \theta$. We also assume that the pulse propagates along the string at a *constant* wave velocity $\vec{v}_w$. As is often the case in physics we can test the validity of our assumptions by seeing if the expression we derive for the wave speed is consistent with observation.

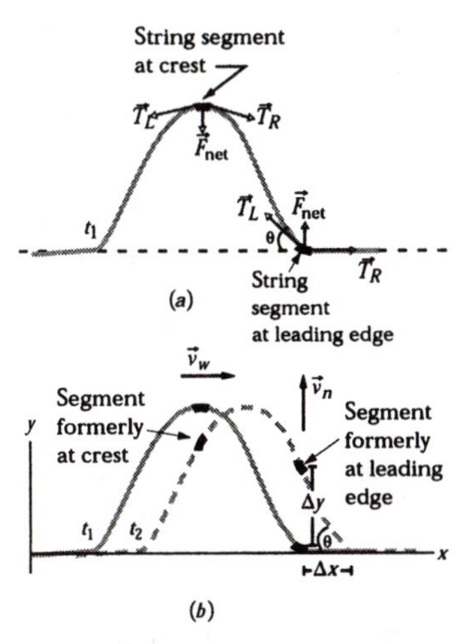

Fig. 17-10: Depictions of a traveling wave pulse. The amplitude is exaggerated for clarity. (*a*) A single pulse travels to the right along a string under constant tension. The string element at the wave crest experiences a net force downward while the string element at the leading edge experiences a net upward force. (*b*) Two snapshots of the wave at times t_1 and t_2 show the wave form traveling with velocity $\vec{v}_w$, so it shifts right a distance Δx during a time interval $\Delta t = t_2 - t_1$. As a result of the net forces on the string elements, the string segment at the wave crest moves down during Δt while the string element at the leading edge moves up.

Let us consider how a segment of string moves as a wave pulse travels by it in a short time interval. If we define the mass per unit length of the string as its mass density μ, then the mass of the segment of string that the wave pulse affects in a time interval Δt is given by

$$m = \mu |\vec{v}_w| \Delta t, \qquad (17\text{-}16)$$

where $|\vec{v}_w|$ is the speed of propagation of the wave pulse. We start by examining the tension forces that act at each end of our small string segment. For example, in Fig. 17-10*a* the segment at the crest of the wave experiences a net downward force. But, the segment at the leading edge of the wave pulse experiences a net upward force. Thus a short time Δt later, the front segment has moved up while the segment at the crest has moved down. If we take all the string segments into account, we see that the entire pulse shape "moves" to the right even though each of the string segments has only moved vertically—either up or down. This is shown in Fig. 17-10*b*.

What happens when the leading edge of the pulse encounters a segment of the string? The net force $\vec{F}_{net}$ due to the unbalanced tension forces at the string segment's ends shown in Fig. 17-10*b* cause the segment to accelerate upward. As a result, the string segment undergoes a change in vertical position of Δy and a momentum change, $\Delta \vec{p}_y$ during the time Δt. We can use the impulse-momentum theorem to describe the motion of the string element:

$$\vec{F}_{net}\Delta t = \Delta \vec{p}_y. \tag{17-17}$$

The segment of string has a mass m and we assume the segment has no vertical velocity at time t_1. If we denote the vertical velocity change imparted to the segment of string in the short time interval Δt as $\Delta \vec{v}_y$, then

$$\vec{F}_{net}\Delta t = m\Delta \vec{v}_y = m(\vec{v}_y - 0) = m\vec{v}_y.$$

But we can see from Fig. 17-10*a* that the magnitude of the net force on the string segment is approximately $T\sin\theta$ so that the impulse delivered to the string segment by the traveling wave can be related to the newly acquired vertical velocity of the string segment by

$$(T\sin\theta)\Delta t \approx m|\vec{v}_y|.$$

From Figure 17-10 we also see that the ratio of the magnitude of the vertical velocity of the leading edge mass segment and the magnitude of the horizontal wave velocity of the temporary disturbance is given by

$$\tan\theta = \frac{\Delta y}{\Delta x} = \frac{|\vec{v}_y|}{|\vec{v}_w|}. \tag{17-18}$$

Combining the last two expressions we derived gives us

$$(T\sin\theta)\Delta t = m|v_w|\tan\theta \quad \text{or} \quad T = \frac{m}{\Delta t}|\vec{v}_w|\frac{1}{\cos\theta}. \tag{17-19}$$

At this point we need to express the mass of the segment in terms of the length of the string segment and the linear density of the string. Thus the mass of the segment in the string can be determined as the density multiplied by the length of the segment so that $m = \mu\Delta x$. But during the time interval Δt the waveform has moved a distance $\Delta x = |\vec{v}_w|\Delta t$. We can now rewrite Eq. 17-19 as

$$T = \mu|v_w|^2\frac{1}{\cos\theta},$$

but for small angles, $\cos\theta$ is approximately 1, so

$$|v_w|^2 = \frac{T}{\mu}.$$

Therefore, $$|\vec{v}_w| = \sqrt{\frac{T}{\mu}} \quad \text{(speed)}. \tag{17-20}$$

Equation 17-20 gives the speed of the pulse in Fig. 17-10 and the speed of *any* other wave pulse or continuous wave on the same string under the same tension. If we take the square root of the ratio of T (dimension MLT^{-2}) and μ (dimension ML^{-1}) we get a dimension of velocity. Thus Eq. 17-20 is dimensionally correct.

Equation 17-20 tells us:

➤The speed of a small amplitude pulse traveling along a stretched string depends only on the tension and linear density of the string.

If the simplifying assumptions we used in the derivation of Eq. 17-20 are reasonable, then we expect that measurements of wave speed, tension and linear mass density can be used to verify Eq. 17-20. Indeed, this theoretical prediction has been experimentally confirmed many, many times. In fact, it turns out that Eq. 17-20 is just as valid for continuous waves as it is for a single wave pulse, and so the speed of a continuous wave is not a function of frequency. The frequency of the wave is fixed entirely by whatever generates the wave (for example, the person jerking the string up and down in Fig. 17-2*b*). However, once the frequency of a continuous wave is set, the wave speed determines the relationship between frequency and wavelength since $\lambda = |\vec{v}_w|/f$ (Eq. 17-12).

READING EXERCISE 17-3: In the derivation of the wave speed in a string shown above, there are references to two different velocities; one is referred to as v_y and one is referred to as v. As completely as possible, describe the differences between these two velocities. Explain why the velocity v is used in deriving the expression for Δm while v_y is used in the discussion of Δp.

READING EXERCISE 17-4: You send a traveling wave along a particular string by oscillating one end. If you increase the frequency of the oscillations, do (a) the speed of the wave and (b) the wavelength of the wave increase, decrease, or remain the same? If, instead, you increase the tension in the string, do (c) the speed of the wave and (d) the wavelength of the wave increase, decrease, or remain the same?

17-7 Energy and Power Transported by a Traveling Wave in a String

When we start up a wave or pulse in a stretched string, *we* provide the energy for the motion of the string. We impart a momentum to a segment of the string. As the wave or pulse moves away, the momentum is transferred from one segment of the string to the next. In addition, the wave transports energy as both kinetic energy and elastic potential energy. Let us consider each of these forms of energy in turn.

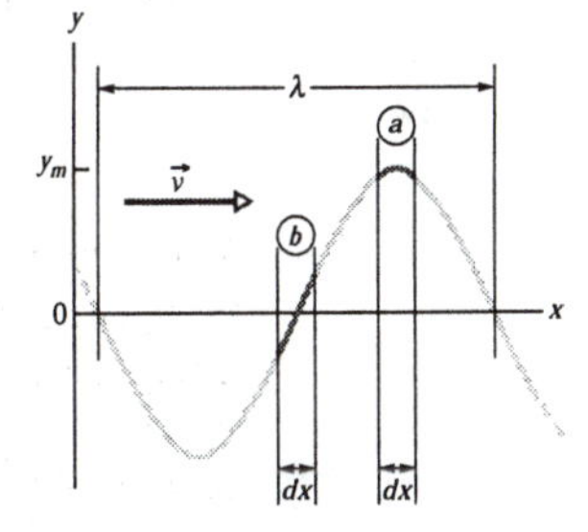

Fig. 17-11: A snapshot of a traveling wave on a string at time $t = 0$. String segment *a* is at displacement $y = y_m$, and string segment *b* is at displacement $y = 0$. The kinetic energy of the string segment at each position depends on the transverse velocity of the segment. The potential energy depends on the amount by which the string is displaced from equilibrium as the wave passes through it.

Kinetic Energy

A segment of the string of mass dm, oscillating transversely in simple harmonic motion as the wave passes through it, has kinetic energy associated with its transverse velocity v_y. When the segment is rushing through its $y = 0$ position (segment *b* in Fig. 17-11), its transverse velocity—and thus its kinetic energy—is a maximum. When the segment is at its extreme position $y = y_m$ (as is segment *a*), its transverse velocity—and thus its kinetic energy—is zero.

Elastic Potential Energy

To send a sinusoidal wave along a previously straight string, the wave must necessarily stretch the string. As a string segment of length *dx* oscillates transversely, its length must increase and decrease in a periodic way if the string segment is to fit the sinusoidal wave form. Elastic potential energy is associated with these length changes, just as for a spring.

When the string segment is at its $y = y_m$ position (segment *b* in Fig. 17-11), its length has its normal undisturbed value *dx*, so its elastic potential energy is zero. However, when the segment is rushing through its $y = 0$ position, it is stretched to its maximum extent, and its elastic potential energy then is a maximum.

Energy Transport

The oscillating string segment thus has both its maximum kinetic energy and its maximum elastic potential energy at $y = 0$. In the snapshot of Fig. 17-11, the regions of the string at maximum displacement have no energy, and the regions at zero displacement have maximum energy. As the wave travels along the string, forces due to the tension in the string continuously do work to transfer energy from regions with energy to regions with no energy.

Suppose we set up a wave on a string stretched along a horizontal x axis so that Eq. 17-3 describes the string's displacement. We might send a wave along the string by oscillating one end of the string, as in Fig. 17-2*b*. In doing so, we provide energy for the motion and stretching of the string—as the string sections oscillate perpendicularly to the x axis, they have kinetic energy and elastic potential energy. As the wave moves into sections that were previously at rest, energy is transferred into those new sections. Thus, we say that the wave *transports* the energy along the string.

The Rate of Energy Transmission

The kinetic energy dK associated with a string segment of mass dm is given by

$$dK = \tfrac{1}{2}\, dm\, {v_y}^2, \tag{17-21}$$

where v_y is the transverse component of velocity of the oscillating string segment. To find v_y, we differentiate Eq. 17-3 with respect to time while holding x constant:

$$v_y = \frac{\partial y}{\partial t} = -\omega y_m \cos(kx - \omega t). \tag{17-22}$$

Using this relation and putting $dm = \mu\, dx$, we rewrite Eq. 17-21 as

$$dK = \tfrac{1}{2}(\mu\, dx)\left[(-\omega y_m)\cos^2(kx - \omega t)\right]^2. \tag{17-23}$$

Dividing Eq. 17-23 by dt gives the rate at which the kinetic energy of a string segment changes, and thus the rate at which kinetic energy is carried along by the wave. The ratio dx/dt that then appears on the right of Eq. 17-23 is the wave speed $|\vec{v}|$, so we obtain

$$\frac{dK}{dt} = \tfrac{1}{2}\mu|\vec{v}|\omega^2 y_m^2 \cos^2(kx - \omega t). \tag{17-24}$$

The *average* rate at which kinetic energy is transported is

$$\begin{aligned}\left(\frac{dK}{dt}\right)_{\text{avg}} &= \tfrac{1}{2}\mu|\vec{v}|\omega^2 y_m^2 \left(\cos^2(kx - \omega t)\right)_{\text{avg}} \\ &= \tfrac{1}{4}\mu|\vec{v}|\omega^2 y_m^2.\end{aligned} \tag{17-25}$$

Here we have taken the average over an integer number of wavelengths and have used the fact that the average value of the square of a cosine function over an integer number of periods is $\tfrac{1}{2}$.

Elastic potential energy is also carried along with the wave, and at the same average rate given by Eq. 17-25. Although we shall not examine the proof, you should recall that, in an oscillating system such as a pendulum or a spring-block system, the average kinetic energy and the average potential energy are indeed equal.

The **average power,** which is the average rate at which energy of both kinds is transmitted by the wave, is then

$$P_{\text{avg}} = 2\left(\frac{dK}{dt}\right)_{\text{avg}}, \tag{17-26}$$

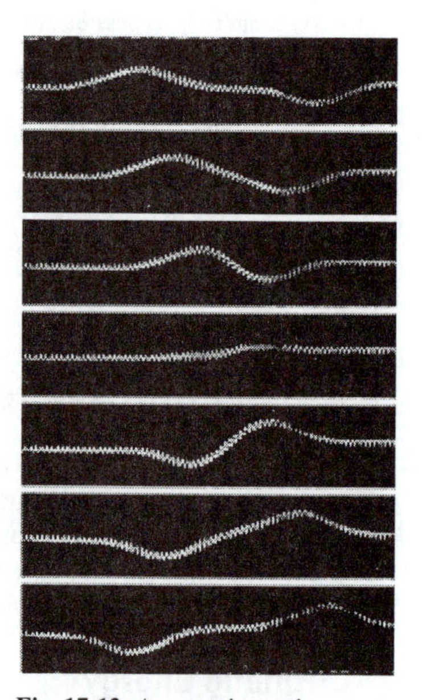

Fig. 17-12: An experimental demonstration of wave pulse superposition. In this sequence of movie frames two pulses are set in motion in opposite directions along a taut spring. The pulses have almost the same shape except that the one starting on the right has positive displacements while the other starting on the left has negative displacements. In the fourth frame the two pulses almost cancel each other. (Movie frames courtesy of the Physical Science Study Committee or PSSC.)

or, from Eq. 17-25,

$$P_{avg} = \tfrac{1}{2}\mu|\vec{v}|\omega^2 y_m^2 \qquad \text{(average power).} \qquad (17\text{-}27)$$

The factors μ and $|\vec{v}|$ in this equation depend on the material and tension of the string. The factors ω and y_m depend on the process that generates the wave. The dependence of the average power of a wave on the square of its amplitude and also on the square of its angular frequency is a general result, true for waves of all types.

17-8 The Principle of Superposition for Waves

It often happens that two or more waves pass simultaneously through the same region. When we listen to a concert, for example, sound waves from many instruments fall simultaneously on our eardrums. The electrons in the antennas of our radio and television receivers are set in motion by the net effect of many electromagnetic waves from many different broadcasting centers. The water of a lake or harbor may be churned up by waves in the wakes of many boats.

In many real world cases, we find the interaction between two overlapping waves to be quite complex. However, we also observe that if the amplitudes of the two waves are small, like those shown in Fig. 17-12, their interaction is well behaved. That is, we observe that if the waves have small amplitudes, the two waves interact so as to produce a resultant wave with an amplitude equal to the sum of the amplitudes of the two original waves. This effect is seen in Figure 17-12 which shows a sequence of snapshots of two pulses traveling in opposite directions on the same stretched string.

➤ When the waves overlap, the displacement of each point on the string is the sum of the two displacements it would have had from each wave independently.

Moreover, we observe that each pulse moves through the other, as if the other were not present:

➤Overlapping waves do not in any way alter the travel of each other.

In order to make these two important observations quantitative, let $y_1(x,t)$ and $y_2(x,t)$ be the displacements associated with two waves traveling simultaneously along the same stretched string. Since the result of the overlapping waves is a resultant wave which is the sum of the two, the displacement of the string when the waves overlap is given by the algebraic sum

$$y'(x,t) = y_1(x,t) + y_2(x,t). \qquad (17\text{-}28)$$

This is an example of the **principle of superposition,** which says that when several effects occur simultaneously, their net effect is the sum of the individual effects.

We can look in more detail at how the sum of two waves passing through each other can look rather odd while the waves overlap. But after they pass each other they look more normal again. This situation is shown in the visualization of the superposition of two waves having different amplitudes in Figs. 17-13 and 17-14.

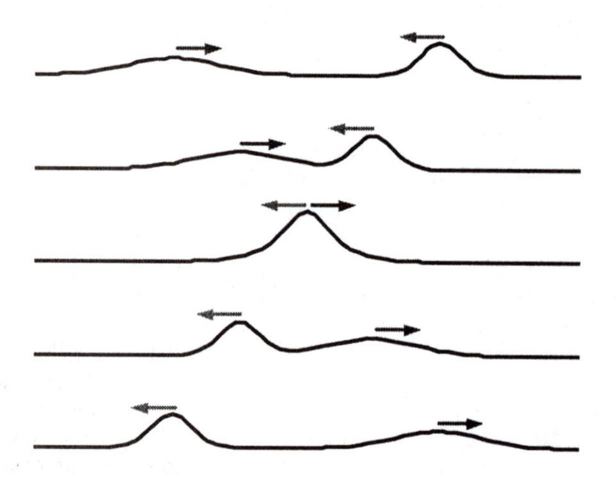

Fig. 17-13: A visualization of two pulses on a taut string passing through each other. A short broad pulse moves from left to right while a sharp tall pulse moves from right to left. When they overlap at times t_2, t_3, and t_4, the wave form is the sum or superposition of the displacement of each pulse at each location along the string. A closeup of this type of superposition is shown in Fig. 17-14.

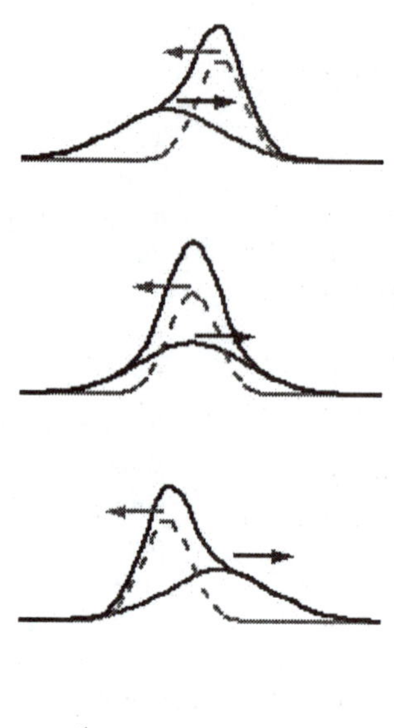

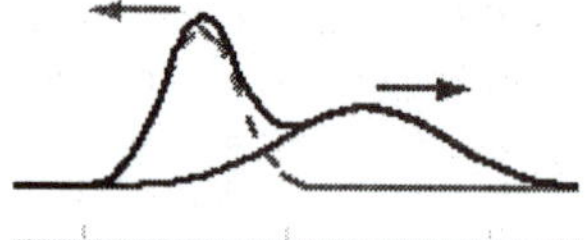

Fig. 17-14: Closeup view of a short, broad pulse and a tall sharp pulse passing through each other. The vertical scale relative to the horizontal scale has been enlarged to show more details. *Note that the resultant wave is the sum of the displacements contributed by each of the two pulses at each location along the string.*

17-9 Interference of Waves

Suppose we send two sinusoidal waves of the same wavelength and amplitude in the same direction along a stretched string. The superposition principle applies. What resultant wave does it predict for the string?

The resultant wave depends on the extent to which the waves are *in phase* (in step) with respect to each other—that is, how much one wave form is shifted from the other wave form. (Figure 17-16 shows waves that are "out of phase" to various degrees.) If the waves are exactly in phase (so that the peaks and valleys of one are exactly aligned with those of the other), they combine to double the displacement of either wave acting alone. If they are exactly out of phase (the peaks of one are exactly aligned with the valleys of the other), they combine to cancel everywhere, and the string remains straight. We call this phenomenon of combining waves **interference,** and the waves are said to **interfere.** (These terms refer only to the displacements of the waves; the travel of the waves is unaffected.)

Let one wave traveling along a stretched string be given by

$$y_1(x,t) = y_m \sin(kx - \omega t), \tag{17-29}$$

and another, shifted from the first, by

$$y_2(x,t) = y_m \sin(kx - \omega t + \phi). \tag{17-30}$$

The waves in question have the same angular frequency ω (that is, the same frequency f), the same wave number k (that is, the same wavelength λ), and the same amplitude y_m. They both travel in the positive direction of the x axis, with the same speed, given by $|v_m| = \sqrt{T/\mu}$. They differ only by a constant angle ϕ, which we call the **phase constant.** These waves are said to be *out of phase* by ϕ or to have a *phase difference* of ϕ, or one wave is said to be *phase-shifted* from the other by ϕ.

From the principle of superposition, the resultant wave is the algebraic sum of the two interfering waves and has displacement

$$\begin{aligned} y'(x,t) &= y_1(x,t) + y_2(x,t) \\ &= y_m \sin(kx - \omega t) + y_m \sin(kx - \omega t + \phi). \end{aligned} \tag{17-31}$$

In Appendix E we see that we can write the sum of the sines of two angles α and β as

$$\sin\alpha + \sin\beta = 2\sin\tfrac{1}{2}(\alpha+\beta)\cos\tfrac{1}{2}(\alpha-\beta). \qquad (17\text{-}32)$$

Applying this relation to Eq. 17-31 leads to

$$y'(x,t) = [2y_m \cos\tfrac{1}{2}\phi]\sin(kx - \omega t + \tfrac{1}{2}\phi). \qquad (17\text{-}33)$$

As Fig. 17-15 shows, the resultant wave is also a sinusoidal wave traveling in the direction of increasing x. It is the only wave you would actually see on the string (you would *not* see the two interfering waves of Eqs. 17-29 and 17-30).

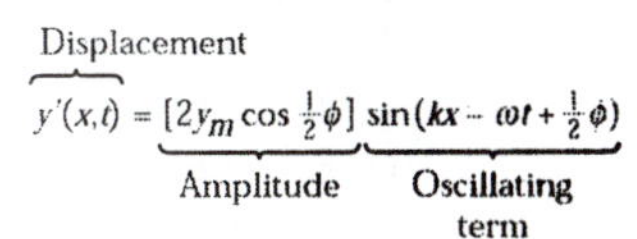

Fig. 17-15: The resultant wave of Eq. 17-33, due to the interference of two sinusoidal transverse waves, is also a sinusoidal wave, with an amplitude and an oscillating term.

➤If two sinusoidal waves of the same amplitude and wavelength travel in the *same* direction along a stretched string, they interfere to produce a resultant sinusoidal wave traveling in that direction.

The resultant wave differs from the interfering waves in two respects: (1) its phase constant is $\frac{1}{2}\phi$, and (2) its amplitude y'_m is the quantity in the brackets in Eq. 17-33:

$$y'_m = 2y_m \cos\tfrac{1}{2}\phi \qquad \text{(amplitude).} \qquad (17\text{-}34)$$

The resultant wave of Eq. 17-33, due to the interference of two sinusoidal transverse waves, is also a sinusoidal transverse wave, with an amplitude and an oscillating term.

Let's consider a couple of special and very important cases. If $\phi = 0$ rad (or 0°), the two interfering waves are exactly in phase, as in Fig. 17-16*a*. Then Eq. 17-33 reduces to

$$y'(x,t) = 2y_m \sin(kx - \omega t) \qquad (\phi = 0). \qquad (17\text{-}35)$$

This resultant wave is plotted in Fig. 17-16*d*. Note from both that figure and $y'(x,t) = 2y_m \sin(kx - \omega t)$ that the amplitude of the resultant wave is twice the amplitude of either interfering wave. That is the greatest amplitude the resultant wave can have, because the cosine term in $y'(x,t) = [2y_m \cos\frac{1}{2}\phi]\sin(kx - \omega t + \frac{1}{2}\phi)$ has its greatest value (unity) when $\phi = 0$. Interference that produces the greatest possible amplitude is called *fully constructive interference*.

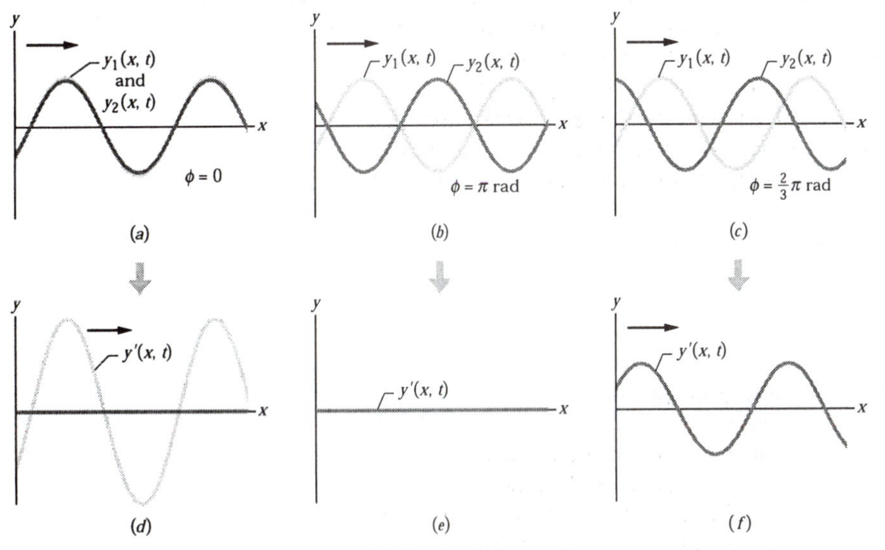

Fig. 17-16: Two identical sinusoidal waves, $y_1(x,t)$ and $y_2(x,t)$, travel along a string in the positive direction of an x axis. They interfere to give a resultant wave $y'(x,t)$. The resultant wave is what is actually seen on the string. The phase difference ϕ between the two interfering waves is (*a*) 0 rad or 0°, (*b*) π rad or 180°, and (*c*) $\frac{2}{3}\pi$ rad or 120°. The corresponding resultant waves are shown in (*d*), (*e*), and (*f*).

If $\phi = \pi$ rad (or 180°), the interfering waves are exactly out of phase as in Fig. 17-16*b*. Then $\cos\frac{1}{2}\phi$ becomes $\cos\pi/2 = 0$, and the amplitude of the resultant wave as given by Eq. 17-34 is zero. We then have, for all values of x and t,

$$y'(x,t) = 0 \qquad (\phi = \pi \text{ rad}). \qquad (17\text{-}36)$$

The resultant wave is plotted in Fig. 17-16*e*. Although we sent two waves along the string, we see no motion of the string. This type of interference is called *fully destructive interference*.

Because a sinusoidal wave repeats its shape every 2π rad, a phase difference $\phi = 2\pi$ rad (or 360°) corresponds to a shift of one wave relative to the other wave by a distance equivalent to one wavelength. Thus, phase differences can be described in terms of wavelengths as well as angles. For example, in Fig. 17-16*b* the waves may be said to be 0.50 wavelength out of phase. Table 17-1 shows some other examples of phase differences and the interference they produce. Note that when interference is neither fully constructive nor fully destructive, it is called *intermediate interference.* The amplitude of the resultant wave is then intermediate between 0 and $2y_m$. For example, from Table 17-1, if the interfering waves have a phase difference of 120° ($\phi = \frac{2}{3}\pi$ rad = 0.33 wavelength), then the resultant wave has an amplitude of y_m, the same as the interfering waves (see Figs. 17-16*c* and *f*).

TABLE 17-1: Phase Differences and Resulting Interference Types[a]

Phase Difference, in			Amplitude of Resultant Wave	Type of Interference
Degrees	Radians	Wavelengths		
0	0	0	$2y_m$	Fully constructive
120	$\frac{2}{3}\pi$	0.33	y_m	Intermediate
180	π	0.50	0	Fully destructive
240	$\frac{4}{3}\pi$	0.67	y_m	Intermediate
360	2π	1.00	$2y_m$	Fully constructive
865	15.1	2.40	$0.60y_m$	Intermediate

[a]The phase difference is between two otherwise identical waves, with amplitude y_m, moving in the same direction.

Touchstone Example 17-9-1, at the end of this chapter, illustrates how to use what you learned in this section.

TE

17-10 Reflections at a Boundary and Standing Waves

In the preceding two sections, we discussed two sinusoidal waves of the same wavelength and amplitude traveling *in the same direction* along a stretched string. What if they travel in opposite directions? We can again find the resultant wave by applying the superposition principle. Figure 17-17 suggests the situation graphically. It shows the two combining waves, one traveling to the left in Fig. 17-17*a*, the other to the right in Fig. 17-17*b*. Figure 17-17*c* shows their sum, obtained by applying the superposition principle graphically.

The outstanding feature of the resultant wave is that there are places along the string, called **nodes**, where the string never moves. Four such nodes are marked by dots in Fig. 17-17*c*. Halfway between adjacent nodes are **antinodes,** where the amplitude of the resultant wave is a maximum. Wave patterns such as that of Fig. 17-17*c* are called **standing waves** because the wave patterns do not move left or right; the locations of the maxima and minima do not change.

➤If two sinusoidal waves of the same amplitude and wavelength travel in *opposite* directions along a stretched string, their interference with each other produces a standing wave.

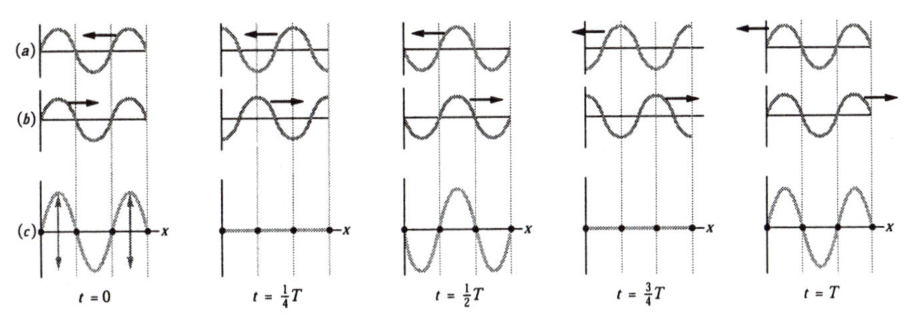

Fig. 17-17: (*a*) Five snapshots of a wave traveling to the left, at the times *t* indicated below part (*c*) (*T* is the period of oscillation). (*b*) Five snapshots of a wave identical to that in (*a*) but traveling to the right, at the same times *t*. (*c*) Corresponding snapshots for the superposition of the two waves on the same string. At $t = 0,\ \frac{1}{2}T,$ and T, fully constructive interference occurs because of the alignment of peaks with peaks and valleys with valleys. At $t = \frac{1}{4}T$ and $\frac{3}{4}T$, fully destructive interference occurs because of the alignment of peaks with valleys. Some points (the nodes, marked with dots) never oscillate; some points (the antinodes) oscillate the most.

To analyze a standing wave, we represent the two combining waves with the equations

$$y_1(x,t) = y_m \sin(kx - \omega t) \qquad (17\text{-}37)$$

and

$$y_2(x,t) = y_m \sin(kx + \omega t). \qquad (17\text{-}38)$$

The principle of superposition gives, for the combined wave,

$$y'(x,t) = y_1(x,t) + y_2(x,t) = y_m \sin(kx - \omega t) + y_m \sin(kx + \omega t).$$

Applying the trigonometric relation of $\sin\alpha + \sin\beta = 2\sin\frac{1}{2}(\alpha+\beta)\cos\frac{1}{2}(\alpha-\beta)$ leads to

$$y'(x,t) = [2y_m \sin kx]\cos\omega t, \qquad (17\text{-}39)$$

which is displayed in Fig. 17-18. This equation does not describe a traveling wave because it is not of the form of $y(x,t) = h(kx \pm \omega t)$. Instead, it describes a standing wave.

The quantity $2y_m \sin kx$ in the brackets of $y'(x,t) = [2y_m \sin kx]\cos\omega t$ can be viewed as the amplitude of oscillation of the string segment that is located at position *x*. However, since an amplitude is always positive and $\sin kx$ can be negative, we take the absolute value of the quantity $2y_m \sin kx$ to be the amplitude at *x*.

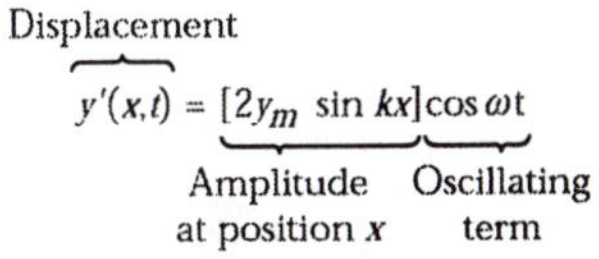

Displacement

$$y'(x,t) = [2y_m \sin kx]\cos\omega t$$

Amplitude at position *x* — Oscillating term

Fig. 17-18: The resultant wave of Eq. 17-39 is a standing wave and is due to the interference of two sinusoidal waves of the same amplitude and wavelength that travel in opposite directions.

In a traveling sinusoidal wave, the amplitude of the wave is the same for all string segments. That is not true for a standing wave, in which the amplitude *varies with position.* In the standing wave of $y'(x,t) = [2y_m \sin kx]\cos\omega t$, for example, the amplitude is zero for values of *kx* that give $\sin kx = 0$. Those values are

$$kx = n\pi, \quad \text{for } n = 0,\ 1,\ 2.... \qquad (17\text{-}40)$$

Substituting $k = 2\pi/\lambda$ in this equation and rearranging, we get

$$x = n\frac{\lambda}{2}, \quad \text{for } n = 0,\ 1,\ 2,... \quad \text{(nodes)}, \qquad (17\text{-}41)$$

as the positions of zero amplitude—the nodes—for the standing wave of Eq. 17-39. Note that adjacent nodes are separated by $\lambda/2$, half a wavelength.

The amplitude of the standing wave of $y'(x,t) = [2y_m \sin kx]\cos\omega t$ has a maximum value of $2y_m$, which occurs for values of *kx* that give $|\sin kx| = 1$. Those values are

$$\begin{aligned} kx &= \tfrac{1}{2}\pi, \tfrac{3}{2}\pi, \tfrac{5}{2}\pi, \ldots \\ &= (n+\tfrac{1}{2})\pi, \quad \text{for } n = 0,\ 1,\ 2,.... \end{aligned} \qquad (17\text{-}42)$$

Substituting $k = 2\pi/\lambda$ in Eq. 17-39 and rearranging, we get

$$x = \left(n + \frac{1}{2}\right)\frac{\lambda}{2} \quad \text{for } n = 0,\ 1,\ 2,... \quad \text{(antinodes)}, \qquad (17\text{-}43)$$

as the positions of maximum amplitude—the antinodes—of the standing wave of Eq. 17-39. The antinodes are separated by $\lambda/2$ and are located halfway between pairs of nodes.

Reflections at a Boundary

We can set up a standing wave in a stretched string by allowing a traveling wave to be reflected from the far end of the string so that it travels back through itself. The incident (original) wave and the reflected wave can then be described by Eqs. 17-37 and 17-38, respectively, and they can combine to form a pattern of standing waves.

In Fig. 17-19, we use a single pulse to show how such reflections take place. In Fig. 17-19*a*, the string is fixed at its left end. When the pulse arrives at that end, it exerts an upward force on the support (the wall). By Newton's Third Law, the support exerts an opposite force of equal magnitude on the string. This force generates a pulse at the support, which travels back along the string in the direction opposite that of the incident pulse. In a "hard" reflection of this kind, there must be a node at the support because the string is fixed there. The reflected and incident pulses must have opposite signs, so as to cancel each other at that point.

In Fig. 17-19*b*, the left end of the string is fastened to a light ring that is free to slide without friction along a rod. When the incident pulse arrives, the ring moves up the rod. As the ring moves, it pulls on the string, stretching the string and producing a reflected pulse with the same sign and amplitude as the incident pulse. Thus, in such a "soft" reflection, the incident and reflected pulses reinforce each other, creating an antinode at the end of the string; the maximum displacement of the ring is twice the amplitude of either of these pulses.

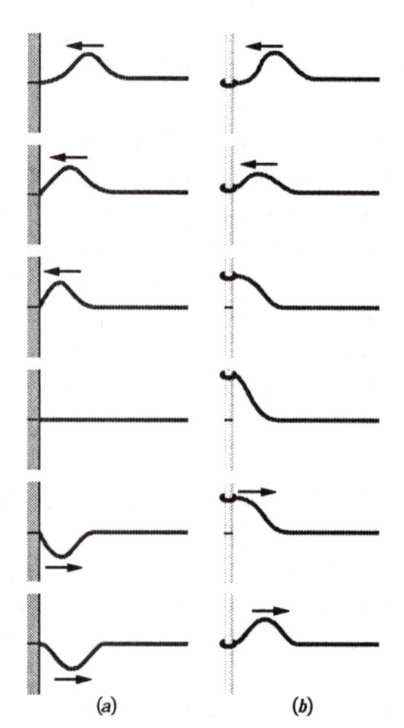

Fig. 17-19: (*a*) A pulse incident from the right is reflected at the left end of the string, which is tied to a wall. Note that the reflected pulse is inverted from the incident pulse. (*b*) Here the left end of the string is tied to a ring that can slide without friction up and down the rod. Now the pulse is not inverted by the reflection.

READING EXERCISE 17-5: Two waves with the same amplitude and wavelength interfere in three different situations to produce resultant waves with the following equations:
(1) $y'(x,t) = (4\text{ m})\sin((5\text{ rad/m})x - (4\text{ rad/s})t)$
(2) $y'(x,t) = (4\text{ m})\sin((5\text{ rad/m})x\cos((4\text{ rad/s})t)$
(3) $y'(x,t) = (4\text{ m})\sin((5\text{ rad/m})x + (4\text{ rad/s})t)$
In which situation are the two combining waves traveling (a) toward positive x, (b) toward negative x, and (c) in opposite directions?

17-11 Standing Waves and Resonance

Consider a string, such as a guitar string, that is stretched between two clamps. Suppose we send a sinusoidal wave of a certain frequency along the string, say, toward the right. When the wave reaches the right end, it reflects and begins to travel back to the left. That left-going wave then overlaps the wave that is still traveling to the right. When the left-going wave reaches the left end, it reflects again and the newly reflected wave begins to travel to the right, overlapping the left-going and right-going waves. In short, we very soon have many overlapping traveling waves, which interfere with one another.

Fig. 17-20: Stroboscopic photographs reveal (imperfect) standing wave patterns on a string being made to oscillate by a vibrator at the left end. The patterns occur at certain frequencies of oscillation.

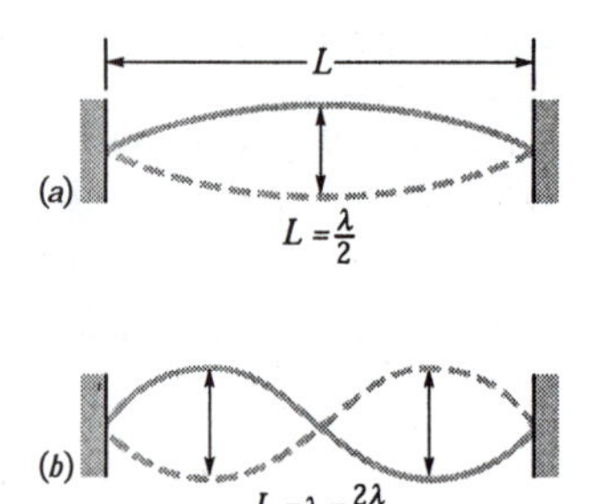

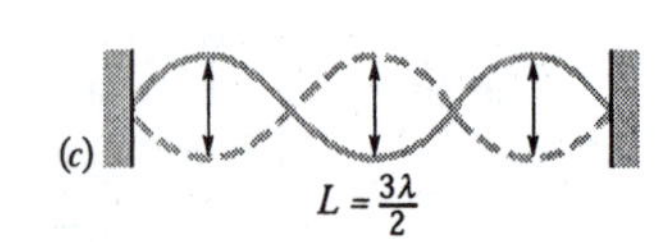

Fig. 17-21: A string, stretched between two clamps, is made to oscillate in standing wave patterns. (*a*) The simplest possible pattern consists of one *loop,* which refers to the composite shape formed by the string in its extreme displacements (the solid and dashed lines). (*b*) The next simplest pattern has two loops. (*c*) The next has three loops.

For certain frequencies, the interference produces a standing wave pattern (or **oscillation mode**) with nodes and large antinodes like those in Fig. 17-20. Such a standing wave is said to be produced at **resonance,** and the string is said to *resonate* at these certain frequencies, called **resonant frequencies.** If the string is oscillated at some frequency other than a resonant frequency, a standing wave is not set up. Then the interference of the right-going and left-going traveling waves results in only small (perhaps imperceptible) oscillations of the string.

Let a string be stretched between two clamps separated by a fixed distance L. To find expressions for the resonant frequencies of the string, we note that a node must exist at each of its ends, because each end is fixed and cannot oscillate. The simplest pattern that meets this key requirement is that in Fig. 17-21*a*, which shows the string at both its extreme displacements (one solid and one dashed, together forming a single "loop"). There is only one antinode, which is at the center of the string. Note that half a wavelength spans the length L, which we take to be the string's length. Thus, for this pattern, $\lambda/2 = L$. This condition tells us that if the left-going and right-going traveling waves are to set up this pattern by their interference, they must have the wavelength $\lambda = 2L$.

A second simple pattern meeting the requirement of nodes at the fixed ends is shown in Fig. 17-21*b*. This pattern has three nodes and two antinodes and is said to be a two-loop pattern. For the left-going and right-going waves to set it up, they must have a wavelength $\lambda = L$. A third pattern is shown in Fig. 17-21*c*. It has four nodes, three antinodes, and three loops, and the wavelength is $\lambda = \frac{2}{3}L$. We could continue this progression by drawing increasingly more complicated patterns. In each step of the progression, the pattern would have one more node and one more antinode than the preceding step, and an additional $\lambda/2$ would be fitted into the distance L.

Thus, a standing wave can be set up on a string of length L by a wave with a wavelength equal to one of the values

$$\lambda = \frac{2L}{n}, \quad \text{for } n = 1, 2, 3.... \tag{17-44}$$

The resonant frequencies that correspond to these wavelengths follow from Eq. 17-12:

$$f = \frac{|\vec{v}_w|}{\lambda} = n\frac{|\vec{v}_w|}{2L}, \quad \text{for } n = 1, 2, 3,.... \tag{17-45}$$

Here $|\vec{v}_w|$ is the speed of traveling waves on the string.

Equation 17-45 tells us that the resonant frequencies are integer multiples of the lowest resonant frequency, $f = |\vec{v}_w|/2L$, which corresponds to $n = 1$. The oscillation mode with that lowest frequency is called the *fundamental mode* or the *first harmonic*. The *second harmonic* is the oscillation mode with $n = 2$, the *third harmonic* is that with $n = 3$, and so on. The frequencies associated with these modes are often labeled f_1, f_2, f_3 and so on. The collection of all possible oscillation modes is called the **harmonic series,** and n is called the **harmonic number** of the nth harmonic.

The phenomenon of resonance is common to all oscillating systems and can occur in two and three dimensions. For example, Fig. 17-22 shows a two-dimensional standing wave pattern on the oscillating head of a kettledrum.

Fig. 17-22: One of many possible standing wave patterns for a kettledrum head, made visible by dark powder sprinkled on the drumhead. As the head is set into oscillation at a single frequency by a mechanical vibrator at the upper left of the photograph, the powder collects at the nodes, which are circles and straight lines (rather than points) in this two-dimensional example.

READING EXERCISE 17-6: In the following series of resonant frequencies, one frequency (lower than 400 Hz) is missing 150, 225, 300, 375 Hz. (a) What is the missing frequency? (b) What is the frequency of the seventh harmonic?

Touchstone Example 17-11-1, at the end of this chapter, illustrates how to use what you learned in this section.

17-12 Phasors

We can represent a string wave (or any other type of wave) vectorially with a **phasor.** In essence, a phasor is a vector that has a magnitude equal to the amplitude of the wave and that rotates around an origin; the angular speed of the phasor is equal to the angular frequency ω of the wave. For example, the wave

$$y_1(x,t) = y_{m1}\sin(kx - \omega t) \qquad (17\text{-}46)$$

is represented by the phasor shown in Fig. 17-23*a*. The magnitude of the phasor is the amplitude y_{m1} of the wave. As the phasor rotates around the origin at angular speed ω, its projection y_1 on the vertical axis varies sinusoidally, from a maximum of y_{m1} through zero to a minimum of $-y_{m1}$ and then back to y_{m1}. This variation corresponds to the sinusoidal variation in the displacement y_1 of any point along the string as the wave passes through it.

When two waves travel along the same string in the same direction, we can represent them and their resultant wave in a *phasor diagram.* The phasors in Fig. 17-23*b* represent the wave of Eq. 17-46 and a second wave given by

$$y_2(x,t) = y_{m2}\sin(kx - \omega t + \phi). \qquad (17\text{-}47)$$

This second wave is phase shifted from the first wave by phase constant ϕ. Because the phasors rotate at the same angular speed ω, the angle between the two phasors is always ϕ. If ϕ is a *positive* quantity, then the phasor for wave 2 *lags* the phasor for wave 1 as they rotate, as drawn in Fig. 17-23*b*. If ϕ is a negative quantity, then the phasor for wave 2 *leads* the phasor for wave 1.

Because waves y_1 and y_2 have the same wave number k and angular frequency ω, we know from Eq. 17-33 that their resultant is of the form

$$y'(x,t) = y'_m\sin(kx - \omega t + \beta), \qquad (17\text{-}48)$$

where y'_m is the amplitude of the resultant wave and β is its phase constant. To find the values of y'_m and β, we would have to sum the two combining waves, as we did to obtain Eq. 17-33.

To do this on a phasor diagram, we vectorially add the two phasors at any instant during their rotation, as in Fig. 17-23*c* where phasor y_{m2} has been shifted to the head of phasor y_{m1}. The magnitude of the vector sum equals the amplitude y'_m in Eq. 17-48. The angle between the vector sum and the phasor for y_1 equals the phase constant β in Eq. 17-48.

Note that, in contrast to the method of Section 17-9,

➤We can use phasors to combine waves *even if their amplitudes are different.*

Touchstone Example 17-12-1, at the end of this chapter, illustrates how to use what you learned in this section.

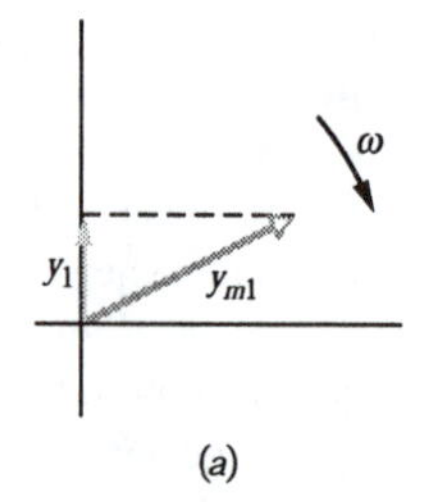

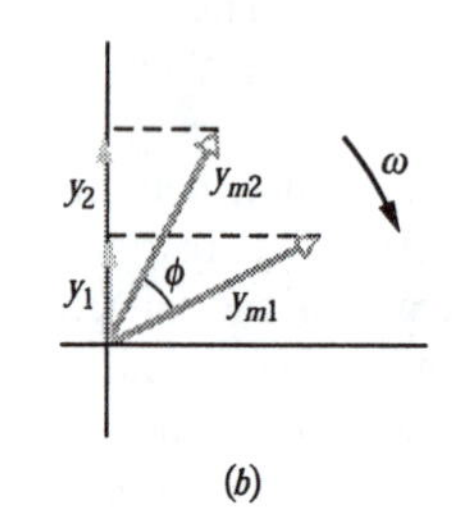

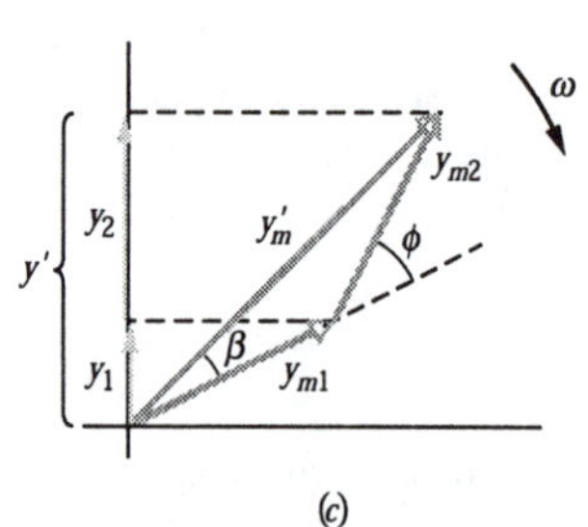

Fig. 17-23: (*a*) A phasor of magnitude y_{m1} rotating about an origin at angular speed ω represents a sinusoidal wave. The phasor's projection y_1 on the vertical axis represents the displacement of a point through which the wave passes. (*b*) A second phasor, also of angular speed ω but of magnitude y_{m2} and rotating at a constant angle ϕ from the first phasor, represents a second wave, with a phase constant ϕ. (*c*) The resultant wave of the two waves is represented by the vector sum y'_m of the two phasors. The projection y' on the vertical axis represents the displacement of a point as that resultant wave passes through it.

Touchstone Example 17-5-1

A wave traveling along a string is described by

$$\mathbf{y(x,t) = (0.00327\ m)\sin((72.1\ rad/m)\,x - (2.72\ rad/s)t)} \qquad \text{(TE17-1)}$$

(a) What is the amplitude of this wave?

SOLUTION: The **Key Idea** is that Eq. TE17-1 is of the same form as Eq. 17-3,

$$y = y_m \sin(kx - \omega t), \qquad \text{(TE17-2)}$$

so we have a sinusoidal wave. By comparing the two equations, we see that the amplitude is

$$y_m = 0.00327\ \text{m} = 3.27\ \text{mm}. \qquad \text{(Answer)}$$

(b) What are the wavelength, period, and frequency of this wave?

SOLUTION: By comparing Eqs. TE17-1 and TE17-2, we see that the angular wave number and angular frequency are

$$k = 72.1\ \text{rad/m} \quad \text{and} \quad \omega = 2.72\ \text{rad/s}.$$

We then relate wavelength λ to k via Eq. 17-6:

$$\lambda = \frac{2\pi}{k} = \frac{2\pi\ \text{rad}}{72.1\ \text{rad/m}}$$
$$= 0.0871\ \text{m} = 8.71\ \text{cm}. \qquad \text{(Answer)}$$

Next, we relate T to ω with Eq. 17-9:

$$T = \frac{2\pi}{\omega} = \frac{2\pi\ \text{rad}}{2.72\ \text{rad/s}} = 2.31\ \text{s}, \qquad \text{(Answer)}$$

and from Eq. 17-10 we have

$$f = \frac{1}{T} = \frac{1}{2.31\ \text{s}} = 0.433\ \text{Hz}. \qquad \text{(Answer)}$$

(c) What is the velocity of this wave?

SOLUTION: The speed of the wave is given by Eq. 17-12:

$$|\vec{v}| = \frac{\omega}{k} = \frac{2.72\ \text{rad/s}}{72.1\ \text{rad/m}} = 0.0377\ \text{m/s}$$
$$= 3.77\ \text{cm/s}. \qquad \text{(Answer)}$$

Because the phase in Eq. TE17-1 contains the position variable x, the wave is moving along the x axis. Also, because the wave equation is written in the form of Eq. 17-3, the *minus* sign in front of the ωt term indicates that the wave is moving in the *positive* direction of the x axis. (Note that the quantities calculated in (b) and (c) are independent of the amplitude of the wave.)

(d) What is the displacement y at $x = 22.5$ cm and $t = 18.9$ s?

SOLUTION: The **Key Idea** here is that Eq. TE17-1 gives the displacement as a function of position x and time t. Substituting the given values into the equation yields

$$y = (0.00327\ \text{m})\sin((72.1\ \text{rad/m}) \times 0.225\ \text{m} - (2.72\ \text{rad/s}) \times 18.9\ \text{s})$$
$$= (0.00327\ \text{m})\sin(-35.1855\ \text{rad})$$
$$= (0.00327\ \text{m})(0.588)$$
$$= 0.00192\ \text{m} = 1.92\ \text{mm}. \qquad \text{(Answer)}$$

Thus, the displacement is positive. (Be sure to change your calculator mode to radians before evaluating the sine.)

Touchstone Example 17-5-2

In Touchstone Example 17-5-1d, we showed that at $t = 18.9$ s the transverse displacement component y of the element of the string at $x = 0.255$ m due to the wave of Eq. TE17-1 is 1.92 mm.

(a) What is v_y, the transverse velocity component of the same element of the string, at that time? (This speed, which is associated with the transverse oscillation of an element of the string, is in the y direction. Do not confuse it with $\vec{v}_w$, the constant velocity at which the *wave form* travels along the x axis.)

SOLUTION: The **Key Idea** here is that the transverse velocity component v_y is the rate at which the displacement y of the element is changing. In general, that displacement is given by

$$y(x, t) = y_m \sin(kx - \omega t). \qquad \text{(TE17-3)}$$

For an element at a certain location x, we find the rate of change of y by taking the derivative of Eq. TE17-3 with respect to t while treating x as a constant. A derivative taken while one (or more) of the variables is treated as a constant is called a *partial derivative* and is represented by the symbol $\partial/\partial x$ rather than d/dx. Here we have

$$v_y = \frac{\partial y}{\partial t} = -\omega y_m \cos(kx - \omega t), \qquad \text{(TE17-4)}$$

where y_m is always taken as positive. Next, substituting numerical values from Touchstone Example 17-5-1 we obtain

$$\begin{aligned} v_y &= (-2.72 \text{ rad/s})(3.27 \text{ mm}) \cos(-35.1855 \text{ rad}) \\ &= 7.20 \text{ mm/s}. \end{aligned} \qquad \text{(Answer)}$$

Thus, at $t = 18.9$ s, the element of string at $x = 22.5$ cm is moving in the positive direction of y, with a velocity of 7.20 mm/s.

(b) What is the transverse acceleration component a_y of the same element at that time?

SOLUTION: The **Key Idea** here is that the transverse acceleration component a_y is the rate at which the transverse velocity of the element is changing. From Eq. TE17-4, again treating x as a constant but allowing t to vary, we find

$$a_y = \frac{\partial v_y}{\partial t} = -\omega^2 y_m \sin(kx - \omega t).$$

Comparison with Eq. TE 17-3 shows that we can write this as

$$a_y = -\omega^2 y.$$

We see that the transverse acceleration component of an oscillating string element is proportional to its transverse component of displacement but opposite in sign. This is completely consistent with the action of the element itself—namely, that it is moving transversely in simple harmonic motion. Substituting numerical values yields

$$\begin{aligned} a_y &= -(2.72 \text{ rad/s})^2(1.92 \text{ mm}) \\ &= -14.2 \text{ mm/s}^2. \end{aligned} \qquad \text{(Answer)}$$

Thus, at $t = 18.9$ s, the element of string at $x = 22.5$ cm is displaced from its equilibrium position by 1.92 mm in the positive y direction and has an acceleration of magnitude 14.2 mm/s^2 in the negative y direction.

Touchstone Example 17-9-1

Two identical sinusoidal waves, moving in the same direction along a stretched string, interfere with each other. The amplitude y_m of each wave is 9.8 mm, and the phase difference ϕ between them is 100°.

(a) What is the amplitude y'_m of the resultant wave due to the interference of these two waves, and what type of interference occurs?

SOLUTION The **Key Idea** here is that these are identical sinusoidal waves traveling in the *same direction* along a string, so they interfere to produce a sinusoidal traveling wave. Because they are identical, they have the *same amplitude.* Thus, the amplitude y'_m of the resultant wave is given by Eq. 17-34:

$$y'_m = 2y_m \cos \tfrac{1}{2}\phi = (2)(9.8 \text{ mm}) \cos(100°/2)$$
$$= 13 \text{ mm.} \qquad \text{(Answer)}$$

We can tell that the interference is *intermediate* in two ways. The phase difference is between 0 and 180° and, correspondingly, amplitude y'_m is between 0 and $2y_m$ (= 19.6 mm).

(b) What phase difference, in radians and wavelengths, will give the resultant wave an amplitude of 4.9 mm?

SOLUTION: The same **Key Idea** applies here as in part (a), but now we are given y'_m and seek ϕ. From Eq. 17-34,

$$y'_m = 2y_m \cos \tfrac{1}{2}\phi.$$

We now have

$$4.9 \text{ mm} = (2)(9.8 \text{ mm}) \cos \tfrac{1}{2}\phi,$$

which gives us (with a calculator in the radian mode)

$$\phi = 2 \cos^{-1} \frac{4.9 \text{ mm}}{(2)(9.8 \text{ mm})}$$
$$= \pm 2.636 \text{ rad} \approx \pm 2.6 \text{ rad.} \qquad \text{(Answer)}$$

There are two solutions because we can obtain the same resultant wave by letting the first wave *lead* (travel ahead of) or *lag* (travel behind) the second wave by 2.6 rad. In wavelengths, the phase difference is

$$\frac{\phi}{2\pi \text{ rad/wavelength}} = \frac{\pm 2.636 \text{ rad}}{2\pi \text{ rad/wavelength}}$$
$$= \pm 0.42 \text{ wavelength.} \qquad \text{(Answer)}$$

Touchstone Example 17-11-1

In Fig. TE17-1, a string, tied to a sinusoidal vibrator at P and running over a support at Q, is stretched by a block of mass m. The separation L between P and Q is 1.2 m, the linear density of the string is 1.6 g/m, and the frequency f of the vibrator is fixed at 120 Hz. The amplitude of the motion at P is small enough for that point to be considered a node. A node also exists at Q.

(a) What mass m allows the vibrator to set up the fourth harmonic on the string?

SOLUTION: One **Key Idea** here is that the string will resonate at only certain frequencies, determined by the wave speed $|\vec{v}_w|$ on the string and the length L of the string. From Eq. 17-45,

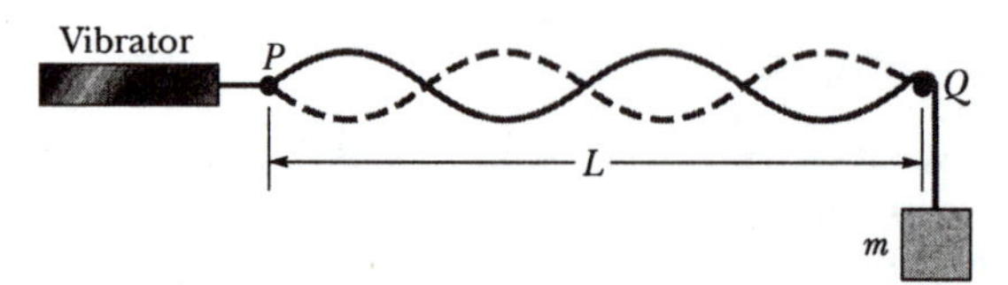

Fig. TE17-1 Touchstone Example 17-11-1. A string under tension connected to a vibrator. For a fixed vibrator frequency, standing wave patterns will occur for certain values of the string tension.

these resonant frequencies are

$$f = n\frac{|\vec{v}_w|}{2L}, \quad \text{for } n = 1, 2, 3, \ldots. \tag{TE17-5}$$

To set up the fourth harmonic (for which $n = 4$), we need to adjust the right side of this equation, with $n = 4$, so that the left side equals the frequency of the vibrator (120 Hz).

We cannot adjust L in Eq. TE17-5; it is set. However, a second **Key Idea** is that we *can* adjust $|\vec{v}_w|$ because it depends on how much mass m we hang on the string. According to Eq. 17-20, wave speed $|v_w| = \sqrt{T/\mu}$. Here the tension T in the string is equal to the magnitude of the weight mg of the block. Thus,

$$|\vec{v}_w| = \sqrt{\frac{T}{\mu}} = \sqrt{\frac{mg}{\mu}}. \tag{TE17-6}$$

Substituting $|\vec{v}_w|$ from Eq. TE17-6 into Eq. TE17-5, setting $n = 4$ for the fourth harmonic, and solving for m give us

$$m = \frac{4L^2f^2\mu}{n^2g} \tag{TE17-7}$$

$$= \frac{(4)(1.2\text{ m})^2(120\text{ Hz})^2(0.0016\text{ kg/m})}{(4)^2(9.8\text{ m/s}^2)}$$

$$= 0.846\text{ kg} \approx 0.85\text{ kg}. \quad \text{(Answer)}$$

(b) What standing wave mode is set up if $m = 1.00$ kg?

SOLUTION: If we insert this value of m into Eq. TE17-7 and solve for n, we find that $n = 3.7$. A **Key Idea** here is that n must be an integer, so $n = 3.7$ is impossible. Thus, with $m = 1.00$ kg, the vibrator cannot set up a standing wave on the string, and any oscillation of the string will be small, perhaps even imperceptible.

Touchstone Example 17-12-1

Two sinusoidal waves $y_1(x, t)$ and $y_2(x, t)$ have the same wavelength and travel together in the same direction along a string. Their amplitudes are $y_{m1} = 4.0$ mm and $y_{m2} = 3.0$ mm, and their phase constants are 0 and $\pi/3$ rad, respectively. What are the amplitude y'_m and phase constant β of the resultant wave? Write the resultant wave in the form of Eq. 17-48.

SOLUTION: One **Key Idea** here is that the two waves have a number of properties in common: Because they travel along the same string, they must have the same speed $|\vec{v}_w|$, as set by the tension and linear density of the string according to Eq. 17-20. With the same wavelength λ, they must have the same angular wave number k $(= 2\pi/\lambda)$. Also, with the same wave number k and speed $|\vec{v}_w|$, they must have the same angular frequency ω $(= k|\vec{v}_w|)$.

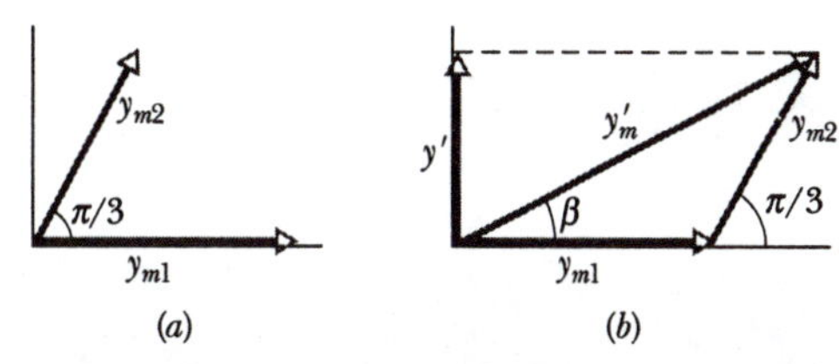

Fig. TE17-2 Touchstone Example 17-12-1. (*a*) Two phasors of magnitudes y_{m1} and y_{m2} and with phase difference $\pi/3$. (*b*) Vector addition of these phasors at any instant during their rotation gives the magnitude y'_m of the phasor for the resultant wave.

A second **Key Idea** is that the waves (call them waves 1 and 2) can be represented by phasors rotating at the same angular speed ω about an origin. Because the phase constant for wave 2 is *greater* than that for wave 1 by $\pi/3$, phasor 2 must *lag* phasor 1 by $\pi/3$ rad in their clockwise rotation, as shown in Fig. TE17-2. The resultant wave due to the interference of waves 1 and 2 can then be represented by a phasor that is the vector sum of phasors 1 and 2.

To simplify the vector summation, we drew phasors 1 and 2 in Fig. TE17-2*a* at the instant when phasor 1 lies along the horizontal axis. We then drew lagging phasor 2 at positive angle $\pi/3$ rad. In Fig. TE17-2*b* we shifted phasor 2 so its tail is at the head of phasor 1. Then we can draw the phasor y'_m of the resultant wave from the tail of phasor 1 to the head of phasor 2. The phase constant β is the angle it makes with phasor 1.

To find values for y'_m and β, we can sum phasors 1 and 2 directly on a vector-capable calculator, by adding a vector of magnitude 4.0 and angle 0 rad to a vector of magnitude 3.0 and angle $\pi/3$ rad, or we can add the vectors by components. For the horizontal components we have

$$\begin{aligned} y'_{mh} &= y_{m1}\cos 0 + y_{m2}\cos \pi/3 \\ &= 4.0\text{ mm} + (3.0\text{ mm})\cos \pi/3 = 5.50\text{ mm}. \end{aligned}$$

For the vertical components we have

$$\begin{aligned} y'_{mv} &= y_{m1}\sin 0 + y_{m2}\sin \pi/3 \\ &= 0 + (3.0\text{ mm})\sin \pi/3 = 2.60\text{ mm}. \end{aligned}$$

Thus, the resultant wave has an amplitude of

$$\begin{aligned} y'_m &= \sqrt{(5.50\text{ mm})^2 + (2.60\text{ mm})^2} \\ &= 6.1\text{ mm} \end{aligned} \qquad \text{(Answer)}$$

and a phase constant of

$$\beta = \tan^{-1}\frac{2.60\text{ mm}}{5.50\text{ mm}} = 0.44\text{ rad}. \qquad \text{(Answer)}$$

From Fig. TE17-2*b*, phase constant β is a *positive* angle relative to phasor 1. Thus, the resultant wave *lags* wave 1 in their travel by phase constant $\beta = +0.44$ rad. From Eq. 17-48, we can write the resultant wave as

$$y'(x, t) = (6.1\text{ mm})\sin(kx - \omega t + 0.44\text{ rad}). \qquad \text{(Answer)}$$

18 Waves—II

This horseshoe bat not only can locate a moth flying in total darkness but can also determine the moth's relative speed, to home in on the insect.

How does the bat's detection system work, and how can a moth "jam" the system or otherwise reduce its effectiveness?

The answer is in this chapter.

18-1 Sound Waves

What is a Sound Wave?

As we saw in Chapter 17, mechanical waves are waves that require a material medium to exist. There are two types of mechanical waves: *Transverse waves* involve oscillations perpendicular to the direction in which the wave travels; *longitudinal waves* involve oscillations parallel to the direction of wave travel. Examples of longitudinal waves given in the last chapter included the expansion/compression waves in the Slinky and the air-filled pipe (Figs. 17-1 and 17-3 respectively).

When we hear sounds, we are detecting longitudinal waves passing through air that have frequencies within the range of human hearing. A **sound wave** can be defined more broadly as a longitudinal wave of any frequency passing through a medium. The medium can be a solid, liquid or gas. For example, geological prospecting teams use sound waves to probe Earth's crust for oil. Ships carry sound-emitting gear (sonar) to detect underwater obstacles. Medical doctors use high-frequency sound waves (ultrasound) to create computer-processed images of soft tissues (as shown in Fig. 18-1). Physics students use ultrasound pulses to track the motion of objects in the laboratory.

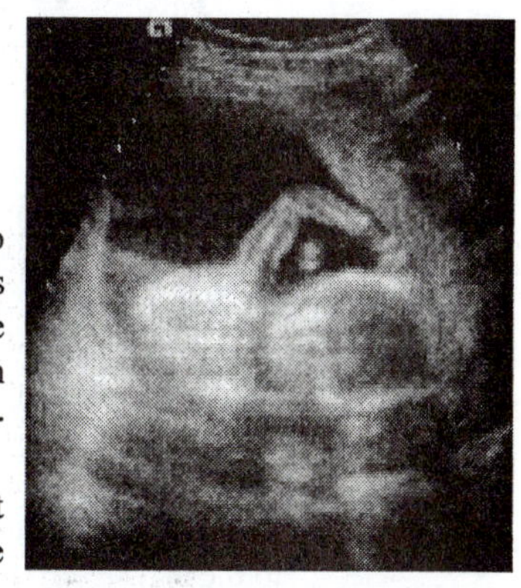

Fig. 18-1: An image of a fetus flexing an arm. This image is made with ultrasound waves that have a frequency of 4 MHz—two-hundred times higher than the threshold of human hearing.

In this chapter we shall focus on the characteristics of audible sound waves that travel through air. We start by considering some differences in how waves propagate in one-, two-, and three-dimensional spaces.

Wave Dimensions

Unlike the wave pulses that travel in a straight line along a string, a sound wave from a small "point-like" source is usually not constrained to travel in only one direction. The same is true for the electromagnetic waves that we will study in Chapters 34-38. In order to understand how sound waves and electromagnetic waves travel, we need to consider how the dimensionality of the space through which a wave propagates affects it.

In the last chapter we observed that the crest of a wave passing along a one-dimensional string moves along a line. If we constrain a sound wave to travel in a long tube, its compression wave crests would lie in a plane perpendicular to the axis of the tube as shown in Fig. 18-4. Such a one-dimensional wave is know as a **plane wave**.

On the other hand, when you see a raindrop fall into the surface of a pond, a wave crest propagates out from the raindrop in an expanding circle. The crest of a water wave is constrained to move along the two-dimensional surface of the water. This two-dimensional wave is known as a **circular wave**.

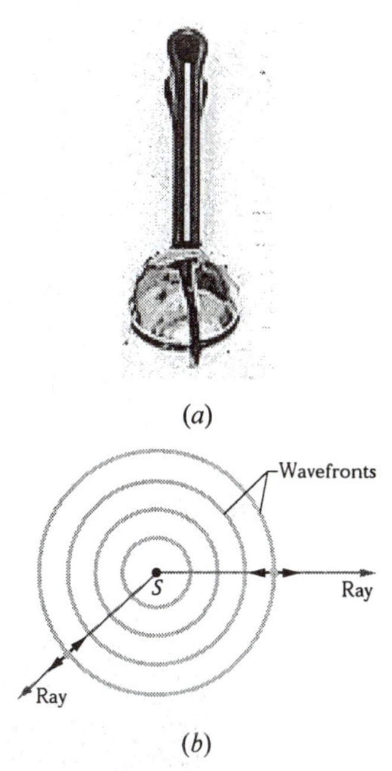

Fig. 18-3: (*a*) A small sleigh bell that is rung at a steady rate. (*b*) If the distances from the bell to its listeners are large compared to the size of the bell, then it acts as a three-dimensional point source of sound. Each compression wave crest moves out in an expanding sphere. The two-dimensional cross section of each of four wave crest spheres is shown. The short double arrows signify that the air particles oscillate parallel to the direction of motion of the wave crests. The lines drawn perpendicular to the wave crests are *rays*.

Fig 18-2: Water on the surface of a shallow tank of water (known as a ripple tank) is disturbed with a pencil. The wave crest propagates radially outward from the source of the disturbance in a widening circle. According to the ruler placed over the tank, the size of a highlighted circular wave crest has increased from 5 cm to 10 cm in 1/6th of a second. Then it increased from 10 cm to 15 cm in the next 1/6th of a second. Thus, the wave crest appears to be moving at a constant speed of 30 cm/s.

If you snap your fingers or ring a tiny sleigh bell, a compression wave is created. At distances that are large relative to the size of the source, compression wave crest travels out in three dimensions as an expanding sphere (Fig. 18-3). This three-dimensional wave is defined as **a spherical wave,** provided there is no preferred direction for the propagation of the wave energy.

It is important to understand that the dimensionality of a sound source (such as a one-dimensional wave in a guitar string) and the dimensionality of sound wave produced by the source are not necessarily the same. We have associated the dimensionality of a wave with the curvature of its wave crest as it spreads not the dimensionality of the source. However, we can actually define the dimensionality of a wave in terms of any point about a propagating wave form. This is because for a wave of a given dimension, the *shape* of the wave crests (maxima) and wave troughs (minima) and points of no deflection (nodes) are the same. For example, in a two-dimensional wave the crests, troughs, and points of zero deflection are all circular. It is useful for us to define a **wavefront** as the collection of all adjacent points on an expanding wave that have the same phase. In other words, a wavefront can be the collection of all adjacent crest points. Or it can be taken to be a collection of all adjacent nodes, or all adjacent troughs. **Rays** are defined as lines directed perpendicular to the wavefronts that indicate the direction of travel of the wavefronts. The short double arrows superimposed on the rays of Fig. 18-3 indicate that the longitudinal oscillations of the air, which transmits the wave, are parallel to the rays.

A one-dimensional or plane wave is defined as any wave that has a wavefront that lies in a plane. The wavefront of a two-dimensional or circular wave lies along an expanding circle. The wavefront of a three-dimensional or spherical wave lies along an expanding sphere. These definitions are often still useful in situations where we do not have ideal point sources. For instance, a flat speaker mounted in the door of a car only emits waves in a "forward" direction. In this case we would have half-spherical three dimensional wave fronts.

As the wavefronts move outward and their radii become larger, their curvatures decrease. Far from the source, these spheres associated with a three-dimensional wave are so large that we lose track of their curved nature all together. For example, when you listen to a singer in a concert hall, your ear is so far away from her that you cannot detect any curvature in the wavefronts she sends out. In such cases we can treat the *local* portion of a wavefront as if it lies in a plane. Thus, there are times in our study of sound when we will treat sound waves as if they are one-dimensional plane waves. Similarly, in Chapter 36 when we study how light waves interfere when passing through slits and traveling parallel to a two-dimensional surface, we can treat light waves as if they were two-dimensional rather than three-dimensional. Let's return to our primary task in this chapter, which is to learn more about the nature of sound waves.

18-2 The Speed of Sound

The ability to calculate the speed of a sound wave as a function of the properties of the medium through which it travels is of great practical importance. For example, we can use our knowledge of the speed of sound in air to estimate how far away a lightning bolt is. Since sound travels approximately 0.2 mi/s in air, it will take about 5 seconds to travel the one mile. Thus a five-second interval between a lightning flash and a thunderclap tells us that the electrical storm is about a mile away. The ultrasonic motion detectors used in many physics laboratories bounce sound pulses off objects in their surroundings. The time of travel of the sound pulses emitted and then reflected from the motion detector is used to measure the distance between it and the reflecting object.

What properties of a medium does the speed of sound depend on? Let's draw an analogy between the speed of sound and the speed of a wave traveling along a string (Section 17-6). We found that the wave speed increases with the tension, T, in the string. But, the tension determines the magnitude of the restoring force that brings a displaced section of string back toward its equilibrium position. We also found, as expected, that the wave speed decreases with the mass of each disturbed length of the string. This make sense since the linear mass density of the string, μ, is an inertial property that determines how rapidly, or slowly, the string can respond to the restoring forces acting on it. Thus, we can generalize (Eq. 17-20), which we derived for the wave speed along a stretched string, to:

$$|\vec{v}_w| = \sqrt{\frac{T}{|\mu|}} = \sqrt{\frac{\text{restoring property}}{\text{inertial property}}}. \qquad (17\text{-}20)$$

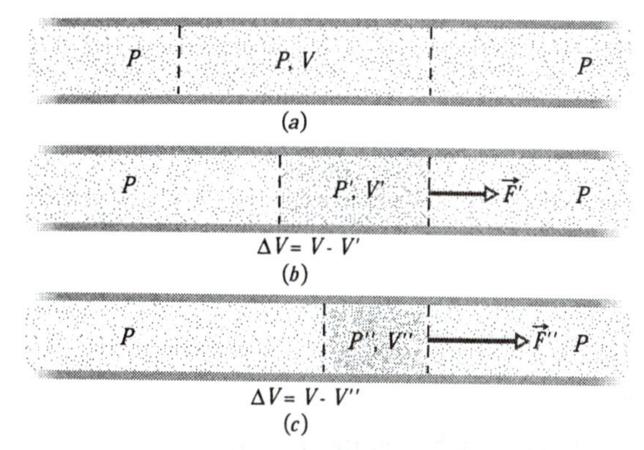

Fig. 18-4: (*a*) An undisturbed tube of fluid at pressure P. (*b*) A compression wave passing through a fluid with a small bulk modulus causes a mass element to be compressed to a smaller volume, V'. (*c*) The same type of compression wave with a larger bulk modulus. It causes a mass element to undergo a greater volume reduction.

If the medium is a fluid (such as air or water) and the wave is longitudinal, we can guess that the inertial property, corresponding to the linear density of the string μ, is the volume density ρ of the fluid. What shall we define as its restoring property? As a sound wave passes through a fluid, elements of mass in the fluid undergo compressions and expansions due to pressure differences within the fluid. When a mass element in the fluid is compressed it has a higher pressure and pushes on additional fluid. This new mass element of fluid becomes compressed into a smaller volume. Then it pushes in turn on another mass element and so on. This is how the compression wave travels through a fluid.

As shown in Fig. 18-4, the restoring property of the fluid is determined by the extent to which an element of mass changes volume when it experiences a difference in pressure (force per unit area). As a compression travels along the density of each mass element increases temporarily. As the fluid mass element becomes more dense its pressure goes up causing a difference in pressure between it and the mass element down the line. The ratio of the pressure difference and the relative volume change is a property of the fluid, defined as its bulk modulus:

$$B \equiv -\frac{\Delta P}{\Delta V/V}, \qquad (18\text{-}1)$$

where $\Delta V/V$ is the fractional change in volume produced by a change in pressure ΔP. (As explained in Section 15-3, the SI unit for pressure is the newton per square meter, which is given a special name, the *pascal* (Pa).) Substituting B for $|\vec{T}|$ and ρ for μ in

$$|\vec{v}_w| = \sqrt{\frac{|\vec{T}|}{\mu}}$$

gives us

$$|\vec{v}_w| = \sqrt{\frac{B}{\rho}}. \qquad \text{(speed of sound in a fluid).} \qquad (18\text{-}2)$$

It can be easily shown that the dimensions of $\sqrt{B/\rho}$ are those of a velocity. It is possible to derive Eq. 18-2 mathematically using methods similar to those used in Section 17-6 to find the expression for the wave speed along a stretched string. Once again experimental results confirm the validity of Eq. 18-2.

Table 18-1 lists the speed of sound in various media.

Table 18-1: The Speed of Sound[a]

Medium	Speed (m/s)
Gases	
Air (0°C)	331
Air (20°C)	343
Helium	965
Hydrogen	1284
Liquids	
Water (0°C)	1402
Water (20°C)	1482
Seawater[b]	1522
Solids	
Aluminum	6420
Steel	5941
Granite	6000

[a]At 0°C and 1 atm pressure, except where noted.
[b]At 20° C and 3.5% salinity.

The density of water is almost 1000 times greater than the density of air. If this were the only relevant factor, we would expect from Eq. 18-2 that the speed of sound in water would be considerably less than the speed of sound in air. However, Table 18-1 shows us that the reverse is true. We conclude (again from Eq. 18-2) that the bulk modulus of water must be more than 1000 times greater than that of air. This is indeed the case. Water is less compressible than air, which (see Eq. 18-1) is another way of saying that its bulk modulus is much greater.

Sound Pressure Variation vs. Time

0.000 0.005 0.010 0.015 0.020

Time (s)

Fig. 18-5: Measurement of the speed of sound. The sound pressure variations from a finger snap travel down a 2.4 m tube and back. A microphone sensor connected to an MBL system is used to measure the time between the initial sound pulses and the reflected pulses.

More About Traveling Sound Waves

Consider for a moment the pressure variations and air molecule displacements associated with a sinusoidal sound wave traveling through air. Figure 18-6 displays such a wave traveling rightward through a long air-filled tube. Recall from Chapter 17 that we can produce such a wave by moving a piston at the left end of the tube back and forth sinusoidally (as in Fig. 17-3). The piston's rightward motion moves the element of air next to it and compresses that air; the piston's leftward motion allows the element of air to move back to the left and the pressure to decrease. As each element of air pushes on the next element in turn, the right-left motion of the air and the change in its pressure is passed to the next bit of air along the tube.

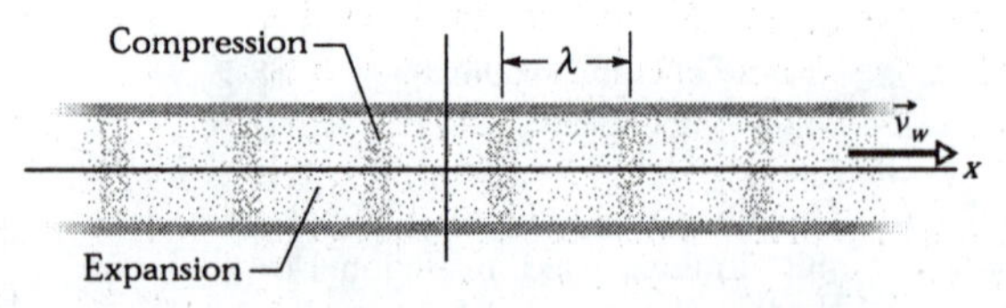

Fig. 18-6: A sound wave, traveling through a long air-filled tube with speed $\vec{v}_w$, consists of a moving, periodic pattern of expansions and compressions of the air. The wave is shown at an arbitrary instant. As the wave passes, a fluid element of thickness Δx oscillates left and right in simple harmonic motion about its equilibrium position.

The alternating compressions and rarefactions (reductions in pressure) propagate as a sound wave. As a reminder of what we learned in Chapter 17, note that it is the oscillations of pressure that propagate, not the air molecules. However, the air molecules (or molecules of another fluid) do move. They oscillate back and forth about their initial position.

As the wave moves, the air pressure changes at any position x in Fig. 18-6 in a sinusoidal fashion, like the displacement of a string element in a transverse wave, except the pressure variations are longitudinal rather than transverse. To describe this pressure variation we write

$$\Delta P(x,t) = \Delta P_m \sin(kx - \omega t). \quad (18\text{-}3)$$

The wave number k, angular frequency ω, frequency f, wavelength λ, speed $|v_w|$, and period T for a longitudinal sound wave are defined and interrelated exactly as for a transverse wave, except that λ is now the distance (again along the direction of travel) in which the pattern of compression and expansion due to the wave begins to repeat itself (see Fig. 18-6). Also note that Eq. 17-12 still holds so that the wave speed is given by $|v_w| = \lambda f = \omega/k$.

A negative value of ΔP in Eq. 18-3 corresponds to an expansion of the air, and a positive value to a compression. Here ΔP_m is the **pressure amplitude,** which is the maximum increase or decrease in pressure due to the wave; ΔP_m is normally very much less than the pressure P present when there is no wave.

READING EXERCISE 18-1: Verify that the dimensions of the expression in Eq. 18-2 are dimensionally correct. In other words, show that the term $\sqrt{B/\rho}$ has the dimensions of m/s which are the dimensions of velocity in SI units.

READING EXERCISE 18-2: Examine Fig. 18-5 shown above. The maximum of the finger snap pulse set is at 0.0002s and the maximum of the reflected pulse is at 0.0133 s. Use these times to calculate the measured value of the wave velocity for the set of sound pulses as they travel back and forth through the air inside the 2.4 m long tube. Is your calculated speed approximately the same as the speed of sound of air at room temperature as reported in Table 18-1? Note: The air temperature in the room was not recorded.

READING EXERCISE 18-3: Use the fact that people float much more easily in seawater than in fresh water to give a plausible explanation of the fact that the measured speed of sound in seawater is larger than that for fresh water.

Touchstone Example 18-2-1 at the end of this chapter, illustrates how to use what you learned in this section.

TE

18-3 Interference

Like transverse waves, sound waves undergo interference when two waves pass through the same point at the same time. Let us consider, in particular, the interference between two identical sound waves traveling in slightly different directions. Figure 18-7 shows how we can set up such a situation: Two point sources S_1 and S_2 emit sound waves that are in phase and of identical wavelength λ. Thus, the sources themselves are said to be in phase; that is, as the waves emerge from the sources, their displacements are always identical. We are interested in the waves that then travel through point P in Fig. 18-7. We assume that the distance to P is much greater than the distance between the sources so that we can approximate the waves as traveling in the same direction at P.

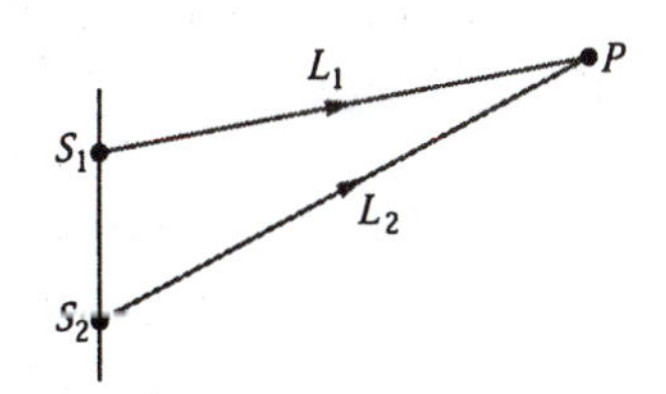

Fig. 18-7: Two point sources S_1 and S_2 emit spherical sound waves in phase. The rays indicate that the waves pass through a common point P.

If the waves (which both start out at the same point in their pressure oscillation, i.e., *in phase*) traveled along paths with identical lengths to reach point P, they would still be in phase there. In this case, the displacements of the two waves would add. As with transverse waves, this means that they would undergo fully constructive interference there. However, in Fig. 18-7, path L_2 traveled by the wave from S_2 is longer than path L_1 traveled by the wave from S_1. The difference in path lengths means that the waves may not be in phase at point P and so might be at different points in their oscillations. In other words, their phase difference ϕ at P depends on their **path length difference** $\Delta L = |L_2 - L_1|$.

To relate phase difference ϕ to path length difference ΔL, we use the definition of phase (from Section 17-4) to determine that a phase difference of 2π rad corresponds to one wavelength. Thus, we can write the proportion

$$\frac{\phi}{2\pi} = \frac{\Delta L}{\lambda}, \qquad (18\text{-}4)$$

from which

$$\phi = \frac{\Delta L}{\lambda} 2\pi. \qquad (18\text{-}5)$$

Fully constructive interference occurs when ϕ is zero, 2π, or any integer multiple of 2π. We can write this condition as

$$\phi = m(2\pi), \quad \text{for } m = 0, 1, 2, \ldots \quad \text{(fully constructive interference)}. \qquad (18\text{-}6)$$

From Eq. 18-5, this occurs when the ratio $\Delta L/\lambda$ is

$$\frac{\Delta L}{\lambda} = 0, 1, 2, \ldots \quad \text{(fully constructive interference)}. \qquad (18\text{-}7)$$

For example, if the path length difference $\Delta L = |L_2 - L_1|$ in Fig. 18-7 is equal to 2λ, then $\Delta L/\lambda = 2$ and the waves undergo fully constructive interference at point P. The interference is fully constructive because the wave from S_2 is phase-shifted relative to the wave from S_1 by 2λ, putting the two waves *exactly in phase* at P.

Fully destructive interference occurs when ϕ is an odd multiple of π, a condition we can write as

$$\phi = (2m+1)\pi, \quad \text{for } m = 0, 1, 2, \ldots \quad \text{(fully destructive interference)}. \qquad (18\text{-}8)$$

From $\phi = \frac{\Delta L}{\lambda} 2\pi$ (Eq. 18-5), this occurs when the ratio $\Delta L/\lambda$ is

$$\frac{\Delta L}{\lambda} = 0.5, 1.5, 2.5, \ldots \quad \text{(fully destructive interference)}. \qquad (18\text{-}9)$$

For example, if the path length difference $\Delta L = |L_2 - L_1|$ in Fig. 18-7 is equal to 2.5λ, then $\Delta L/\lambda = 2.5$ and the waves undergo fully destructive interference at point P. The interference is fully destructive because the wave from S_2 is phase-shifted relative to the wave from S_1 by 2.5 wavelengths, which puts the two waves *exactly out of phase* at P.

Of course, two waves could produce intermediate interference as, say, when $\Delta L/\lambda = 1.2$. This would be closer to fully constructive interference $(\Delta L/\lambda = 1.0)$ than to fully destructive interference $(\Delta L/\lambda = 1.5)$.

Touchstone Example 18-3-1, at the end of this chapter, illustrates how to use what you learned in this section.

TE

18-4 Intensity and Sound Level

If you have ever tried to sleep while someone played loud music nearby, you are well aware that there is more to sound than frequency, wavelength, and speed. Humans also detect how *loud* a sound is. Although the human ear does detect pressure amplitudes, it turns out that we are more sensitive to the *energy fluctuations* in a propagating wave than we are to the pressure alone. Hence, we will define a new energy related quantity associated with waves. The **intensity** I of a sound wave at a surface is the average rate per unit area at which energy is transferred by the wave through or onto the surface. This is what we commonly refer to as loudness. By definition, the intensity of a wave is

$$I = \frac{Power}{A}, \tag{18-10}$$

where "*Power*" is the time rate of energy transfer (the power) of the sound wave, and A is the area of the surface intercepting the sound.

There is a relationship between the energy-related quantity ***intensity*** and the variations in pressure associated with a sound wave. The intensity I is related to the change in pressure ΔP of the sound wave by

$$I = \frac{(\Delta P)^2}{\rho|\vec{v}|} = \frac{\Delta P_m^{\ 2}}{\rho|\vec{v}|}, \tag{18-11}$$

where ΔP_m is the pressure amplitude of the wave, ρ is the density of the propagation medium and $|\vec{v}|$ is the speed of the wave. A related expression in terms of a (particle) displacement amplitude can also be used. Because this is a general result which is seen over and over again in many different areas of physics that involve waves, note that

➤The intensity of a (sound) wave varies as the square of its amplitude.

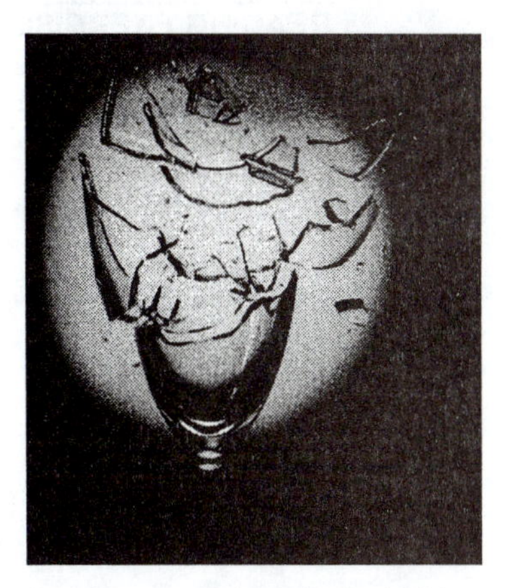

Sound can cause the wall of a drinking glass to oscillate. If the sound produces a standing wave of oscillations and if the intensity of the sound is large enough, the glass will shatter.

Variation of Intensity with Distance

How intensity varies with distance from a real sound source is often complex. Some real sources (like loudspeakers) may transmit sound only in particular directions, and the environment usually produces echoes (reflected sound waves) that overlap the direct sound waves. In some situations, however, we can ignore echoes and assume that the sound source is a point source that emits the sound *isotropically*—that is, with equal intensity in all directions. The wavefronts spreading from such an isotropic point source S at a particular instant are shown in Fig. 18-8.

Let us assume that the mechanical energy of the sound waves is conserved as they spread from this source. Let us also center an imaginary sphere of radius r on the source, as shown in Fig. 18-8. All the energy emitted by the source must pass through the surface of the sphere. Thus, the time rate at which energy is transferred through the surface by the sound waves must equal the time rate at which energy is emitted by the source (that is, the power P_s of the source). From the fact that intensity is equal to the ratio of power to area (Eq. 18-10), the intensity I at the sphere must then be

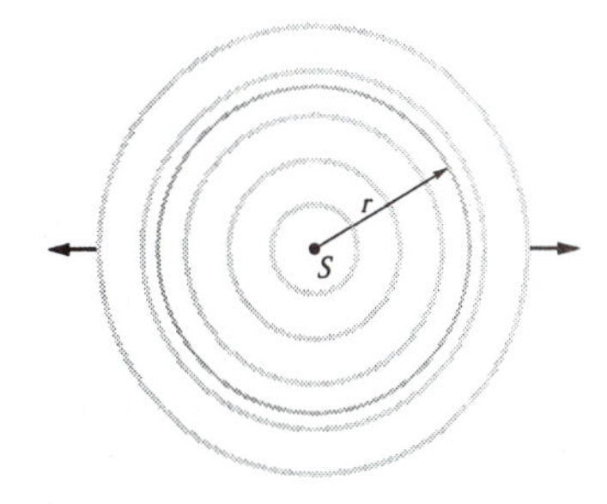

Fig. 18-8: A point source S emits sound waves uniformly in all directions. The waves pass through an imaginary sphere of radius r that is centered on S.

$$I = \frac{P_s}{4\pi r^2}, \tag{18-12}$$

where $4\pi r^2$ is the area of the sphere. Equation 18-12 tells us that the intensity of sound from an isotropic point source decreases with the square of the distance r from the source.

READING EXERCISE 18-4: The figure indicates three small patches 1, 2, and 3 that lie on the surfaces of two imaginary spheres; the spheres are centered on an isotropic point source S of sound. The rates at which energy is transmitted through the three patches by the sound waves are equal. Rank the patches according to (a) the intensity of the sound on them and (b) their area, greatest first.

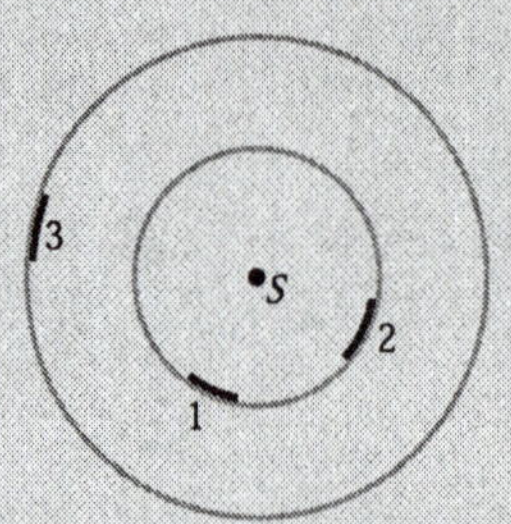

The Decibel Scale

The displacement amplitude at the human ear ranges from about 10^{-5} m for the loudest tolerable sound to about 10^{-11} m for the faintest detectable sound, a ratio of 10^6. From our discussions above, we know that the intensity of a sound varies as the *square* of its amplitude, so the ratio of intensities at these two limits of the human auditory system is 10^{12}. Humans can hear over an enormous range of intensities.

We deal with such an enormous range of values by using logarithms. Consider the relation

$$y = \log x,$$

in which x and y are variables. It is a property of this equation that if we *multiply* x by 10, then y increases by 1. To see this, we write

$$y' = \log(10x) = \log 10 + \log x = 1 + y.$$

Similarly, if we multiply x by 10^{12}, y increases by only 12.

Thus, instead of speaking of the intensity I of a sound wave, it is much more convenient to speak of its **sound level** β, defined as

$$\beta = (10\text{ dB})\log\frac{I}{I_0}. \qquad (18\text{-}13)$$

Here dB is the abbreviation for **decibel,** the unit of sound level, a name that was chosen to recognize the work of Alexander Graham Bell. I_0 in Eq. 18-13 is a standard reference intensity $(= 10^{-12}\ \text{W/m}^2)$, chosen because it is near the lower limit of the human range of hearing. For $I = I_0$, Eq. 18-13 gives $\beta = 10 \log 1 = 0$, so our standard reference level corresponds to zero decibels. Then β increases by 10 dB every time the sound intensity increases by an order of magnitude (a factor of 10). Thus, $\beta = 40$ corresponds to an intensity that is 10^4 times the standard reference level. Table 18-2 lists the sound levels for a variety of environments.

Table 18-2: Some Sound Levels (dB)

Hearing threshold	0	Rock concert	110
Rustle of leaves	10	Pain threshold	120
Conversation	60	Jet engine	130

Touchstone Examples 18-4-1 and 18-4-2, at the end of this chapter, illustrate how to use what you learned in this section.

TE

18-5 Sources of Musical Sound

Musical sounds can be set up by oscillating strings (guitar, piano, violin), membranes (kettledrum, snare drum), air columns (flute, oboe, pipe organ, and the fujara of Fig. 18-9), wooden blocks or steel bars (marimba, xylophone), and many other oscillating bodies. Most instruments involve more than a single oscillating part. In the violin, for example, both the strings and the body of the instrument participate in producing the music.

Fig. 18-9: The air column within a fujara oscillates when that traditional Slovakian instrument is played.

Recall from Chapter 17 that standing waves can be set up on a stretched string that is fixed at both ends. They arise because waves traveling along the string are reflected back onto the string at each end. If the wavelength of the waves is suitably matched to the length of the string, the superposition of waves traveling in opposite directions produces a standing wave pattern (or oscillation mode). The wavelength required of the waves for such a match is one that corresponds to a *resonant frequency* of the string. The advantage of setting up standing waves is that the string then oscillates with a large, sustained amplitude, pushing back and forth against the surrounding air and thus generating a noticeable sound wave with the same frequency as the oscillations of the string. This production of sound is of obvious importance to, say, a guitarist.

We can set up standing waves of sound in an air-filled pipe in a similar way. As sound waves travel through the air in the pipe, they are reflected at each end and travel back through the pipe. (The reflection occurs even if an end is open, but the reflection is not as complete as when the end is closed.) If the wavelength of the sound waves is suitably matched to the length of the pipe, the superposition of waves traveling in opposite directions through the pipe sets up a standing wave pattern. The wavelength required of the sound waves for such a match is one that corresponds to a resonant frequency of the pipe. The advantage of such a standing wave is that the air in the pipe oscillates with a large, sustained amplitude, emitting at any open end a sound wave that has the same frequency as the oscillations in the pipe. This emission of sound is of obvious importance to, say, an organist.

Many other aspects of standing sound wave patterns are similar to those of string waves: The closed end of a pipe is like the fixed end of a string in that there must be a displacement node (zero displacement in a string, zero motion of molecules in a fluid like air) located there. A zero particle velocity immediately against the closed end of the pipe leads to a large change in pressure (a pressure antinode). The open end of a pipe is like the end of a string attached to a freely moving ring, as in Fig. 17-19*b*, in that there must be a displacement antinode (maximum displacement in a string, maximum motion of molecules in a fluid like air) located there. This too makes sense, as the molecules of air (or other fluid) are completely unconstrained once they leave the end of the pipe. The large particle velocities here lead to very small changes in pressure (a pressure antinode). To be perfectly precise, the antinode for the open end of a pipe is located slightly beyond the end, but this difference is so minor that it is irrelevant in our discussions.

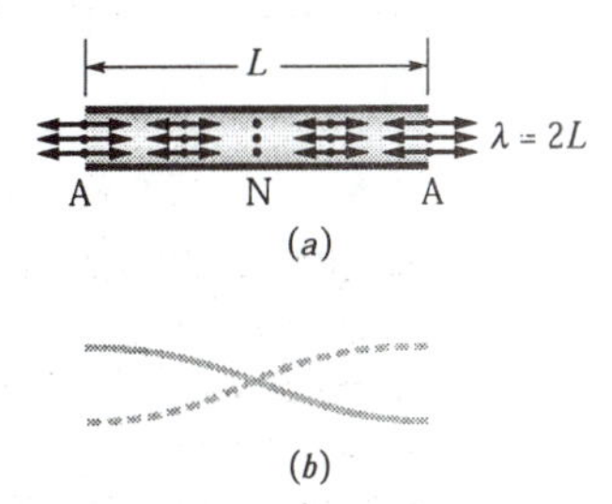

Fig. 18-10: (*a*) The simplest standing wave pattern of displacement for (longitudinal) sound waves in a pipe with both ends open has an antinode (A) across each end and a node (N) across the middle. (The longitudinal displacements represented by the double arrows are greatly exaggerated.) (*b*) The corresponding standing wave pattern for transverse waves.

So, the simplest standing wave pattern that can be set up in a pipe with two open ends is one with a particle displacement antinode (maximum particle velocities and zero change in pressure) at both ends and no other antinodes between them. Under these conditions, the only way that there can be a standing wave pattern at all is for there to be a displacement node (zero particle velocity and maximum change in pressure) in the middle of the pipe. This is shown in Fig. 18-10*a*. An easier way of representing standing longitudinal sound waves is shown in Fig. 18-10*b*—by drawing them as standing transverse waves. This representation works because it mirrors the particle velocities as a function of position within the pipe.

The standing wave pattern of Fig. 18-10*a* is called the *fundamental mode* or *first harmonic*. For it to be set up, the sound waves in a pipe of length L must have a wavelength given by $L = \lambda/2$, so that $\lambda = 2L$. Several more standing sound wave patterns for a pipe with two open ends are shown in Fig. 18-11*a* using transverse wave representations. The *second harmonic* requires sound waves of wavelength $\lambda = L$, the *third harmonic* requires wavelength $\lambda = 2L/3$, and so on.

More generally, the resonant frequencies for a pipe of length L with two open ends correspond to the wavelengths

$$\lambda = \frac{2L}{n}, \quad \text{for } n = 1, 2, 3,\ldots, \tag{18-14}$$

where n is called the *harmonic number*. The resonant frequencies for a pipe with two open ends are then given by

$$f = \frac{|\vec{v}|}{\lambda} = \frac{n|\vec{v}|}{2L}, \quad \text{for } n = 1, 2, 3,\ldots \quad \text{(pipe, two open ends)}, \tag{18-15}$$

where v is the speed of sound.

Figure 18-11*b* shows (using transverse wave representations) some of the standing sound wave patterns that can be set up in a pipe with only one open end. As required, across the open end there is a displacement antinode (maximum particle velocities and zero pressure change) and across the closed end there is a displacement node (zero particle velocities and maximum pressure change). The simplest pattern requires sound waves having a wavelength given by $L = \lambda/4$, so that $\lambda = 4L$. The next simplest pattern requires a wavelength given by $L = 3\lambda/4$, so that $\lambda = 4L/3$, and so on.

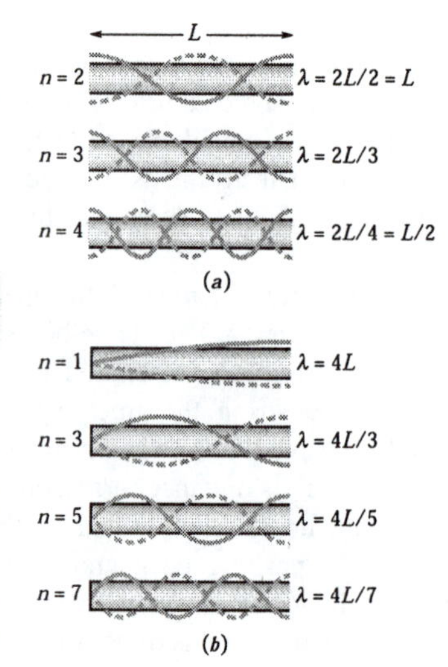

Fig. 18-11: Standing wave patterns for transverse waves superimposed on pipes to represent standing sound wave patterns in the pipes. (*a*) With *both* ends of the pipe open, any harmonic can be set up in the pipe. (*b*) With only *one* end open, only odd harmonics can be set up.

More generally, the resonant frequencies for a pipe of length L with only one open end correspond to the wavelengths

$$\lambda = \frac{4L}{n}, \quad n = 1, 3, 5,\ldots, \tag{18-16}$$

in which the harmonic number n *must be an odd number*. The resonant frequencies are then given by

$$f = \frac{|\vec{v}|}{\lambda} = \frac{n|\vec{v}|}{4L}, \quad \text{for } n = 1, 3, 5\ldots \quad \text{(pipe, one open end)}. \tag{18-17}$$

Note again that only odd harmonics can exist in a pipe with one open end. For example, the second harmonic, with $n = 2$, cannot be set up in such a pipe. Note also that for such a pipe the numeric adjective (e.g. first, second, third...) before the word harmonic in a phrase such as "the third harmonic" always refers to the harmonic number n and not to the nth *possible* harmonic.

The length of a musical instrument reflects the range of frequencies over which the instrument is designed to function, and smaller length implies higher frequencies. Figure 18-12, for example, shows the saxophone and violin families, with their frequency ranges suggested by the piano keyboard. Note that, for every instrument, there is overlap with its higher- and lower-frequency neighbors.

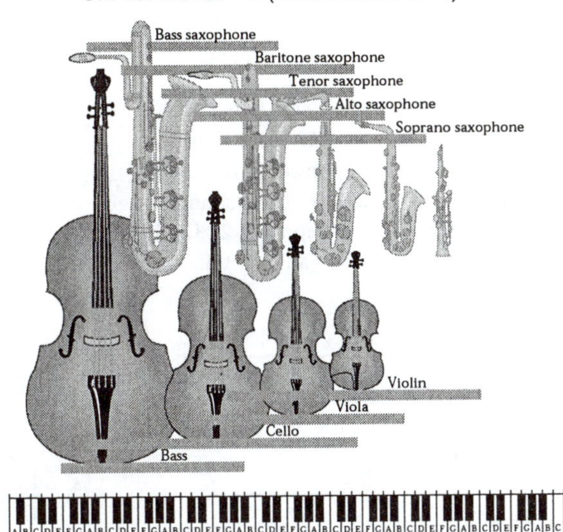

Fig. 18-12: The saxophone and violin families, showing the relations between instrument length and frequency range. The frequency range of each instrument is indicated by a horizontal bar along a frequency scale suggested by the keyboard at the bottom; the frequency increases toward the right.

In any oscillating system that gives rise to a musical sound, whether it is a violin string or the air in an organ pipe, the fundamental and one or more of the higher harmonics are usually generated simultaneously. Thus, you hear them together—that is, superimposed as a net wave. When different instruments are played at the same note, they produce the same fundamental frequency but different intensities for the higher harmonics. For example, the fourth harmonic of middle C might be relatively loud on one instrument and relatively quiet or even missing on another. Thus, because different instruments produce different net waves, they sound different to you even when they are played at the same note. That would be the case for the three net waves shown in Fig. 18-13, which were produced at the same note by different instruments.

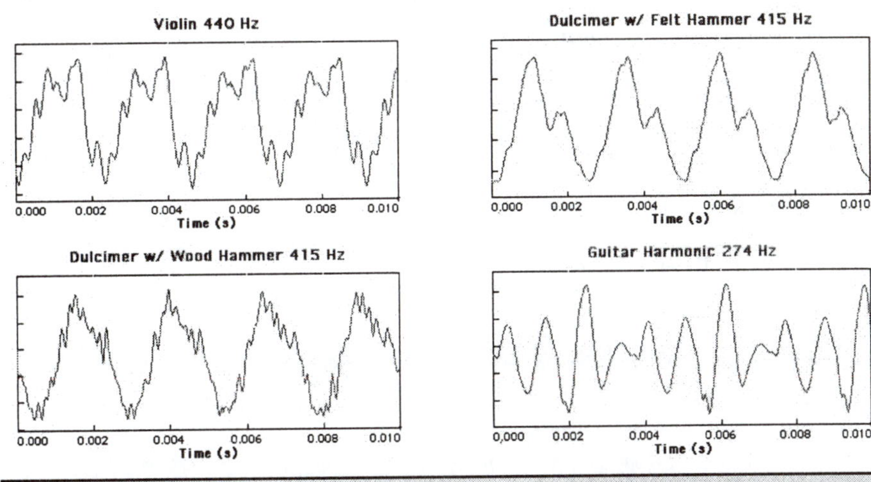

Fig. 18-13: The wave forms produced by a violin, a hammer dulcimer, and a guitar. These graphs of the relative sound pressure variation vs. time are produced using an MBL system with a microphone sensor attached.

READING EXERCISE 18-5: Pipe A, with length L, and pipe B, with length $2L$, both have two open ends. Which harmonic of pipe B has the same frequency as the fundamental of pipe A?

Touchstone Example 18-5-1 at the end of this chapter, illustrates how to use what you learned in this section.

TE

18-6 Beats

If we listen, a few minutes apart, to two sounds whose frequencies are, say, 8 and 10 Hz, most of us cannot tell one from the other. However, if the sounds reach our ears simultaneously, what we hear is a sound whose frequency is 9 Hz, the *average* of the two combining frequencies. We also hear a striking variation in the intensity of this sound—it increases and decreases in slow, wavering **beats** that repeat at a frequency of 2 Hz, the *difference* between the two combining frequencies. Figure 18-14 shows this beat phenomenon.

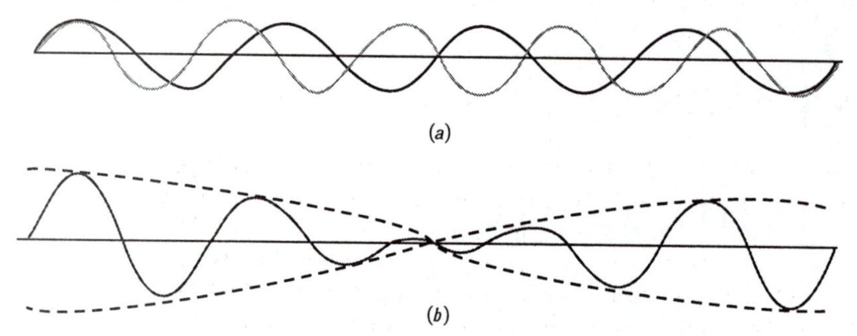

Fig. 18-14: (*a*) The pressure variations ΔP of two sound waves as they would be detected separately are plotted as a function of time. The frequencies of the waves are nearly equal. (*b*) If the two waves are detected simultaneously the resultant pressure variation is the superposition of the two waves. Notice how the waves cancel each other at the center of the plot and reinforce each other at the two ends of the plot.

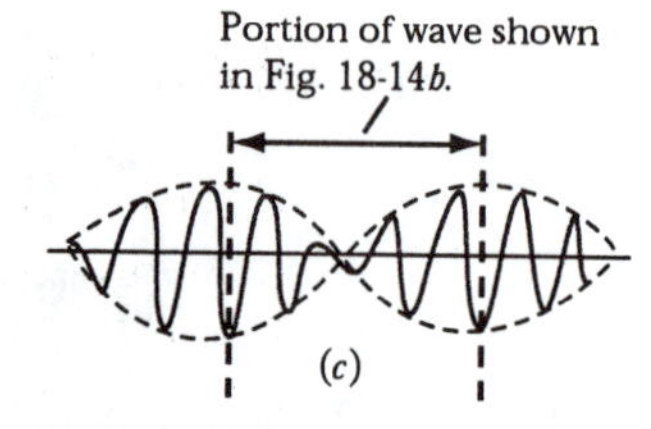

Fig. 18-14(*c*): When plotted over a longer time period the two waves shown in Fig. 18-14 show a beat pattern of 2 Hz if the frequencies of the original waves are 10 Hz and 12 Hz.

Let the time-dependent variations of pressure due to two sound waves at a particular location be

$$\Delta P_1 = \Delta P_m \cos\omega_1 t \quad \text{and} \quad \Delta P_2 = \Delta P_m \cos\omega_2 t, \tag{18-18}$$

where $\omega_1 > \omega_2$. We have assumed, for simplicity, that the waves have the same amplitude. According to the superposition principle, the resultant pressure variation is

$$\Delta P = \Delta P_1 + \Delta P_2 = \Delta P_m(\cos\omega_1 t + \cos\omega_2 t).$$

Using the trigonometric identity (see Appendix E)

$$\cos\alpha + \cos\beta = 2\cos\tfrac{1}{2}(\alpha - \beta)\cos\tfrac{1}{2}(\alpha + \beta)$$

allows us to write the resultant pressure variation as

$$\Delta P = 2\Delta P_m \cos\tfrac{1}{2}(\omega_1 - \omega_2)t\cos\tfrac{1}{2}(\omega_1 + \omega_2)t. \tag{18-19}$$

If we write

$$\omega' = \tfrac{1}{2}(\omega_1 - \omega_2) \text{ and } \omega = \tfrac{1}{2}(\omega_1 + \omega_2) \tag{18-20}$$

we can then write Eq. 18-19 as

$$\Delta P(t) = [2\Delta P_m \cos\omega' t]\cos\omega t. \tag{18-21}$$

We now assume that the angular frequencies ω_1 and ω_2 of the combining waves are almost equal, which means that $\omega >> \omega'$ in Eq. 18-20. We can then regard Eq. 18-21 as a cosine function whose angular frequency is ω and whose amplitude (which is not constant but varies with angular frequency ω') is the quantity in the brackets.

A maximum amplitude will occur whenever $\cos\omega' t$ in Eq. 18-21 has the value +1 or −1, which happens twice in each repetition of the cosine function. Because $\cos\omega' t$ has angular frequency ω', the angular frequency ω_{beat} at which beats occur is $\omega_{\text{beat}} = 2\omega'$. Then, with the aid of Eq. 18-20, we can write

$$\omega_{\text{beat}} = 2\omega' = (2)(\tfrac{1}{2})(\omega_1 - \omega_2) = \omega_1 - \omega_2.$$

Because $\omega = 2\pi f$, we can recast this as

$$f_{\text{beat}} = f_1 - f_2 \qquad \text{(beat frequency).} \qquad (18\text{-}22)$$

Musicians use the beat phenomenon in tuning their instruments. If an instrument is sounded against a standard frequency (for example, the lead oboe's reference A) and tuned until the beat disappears, then the instrument is in tune with that standard. In musical Vienna, concert A (440 Hz) is available as a telephone service for the benefit of the city's many professional and amateur musicians.

18-7 The Doppler Effect

An ambulance is parked by the side of the highway, sounding its 1000 Hz siren. If you are also parked by the highway, you will hear that same frequency. However, if there is relative motion between you and the ambulance, either toward or away from each other, you will hear a different frequency. For example, if you are driving *toward* the ambulance at 120 km/h (about 75 mi/h), you will hear a *higher* frequency (1096 Hz, an *increase* of 96 Hz). If you are driving *away from* the ambulance at that same speed, you will hear a *lower* frequency (904 Hz, a *decrease* of 96 Hz).

These motion-related frequency changes are examples of the **Doppler effect.** The effect was proposed (although not fully worked out) in 1842 by Austrian physicist Johann Christian Doppler. It was tested experimentally in 1845 by Buys Ballot in Holland, "using a locomotive drawing an open car with several trumpeters."

The Doppler effect holds not only for sound waves but also for electromagnetic waves, including microwaves, radio waves, and visible light. Here, however, we shall consider only sound waves, and we shall take as a reference frame the body of air through which these waves travel. This means that we shall measure the speeds of a source S of sound waves and a detector D of those waves *relative to that body of air.* (Unless otherwise stated, the body of air is stationary relative to the ground, so the speeds can also be measured relative to the ground.) We shall assume that S and D move either directly toward or directly away from each other, at speeds less than the speed of sound.

If either the detector or the source are moving, or both is moving, the emitted frequency f and the detected frequency f' are related by

$$f' = f\,\frac{|\vec{v}| \pm |\vec{v}_D|}{|\vec{v}| \pm |\vec{v}_S|} \qquad \text{(general Doppler effect),} \qquad (18\text{-}23)$$

where $|\vec{v}|$ is the speed of sound through the air, $|\vec{v}_D|$ is the detector's speed relative to the air, and $|\vec{v}_S|$ is the source's speed relative to the air. The choice of plus or minus signs is set by this rule:

➤When the motion of detector or source is toward the other, the sign on its speed must give an upward shift in frequency. When the motion of detector or source is away from the other, the sign on its speed must give a downward shift in frequency.

In short, toward means shift up, and away means shift down.

Here are some examples of the rule. If the detector moves toward the source, use the plus sign in the numerator of Eq. 18-23 to get a shift up in the frequency. If it moves away, use the minus sign in the numerator to get a shift down. If it is stationary, substitute 0 for $|\vec{v}_D|$. If the source moves toward the detector, use the minus sign in the denominator of Eq. 18-23 to get a shift up in the frequency. If it moves away, use the plus sign in the denominator to get a shift down. If the source is stationary, substitute 0 for $|\vec{v}_S|$.

Next, we derive equations for the Doppler effect for two specific situations and then derive Eq. 18-23 for the general situation.

1. When the detector moves relative to the air and the source is stationary relative to the air, the motion changes the frequency at which the detector intercepts wavefronts and thus the detected frequency of the sound wave.
2. When the source moves relative to the air and the detector is stationary relative to the air, the motion changes the wavelength of the sound wave and thus the detected frequency (recall that frequency is related to wavelength).

Detector Moving; Source Stationary

In Fig. 18-15, a detector *D* (represented by an ear) is moving at speed $|\vec{v}_D|$ toward a stationary source *S* that emits spherical wavefronts, of wavelength λ and frequency *f*, moving at the speed $|\vec{v}|$ of sound in air. The wavefronts are drawn one wavelength apart. The frequency detected by detector *D* is the rate at which *D* intercepts wavefronts (or individual wavelengths). If *D* were stationary, that rate would be *f*, but since *D* is moving into the wavefronts, the rate of interception is greater, and thus the detected frequency f' is greater than *f*.

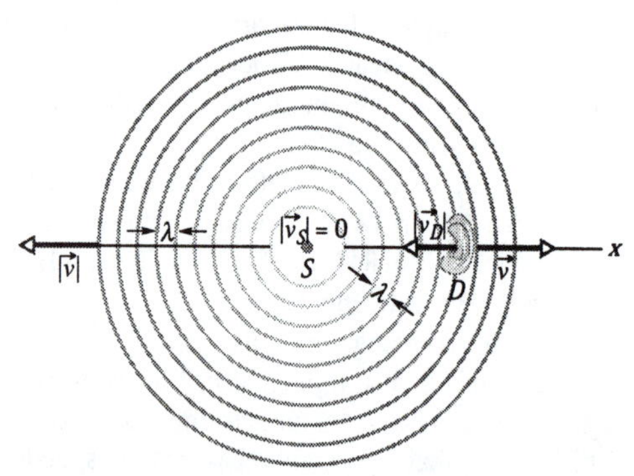

Fig. 18-15: A stationary source of sound *S* emits spherical wavefronts, shown one wavelength apart, that expand outward at speed $|\vec{v}|$. A sound detector *D*, represented by an ear, moves with velocity $|\vec{v}_D|$ toward the source. The detector senses a higher frequency because of its motion.

Let us for the moment consider the situation in which *D* is stationary (Fig. 18-16). In time *t*, the wavefronts move to the right a distance $|\vec{v}|t$. The number of wavelengths in that distance $|\vec{v}|t$ is the number of wavelengths intercepted by *D* in time *t*, and that number is $|\vec{v}|t/\lambda$. The rate at which *D* intercepts wavelengths, which is the frequency *f* detected by *D*, is

$$f = \frac{|\vec{v}|t/\lambda}{t} = \frac{|\vec{v}|}{\lambda}. \tag{18-24}$$

In this situation, with *D* stationary, there is no Doppler effect—the frequency detected by *D* is the frequency emitted by *S*.

Now let us again consider the situation in which *D* moves opposite the wavefronts (Fig. 18-17). In time *t*, the wavefronts move to the right a distance $|\vec{v}|t$ as previously, but now *D* moves to the left a distance $|\vec{v}_D|t$. Thus, in this time *t*, the distance moved by the wavefronts relative to *D* is $|\vec{v}|t + |\vec{v}_D|t$. The number of wavelengths in this relative distance $|\vec{v}|t + |\vec{v}_D|t$ is the number of wavelengths intercepted by *D* in time *t*, and is $(|\vec{v}|t + |\vec{v}_D|t/\lambda)$. The *rate* at which *D* intercepts wavelengths in this situation is the frequency f', given by

$$f' = \frac{(|\vec{v}|t + |\vec{v}_D|t)/\lambda}{t} = \frac{|\vec{v}| + |\vec{v}_D|}{\lambda}. \tag{18-25}$$

From Eq. 18-24, we have $\lambda = |\vec{v}|/f$. Then Eq. 18-25 becomes

$$f' = \frac{|\vec{v}| + |\vec{v}_D|}{|\vec{v}|/f} = f\frac{|\vec{v}| + |\vec{v}_D|}{|\vec{v}|}. \tag{18-26}$$

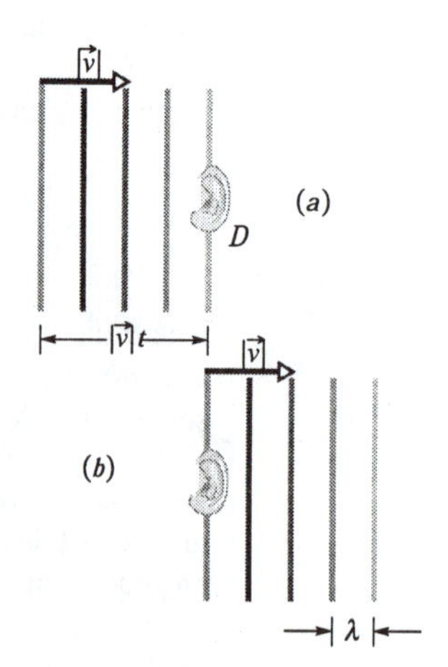

Fig. 18-16: Wavefronts of Fig. 18-15, assumed planar, (*a*) reach and (*b*) pass a stationary detector *D*; they move a distance $|\vec{v}|t$ to the right in time *t*.

Note that in Eq. 18-26, f' must be greater than f unless $|\vec{v}_D| = 0$ (the detector is stationary).

Similarly, we can find the frequency detected by D if D moves away from the source. In this situation, the wavefronts move a distance $|\vec{v}|t - |\vec{v}_D|t$ relative to D in time t, and f' is given by

$$f' = f\frac{|\vec{v}| - |\vec{v}_D|}{|\vec{v}|}. \tag{18-27}$$

In Eq. 18-27, f' must be less than f unless $|\vec{v}_D| = 0$.

We can summarize Eqs. 18-26 and 18-27 with

$$f' = f\frac{|\vec{v}| \pm |\vec{v}_D|}{|\vec{v}|} \quad \text{(detector moving; source stationary).} \tag{18-28}$$

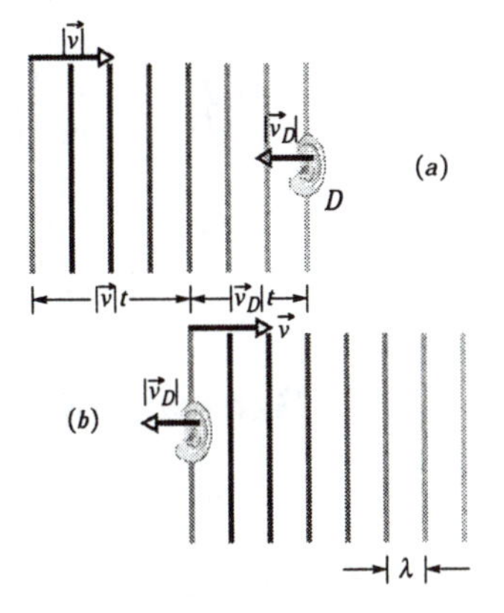

Fig. 18-17: Wavefronts (*a*) reach and (*b*) pass detector *D*, which moves opposite the wavefronts. In time *t*, the wavefronts move a distance $|\vec{v}|t$ to the right and *D* moves a distance $|\vec{v}_D|t$ to the left.

Source Moving; Detector Stationary

Let detector D be stationary with respect to the body of air, and let source S move toward D at speed $|\vec{v}_S|$ (Fig. 18-18). The motion of S changes the wavelength of the sound waves it emits, and thus the frequency detected by D.

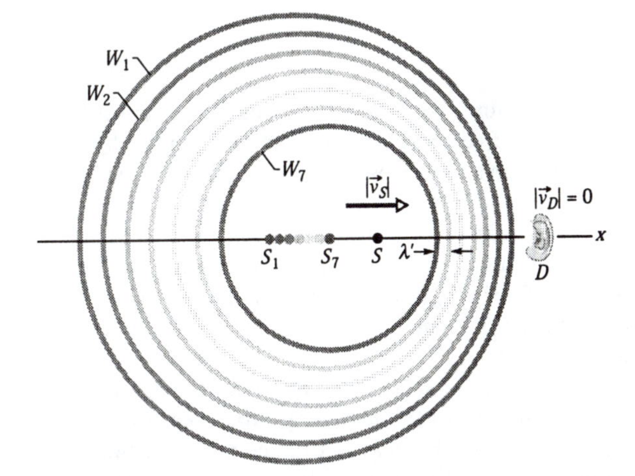

Fig. 18-18: A detector D is stationary, and a source S is moving toward it at speed v_S. Wavefront W_1 was emitted when the source was at S_1, wavefront W_7 when it was at S_7. At the moment depicted, the source is at S. The detector perceives a higher frequency because the moving source, chasing its own wavefronts, emits a reduced wavelength λ' in the direction of its motion.

To see this change, let $T(=1/f)$ be the time between the emission of any pair of successive wavefronts W_1 and W_2. During T, wavefront W_1 moves a distance $|\vec{v}|T$ and the source moves a distance $|\vec{v}_S|T$. At the end of T, wavefront W_2 is emitted. In the direction in which S moves, the distance between W_1 and W_2, which is the wavelength λ' of the waves moving in that direction, is $|\vec{v}|T - |\vec{v}_S|T$. If D detects those waves, it detects frequency f' given by

$$\begin{aligned} f' &= \frac{|\vec{v}|}{\lambda'} = \frac{|\vec{v}|}{|\vec{v}|T - |\vec{v}_S|T} = \frac{|\vec{v}|}{|\vec{v}|/f - |\vec{v}_S|/f} \\ &= f\frac{|\vec{v}|}{|\vec{v}| - |\vec{v}_S|}. \end{aligned} \tag{18-29}$$

Note that f' must be greater than f unless $|\vec{v}_S| = 0$.

In the direction opposite that taken by S, the wavelength λ' of the waves is $|\vec{v}|T - |\vec{v}_S|T$. If D detects those waves, it detects frequency f' given by

$$f' = f\frac{|\vec{v}|}{|\vec{v}|+|\vec{v}_S|}. \qquad (18\text{-}30)$$

Now f' must be less than f unless $|\vec{v}_S| = 0$.

We can summarize Eqs. 18-29 and 18-30 with

$$f' = f\frac{|\vec{v}|}{|\vec{v}| \pm |\vec{v}_S|} \quad \text{(source moving; detector stationary).} \qquad (18\text{-}31)$$

General Doppler Effect Equation

We can now derive the general Doppler effect equation by replacing f in Eq. 18-31 (the frequency of the source) with f' of Eq. 18-28 (the frequency associated with motion of the detector). The result is Eq. 18-23 for the general Doppler effect.

That general equation holds not only when both detector and source are moving but also in the two specific situations we just discussed. For the situation in which the detector is moving and the source is stationary, substitution of $|\vec{v}_S| = 0$ into Eq. 18-23 gives us Eq. 18-28, which we previously found. For the situation in which the source is moving and the detector is stationary, substitution of $|\vec{v}_D| = 0$ into Eq. 18-22 gives us Eq. 18-31, which we previously found. Thus, Eq. 18-23 is the equation to remember.

Bat Navigation

Bats navigate and search out prey by emitting, and then detecting reflections of, ultrasonic waves. These are sound waves with frequencies greater than can be heard by a human. For example, a horseshoe bat emits ultrasonic waves at 83 kHz, well above the 20 kHz limit of human hearing.

After the sound is emitted through the bat's nostrils, it might reflect (echo) from a moth, and then return to the bat's ears. The motions of the bat and the moth relative to the air cause the frequency heard by the bat to differ by a few kilohertz from the frequency it emitted. The bat automatically translates this difference into a relative speed between itself and the moth, so it can zero in on the moth.

Some moths evade capture by flying away from the direction in which they hear ultrasonic waves. That choice of flight path reduces the frequency difference between what the bat emits and what it hears, and then the bat may not notice the echo. Some moths avoid capture by clicking to produce their own ultrasonic waves, thus "jamming" the detection system and confusing the bat. (Surprisingly, moths and bats do all this without first studying physics.)

It is interesting to note that the ultrasonic motion detectors described in Section 1-8 use the same technology as bats to measure the velocity of objects in physics laboratories. They do this by alternately emitting and receiving 50 KHz sound pulses. For this reason some physics instructors have been known to describe motion detectors as "dumb bats."

READING EXERCISE 18-6: The figure indicates the directions of motion of a sound source and a detector for six situations in stationary air. For each situation, is the detected frequency greater than or less than the emitted frequency, or can't we tell without more information about the actual speeds?

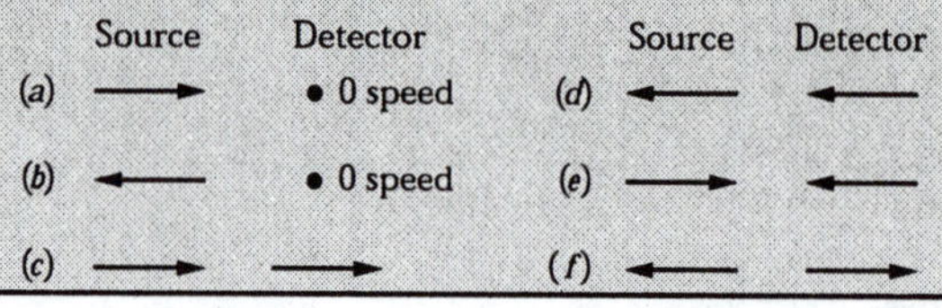

	Source	Detector		Source	Detector
(a)	→	• 0 speed	(d)	←	←
(b)	←	• 0 speed	(e)	→	←
(c)	→	→	(f)	←	→

Touchstone Example 18-7-1, at the end of this chapter, illustrates how to use what you learned in this section.

TE

18-8 Supersonic Speeds; Shock Waves

If a source is moving toward a stationary detector at a speed equal to the speed of sound—that is, if $|\vec{v}_S| = |\vec{v}|$—Eqs. 18-23 and 18-31 predict that the detected frequency f' will be infinitely great. This means that the source is moving so fast that it keeps pace with its own spherical wavefronts, as Fig. 18-19*a* suggests. What happens when the speed of the source exceeds the speed of sound?

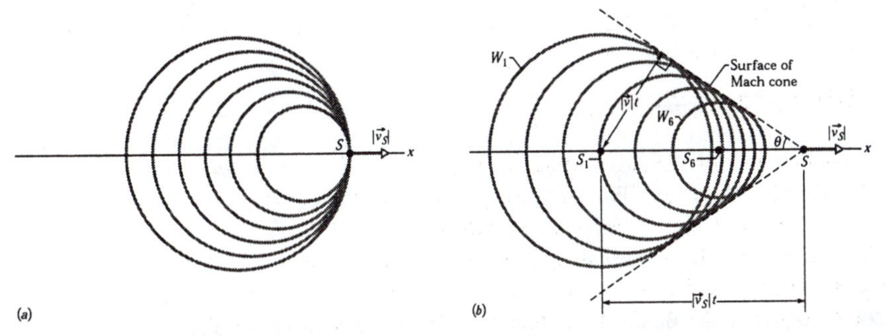

Fig. 18-19: (*a*) A source of sound *S* moves at speed $|\vec{v}_S|$ equal to the speed of sound and thus as fast as the wavefronts it generates. (*b*) A source *S* moves at speed $|\vec{v}_S|$ faster than the speed of sound and thus faster than the wavefronts. When the source was at position S_1 it generated wavefront W_1, and at position S_6 it generated W_6. All the spherical wavefronts expand at the speed of sound $|\vec{v}|$ and bunch along the surface of a cone called the Mach cone, forming a shock wave. The surface of the cone has half-angle θ and is tangent to all the wavefronts.

For such supersonic speeds, Eqs. 18-23 and 18-31 no longer apply. Figure 18-19*b* depicts the spherical wavefronts that originated at various positions of the source. The radius of any wavefront in this figure is $|\vec{v}|t$, where $|\vec{v}|$ is the speed of sound and t is the time that has elapsed since the source emitted that wavefront. Note that all the wavefronts bunch along a V-shaped envelope in the two-dimensional drawing of Fig. 18-19*b*. The wavefronts actually extend in three dimensions, and the bunching actually forms a cone called the *Mach cone*. A *shock wave* is said to exist along the surface of this cone, because the bunching of wavefronts causes an abrupt rise and fall of air pressure as the surface passes through any point. From Fig. 18-19*b*, we see that the half-angle θ of the cone, called the *Mach cone angle*, is given by

$$\sin\theta = \frac{|\vec{v}|t}{|\vec{v}_S|t} = \frac{|\vec{v}|}{|\vec{v}_S|} \quad \text{(Mach cone angle).} \quad (18\text{-}32)$$

The ratio $|\vec{v}_S|/|\vec{v}|$ is called the *Mach number*. When you hear that a particular plane has flown at Mach 2.3, it means that its speed was 2.3 times the speed of sound in the air through which the plane was flying. The shock wave generated by a supersonic aircraft (Fig. 18-20) or projectile produces a burst of sound, called a *sonic boom*, in which the air pressure first suddenly increases and then suddenly decreases below normal before returning to normal. Part of the sound that is heard when a rifle is fired is the sonic boom produced by the bullet. A sonic boom can also be heard from a long bullwhip when it is snapped quickly: Near the end of the whip's motion, its tip is moving faster than sound and produces a small sonic boom—the *crack* of the whip.

Fig. 18-20: Shock waves produced by the wings of an F-4 Phantom II jet. They are visible because the sudden decrease in air pressure in the shock waves caused water molecules in the air to condense, forming a fog.

Touchstone Example 18-2-1

One clue used by your brain to determine the direction of a source of sound is the time delay Δt between the arrival of the sound at the car closer to the source and the arrival at the farther car. Assume that the source is distant so that a wavefront from it is approximately planar when it reaches you, and let D represent the separation between your cars.

(a) Find an expression that gives Δt in terms of D and the angle θ between the direction of the source and the forward direction.

SOLUTION: The situation is shown (from an overhead view) in Fig. TE18-1, where wavefronts approach you from a source that is located in front of you and to your right. The **Key Idea** here is that the time delay Δt is due to the distance d that each wavefront must travel to reach your left ear (L) after it reaches your right ear (R). From Fig. TE18-1, we find

$$\Delta t - \frac{d}{|\vec{v}|} = \frac{D \sin \theta}{|\vec{v}|}. \qquad \text{(Answer)} \quad \text{(TE18-1)}$$

where $|\vec{v}|$ is the speed of sound in air. Based on a lifetime of experience, your brain correlates each detected value of Δt (from zero to the maximum value) with a value of θ (from zero to 90°) for the direction of the sound source.

(b) Suppose that you are submerged in water at 20°C when a wavefront arrives from directly to your right. Based on the time-delay clue, at what angle θ from the forward direction does the source seem to be?

SOLUTION: The **Key Idea** here is that the speed of the sound is now the speed $|\vec{v}_w|$ in water, so in Eq. TE18-1 we substitute $|\vec{v}_w|$ for $|\vec{v}|$ and 90° for θ, finding that

$$\Delta t_w = \frac{D \sin 90°}{|\vec{v}_w|} = \frac{D}{|\vec{v}_w|}, \qquad \text{(TE18-2)}$$

Since $|\vec{v}_w|$ is about four times $|\vec{v}|$, delay Δt_w is about one-fourth the maximum time delay in air. Based on experience, your brain will process the water time delay as if it occurred in air. Thus, the sound source appears to be at an angle θ smaller than 90°. To find that apparent angle, we substitute the time delay $D \div |\vec{v}_w|$ from Eq. TE18-2 for Δt in Eq. TE18-1, obtaining

$$\frac{D}{|\vec{v}_w|} = \frac{D \sin \theta}{|\vec{v}|}. \qquad \text{(TE18-3)}$$

Then, to solve for θ we substitute $|\vec{v}| = 343$ m/s and $|\vec{v}_w| = 1482$ m/s (from Table 18-1) into Eq. TE18-2, finding

$$\sin \theta = \frac{|\vec{v}|}{|\vec{v}_w|} = \frac{343 \text{ m/s}}{1482 \text{ m/s}} = 0.231$$

and thus

$$\theta = 13°. \qquad \text{(Answer)}$$

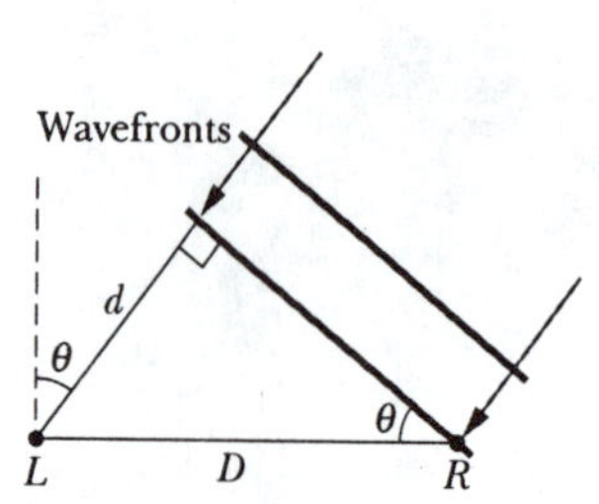

Fig. TE18-1 A wavefront travels a distance d (= $D \sin \theta$) farther to reach the left ear (L) than to reach the right ear (R).

Touchstone Example 18-3-1

In Fig. TE18-2*a*, two point sources S_1 and S_2, which are in phase and separated by distance $D = 1.5\lambda$, emit identical sound waves of wavelength λ.

(a) What is the path length difference of the waves from S_1 and S_2 at point P_1, which lies on the perpendicular bisector of distance D, at a distance greater than D from the sources? What type of interference occurs at P_1?

SOLUTION: The **Key Idea** here is that, because the waves travel identical distances to reach P_1, their path length difference is

$$\Delta L = 0. \qquad \text{(Answer)}$$

From Eq. 18-7, this means that the waves undergo fully constructive interference at P_1.

(b) What are the path length difference and type of interference at point P_2 in Fig. TE18-2*a*?

SOLUTION: Now the **Key Idea** is that the wave from S_1 travels the extra distance D ($= 1.5\lambda$) to reach P_2. Thus, the path length difference is

$$\Delta L = 1.5\lambda. \qquad \text{(Answer)}$$

From Eq. 18-9, this means that the waves are exactly out of phase at P_2 and undergo fully destructive interference there.

(c) Figure TE18-2*b* shows a circle with a radius much greater than D, centered on the midpoint between sources S_1 and S_2. What is the number of points N around this circle at which the interference is fully constructive?

SOLUTION: Imagine that, starting at point a, we move clockwise along the circle to point d. One **Key Idea** here is that as we move to point d, the path length difference ΔL increases and so

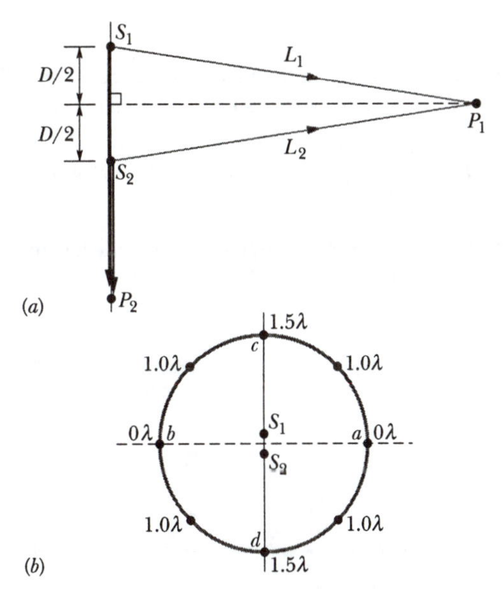

Fig. TE18-2 (*a*) Two point sources S_1 and S_2, separated by distance D, emit spherical sound waves in phase. The waves travel equal distances to reach point P_1. Point P_2 is on the line extending through S_1 and S_2. (*b*) The path length difference (in terms of wavelength) between the waves from S_1 and S_2, at eight points on a large circle around the sources.

the type of interference changes. From (a), we know that the path length difference is $\Delta L = 0\lambda$ at point a. From (b), we know that $\Delta L = 1.5\lambda$ at point d. Thus, there must be one point along the circle between a and d at which $\Delta L = \lambda$, as indicated in Fig. TE18-2*b*. From Eq. 18-7, fully constructive interference occurs at that point. Also, there can be no other point along the way from point a to point d at which fully constructive interference occurs, because there is no other integer than 1 between 0 and 1.5.

Another **Key Idea** here is to use symmetry to locate the other points of fully constructive interference along the rest of the circle. Symmetry about line cd gives us point b, at which $\Delta L = 0\lambda$. Also, there are three more points at which $\Delta L = \lambda$. In all we have

$$N = 6. \qquad \text{(Answer)}$$

Touchstone Example 18-4-1

An electric spark jumps along a straight line of length $L = 10$ m, emitting a pulse of sound that travels radially outward from the spark. (The spark is said to be a *line source* of sound.) The power of the emission is $P_s = 1.6 \times 10^4$ W.

(a) What is the intensity I of the sound when it reaches a distance $r = 6$ m from the spark?

SOLUTION: Let us center an imaginary cylinder of radius $r = 6$ m and length $L = 10$ m (open at both ends) on the spark, as shown in Fig. TE18-3. One **Key Idea** here is that the intensity I at the cylindrical surface is the ratio P/A of the time rate P at which sound energy passes through the surface to the surface area A. Another **Key Idea** is to assume that the principle of conservation of energy applies to the sound energy. This means that the rate P at which energy is transferred through the cylinder must equal the rate P_s at which energy is emitted by the source. Putting these ideas together and noting that the area of the cylindrical surface is $A = 2\pi rL$, and neglecting the relatively small amount of sound energy that flows out through the ends of the cylinder, we have

$$I = \frac{P}{A} = \frac{P_s}{2\pi rL}. \qquad \text{(TE18-4)}$$

This tells us that the intensity of the sound from a line source decreases with distance r (and not with the square of distance r as for a point source). Substituting the given data, we find

$$I = \frac{1.6 \times 10^4 \text{ W}}{2\pi(6 \text{ m})(10 \text{ m})} = 42.4 \text{ W/m}^2 \approx 42 \text{ W/m}^2. \qquad \text{(Answer)}$$

(b) At what time rate P_d is sound energy intercepted by an acoustic detector of area $A_d = 2.0$ cm^2, aimed at the spark and located a distance $r = 6$ m from the spark?

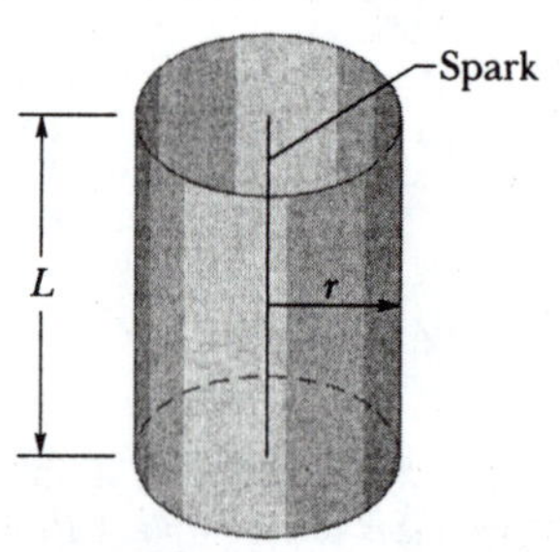

Fig. TE18-3 A spark along a straight line of length L emits sound waves radially outward. The waves pass through an imaginary cylinder of radius r and length L that is centered on the spark.

SOLUTION: Applying the first **Key Idea** of part (a), we know that the intensity of sound at the detector is the ratio of the energy transfer rate P_d there to the detector's area A_d:

$$I = \frac{P_d}{A_d}. \qquad \text{(TE18-5)}$$

We can imagine that the detector lies on the cylindrical surface of (a). Then the sound intensity at the detector is the intensity I (= 42.4 W/m²) at the cylindrical surface. Solving Eq. TE18-5 for P_d gives us

$$P_d = (42.4\ \text{W/m}^2)(2.0 \times 10^{-4}\ \text{m}^2) = 8.4\ \text{mW}. \qquad \text{(Answer)}$$

Touchstone Example 18-4-2

In 1976, the Who set a record for the loudest concert—the sound level 46 m in front of the speaker systems was $\beta_2 = 120$ dB. What is the ratio of the intensity I_2 of the band at that spot to the intensity I_1 of a jackhammer operating at sound level $\beta_1 = 92$ dB?

SOLUTION: The **Key Idea** here is that for both the Who and the jackhammer, the sound level β is related to the intensity by the definition of sound level in Eq. 18-13. For the Who, we have

$$\beta_2 = (10\ \text{dB}) \log \frac{I_2}{I_0},$$

and for the jackhammer, we have

$$\beta_1 = (10\ \text{dB}) \log \frac{I_1}{I_0}.$$

The difference in the sound levels is

$$\beta_2 - \beta_1 = (10\ \text{dB})\left(\log \frac{I_2}{I_0} - \log \frac{I_1}{I_0}\right). \qquad \text{(TE18-6)}$$

Using the identity

$$\log \frac{a}{b} - \log \frac{c}{d} = \log \frac{ad}{bc},$$

we can rewrite Eq. TE18-6 as

$$\beta_2 - \beta_1 = (10\ \text{dB}) \log \frac{I_2}{I_1}. \qquad \text{(TE18-7)}$$

Rearranging and substituting the known sound levels now yield

$$\log \frac{I_2}{I_1} = \frac{\beta_2 - \beta_1}{10\ \text{dB}} = \frac{120\ \text{dB} - 92\ \text{dB}}{10\ \text{dB}} = 2.8.$$

Taking the antilog of the far left and far right sides of this equation (the antilog key on your calculator is probably marked as 10^x), we find

$$\frac{I_2}{I_1} = \log^{-1} 2.8 = 630. \qquad \text{(Answer)}$$

Thus, the Who was *very* loud.

Temporary exposure to sound intensities as great as those of a jackhammer and the 1976 Who concert results in a temporary reduction of hearing. Repeated or prolonged exposure can result in permanent reduction of hearing (Fig. TE18-4). Loss of hearing is a clear risk for anyone continually listening to, say, heavy metal at high volume, especially on headphones.

Fig. TE18-4 Peter Townshend of the Who, playing in front of a speaker system. He suffered a permanent reduction in his hearing ability due to his exposure to high-intensity sound, not so much during on-stage performances as from wearing headphones in recording studios and at home.

Touchstone Example 18-5-1

Weak background noises from a room set up the fundamental standing wave in a cardboard tube of length $L = 67.0$ cm with two open ends. Assume that the speed of sound in the air within the tube is 343 m/s.

(a) What frequency do you hear from the tube?

SOLUTION: The **Key Idea** here is that, with both pipe ends open, we have a symmetric situation in which the standing wave has a displacement antinode at each end of the tube. The standing wave pattern (in string wave style) is that of Fig. 18-10*b*. The frequency is given by Eq. 18-15 with $n = 1$ for the fundamental mode:

$$f = \frac{n|\vec{v}|}{2L} = \frac{(1)(343\text{ m/s})}{2(0.670\text{ m})} = 256\text{ Hz}. \qquad \text{(Answer)}$$

If the background noises set up any higher harmonics, such as the second harmonic, you must also hear frequencies that are *integer* multiples of 256 Hz.

(b) If you jam your ear against one end of the tube, what fundamental frequency do you hear from the tube?

SOLUTION: The **Key Idea** now is that, with your ear effectively closing one end of the tube, we have an asymmetric situation—a displacement antinode still exists at the open end but a displacement node is now at the other (closed) end. The standing wave pattern is the top one in Fig. 18-11*b*. The frequency is given by Eq. 18-17 with $n = 1$ for the fundamental mode:

$$f = \frac{n|\vec{v}|}{4L} = \frac{(1)(343 \text{ m/s})}{4(0.670 \text{ m})} = 128 \text{ Hz}. \qquad \text{(Answer)}$$

If the background noises set up any higher harmonics, they will be *odd* multiples of 128 Hz. That means that the frequency of 256 Hz (which is an even multiple) cannot now occur.

Touchstone Example 18-7-1

A rocket moves at a speed of 242 m/s directly toward a stationary pole (through stationary air) while emitting sound waves at frequency $f = 1250$ Hz.

(a) What frequency f' is measured by a detector that is attached to the pole?

SOLUTION: We can find f' with Eq. 18-23 for the general Doppler effect. The **Key Idea** here is that, because the sound source (the rocket) moves through the air *toward* the stationary detector on the pole, we need to choose the sign on $|\vec{v}_S|$ that gives a *shift up* in the frequency of the sound. Thus, in Eq. 18-23 we use the minus sign in the denominator. We then substitute 0 for the detector speed $|\vec{v}_D|$, 242 m/s for the source speed $|\vec{v}_S|$, 343 m/s for the speed of sound $|\vec{v}|$ (from Table 18-1), and 1250 Hz for the emitted frequency f. We find

$$f' = f\frac{|\vec{v}| \pm |\vec{v}_D|}{|\vec{v}| \pm |\vec{v}_S|} = (1250 \text{ Hz})\frac{343 \text{ m/s} \pm 0}{343 \text{ m/s} - 242 \text{ m/s}}$$

$$= 4245 \text{ Hz} \approx 4250 \text{ Hz}, \qquad \text{(Answer)}$$

which, indeed, is a greater frequency than the emitted frequency.

(b) Some of the sound reaching the pole reflects back to the rocket as an echo. What frequency f'' does a detector on the rocket detect for the echo?

SOLUTION: Two **Key Ideas** here are the following:

1. The pole is now the source of sound (because it is the source of the echo), and the rocket's detector is now the detector (because it detects the echo).
2. The frequency of the sound emitted by the source (the pole) is equal to f', the frequency of the sound the pole intercepts and reflects.

We can rewrite Eq. 18-23 in terms of the source frequency f' and the detected frequency f'' as

$$f'' = f'\frac{|\vec{v}| \pm |\vec{v}_D|}{|\vec{v}| \pm |\vec{v}_S|} \qquad \text{(TE18-8)}$$

A third **Key Idea** here is that, because the detector (on the rocket) moves through the air *toward* the stationary source, we need to use the sign on v_D that gives a *shift up* in the frequency of the sound. Thus, we use the plus sign in the numerator of Eq. TE18-8. Also, we substitute $|\vec{v}_D| = 242$ m/s, $|\vec{v}_S| = 0$, $|\vec{v}| = 343$ m/s, and $f' = 4245$ Hz. We find

$$f'' = (4245 \text{ Hz})\frac{343 \text{ m/s} + 242 \text{ m/s}}{343 \text{ m/s} \pm 0}$$

$$= 7240 \text{ Hz}, \qquad \text{(Answer)}$$

which, indeed, is greater than the frequency of the sound reflected by the pole.

19 Temperature, Heat Energy, and the First Law of Thermodynamics

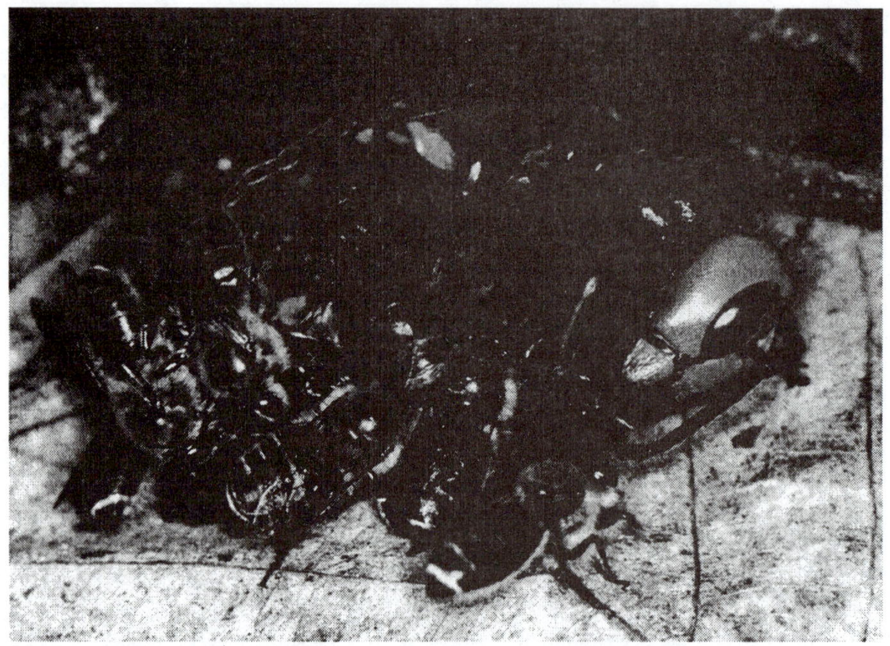

The giant hornet Vespa mandarinia japonica preys on Japanese bees. However, if one of the hornets attempts to invade a bee hive, several hundred of the bees quickly form a compact ball around the hornet to stop it. After about 20 minutes the hornet is dead, although the bees do not sting, bite, crush, or suffocate it.

Why, then, does the hornet die?

The answer is in this chapter.

19-1 Thermodynamics

In the next three chapters we focus on a new subject—thermodynamics. The development of thermodynamic principles is one of humankind's most profound intellectual achievements. Why? Thermodynamics has enriched our fundamental understanding of phenomena ranging from the metabolism of a lizard to the evolution of the universe. The steam engine, that powered the industrial revolution, operates according to thermodynamic principles.

A key idea in thermodynamics is that ordinary matter contains hidden internal energy. This hidden energy is thought of as a type of thermal energy because matter contains more of this energy when it is hot than when it is cold. The nineteenth century discovery of internal thermal energy is similar to Einstein's twentieth century discovery that matter also contains hidden mass-energy by virtue of its mass.

Although we have a natural ability to sense hot and cold, we can only use our sense of touch to tell if a material object is hot or cold over a relatively narrow range of temperatures. But, in order to use thermodynamics, we need to *quantify* our intuitive sense of hotness. Recall that any characteristic of a material or object that is measurable can be referred to as a **quantity** or **measurable property**. A **measurable property** is one that can be *quantified* through physical comparison with a reference (Sections 1-1 and 1-2). Careful observation of our everyday world tells us that some objects such as a balloon full of air or a metal rod have characteristics that change as the object gets hotter. For example, as temperature increases, the pressure and volume of the air in a balloon may increase or a metal rod may grow a little longer. Volume, length, electrical resistance and pressure are examples of *measurable properties* of a material object that can change. We can use any one of the properties of materials that *change* as an object gets hotter or colder to design crude (or not so crude) devices that quantify the hotness of an object. Such a device is called a **thermometer**.

Designing an accurate thermometer is not a trivial task. Nevertheless, thermometers are common devices and so many of us have a basic, "common sense" understanding of what a thermometer is and how to use one. We begin our study of thermodynamics with temperature. However, we will reconsider the topic later in the chapter to refine and expand our understanding of the concept of temperature and its measurement.

(*a*) (*b*)

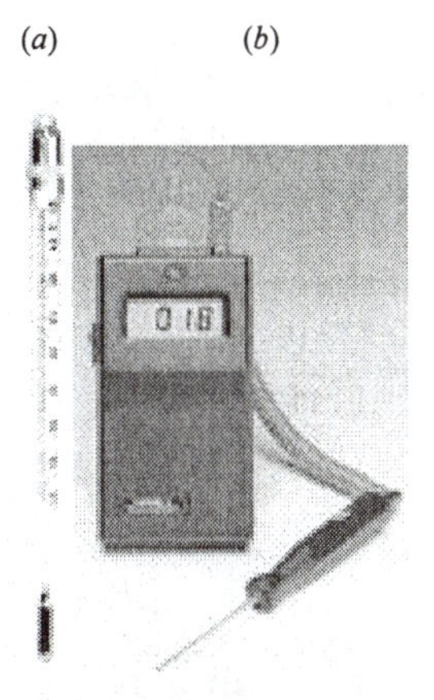

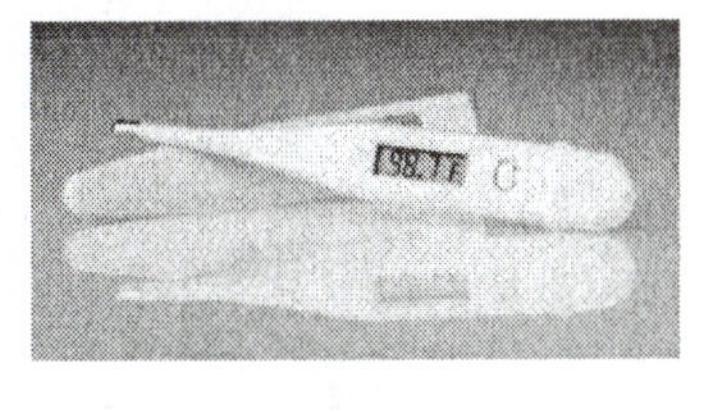

(*c*)

Fig. 19-1: Two types of thermometers based on changes in measurable properties of materials with hotness. (*a*) A liquid-thermometer in which the liquid in a tube expands more than the glass that contains it when placed in hotter surroundings. (*b*) and (*c*) Electronic thermometers in which the electrical resistance of a sensing element embedded at the end of a thin rod changes when placed in hotter surroundings. (*a*) and (*c*) courtesy of Pocket Nurse Enterprises Inc. (*b*) courtesy of Electronic Temperature Instruments, Ltd.

READING EXERCISE 19-1: List several additional measurable properties of an object. List several properties of an object that you believe are not measurable.

19-2 An Introduction to Thermometers and Temperature Scales

Most, but not all, substances expand when heated. You can loosen a tight metal lid on a jar by holding it under a stream of hot water. Both the metal of the lid and the glass of the jar expand as the hot water transfers some of its hidden thermal energy to both the jar and lid. (With the added energy, the atoms can move a bit farther from each other than usual, against the spring-like interatomic forces that hold every solid together.) However, because the metal lid expands more than the glass, the lid is loosened. The familiar sealed liquid-in-glass thermometer (Fig. 19-1*a*) works in a similar way. The mercury or colored alcohol contained in the hollow glass bulb and tube expands more than the glass that surrounds it.

Suppose you dip an unmarked liquid-thermometer filled with red colored alcohol into a cup of cold water. Since the water feels cold to the touch you can make a scratch in the tube where liquid stands and define that height as a cold temperature. Then you can transfer your unmarked thermometer to a cup of water that feels hot to the touch. You make another scratch where the liquid now stands and define that height as a hot temperature. What if you dip the thermometer into a cup of water that feels neither hot, nor cold to your touch and the height of the alcohol is halfway between? Is it reasonable to assume that the new temperature is halfway in between the two original temperatures? This assumption has guided the development of the historical Fahrenheit and Celsius temperature scales. And, it is not too far from the truth. Thus, it is useful to start our study of thermodynamics with a very crude definition of temperature change as *a quantity that is proportional to changes of the height of the liquid inside a thermometer.* In order to quantify temperature, we need to assign numbers to various heights of the liquid in our glass tube. That is, we must set up a *temperature scale*.

There are two temperature scales in common use today. You are probably familiar with the Fahrenheit scale from US weather forecasts and you may have worked with the Celsius scale in other science courses. Each scale is set up using a reproducible low temperature "fixed point" and one at a higher temperature. A temperature is assigned to each fixed point on the thermometer column. Then the distance between the column is divided into equally spaced "degrees" between the lower and higher temperatures.

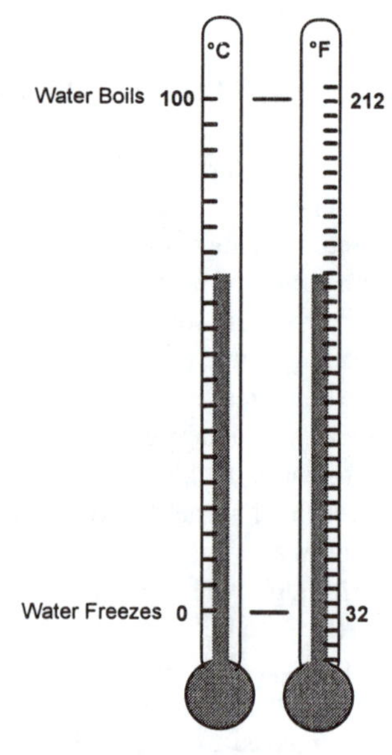

Fig. 19-2: The Celsius, and Fahrenheit temperature scales compared for the freezing and boiling points of water at standard atmospheric pressure at sea level.

The Fahrenheit Scale

The scale set up by Gabriel Fahrenheit in 1714 starts with a zero point (0°F) defined to be the lowest temperature attainable by a mixture of ice and salt. His upper fixed point was set at human body temperature and defined as 96°F. So Fahrenheit put 96 divisions along the glass tube between his two fixed points. The Fahrenheit scale has proven to be quite awkward because the freezing and boiling points of water are so much easier to reproduce when constructing a thermometer than the salt-ice temperature and body temperature. Unfortunately the freezing point of water at sea level turns out to be about 32°F and the boiling point of water at sea level turns out to be about 212°F as shown in Fig. 19-2.

The Celsius Scale

In 1742 a Swedish investigator named Celsius devised a more sensible scale that he called the centigrade scale. Celsius defined the freezing point of water at sea level as 0°C and the boiling point as 100°C. The modern Celsius scale was developed based on a degree that is the same "size" as the centigrade scale. But, it has been adjusted so one of its fixed points is the triple point of water. The triple point of water is the temperature and pressure at which solid ice, liquid water, and water vapor coexist. The triple point temperature has been defined as 0.01°C, so that a modern Celsius thermometer and the historical centigrade thermometer will give essentially the same reading. We discuss the triple point in more detail later in this chapter. Another refinement has been to define a standard value for the atmospheric pressure at sea level to help make the boiling point temperature more stable.

The Kelvin Scale

A nineteenth century British physicist, Lord Kelvin, discovered that there is a natural limit to how cold any object can get. An important temperature scale used in thermodynamics built on this natural zero point for temperature. We discuss the Kelvin scale in more detail in Section 19-9.

Temperature Conversions

The Fahrenheit scale employs a smaller degree than the Celsius scale and a different zero of temperature. You can easily verify both these differences by examining an ordinary room thermometer on which both scales are marked as shown in Fig. 19-2. The equation for converting between these two scales can be derived quite easily by remembering a few corresponding fixed point temperatures for each scale (see Table 19-1 or Fig. 19-2). The equation is

$$T_F = \tfrac{9}{5}T_C + 32°, \tag{19-1}$$

where T_F is Fahrenheit temperature.

TABLE 19-1: Some Corresponding Temperatures

Temperature	°C	°F
Boiling point of water[a]	100	212
Normal body temperature	37.0	98.6
Accepted comfort level	20	68
Freezing point of water[a]	0	32
Zero of Fahrenheit scale	≈–18	0
Scales coincide	–40	–40

[a]Strictly, the boiling point of water on the Celsius scale is 99.975°C, and the freezing point is 0.00°C. Thus there is slightly less than 100° C between those two points.

We use the letters C and F to distinguish measurements and degrees on the two scales. Thus,

$$0°\text{C} = 32°\text{F} \tag{19-2}$$

means that 0° on the Celsius scale measures the same temperature as 32° on the Fahrenheit scale, whereas

$$\Delta T = 5\ \text{C°} = 9\ \text{°F}$$

means that a temperature difference of 5 Celsius degrees (note the degree symbol appears *after* C) is equivalent to a temperature difference of 9 Fahrenheit degrees.

Problems with the Initial Definition of Temperature

There are several problems with basing our definition of temperature on the height of liquid in a scaled liquid-thermometer: (1) the historically chosen fixed points are not highly reproducible since freezing and boiling points depend on pressure; (2) the height range over which liquids can vary in a glass tube is quite limited (liquids freeze when very cold or vaporize when very hot); and (3) the assumption that liquids expand in proportion to temperature is not completely accurate since no two liquid substances give us exactly the same temperature readings at equally spaced points between designated fixed temperatures. We will return to our discussion of how to define temperature and design a better thermometer in Section 19-9. Meanwhile, our initial definition of temperature as the height of a liquid in a liquid-thermometer that has been scaled is an adequate starting point for our study of thermal interactions.

19-3 Thermal Interactions

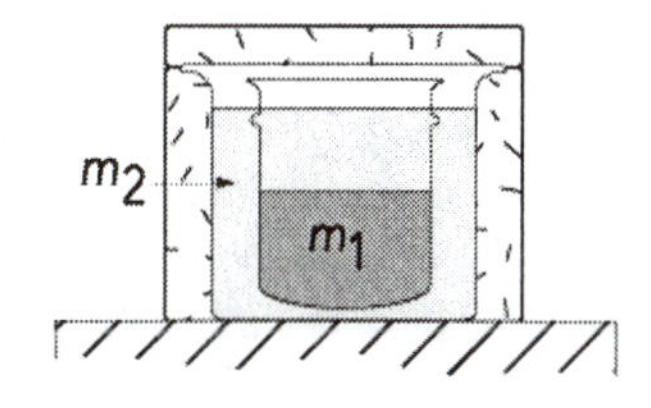

Fig. 19-3: Two vessels of water that have different temperatures at first are placed in thermal contact. They are surrounded by an insulating styrofoam cup with a lid so that they don't interact significantly with their surroundings (room air). What happens to their temperatures?

Now that we have a means to measure temperature we can explore what happens to temperatures when objects come into contact with each other. Consider the following two observations:

1. When a metal bucket of hot water is placed in a room, the reading of a thermometer placed in the water always decreases until, eventually, it matches the reading of a (assumed identical) thermometer on the wall of the room.
2. When a bucket of cold water is placed in a room, the reading of a thermometer in the bucket always increases until, eventually, it matches the reading of a thermometer in the room.

Our common sense tells us these observations indicate that there has been an interaction between the bucket of water and its surroundings. These interactions are examples of **thermal interactions**. We will call the bucket of water our *system*. We first introduced the term "system" back in Chapter 8 in regard to the concept of conservation of momentum. Just as we did there, we will define a **system** to be the object or objects that are the primary focus of our interest. That, of course, means that we can make choices about what objects we include in our system. We will often be considering the interaction between two or more systems that can exchange heat energy or do physical work on each other without exchanging matter. At other times we will consider the interaction between a single system and its surroundings.

The **environment** is defined as a system's surroundings or everything outside of it. Many thermal interactions of interest are interactions between a system (for example the bucket of water) and its environment. For practical purposes, the environment can often be taken as a system's immediate surroundings provided these surroundings are much bigger than the system. For example, the environment for a bucket of water could be the room full of air that surrounds it. Now consider two additional observations:

3. When two objects at different temperature are brought into contact with one another, the thermometer reading for the hotter object decreases and the thermometer reading of the colder object increases until finally the two readings are the same. (See Figs. 19-3 and 19-4).
4. When two objects at the same temperature are brought together, no changes in the thermometer readings occur.

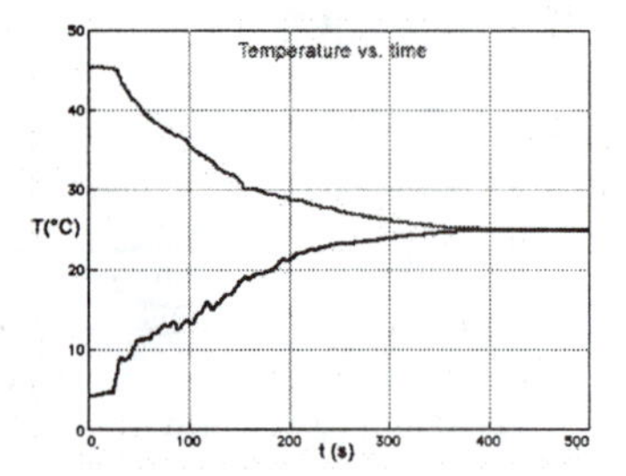

Fig. 19-4: A system consisting of $m_1 = 200$ g of cool water at about 5°C in a thin plastic cup is brought into thermal contact with another system consisting of $m_2 = 200$ g of cold water in a larger insulated container at about 45°C. An MBL system is used to monitor their temperatures. Even though no matter is exchanged between the systems (as shown in Fig. 19-3), the temperature of the cool water rises while that of the warm water falls until they reach thermal equilibrium at about 25°C, 400 seconds later.

All four of the observations discussed above are examples of a condition we will call *thermal equilibrium.* If two objects (for example, our system and the surrounding environment) produce the same thermometer readings (assuming identical thermometers) then the two objects are said to be in **thermal equilibrium.** That is, two objects are in thermal equilibrium if they have the same temperature.

The concept of thermal equilibrium is important in understanding temperature measurements. For example, as you likely know from common experience, it takes some time for a thermometer reading to reach its maximum value and stabilize. If you place a thermometer under your tongue, it does not immediately measure your correct body temperature. This is (at least in part) because it takes some time for a thermometer to reach thermal equilibrium with another object like your body. When a thermometer comes to thermal equilibrium with an object that is not heating up or cooling down, the thermometer reading will reach a constant value. Only then do we have an accurate measurement of the object's temperature.

Thermal equilibrium is important in the measurement of temperature in another way. Consider three objects: object A, object B and object T. Object A and object B are placed in separate, well insulated environments as shown in Fig. 19-5. We measure the temperatures of the three objects, two at time. Suppose that we find that object A and object T are in thermal equilibrium. We then compare the temperatures of object B and object T and find that they are in thermal equilibrium with each other as well (at the same thermometer reading as object A). Are object A and object B then necessarily in thermal equilibrium too? Experimentation (for example, using three identical thermometers to simultaneously measure the temperatures of all three objects) provides the answer which is referred to as the **zeroth law of thermodynamics:**

➤If bodies A and C are each in thermal equilibrium with a third body B, then they are in thermal equilibrium with each other.

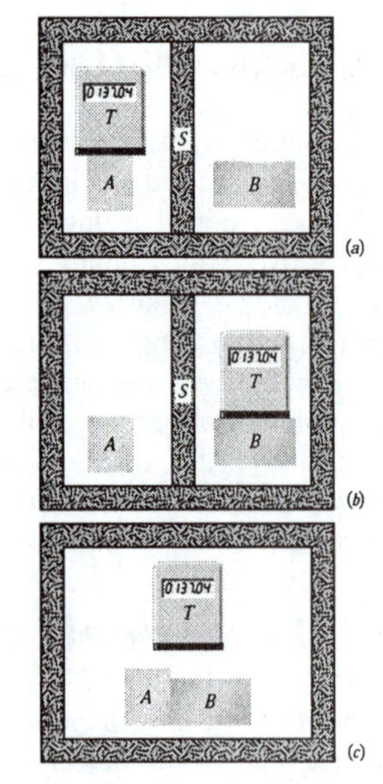

Fig. 19-5: (*a*) Body T (perhaps a thermometer) and body A are in thermal equilibrium. (Body S is a thermally insulating screen.) (*b*) Body T and body B are also in thermal equilibrium, (at the same reading of an electronic (or digital) thermometer. (*c*) If (*a*) and (*b*) are true, the zeroth law of thermodynamics states that body A and body B are also in thermal equilibrium.

If we choose object T to be a thermometer, we see that we use the zeroth law constantly in science laboratories. It is the basis of our acceptance of the use of thermometers to compare temperatures. For example, if we want to know whether the liquids in two beakers are at the same temperature, we can measure the two temperatures separately with a single thermometer and compare them. We do not need to place a separate thermometer in each beaker and look for equal temperature readings.

19-4 Heating, Cooling and Temperature

Is just measuring the initial and final temperature of an object that is changing temperature a good way to learn about how objects heat or cool? In order to answer this question, let us consider several examples of how water cools:

1. Consider what happens when a small cup of very hot water is put in a room that is being maintained at comfortable air temperature of 20°C. Obviously the water will cool down until it reaches thermal equilibrium with the surrounding air.
2. If we place a large bucket of hot water in the same room, it will also cool down, but we know it will take longer than the cup of water does to cool.
3. Also we know that if we first place a small cup of hot water in a sealed thermos bottle and then in the room, the water will still reach thermal equilibrium with the room but it will take much longer to cool down to room temperature than it did before.
4. Now what happens when we put a cup containing water at 30°C in a very cold freezer? We find the water freezes and becomes ice. If during the freezing process we keep the water-ice mixture well-stirred and measure its temperature as a function of time, what do we find? As shown in Fig. 19-6, the ice-water mixture undergoes no change in temperature until all the ice is frozen. In this situation, and in the case of the insulated thermos bottle containing hot water, the measurement of temperature alone does not fully reflect the nature of the thermal interactions that are taking place.

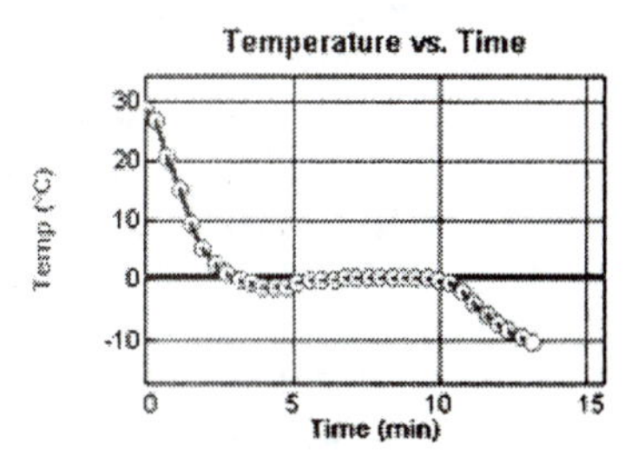

Fig. 19-6: A temperature vs. time graph showing what happens to a cup of water after it is placed in a cold freezer. When the water temperature decreases to 0°C ice begins to form. While ice is forming the temperature does not change. Once all the water is changed to ice the temperature starts decreasing again.

The examples above are evidence that our ability to analyze thermal interactions would be significantly limited if we rely solely on initial and final temperature measurements. Hence, we are motivated to invent a new, broader concept that we can use in discussing thermal interactions—even in cases where temperature changes are not the significant feature. The name of the process we will introduce is "(to) heat," which we can use to describe the interaction between a hotter body and colder body, *even if no temperature change occurs*. For example, if you *heat* ice, it will melt. Or, it takes longer to *heat* cold water if it is insulated from its environment than if it is not. Observations such as those above form the basic operational definition of the *process* called heating.

In order to gain some additional insight into the process of heating, consider two more observations.

1. A container of water is placed on a burner to raise its temperature. It takes fuel (an energy source) for this process to occur.
2. If we start with two containers of water, one large and one small, at the same temperature and place them over identical burners, it takes a longer time (and so more fuel) to elevate the temperature of the larger amount of water to the same final value as that of the smaller.

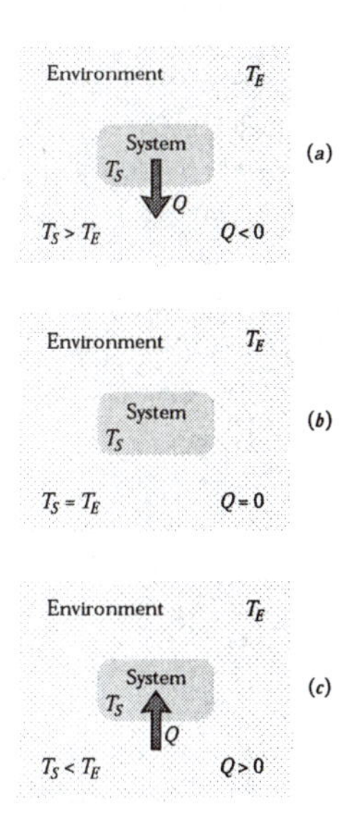

Fig. 19-7: If the temperature of a system exceeds that of its environment as in (*a*), an amount of heat energy Q is lost by the system to the environment until thermal equilibrium as shown in part (*b*) is established. (*c*) If the temperature of the system is below that of the environment, heat energy is absorbed by the system until thermal equilibrium is established.

Both of these observations are direct indications that there is energy involved in heating. Furthermore, the idea that heating is a *transfer of energy* from a hotter object to a colder object is consistent with all the other observations we have considered in this chapter. Energy associated with thermal interactions like heating is often referred to as *heat energy*. We are led then to this important statement:

➤Heating is the transfer of energy between a system and its environment that occurs because a temperature difference exists between them.

The transferred energy is called **heat energy,** and is given the symbol Q.

Unfortunately, the term *heat energy* is sometimes shortened to just "**heat.**" The heat energy Q is taken to be *positive* when the energy is transferred *to a system* from its environment (then we say that heat energy is absorbed). The transferred heat energy is *negative* when energy is transferred *from the system* to its environment (we then say that energy is released or lost from the system). This transfer of energy is shown in Fig. 19-7. In the situation of Fig. 19-7*a*, in which the temperature of the system T_S is greater than the temperature of the environment T_E ($T_S > T_E$), heat energy is transferred from the system to the environment, so Q is negative. In Fig. 19-7*b*, in which $T_S = T_E$, there is no such transfer, Q is zero, and heat energy is neither released nor absorbed. In Fig. 19-7*c*, in which $T_S < T_E$, the transfer is to the system from the environment, so Q is positive.

It took scientists a while to realize that "heating" was associated with the transfer of energy from one system to another. Hence, heat energy ended up with its own unit. Since heating was initially considered strictly in terms of temperature change, the **calorie** (cal) was defined as the amount of heating that would raise the temperature of 1 g of water from 14.5°C to 15.5°C. In the British system, the corresponding unit of heat energy was the **British thermal unit** (Btu), defined as the amount of heat energy that would raise the temperature of 1 lb of water from 63°F to 64°F.

In 1948, the scientific community decided that since heating (like work) is an energy transfer process, the SI unit for heat energy should be the one we use for all other energy, namely, the **joule**. Joules are used instead of calories in most countries, as seen on the can of soda from Australia in Fig. 19-8. The calorie is now defined to be 4.1860 J (exactly), with no reference to the heating of water. (The "calorie" used in nutrition, sometimes called the Calorie (Cal), is really a kilocalorie or 1000 calories.) The relations among the various heat energy units are

$$1 \text{ cal} = 3.969 \times 10^{-3} \text{ Btu} = 4.1860 \text{ J} = 0.001 \text{ Cal (or food calorie)}. \quad (19\text{-}3)$$

Fig. 19-8: In the US we have not adopted the international system of units when referring to food energy so the food calorie (a kilocalorie) is used in diet books and in all government regulated food labels. Most other countries use *Joules* (the accepted SI unit for energy) on food labels. The can in the photo above was sold in Australia.

Mechanisms for Transfer of Heat Energy

We have discussed the transfer of heat energy between a system and its environment, but we have not yet described how that transfer takes place. We will briefly describe the three mechanisms of transferring heat energy here and return to discuss them more fully at the end of this chapter.

If you leave the end of a metal poker in a fire for enough time, its handle will get hot. There are large vibrations of the atoms and electrons of the metal at the fire end of the poker because of the high temperature of their environment. These increased vibrational amplitudes, and thus the associated energy, are passed along the poker, from atom to atom, during collisions between adjacent atoms. In this way, a region of rising temperature extends itself along the poker to the handle even though there has been no flow of matter. In this case, we say that heat energy is transferred from the fire to your hand by (thermal) **conduction** along the length of the poker.

Conduction is, by definition, a transfer mechanism that requires direct contact between two objects at different temperatures. If there is not direct contact between the colder object and hotter object, there can not be a transfer of heat energy by conduction. However, it is also important to note that there is *no flow or transfer matter* involved in heating by conduction. In some real situations, it is sometimes impossible to avoid the transfer of material from one object to the other during heating. However, in such cases, the process is no longer one of "pure" conduction.

When you open a low and high window in a heated house, you can feel cold outside air rushing into the room through the low window and warm room air rushing outside through the high window. This is because the cold air is more dense than the warm air so it displaces the warm room at the bottom of the room. The warm air rises as a result of buoyant forces and flows out the top window. The room gets colder because of the exchange of cold and warm air. In this situation, heat energy is being transported by the flow of matter (air currents in this case). The transfer by the exchange of hotter and cooler fluids is known as **convection.** Examples of heating by convection are everywhere. This is how the warmth spreads through a room when a radiator or heater gets hot. It is why *all* the water in a tea kettle gets hot (as opposed to only the water in contact with the hot kettle surface) when the kettle is placed on a hot stove. Convection is part of many other natural processes. Atmospheric convection plays a fundamental role in determining global climate patterns and daily weather variations. Glider pilots and birds alike seek rising thermals (convection currents of warm air) that keep them aloft. Huge energy transfers take place within the oceans by the same process. Finally, energy is transported to the surface of the sun from the nuclear furnace at its core by enormous cells of convection, in which hot gas rises to the surface along the cell core and cooler gas around the core descends below the surface.

When you sit in the sun you can feel your skin getting warmer. If you put a shield between you and the sun, the sensation of warmth immediately disappears. How is this energy transfer taking place? We know that the light from the sun has to pass through over a hundred million kilometers of almost empty space. So, this heat energy transfer can't be attributed to either convection or conduction. Solar energy transfer is attributed to a third transfer process—the absorption of **electromagnetic radiation.** Although visible light is one kind of electromagnetic radiation the Sun and a hot fire that can also warm you emit both visible light and invisible infrared radiation that has a longer characteristic wavelength than light. (See Chapter 34 for more details on electromagnetic waves.) No medium is required for energy transfer via electromagnetic radiation. Heat energy transferred by infrared electromagnetic waves is often called **thermal radiation** or **radiant energy.**

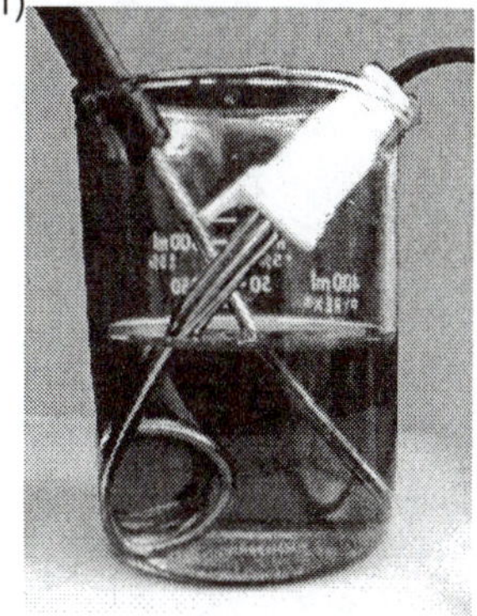

Fig. 19-9(*a*): An MBL system outfitted with a digital temperature sensor can be used to monitor the temperature rise in a liquid while heat energy is transferred to it at a constant rate by an immersion heater. The liquid in this photo is motor oil.

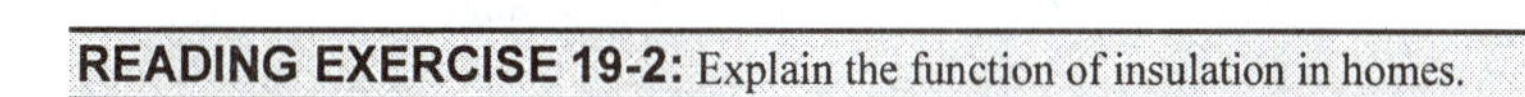

READING EXERCISE 19-2: Explain the function of insulation in homes.

19-5 Heat Energy Transfer to Solids and Liquids

If we start with two containers of water, one large and one small, at the same temperature and place them to heat over identical burners, it takes a longer time (and so more fuel and thus more heat energy) to elevate the temperature of the larger amount of water to the same final value as that of the smaller. It takes almost twice as much energy to heat a cup of water as it does to heat a cup of motor oil to the same temperature. This is shown in Figs. 19-9*a* and 19-9*b*. A common electric immersion heater is placed in a container of motor oil and plugged in. It puts out a bit less than 200W of power at a constant rate. Since energy is power × time the total amount of heat energy transferred to the oil is directly proportional to the time the heater has been on. The graph in Fig. 19-9*b* shows that after 70 seconds, the change in water temperature, $T_f - T_i$, is about 15°C while the oil temperature has risen about 30°C. We can say the "water holds its heat" twice as well as motor oil.

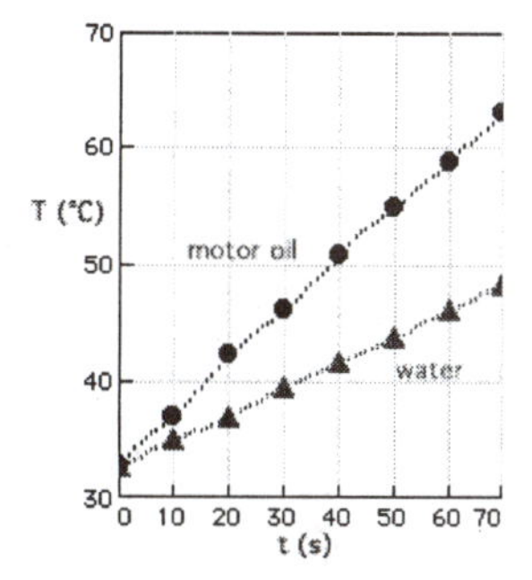

Fig. 19-9(*b*): If heat energy is transferred to two different liquids that have the same mass, the temperature does not usually rise at the same rate. So we define a different specific heat for each liquid. For example a small immersion heater shown in Fig. 19-9*a* "heats up" 175 g of motor oil faster than 175 g of water. Data were taken once each second for 70 seconds with an MBL system.

If we look at heating and cooling curves for a large number of different objects and different substances, (see Figs. 19-9*a* and *b* for an example) we find that the relationship between heat energy transfer and temperature is linear (so long as the material doesn't melt, freeze or vaporize). However, the scaling factor (proportionality constant) changes. The **heat capacity** C of an object is the name that we give to the proportionality constant between the heat energy Q that the object absorbs or loses and the resulting temperature change ΔT of the object; that is,

$$Q = C\,\Delta T = C(T_f - T_i), \tag{19-4}$$

in which T_i and T_f are the initial and final temperatures of the object. Heat capacity C has the unit of energy per degree Celsius. The heat capacity C of, say, a marble slab used in a bun warmer might be 179 cal/C°.

The word "capacity" in this context is really misleading in that it suggests analogy with the capacity of a bucket to hold water. Since heat is not a substance *that analogy is misleading,* and you should not think of the object as "containing" heat energy or being limited in its ability to absorb heat energy. Heat energy transfer can proceed without limit as long as the necessary temperature difference between the object and its surroundings is maintained. The object may, of course, melt or vaporize during the process.

Specific Heat

Two objects made of the same material—say, marble—will have heat capacities proportional to their masses. It is therefore convenient to define a "heat capacity per unit mass" or **specific heat** c that refers not to an object but to a unit mass of the material of which the object is made. Equation 19-4 then becomes

$$Q = cm\,\Delta T = cm(T_f - T_i). \tag{19-5}$$

Through experiment we would find that although the heat capacity of a particular marble slab might be $179\,\text{cal/C}°$, the specific heat of marble itself (in that slab or in any other marble object) is $0.21\,\text{cal/g}\cdot\text{C}°$.

From the way the calorie and the British thermal unit were initially defined, the specific heat of water is

$$c = 1\,\text{cal/g}\cdot\text{C}° = 1\ \text{Btu/lb}\cdot\text{F}° = 4190\ \text{J/kg}\cdot\text{C}°. \tag{19-6}$$

Table 19-2 shows the specific heats of some substances at room temperature. Note that the value for water is relatively high. The specific heat of any substance actually depends somewhat on temperature, but the values in Table 19-2 apply reasonably well in a range of temperatures near room temperature.

TABLE 19-2: Specific Heats of Some Substances at Room Temperature

Substance	Specific Heat $\frac{\text{cal}}{\text{g}\cdot\text{C}°}$	Specific Heat $\frac{\text{J}}{\text{kg}\cdot\text{C}°}$	Molar Specific Heat $\frac{\text{J}}{\text{mol}\cdot\text{C}°}$
Elemental Solids			
Lead	0.0305	128	26.5
Tungsten	0.0321	134	24.8
Silver	0.0564	236	25.5
Copper	0.0923	386	24.5
Aluminum	0.215	900	24.4
Other Solids			
Brass	0.092	380	
Granite	0.19	790	
Glass	0.20	840	
Ice (-10°C)	0.530	2220	
Liquids			
Mercury	0.033	140	
Ethyl alcohol	0.58	2430	
Seawater	0.93	3900	
Water	1.00	4190	

READING EXERCISE 19-3: If an amount of heat energy Q is transferred to object A it will cause each gram of A to rise in temperature by 3 C°. If the same amount of energy is transferred to object B then the temperature of each gram of B will rise by 4 C°. Which object has the greater specific heat?

Molar Specific Heat

In many instances the most convenient unit for specifying the amount of a substance is the mole (mol), where

$$1\ \text{mol} = 6.02 \times 10^{23}\ \text{elementary units}$$

of *any* substance. Thus 1 mol of aluminum means 6.02×10^{23} atoms (the atom being the elementary unit), and 1 mol of aluminum oxide means 6.02×10^{23} molecules of the oxide (because the molecule is the elementary unit of a compound).

When quantities are expressed in moles, specific heats must also involve moles (rather than a mass unit); they are then called **molar specific heats**. Table 19-2 shows the values for some elemental solids (each consisting of a single element) at room temperature.

An Important Point

In determining and then using the specific heat of any substance, we need to know the conditions under which heat energy is transferred. For solids and liquids, we usually assume that the sample is under constant pressure (usually atmospheric) during the transfer. It is also conceivable that the sample is held at constant volume while the heat energy is absorbed. This means that thermal expansion of the sample is prevented by applying external pressure. For solids and liquids, this is very hard to arrange experimentally but the effect can be calculated, and it turns out that the specific heats under constant pressure and constant volume for any solid or liquid differ usually by no more than a few percent. Gases, as you will see, have quite different values for their specific heats under constant-pressure conditions and under constant-volume conditions.

Heats of Transformation

Recall our example above of the ice (solid water). "Solid" is a description of the water which we will call the water's **phase**. There are, in general, three phases of matter: solid, liquid and vapor. As a *solid*, the molecules of a sample are locked into a fairly rigid structure by their mutual attraction. As a *liquid*, the molecules have more energy and move about more. They may form brief clusters, but the sample does not have a rigid structure and can flow or settle into a container. As a *gas* or *vapor*, the molecules have even more energy, are free of one another, and can fill up the full volume of a container. To fully describe a material for thermodynamic purposes we must specify not only the phase (solid, liquid, or vapor) but also the temperature, pressure, and volume. The phase of a material along with the temperature, pressure and volume of the material specify the **state** of the material.

As we all know, we find that ice melts when exposed to a warm room and becomes liquid water. The melting is called a **change of phase.** When heat energy is absorbed or lost by a solid, liquid or vapor the temperature of the sample does not necessarily change. Instead, the sample may change from one phase to another. Through experiments we have found that while a material is undergoing a change in phase additional transfers of heat energy do not change the temperature of the material. We saw one example of this in Fig. 19-6.

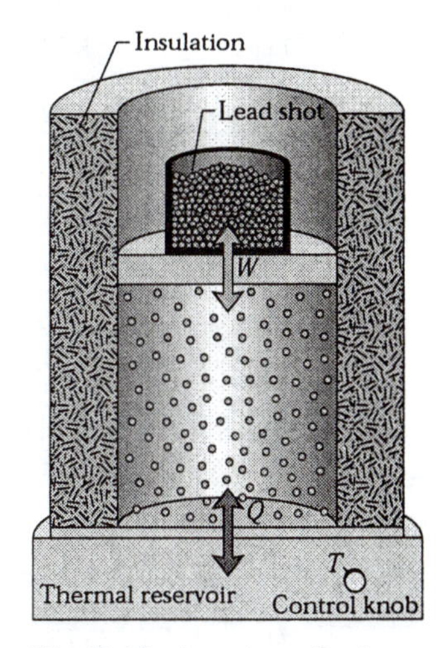

Fig. 19-10 : A gas is confined to a cylinder with a movable piston. Heat energy ΔQ can be added to, or withdrawn from, the gas by regulating the temperature T of the adjustable thermal reservoir. Work W can be done by the gas by raising or lowering the piston.

To *melt* a solid means to change it from the solid phase to the liquid phase. The process requires energy because the molecules of the solid must be freed from their rigid structure. Melting an ice cube to form liquid water is a common example. To *freeze* a liquid to form a solid is the reverse of melting and requires that energy be removed from the liquid, so that the molecules can settle into a rigid structure. To *vaporize* a liquid means to change it from the liquid phase to the vapor or gas phase. This process, like melting, requires energy because the molecules must be freed from their clusters. Boiling liquid water to transfer it to water vapor (or steam—a gas of individual water molecules) is a common example. *Condensing* a gas to form a liquid is the reverse of vaporizing; it requires that energy be removed from the gas, so that the molecules can cluster instead of flying away from one another.

The amount of energy per unit mass that must be transferred as heat energy when a sample completely undergoes a phase change is called the heat of transformation L. Thus, when a sample of mass m completely undergoes a phase change, the total energy transferred is

$$Q = Lm. \tag{19-7}$$

When the phase change is from liquid to gas (then the sample must absorb heat energy) or from gas to liquid (then the sample must release heat energy), the heat energy of transformation is called the **heat of vaporization** L_V. For water at its normal boiling or condensation temperature,

$$L_V = 539 \text{ cal/g} = 40.7 \text{ kJ/mol} = 2256 \text{ kJ/kg}. \tag{19-8}$$

When the phase change is from solid to liquid (then the sample must absorb heat energy) or from liquid to solid (then the sample must release heat energy), the heat of transformation is called the **heat of fusion** L_F. For water at its normal freezing or melting temperature,

$$L_F = 79.5 \text{ cal/g} = 6.01 \text{ kJ/mol} = 333 \text{ kJ/kg}. \tag{19-9}$$

Table 19-3 shows the heats of transformation for some substances.

TABLE 19-3: Some Heats of Transformation

Substance	Melting		Boiling	
	Melting Point (°C)	Heat of Fusion L_F (kJ/kg)	Boiling Point (°C)	Heat of Vaporization L_V (kJ/kg)
Hydrogen	-259.2	58.0	-252.9	455
Oxygen	-218.4	13.9	-183.0	213
Mercury	-39.2	11.4	356.9	296
Water	-0.1	333	99.9	2256
Lead	327.9	23.2	1743.9	858
Silver	961.9	105	2049.9	2336
Copper	1082.9	207	2594.9	4730

READING EXERCISE 19-4: Notice in Table 19-3 above that water has a very large heat of vaporization. What is the significance of this to a fire fighter who finds the water sprayed on a very hot fire converted to steam? Suppose that 1 ml of steam comes in contact with 1 ml of a firefighters flesh and condenses. Assuming that flesh has the same heat capacity as water, what would the temperature rise in 1 ml of the firefighter's flesh be?

Touchstone Examples 19-5-1 and 19-5-2, at the end of this chapter, illustrate how to use what you learned in this section.

TE

19-6 A Closer Look at Heat Energy and Work

Here we look in some detail at how energy can be transferred both as heat energy and as work between a system and its environment. Let us take as our system a gas confined to a cylinder with a movable piston, as in Fig. 19-10. This is the kind of device that is used to drive a steam engine, the engine of an automobile, and many other tools—all of which convert heat energy into work.

In our piston-cylinder system, the upward force on the piston due to the pressure of the confined gas is equal to the weight of lead shot loaded onto the top of the piston. The walls of the cylinder are made of insulating material that does not allow any transfer of heat energy. The bottom of the cylinder however rests on a reservoir of heat energy, a *thermal reservoir* (perhaps a hot plate) whose temperature T you can control by turning a knob.

The system (the gas) starts from an *initial state* i, described by a pressure p_i, a volume V_i, and a temperature T_i. You want to change the system to a final state f, described by a pressure p_f, a volume V_f, and a temperature T_f. The procedure by which you change the system from its initial state to its final state is called a *thermodynamic process*. During such a process, energy may be transferred into the system from the thermal reservoir ($+Q$) or vice versa ($-Q$). Also, work can be done by the system to raise the loaded piston (positive work) or lower it (negative work). We assume that all such changes occur slowly, with the result that the system is always in (approximate) thermal equilibrium (that is, every part of the system is always in thermal equilibrium with every other part).

Suppose that you remove a few of the lead shot from the piston of Fig. 19-10, allowing the gas to push the piston and remaining shot upward through a differential displacement $d\vec{s}$ with an upward force $\vec{F}$. Since the displacement is tiny, we can assume that $\vec{F}$ is constant during the displacement. Then $\vec{F}$ has a magnitude that is equal to PA, where P is the pressure of the gas and A is the face area of the piston. The differential work dW done by the gas during the displacement is

$$dW = \vec{F} \cdot d\vec{s} = (P|A|)(d|s|) = P(|A|d|s|) = PdV, \tag{19-10}$$

in which dV is the differential change in the volume of the gas owing to the movement of the piston. When you have removed enough shot to allow the gas to change its volume from V_i to V_f, the total work done by the gas is

$$W = \int dW = \int_{V_i}^{V_f} P\,dV. \tag{19-11}$$

During the change in volume, the pressure and temperature of the gas may also change. To evaluate the integral in this expression directly, we need to know how pressure varies with volume for the actual process by which the system changes from state i to state f.

There are actually many ways to take the gas from state i to state f. One way is shown in Fig. 19-11*a*, which is a plot of the pressure of the gas versus its volume and which is called a *P-V* diagram. In Fig. 19-11*a*, the curve indicates that the pressure decreases as the volume increases. The integral in Eq. 19-11 (and thus the work W done by the gas) is represented by the shaded area under the curve between points i and f. Regardless of what exactly we do to take the gas along the curve, that work is positive, owing to the fact that the gas increases its volume by forcing the piston upward.

Another way to get from state i to state f is shown in Fig. 19-11*b*: the change takes place in two steps—the first from state i to state a, and the second from state a to state f.

Step *ia* of this process is carried out at constant pressure, which means that you leave undisturbed the lead shot that ride on top of the piston in Fig. 19-10. You cause the volume to increase (from V_i to V_f) by slowly turning up the temperature control knob, raising the temperature of the gas to some higher value T_a. (Increasing the temperature increases the force from the gas on the piston, moving it upward.) During this step, positive work is done by the expanding gas (to lift the loaded piston) and heat energy is absorbed by the system from the thermal reservoir (in response to the arbitrarily small temperature differences that you create as you turn up the temperature). The heat energy (Q) is positive because it is added to the system.

Step *af* of the process of Fig. 19-11*b* is carried out at constant volume, so you must wedge the piston, preventing it from moving. Then as you use the control knob to decrease the temperature, you find that the pressure drops from P_a to its final value P_f. During this step, heat energy is lost by the system to the thermal reservoir.

For the overall process *iaf*, the work W, which is positive and is carried out only during step *ia*, is represented by the shaded area under the curve. Energy is transferred as heat energy during both steps *ia* and *af*, with a net energy transfer Q.

Fig. 19-11*c* shows a process in which the previous two steps are carried out in reverse order. The work W in this case is smaller than for Fig. 19-11*b*, as is the net heat energy absorbed. Fig. 19-11*d* suggests that you can make the work done by the gas as small as you want (by following a path like *icdf*) or as large as you want (by following a path like *ighf*).

To sum up: A system can be taken from a given initial state to a given final state by an infinite number of processes. Heat energy transfers may or may not be involved, and in general, the work W and the heat energy transfer ΔQ will have different values for different processes. We say that heat energy and work are *path-dependent* quantities.

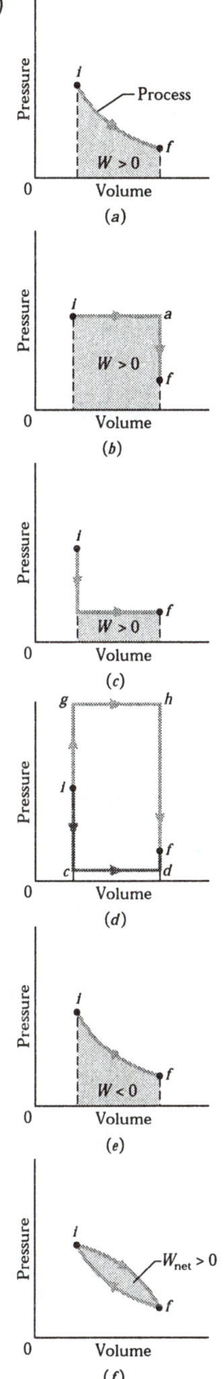

Fig. 19-11: (*a*) The shaded area represents the work W done by a system as it goes from an initial state i to a final state f. Work W is positive because the system's volume increases. (*b*) W is still positive, but now greater. (*c*) W is still positive, but now smaller. (*d*) W can be even smaller (path *icdf*) or larger (path *ighf*). (*e*) Here the system goes from state f to state i, as the gas is compressed to less volume by an external force. The work W done by the system is now negative. (*f*) The net work W_{net} done by the system during a complete cycle is represented by the shaded area.

Fig. 19-11*e* shows an example in which negative work is done by a system as some external force compresses the system, reducing its volume. The absolute value of the work done is still equal to the area beneath the curve, but because the gas is compressed, the work done by the gas is negative.

Fig. 19-11*f* shows a thermodynamic cycle in which the system is taken from some initial state *i* to some other state *f* and then back to *i*. The net work done by the system during the cycle is the sum of the positive work done during the expansion and the negative work done during the compression. In Fig. 19-11*f*, the net work is positive because the area under the expansion curve (*i* to *f*) is greater than the area under the compression curve (*f* to *i*).

READING EXERCISE 19-5: The *P*-*V* diagram here shows six curved paths (connected by vertical paths) that can be followed by a gas. Which two of them should be part of a closed cycle if the net work done by the gas is to be at its maximum positive value?

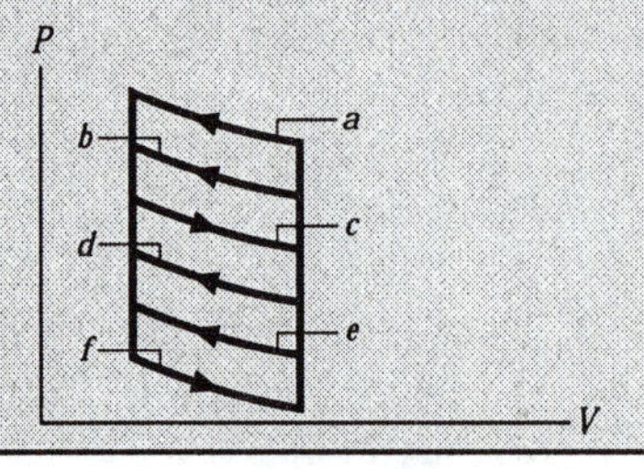

19-7 The First Law of Thermodynamics

You have just seen that when a system changes from a given initial state to a given final state, both the work W done by the system and the heat energy Q transferred to the system depend on the nature of the process. Experimentally, however, we find a surprising thing. The quantity $Q-W$ *is the same for all processes*. It depends only on the initial and final states and does not depend at all on how the system gets from one to the other. All other combinations of Q and W, including Q alone, W alone, $Q+W$, and $Q-2W$, are path dependent; only the quantity $Q-W$ is not.

The quantity Q must represent a change in some intrinsic property of the system. We call this property the internal energy E_{int} and we write

$$\Delta E_{int} = E_{int,f} - E_{int,i} = Q - W \qquad \text{(first law)} \qquad (19\text{-}12)$$

Equation 19-12 above is called the **first law of thermodynamics.** If the thermodynamic system undergoes only a differential change, we can write the first law as*

$$dE_{int} = dQ - dW \qquad \text{(first law)} \qquad (19\text{-}13)$$

➤The internal energy E_{int} of a system tends to increase if heat energy ΔQ is added and tends to decrease if energy is lost when the system does an amount of work W on its surroundings.

In Chapter 10, we discussed the principle of energy conservation as it applies to isolated systems—that is, to systems in which no energy enters or leaves the system. The first law of thermodynamics is an extension of that principle to systems that are ***not*** isolated. In such cases, energy may be transferred into or out of the system as either work W or heat energy Q. In our statement of the first law of thermodynamics above, we assume that there are no changes in the kinetic energy or the potential energy of the system as a whole; that is, $\Delta K = \Delta U = 0$.

*Here dQ and dW, unlike dE_{int}, are not true differentials; that is, there are no such functions as $Q(P, V)$ and $W(P, V)$ that depend only on the state of the system. The quantities dQ and dW are inexact differentials and are usually represented by the symbols $đQ$ and $đW$. For our purposes, we can treat them simply as infinitesimally small energy transfers.

Before this chapter, the term *work* and the symbol W always meant the work done on a system. However, starting now (actually, starting with Eq. 19-10 above) and continuing through the next two chapters about thermodynamics, we focus on the work done *by* a system, such as the gas in Fig. 19-10.

The work done *on* a system is always the negative of the work done by the system, so if we rewrite $\Delta E_{\text{int}} = E_{\text{int},f} - E_{\text{int},i} = Q - W$ in terms of the work W_{on} done *on* the system, we have $\Delta E_{\text{int}} = Q + W_{\text{on}}$. This tells us the following: The internal energy of a system tends to increase if heat energy is absorbed by the system or if positive work is done *on* the system. Conversely, the internal energy tends to decrease if heat energy is lost by the system or if negative work is done *on* the system.

READING EXERCISE 19-6: The figure here shows four paths on a P-V diagram along which a gas can be taken from state i to state f. Rank the paths according to (a) the change ΔE_{int}, (b) the work W done by the gas, and (c) the magnitude of the heat energy Q transferred to the gas, greatest first.

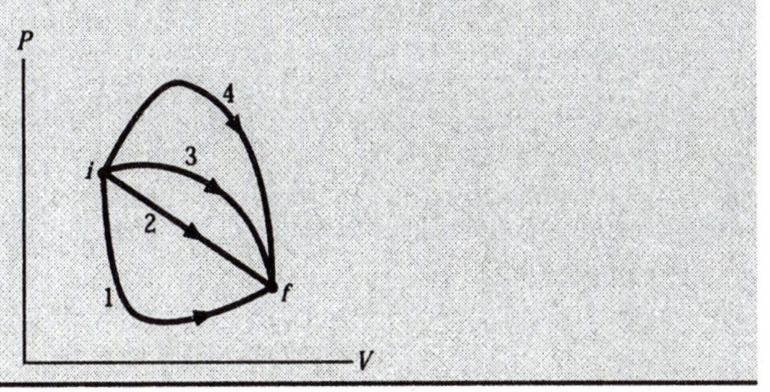

19-8 Some Special Cases of the First Law of Thermodynamics

Here we look at four different thermodynamic processes, in each of which a certain restriction is imposed on the system. We then see what consequences follow when we apply the first law of thermodynamics to the process. The results are summarized in Table 19-4.

TABLE 19-4: The First Laws Of Thermodynamics: Four Special Cases

The 1st Law: $\Delta E_{\text{int}} = Q - W$ (Eq. 19-12)

Process	Restriction	Consequence
Adiabatic	$Q = 0$	$\Delta E_{\text{int}} = -W$
Constant volume	$W = 0$	$\Delta E_{\text{int}} = Q$
Closed cycle	$\Delta E_{\text{int}} = 0$	$Q = W$
Free expansion	$Q = W = 0$	$\Delta E_{\text{int}} = 0$

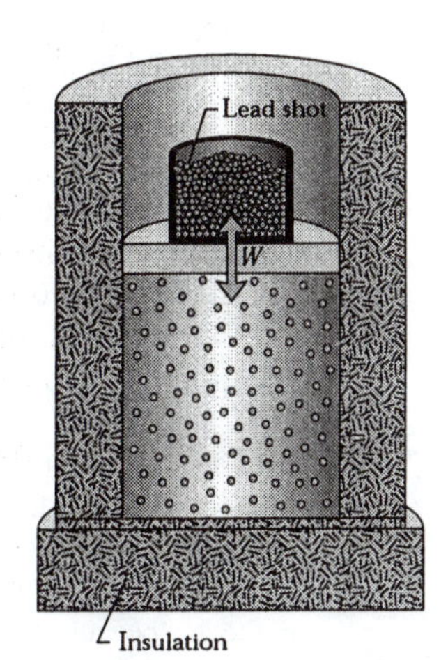

Fig. 19-12: An adiabatic expansion can be carried out by slowly removing lead shot from the top of the piston. Adding lead shot reverses the process at any stage.

1. ***Adiabatic processes.*** An adiabatic process is one that occurs so rapidly or occurs in a system that is so well insulated that *no transfer of heat energy* occurs between the system and its environment. Putting $Q = 0$ in the first law ($\Delta E_{\text{int}} = E_{\text{int},f} - E_{\text{int},i} = Q - W$) yields

$$\Delta E_{\text{int}} = -W \quad \text{(adiabatic process).} \tag{19-14}$$

This tells us that if work is done *by* the system (that is, if W is positive), the internal energy of the system decreases by the amount of work. Conversely, if work is done *on* the system (that is, if W is negative), the internal energy of the system increases by that amount.

Fig. 19-12 shows an idealized adiabatic process. Heat energy cannot be transferred to or from a system because it is insulated from its surrounding. Thus, the only way energy can be transferred between the system and its environment is if work is either by the system or on the system. If we remove shot from the piston and allow the gas to expand, work is done by the system (the gas) and thus is positive. Thus the internal energy of the gas must decrease. If, instead, we add shot and compress the gas, the work done by the system is negative and the internal energy of the gas increases.

2. ***Constant-volume processes.*** If the volume of a system (such as a gas) is held constant, that system can do no work. Putting $W=0$ in the first law ($\Delta E_{\text{int}} = E_{\text{int},f} - E_{\text{int},i} = Q - W$) yields

$$\Delta E_{\text{int}} = Q \qquad \text{(constant-volume process).} \qquad (19\text{-}15)$$

Thus, if heat energy is transferred to a system (so that ΔQ is positive), the internal energy of the system increases. Conversely, if heat energy is transferred from the system to its surroundings so that ΔQ is negative, the internal energy of the system must decrease.

3. ***Cyclical processes.*** There are processes in which, after certain interchanges of heat energy and work, the system is restored to its initial state. Engines undergo this type of process. Since the system is restored to its initial state, no intrinsic property of the system—including its internal energy—can possibly change. Putting $\Delta E_{\text{int}} = 0$ in the first law ($\Delta E_{\text{int}} = E_{\text{int},f} - E_{\text{int},i} = Q - W$) yields

$$Q = W \qquad \text{(cyclical process).} \qquad (19\text{-}16)$$

Thus, the net work done during the process must exactly equal the net amount of heat energy transferred to the system. So the store of internal energy hidden in the system remains unchanged. Cyclical processes form a closed loop on a *P-V* plot, as shown in Fig. 19-11*f*.

4. ***Free expansions.*** These are adiabatic processes in which no transfer of heat energy occurs between the system and its environment and no work is done on or by the system. Thus, $Q = W = 0$ and the first law requires that

$$\Delta E_{\text{int}} = 0 \qquad \text{(free expansion).} \qquad (19\text{-}17)$$

Fig. 19-13 shows how such an expansion can be carried out. A gas, which is in thermal equilibrium within itself, is initially confined by a closed stopcock to one half of an insulated double chamber; the other half is evacuated. The stopcock is opened, and the gas expands freely to fill both halves of the chamber. No heat energy is transferred to or from the gas because of the insulation. No work is done by the gas because it rushes into a vacuum and thus does not meet any pressure.

A free expansion differs from all other processes we have considered because it cannot be done slowly and in a controlled way. As a result, at any given instant during the sudden expansion, the gas is not in thermal equilibrium and its pressure is not the same everywhere. Therefore, although we can plot the initial and final states on a *P-V* diagram, we cannot plot the expansion itself.

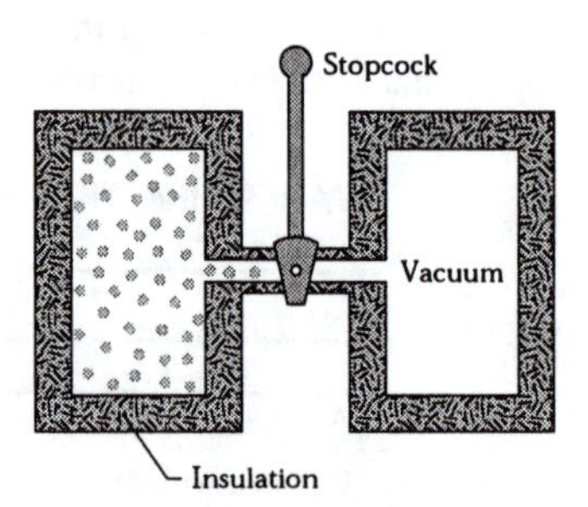

Fig. 19-13: The initial stage of a free-expansion process. After the stopcock is opened, the gas fills both chambers and eventually reaches an equilibrium state.

READING EXERCISE 19-7: For one complete cycle as shown in the *P-V* diagram here, (a) is the internal energy change ΔE_{int} of the gas positive, negative, or zero? and (b) what about the net heat energy, ΔQ, transferred to the gas?

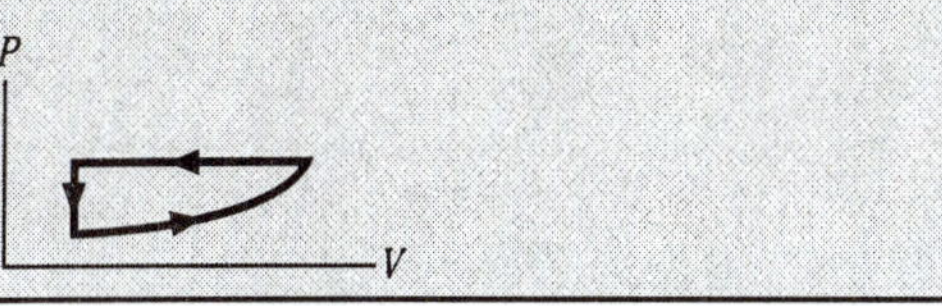

Touchstone Example 19-8-1, at the end of this chapter, illustrates how to use what you learned in this section.

TE

19-9 More on Temperature Measurement

In Section 1-4 we discussed the fact that there are only seven fundamental quantities in physics that serve as base units for the entire international system (or SI) of units. Temperature in kelvins is one of them. How is the kelvin unit defined? Why has a gas thermometer operating under constant volume become the standard thermometer? What are its fixed points? Why is it superior to the liquid-thermometers we discussed in Section 19-2?

As we discussed in Section 19-2 there are several difficulties with defining temperature in terms of the height of a liquid column in a liquid-thermometer. First, the historical fixed points (such as freezing and boiling points at sea level, body temperature, and lowest ice/salt mixture temperature) are not reproducible to a high accuracy. Second liquid-thermometers are only usable in a narrow range of temperatures between the freezing and boiling points of the liquid. Third, no two liquids expand and contract in exactly the same way as their temperatures change.

The aforementioned difficulties prompted a search for highly reproducible fixed points and a way to measure temperature that is independent of the behavior of any one particular substance. Gases are more promising substances for temperature measurements since they are already "boiling" and have no upper limit except at the melting point of their container. Also gases tend to liquefy at very low temperatures. For example air liquefies at about –200°C. Generally a gas volume (at constant pressure) or a gas pressure (at constant volume) can be measured between two fixed points. The scales can be determined in the same way as they are for liquids. The various gas scales have been found to agree among themselves better than liquid scales do. Other thermometers are based on changes of the electrical properties or materials or changes in the light given off by glowing substances and so on. Although we have dozens of thermometric scales based on the behaviors of various substances, none of them can be proven exactly true. Here we present some methods for identifying better fixed points and designing more accurate thermometers.

Defining Standard Fixed Points

Using a low-density gas instead of liquid to measure low temperatures gives very interesting results. As we suggested in Section 19-2, when more and more heat energy is extracted from most substances including any gas held at a constant pressure, its temperature drops and so does its volume. A simple apparatus for doing such an experiment over a limited range of temperature is shown in Fig. 19-14*a*.

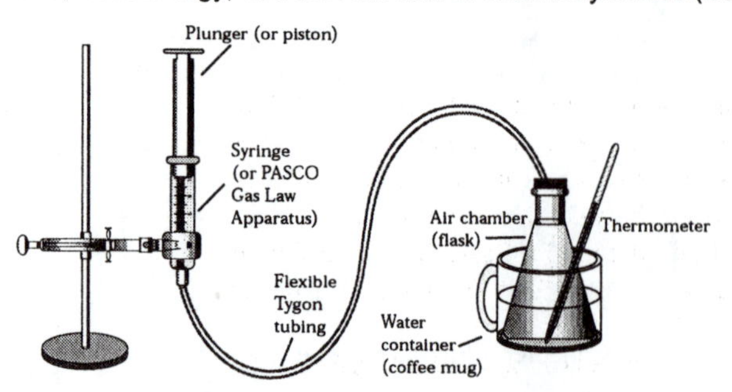

Fig. 19-14(*a*): Diagram of apparatus that can be used to measure volume changes as a function of the temperature of air or some other gas trapped at constant pressure. The pressure is a combination of atmospheric pressure plus the pressure exerted by the weight of the plunger of cross-sectional area A, so that $P = P_{atm} + mg/A$.

If you plot data for the volume of any low-density gas as a function of temperature, the graph is linear. By extrapolating the graph to zero volume, you can predict that the volume will go to zero at a temperature of approximately –273°C. In reality any gas will liquefy before its temperature gets that low. An alternative approach is to hold the volume of a gas constant and observe that its pressure will also approach zero at a temperature of about –273°C. A simple apparatus for doing this experiment is shown in Fig. 19-14*b*.

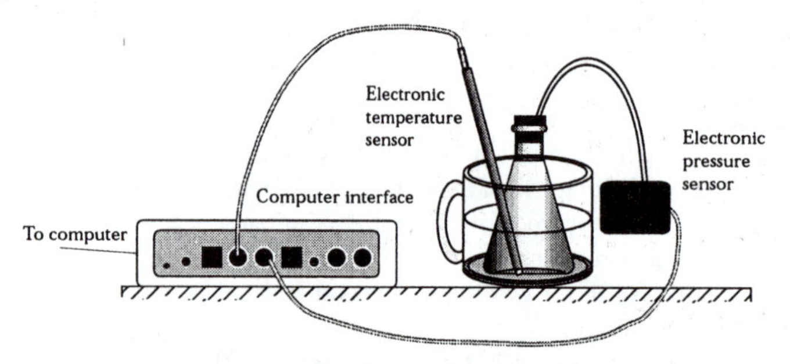

Fig. 19-14(*b*): Diagram of apparatus that uses an MBL system to measure pressure changes as a function of the temperature of air or some other gas trapped at constant volume.

Actually the extrapolation to zero pressure gives a slightly different result depending on how much gas is placed in the flask that holds it to a constant volume. However, as we place less and less gas in the flask and retake the data we find that the temperature at zero pressure converges to a lower limit of –273.16°C. Student data for this experiment using different apparatus is shown in Fig. 19-15.

Since vanishingly small samples of all gases appear to approach the same minimum temperature regardless of their chemical composition, this temperature seems to be a fundamental property of nature. We call this minimum temperature **absolute zero.** We define absolute zero as the temperature of a body when it has the minimum possible internal energy. *Because of the universality of this minimum temperature of –273.16°C it has become our standard low temperature fixed point.*

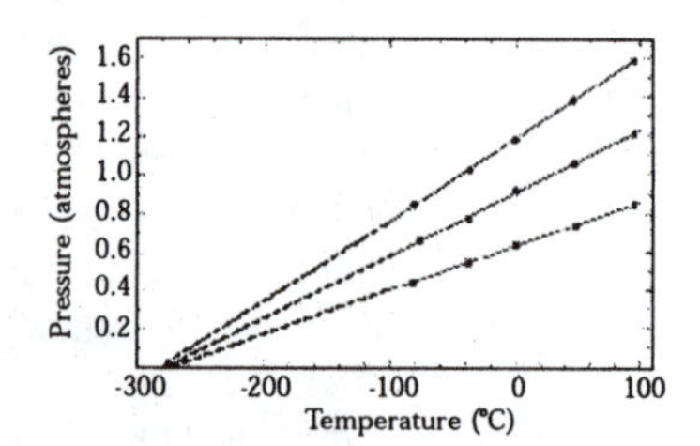

Fig. 19-15: Data from a student experiment measuring the pressure of a fixed volume of gas at various temperatures. The three data sets are for three different amounts of gas (air) in a container. Regardless of the amount of gas, the pressure is a linear function of temperature that extrapolates to zero at approximately –280°C. (More precise measurements show that the zero-point does depend slightly on the amount of gas, but has a well-defined limit of –273.16°C as the density of the gas goes to zero.)

But what should we use if we want a second standard fixed point? We could, for example, select the freezing point or the boiling point of water. However, water's change of phase, for example from liquid to vapor (boiling), depends not just on temperature, but also on pressure. This introduces some technical difficulties and reduces our confidence in using boiling or freezing as fixed, reproducible thermal phenomena. As we suggested in Section 19-2 a state known as the triple point of water is a good candidate for a fixed point.

We have found through extensive experimentation that three phases—liquid water, solid ice, and water vapor (gaseous water)—can coexist, in thermal equilibrium, at one and only one set of values of pressure and temperature. Thus, this temperature is called the **triple point of water**. Since both the temperature and pressure are well defined at this point, *the triple point of water is often used as a standard high temperature fixed point in designing a thermometer.*

Fig. 19-16 shows a triple-point cell, in which the triple point of water can be set up in a laboratory. For reasons that we will explain when introducing the Kelvin temperature scale, the triple point of water has been assigned a value of 0.01°C.

The Kelvin Temperature Scale

The accepted SI unit of temperature is the kelvin. The kelvin is named after Lord Kelvin who first proposed that there is a natural limit to how cold any object can get. It is defined by using absolute zero and the triple point of water as its two fixed points. In order to tie in with the popular Celsius scale, the kelvin was set to have the same "size" degree as the Celsius scale. This yields a rather odd definition. Basically the minimum possible temperature is defined as zero kelvin or 0 K. And, the triple point of water is defined as exactly +273.16 K.

Since the triple point of water is 0.01°C, the conversion between a Celsius scale temperature and the Kelvin scale temperature is very simple since one merely needs to add 273.15 to the Celsius temperature, T_C, to get the Kelvin temperature T. In other words,

$$T = T_C + 273.15 \tag{19-18}$$

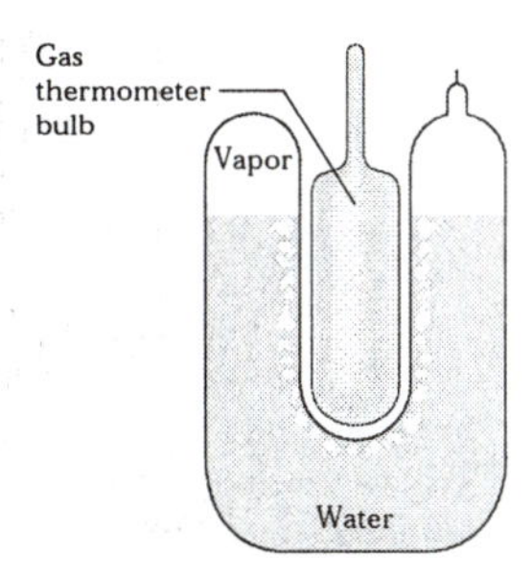

Fig. 19-16: A triple-point cell, in which solid ice, liquid water, and water vapor coexist in thermal equilibrium. By international agreement, the temperature of this mixture has been defined to be 273.16 K. The bulb of a constant-volume gas thermometer is shown inserted into the well of the cell.

Figure 19-17 shows a wide range of temperatures in Kelvin, either measured or conjectured. When expressing temperatures on the Fahrenheit or Celsius scales, we commonly say degrees Fahrenheit (°F) or degrees Celsius (°C). But, Kelvin temperatures are measured in a unit simply called "kelvin." Thus, it is not customary to say "degrees kelvin." Figure 19-18 compares the Kelvin, Celsius, and Fahrenheit scales.

When the universe began, some 10 to 20 billion years ago, its temperature was about 10^{39} K. As the universe expanded it cooled, and it has now reached an average temperature of about 3 K. We on Earth are a little warmer than that because we happen to live near a star. Without our sun, we too would be at 3 K (or, rather, we could not exist).

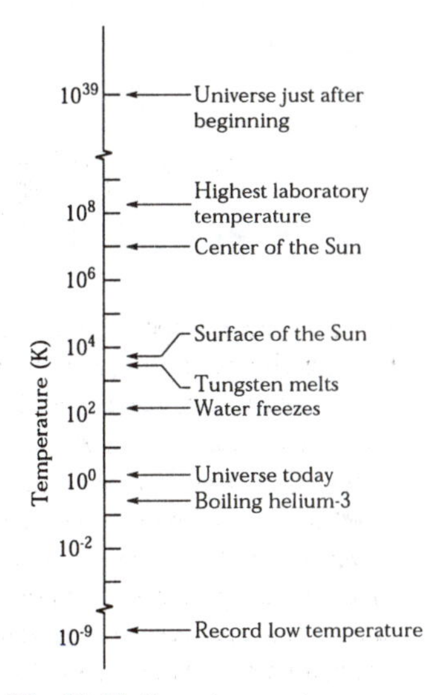

Fig. 19-17: Some temperatures on the Kelvin scale. Temperature $T = 0$ corresponds to $10^{-\infty}$ and cannot be plotted on this logarithmic scale.

The Constant-Volume Gas Thermometer

Now that we have better fixed points we can use them as part of a standard thermometer which uses gas rather than a liquid as its medium. The standard thermometer, against which all other thermometers are calibrated, is based on the effect of temperature changes on the pressure of a gas occupying a fixed volume. Fig. 19-19 shows such a **constant-volume gas thermometer**; it consists of a gas-filled bulb connected by a tube to a mercury manometer. By raising and lowering reservoir R, the mercury level on the left can always be brought to the zero of the scale to keep the gas volume constant (variations in the gas volume can affect temperature measurements).

The temperature of any body in thermal contact with the bulb (like the liquid in Fig. 19-19) is then defined to be

$$T = CP, \tag{19-19}$$

in which P is the pressure within the gas and C is a constant. From Eq. 15-10, the pressure P is

$$P = P_{atm} - \rho_{Hg} gh, \tag{19-20}$$

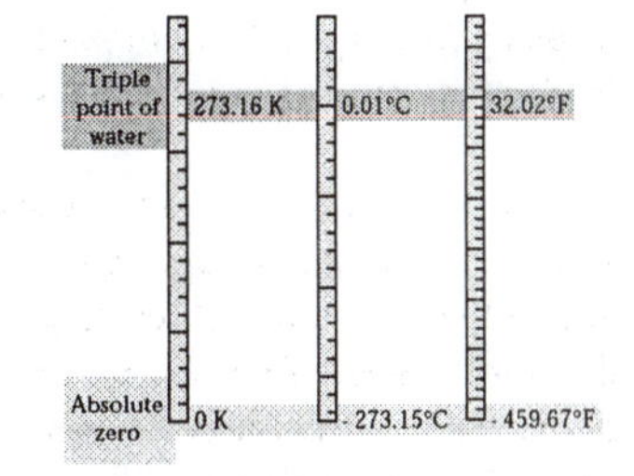

Fig. 19-18: A comparison of the Kelvin, Celsius, and Fahrenheit temperature scales.

in which P_{atm} is the atmospheric pressure, ρ_{Hg} is the density of the mercury in the open-tube manometer (like that described in Section 15-5), and h is the measured difference between the mercury levels in the two arms of the tube.*

If we next put the bulb in a triple-point cell (Fig. 19-16), the temperature now being measured is

$$T_3 = CP_3 \tag{19-21}$$

*For pressure units, we shall use units introduced in Section 15-3. The SI unit for pressure is the newton per square meter, which is called the pascal (Pa). The pascal is related to other common pressure units by

$$1 \text{ atm} = 1.01 \times 10^5 \text{ Pa} = 760 \text{ torr} = 14.7 \text{ lb/in.}^2 .$$

in which P_3 is the gas pressure now. Eliminating C between Eqs. 19-19 and 19-21 gives us the temperature as

$$T = T_3\left(\frac{P}{P_3}\right) = (273.16\text{ K})\left(\frac{P}{P_3}\right) \qquad \text{(provisional).} \qquad (19\text{-}22)$$

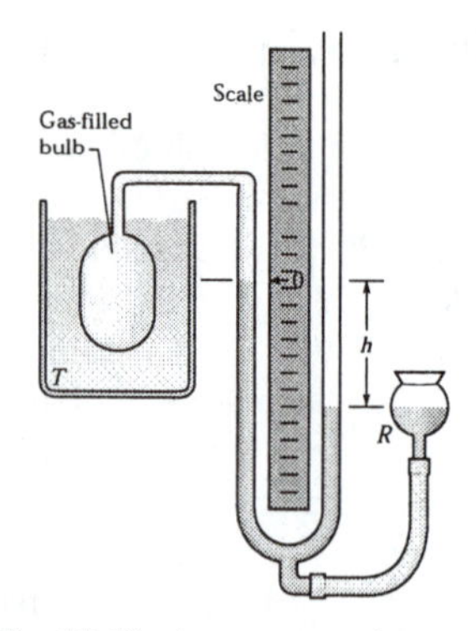

Fig. 19-19: A constant-volume gas thermometer, the gas-filled bulb on the left is immersed in a liquid whose temperature T is being measured. The mercury-filled bulb on the right is raised or lowerd as P changes to keep the left-hand column of mercury at the zero-point on the scale so the gas volume stays constant.

We still have a problem with this thermometer. If we use it to measure a given temperature, we find that different gases in the bulb give slightly different results. However, as we use smaller and smaller amounts of gas to fill the bulb, the readings converge nicely to a single temperature, no matter what gas we use.

Thus the recipe for measuring a temperature with a gas thermometer, is

$$T = (273.16\text{ K})\left(\lim_{\text{gas}\to 0}\frac{P}{P_3}\right). \qquad (19\text{-}23)$$

The recipe instructs us to measure an unknown temperature T as follows: Fill the thermometer bulb with an arbitrary amount of *any* gas (for example, nitrogen) and measure P_3 (using a triple-point cell) and P, the gas pressure at the temperature being measured. (Keep the gas volume the same.) Calculate the ratio P/P_3. Then repeat both measurements with a smaller amount of gas in the bulb, and again calculate this ratio. Continue this way, using smaller and smaller amounts of gas, until you can extrapolate to the ratio P/P_3 that you would find if there were approximately no gas in the bulb. Calculate the temperature T by substituting that extrapolated ratio into Eq. 19-23. (The temperature is called the *ideal gas temperature*.)

19-10 Thermal Expansion

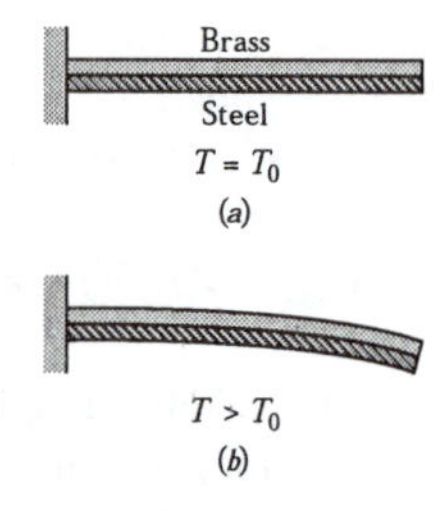

Fig. 19-20: (*a*) A bimetal strip, consisting of a strip of brass and a strip of steel welded together, at temperature T_0. (*b*) The strip bends as shown at temperatures above this reference temperature. Below the reference temperature the strip bends the other way. Many thermostats operate on this principle, making and breaking an electrical contact as the temperature rises and falls.

We already know that some materials expand when heated and contract when cooled. Indeed, we used this principle in our first definition of temperature. The design of thermometers and thermostats is often based on the differences in expansion between the components of a *bimetal strip* (Fig. 19-20). In aircraft manufacture, rivets and other fasteners are often cooled in dry ice before insertion and then allowed to expand to a tight fit. However, such **thermal expansion** is not always desirable, as Fig. 19-21 suggests. To preclude buckling, expansion slots must be placed in bridges to accommodate roadway expansion on hot days. Dental materials used for fillings must be matched in their thermal expansion properties to those of tooth enamel (otherwise consuming hot coffee or cold ice cream would be quite painful). Regardless of whether we might wish to exploit or avoid thermal expansion, this property has many important implications and we need to consider how it works in detail.

Linear Expansion

If the temperature of a metal rod of length L is raised by an amount ΔT, its length is found to increase by an amount ΔL according to

$$\frac{\Delta L}{L} = \alpha \Delta T, \qquad (19\text{-}24)$$

in which α is a constant called the coefficient of linear expansion. For small changes in temperature the fractional length change $(\Delta L/L)$ is proportional to the change in temperature. The coefficient α should be thought of as the ratio of how much the length changes per unit length given a certain change in temperature. It has the unit "per degree" or "per kelvin" and depends on the material. Although α varies somewhat with temperature, for most practical purposes it can be taken as constant for a particular material. Table 19-5 shows some coefficients of linear expansion. Note that the unit C° there could be replaced with the unit K.

Fig. 19-21: Railroad tracks in Asbury Park, New Jersey, distorted because of thermal expansion on a very hot July day.

TABLE 19-5: Some Coefficients of Linear Expansion[a]

Substance	α (10^{-6}/C°)	Substance	α (10^{-6}/C°)
Ice (at 0°C)	51	Steel	11
Lead	29	Glass (ordinary)	9
Aluminum	23	Glass (Pyrex)	3.2
Brass	19	Diamond	1.2
Copper	17	Invar[b]	0.7
Concrete	12	Fused quartz	0.5

[a]Room temperature values except for the listing for ice.
[b]This alloy was designed to have a low coefficient of expansion. The word is a shortened form of "invariable."

The thermal expansion of a solid is like (three-dimensional) photographic enlargement. Fig. 19-22*b* shows the (exaggerated) expansion of a steel ruler after its temperature is increased from that of Fig. 19-22*a*. Equation 19-24 applies to every linear dimension of the ruler, including its edge, thickness, diagonals, and the diameters of the circle etched on it and the circular hole cut in it. If the disk cut from that hole originally fits snugly in the hole, it will continue to fit snugly if it undergoes the same temperature increase as the ruler.

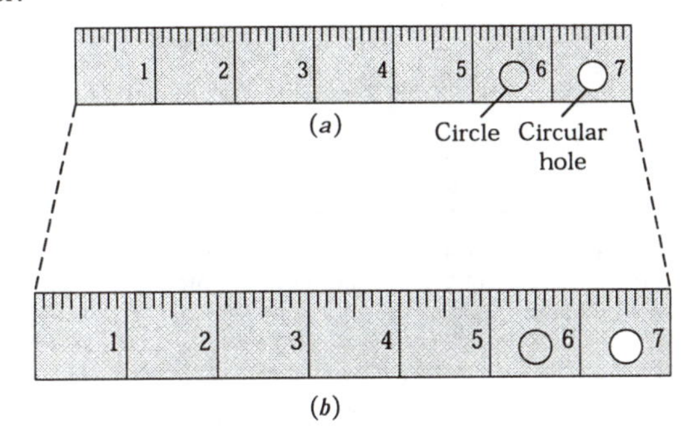

Fig. 19-22: The same steel ruler at two different temperatures. When it expands, the scale, the numbers, the thickness, and the diameters of the circle and circular hole are all increased by the same factor. (The expansion has been exaggerated for clarity.)

Volume Expansion

If all dimensions of a solid expand with temperature, the volume of that solid must also expand. For liquids, volume expansion is the only meaningful expansion parameter. If the temperature of a solid or liquid whose volume is V is increased by an amount ΔT, the increase in volume is found to be ΔV where

$$\frac{\Delta V}{V} = \beta\, \Delta T, \qquad (19\text{-}25)$$

and β is the **coefficient of volume expansion** of the solid or liquid.

The coefficients of volume expansion β and linear expansion α for a solid are related. In order to see how, consider a rectangular solid of height h, width w and length l. The volume of the rectangular solid is hwl. If the solid is heated, each dimension expands linearly so that the height becomes $h + \Delta h$, the width becomes $w+\Delta w$ and the length becomes $l + \Delta l$. Hence, the new volume is $(h + \Delta h)(w + \Delta w)(l + \Delta l)$. If we multiply this out, ignoring all terms with two or more deltas (Δ) because those terms will be very small, we get $hwl + hw\Delta l + h\Delta wl + \Delta hwl$. Since each dimension expands linearly, this means that our new volume is equal to $hwl + hwl(3\alpha\Delta T)$. Hence,

$$\beta = 3\alpha. \qquad (19\text{-}26)$$

The most common liquid, water, does not behave like other liquids. Above about 4°C, water expands as the temperature rises, as we would expect. Between 0 and about 4°C, however, water *contracts* with increasing temperature. Thus, at about 4°C, the density of water passes through a maximum. At all other temperatures, the density of water is less than this maximum value.

This behavior of water is the reason why lakes freeze from the top down rather than from the bottom up. As water on the surface is cooled from, say, 10°C toward the freezing point, it becomes denser ("heavier") than lower water and sinks to the bottom. Below 4°C, however, further cooling makes the water then on the surface *less* dense ("lighter") than the lower water, so it stays on the surface until it freezes. Thus the surface freezes while the lower water is still liquid. If lakes froze from the bottom up, the ice so formed would tend not to melt completely during the summer, because it would be insulated by the water above. After a few years, many bodies of open water in the temperate zones of Earth would be frozen solid all year round—and aquatic life as we know it could not exist.

READING EXERCISE 19-8: The figure below shows four rectangular metal plates, with sides of L, $2L$, or $3L$. They are all made of the same material, and their temperature is to be increased by the same amount. Rank the plates according to the expected increase in (a) their vertical heights and (b) their areas, greatest first.

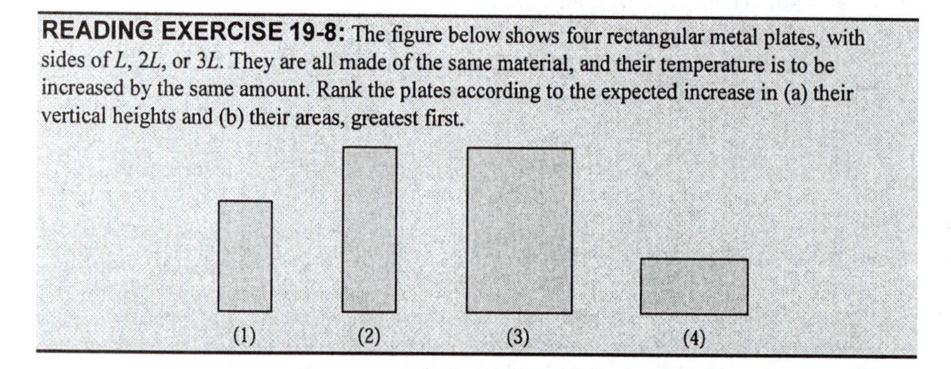

READING EXERCISE 19-9: Suppose that one of the plates shown above has a round hole cut out of its center. The temperature is increased. Does the hole in the center get larger, smaller, or remain unchanged in size? Explain your reasoning.

READING EXERCISE 19-10: Consider a cylindrical metal rod that stands on its base in a vertical orientation. The temperature increases. What happens to the pressure at the base of the rod? Does it increase, decrease, or remain unchanged? Explain your reasoning.

Touchstone Example 19-10-1, at the end of this chapter, illustrates how to use what you learned in this section.

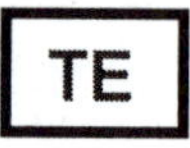

19-11 More on Heat Energy Transfer Mechanisms

In Section 19-4, during our initial discussion of the transfer of heat energy between a system and its environment, we qualitatively discussed the three transfer mechanisms (conduction, convection, and radiation). In order to expand and refine our understanding of these transfer mechanisms, we will return to that discussion now.

Conduction

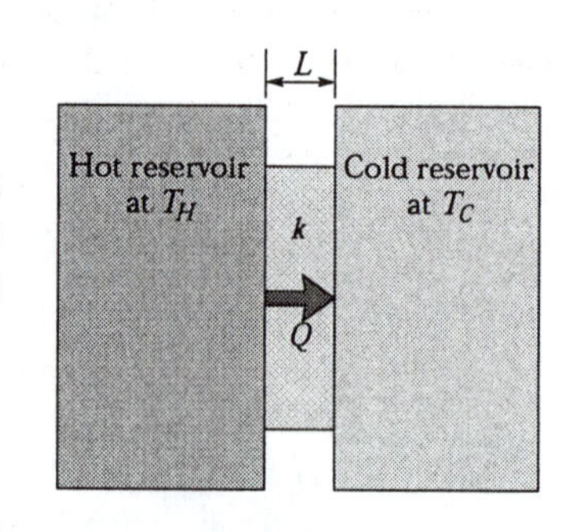

Fig. 19-23: Thermal conduction. Energy is transferred as heat from a reservoir at temperature T_H to a cooler reservoir at temperature T_C through a conducting slab of thickness L and thermal conductivity k.

We all have a natural ability to sense hot and cold. But unfortunately, our "temperature sense" is in fact not always reliable. On a cold winter day, for example, why does an iron railing seem much colder to the touch than a wooden fence post when both are at the same temperature? Why are frying pans made out of metal while pot holders are made out of cloth and other fibers? Because some materials are much more effective than others at transferring heat energy via conduction.

Consider a slab of face area A and thickness L, whose faces are maintained at temperatures T_H and T_C by a hot reservoir and a cold reservoir, as in Fig. 19-23. Let ΔQ be the energy that is transferred as heat energy through the slab, from its hot face to its cold face, in a time interval Δt. Experiment shows that the heat energy *conduction rate* P_{cond} (the power or heat energy transferred per unit time—not a pressure) is

$$H_{\text{cond}} = \frac{\Delta Q}{\Delta t} = kA\frac{T_H - T_C}{L}, \tag{19-27}$$

in which k, called the *thermal conductivity*, is a constant that depends on the material of which the slab is made. A material that readily transfers energy by conduction is a *good thermal conductor* and has a high value of k. That is, the thermal conductivity k is a *property* of the material. Table 19-6 gives the thermal conductivities of some common metals, gases, and building materials.

TABLE 19-6: Some Thermal Conductivities[a]

Substance	$k\ (W/m\cdot K)$
Metals	
Stainless steel	14
Lead	35
Aluminum	235
Copper	401
Silver	428
Gases	
Air (dry)	0.026
Helium	0.15
Hydrogen	0.18
Building Materials	
Polyurethane foam	0.024
Rock wool	0.043
Fiberglass	0.048
White pine	0.11
Window glass	1.0

[a]Conductivities change somewhat with temperature. The given values are at room temperature.

Thermal Resistance to Conduction (*R*-Value)

If you are interested in insulating your house or in keeping cola cans cold on a picnic, you are more concerned with poor conductors of heat energy than with good ones. For this reason, the concept of *thermal resistance* R has been introduced into engineering practice. The R-value of a slab of thickness L is defined as

$$R = \frac{L}{k}. \tag{19-28}$$

The lower the thermal conductivity of the material of which a slab is made, the higher the R-value of the slab, so something that has a high R-value is a *poor thermal conductor* and thus a *good thermal insulator*.

Note that R is a property attributed to a slab of a specified thickness, not to a material. That is, R is *not* a property of a material. The commonly used unit for R (which, in the United States at least, is almost never stated) is the square foot-fahrenheit degree-hour per British thermal unit $(\text{ft}^2\cdot \text{F}^\circ\cdot\text{h/Btu})$. (The unit is rarely stated!)

Conduction Through a Composite Slab

Fig. 19-24 shows a composite slab, consisting of two materials having different thicknesses L_1 and L_2 and different thermal conductivities k_1 and k_2. The temperatures of the outer surfaces of the slab are T_H and T_C. Each face of the slab has area A. Let us derive an expression for the conduction rate through the slab under the assumption that the transfer is a *steady-state* process; that is, the temperatures everywhere in the slab and the rate of energy transfer do not change with time.

In the steady state, the conduction rates through the two materials must be equal. This is the same as saying that the energy transferred through one material in a certain time must be equal to that transferred through the other material in the same time. If this were not true, temperatures in the slab would be changing and we would not have a steady-state situation. Letting T_X be the temperature of the interface between the two materials, we can now use Eq. 19-27 to express the rate of heat energy transfer as

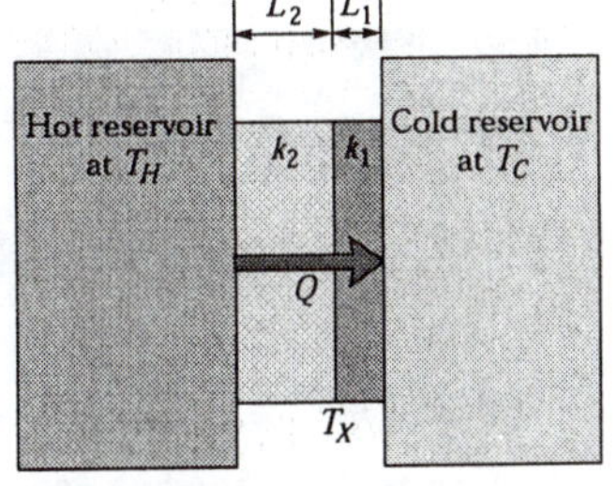

Fig. 19-24: Heat energy is transferred from a hot reservior to a cold reservoir at a steady rate through a composite slab made up of two different materials with different thicknesses and different thermal conductivities. The steady-state temperature at the interface of the two materials is denoted T_X.

$$P_{\text{cond}} = \frac{k_2A(T_H - T_X)}{L_2} = \frac{k_1A(T_X - T_C)}{L_1}. \tag{19-29}$$

Solving for T_X yields, after a little algebra,

$$T_X = \frac{k_1L_2T_C + k_2L_1T_H}{k_1L_2 + k_2L_1}. \tag{19-30}$$

Substituting this expression for T_X into either equality of Eq. 19-29 yields

$$P_{\text{cond}} = \frac{A(T_H - T_C)}{L_1/k_1 + L_2/k_2}. \tag{19-31}$$

We can extend Eq. 19-31 to apply to any number n of materials making up a slab:

$$P_{\text{cond}} = \frac{A(T_H - T_C)}{\sum_{i=1}^{n}(L_i/k_i)}. \tag{19-32}$$

The summation sign in the denominator tells us to add the values of L/k for all the materials.

READING EXERCISE 19-11: The figure shows the face and interface temperatures of a composite slab consisting of four materials, of identical thicknesses, through which the heat energy transfer is steady. Rank the materials according to their thermal conductivities, greatest first.

25°C 15°C 10°C - 5.0°C - 10°C

a *b* *c* *d*

Radiation

The rate P_{rad} at which an object emits energy via electromagnetic radiation depends on the object's surface area A and the temperature T of that area in kelvins and is given by

$$P_{rad} = \sigma \varepsilon A T^4. \qquad (19\text{-}33)$$

Here $\sigma = 5.6703 \times 10^{-8}\ \mathrm{W/m^2 \cdot K^4}$ is called the *Stefan-Boltzmann constant* after Josef Stefan (who discovered Eq. 19-33 experimentally in 1879) and Ludwig Boltzmann (who derived it theoretically soon after). The symbol ε represents the *emissivity* of the object's surface, which has a value between 0 and 1, depending on the composition of the surface. A surface with the maximum emissivity of 1.0 is said to be a *blackbody radiator*, but such a surface is an ideal limit and does not occur in nature. Note again that the temperature in Eq. 19-33 must be in kelvins so that a temperature of absolute zero corresponds to no radiation. Note also that every object whose temperature is above 0 K—including you—emits thermal radiation. (See Fig. 19-25.)

Fig. 19-25: A false-color thermogram reveals the rate at which energy is radiated by houses along a street. The rates, from largest to smallest, are color coded as white, red, pink, blue, and black. You can tell where there is insulation in the walls, a heavy curtain over a window, and a higher air temperature at the ceiling on the second floor.

The rate P_{abs} at which an object absorbs energy via thermal radiation from its environment, which we take to be at uniform temperature T_{env} (in kelvins), is

$$P_{abs} = \sigma \varepsilon A T_{env}^4. \qquad (19\text{-}34)$$

The emissivity ε in Eq. 19-34 is the same as that in Eq. 19-33. An idealized blackbody radiator, with $\varepsilon = 1$, will absorb all the radiated energy it intercepts (rather than sending a portion back away from itself through reflection or scattering).

Because an object will radiate energy to the environment while it absorbs energy from the environment, the object's net rate P_{net} of energy exchange due to thermal radiation is

$$P_{net} = P_{abs} - P_{rad} = \sigma \varepsilon A (T_{env}^4 - T^4). \qquad (19\text{-}35)$$

P_{net} is positive if net energy is being absorbed via radiation, and negative if it is being lost via radiation.

Touchstone Examples 19-11-1 and 19-11-2, at the end of this chapter, illustrate how to use what you learned in this section.

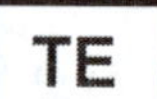

Touchstone Example 19-5-1

(a) How much heat must be absorbed by ice of mass m = 720 g at −10°C to take it to liquid state at 15°C?

SOLUTION: The first **Key Idea** is that the heating process is accomplished in three steps.

Step 1. The **Key Idea** here is that the ice cannot melt at a temperature below the freezing point—so initially, any energy transferred to the ice as heat can only increase the temperature of the ice. The heat Q_1 needed to increase that temperature from the initial value $T_i = -10°C$ to a final value $T_f = 0°C$ (so that the ice can then melt) is given by Eq. 19-5 ($Q = cm\,\Delta T$). Using the specific heat of ice c_{ice} in Table 19-2 gives us

$$\begin{aligned} Q_1 &= c_{ice} m (T_f - T_i) \\ &= (2220\ \text{J/kg}\cdot\text{C°})(0.720\ \text{kg})[0°\text{C} - (-10°\text{C})] \\ &= 15\,984\ \text{J} = 15.98\ \text{kJ}. \end{aligned}$$

Step 2. The next **Key Idea** is that the temperature cannot increase from 0°C until all the ice melts—so any energy transferred to the ice as heat now can only change ice to liquid water. The heat energy Q_2 needed to melt all the ice is given by Eq. 19-7 ($Q = Lm$). Here L is the heat of fusion L_F, with the value given in Eq. 19-9 and Table 19-3. We find

$$Q_2 = L_F m = (333\ \text{kJ/kg})(0.720\ \text{kg}) = 239.8\ \text{kJ}.$$

Step 3. Now we have liquid water at 0°C. The next **Key Idea** is that the energy transferred to the liquid water as heat now can only increase the temperature of the liquid water. The heat energy Q_3 needed to increase the temperature of the water from the initial value $T_i = 0°C$ to the final value $T_f = 15°C$ is given by Eq. 19-5 (with the specific heat of liquid water c_{liq}):

$$\begin{aligned} Q_3 &= c_{liq} m (T_f - T_i) \\ &= (4190\ \text{J/kg}\cdot\text{C°})(0.720\ \text{kg})(15°\text{C} - 0°\text{C}) \\ &= 45\,252\ \text{J} \approx 45.25\ \text{kJ}. \end{aligned}$$

The total required heat energy Q_{tot} is the sum of the amounts required in the three steps:

$$\begin{aligned} Q_{tot} &= Q_1 + Q_2 + Q_3 \\ &= 15.98\ \text{kJ} + 239.8\ \text{kJ} + 45.25\ \text{kJ} \\ &\approx 300\ \text{kJ}. \end{aligned} \qquad \text{(Answer)}$$

Note that the heat energy required to melt the ice is much greater than that required to raise the temperature of either the ice or the liquid water.

(b) If we supply the ice with a total heat energy of only 210 kJ, what then are the final state and temperature of the water?

SOLUTION: From step 1, we know that 15.98 kJ is needed to raise the temperature of the ice to the melting point. The remaining heat energy Q_{rem} is then 210 kJ − 15.98 kJ, or about 194 kJ. From step 2, we can see that this amount of heat energy is insufficient to melt all the ice. Then this **Key Idea** becomes important: Because the melting of the ice is incomplete, we must end up with a mixture of ice and liquid; the temperature of the mixture must be the freezing point, 0°C.

We can find the mass m of ice that is melted by the available heat energy Q_{rem} by using Eq. 19-7 with L_F:

$$m = \frac{Q_{rem}}{L_F} = \frac{194\ \text{kJ}}{333\ \text{kJ/kg}} = 0.583\ \text{kg} \approx 580\ \text{g}.$$

Thus, the mass of the ice that remains is 720 g − 580 g, or 140 g, and we have

$$580\ \text{g water} \quad \text{and} \quad 140\ \text{g ice}, \quad \text{at } 0°\text{C}. \qquad \text{(Answer)}$$

Touchstone Example 19-5-2

A copper slug whose mass m_c is 75 g is heated in a laboratory oven to a temperature T of 312°C. The slug is then dropped into a glass beaker containing a mass $m_w = 220$ g of water. The heat capacity C_b of the beaker is 45 cal/K. The initial temperature T_i of the water and the beaker is 12°C. Assuming that the slug, beaker, and water are an isolated system and the water does not vaporize, find the final temperature T_f of the system at thermal equilibrium.

SOLUTION: One **Key Idea** here is that, with the system isolated, only internal transfers of energy can occur. There are three such transfers, all as heat energy. The slug loses energy, the water gains energy, and the beaker gains energy. Another **Key Idea** is that, because these transfers do not involve a phase change, the energy transfers can only change the temperatures. To relate the transfers to the temperature changes, we can use Eqs. 19-4 and 19-5 to write

$$\text{for the water:} \quad Q_w = c_w m_w (T_f - T_i); \qquad \text{(TE19-1)}$$

$$\text{for the beaker:} \quad Q_b = C_b (T_f - T_f); \qquad \text{(TE19-2)}$$

$$\text{for the copper:} \quad Q_c = c_c m_c (T_f - T). \qquad \text{(TE19-3)}$$

A third **Key Idea** is that, with the system isolated, the total energy of the system cannot change. This means that the sum of these three energy transfers is zero:

$$Q_w + Q_b + Q_c = 0. \qquad \text{(TE19-4)}$$

Substituting Eqs. TE19-1 through TE19-3 into Eq. TE19-4 yields

$$c_w m_w (T_f - T_i) + C_b (T_f - T_i) + c_c m_c (T_f - T) = 0. \qquad \text{(TE19-5)}$$

Temperatures are contained in Eq. TE19-5 only as differences. Solving it for T_f, we obtain

$$T_f = \frac{c_c m_c T + C_b T_i + c_w m_w T_i}{c_w m_w + C_b + c_c m_c}.$$

Using Celsius temperatures and taking values for c_c and c_w from Table 19-2, we find the numerator to be

$$(0.0923 \text{ cal/g}\cdot\text{C°})(75 \text{ g})(312\text{°C}) + (45 \text{ cal/C°})(12\text{°C}) + (1.00 \text{ cal/g}\cdot\text{C°})(220 \text{ g})(12\text{°C}) = 5339.8 \text{ cal},$$

and the denominator to be

$$(1.00 \text{ cal/g}\cdot\text{C°})(220 \text{ g}) + 45 \text{ cal/C°} + (0.0923 \text{ cal/g}\cdot\text{C°})(75 \text{ g}) = 271.9 \text{ cal/C°}.$$

We then have

$$T_f = \frac{5339.8 \text{ cal}}{271.9 \text{ cal/C°}} = 19.6\text{°C} \approx 20\text{°C}. \qquad \text{(Answer)}$$

From the given data you can show that

$$Q_w \approx 1670 \text{ cal}, \qquad Q_b \approx 342 \text{ cal}, \qquad Q_c \approx -2020 \text{ cal}.$$

Apart from rounding errors, the algebraic sum of these three heat transfers is indeed zero, as Eq. TE19-4 requires.

Touchstone Example 19-8-1

Let 1.00 kg of liquid water at 100°C be converted to steam at 100°C by boiling at standard atmospheric pressure (which is 1.00 atm or 1.01×10^5 Pa) in the arrangement of Fig. TE19-1. The volume of that water changes from an initial value of 1.00×10^{-3} m³ as a liquid to 1.671 m³ as steam.

(a) How much work is done by the system during this process?

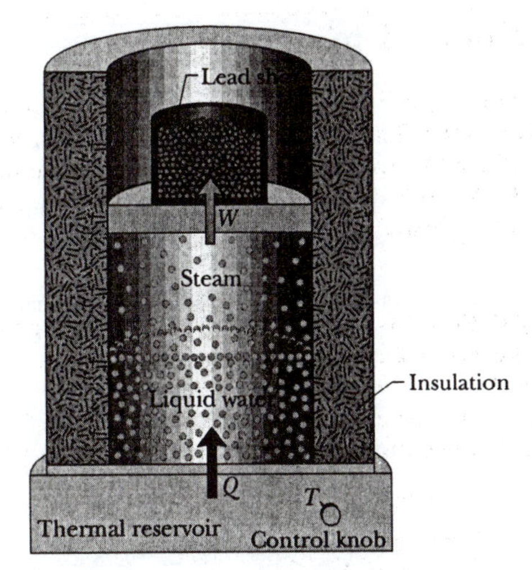

Fig.TE19-1 Water boiling at constant pressure. Energy is transferred from the thermal reservoir as heat until the liquid water has changed completely into steam. Work is done by the expanding gas as it lifts the loaded piston.

SOLUTION: The **Key Idea** here is that the system must do positive work because the volume increases. In the general case we would calculate the work W done by integrating the pressure with respect to the volume (Eq. 19-11). However, here the pressure is constant at 1.01×10^5 Pa, so we can take P outside the integral. We then have

$$\begin{aligned} W &= \int_{V_i}^{V_f} P\,dV = P\int_{V_i}^{V_f} dV = P(V_f - V_i) \\ &= (1.01 \times 10^5\ \text{Pa})(1.671\ \text{m}^3 - 1.00 \times 10^{-3}\ \text{m}^3) \\ &= 1.69 \times 10^5\ \text{J} = 169\ \text{kJ}. \qquad \text{(Answer)} \end{aligned}$$

(b) How much energy is transferred as heat during the process?

SOLUTION: The **Key Idea** here is that the heat energy causes only a phase change and not a change in temperature, so it is given fully by Eq. 19-7 ($Q = Lm$). Because the change is from liquid to gaseous phase, L is the heat of vaporization L_V, with the value given in Eq. 19-8 and Table 19-3. We find

$$\begin{aligned} Q &= L_V m = (2256\ \text{kJ/kg})(1.00\ \text{kg}) \\ &= 2256\ \text{kJ} \approx 2260\ \text{kJ}. \qquad \text{(Answer)} \end{aligned}$$

(c) What is the change in the system's internal energy during the process?

SOLUTION: The **Key Idea** here is that the change in the system's internal energy is related to the heat energy (here, this is energy transferred into the system) and the work (here, this is energy transferred out of the system) by the first law of thermodynamics (Eq. 19-12). Thus, we can write

$$\begin{aligned} \Delta E_{\text{int}} &= Q - W = 2256\ \text{kJ} - 169\ \text{kJ} \\ &\approx 2090\ \text{kJ} = 2.09\ \text{MJ}. \qquad \text{(Answer)} \end{aligned}$$

This quantity is positive, indicating that the internal energy of the system has increased during the boiling process. This energy goes into separating the H_2O molecules, which strongly attract each other in the liquid state. We see that, when water is boiled, about 7.5% (= 169 kJ/2260 kJ) of the heat goes into the work of pushing back the atmosphere. The rest of the heat goes into the system's internal energy.

Touchstone Example 19-10-1

On a hot day in Las Vegas, an oil trucker loaded 37 000 L of diesel fuel. He encountered cold weather on the way to Payson, Utah, where the temperature was 23.0 C° lower than in Las Vegas, and where he delivered his entire load. How many liters did he deliver? The coefficient of volume expansion for diesel fuel is 9.50×10^{-4}/C°, and the coefficient of linear expansion for his steel truck tank is 11×10^{-6}/C°.

SOLUTION: The **Key Idea** here is that the volume of the diesel fuel depends directly on the temperature. Thus, because the temperature decreased, the volume of the fuel did also. From Eq. 19-25, the volume change is

$$\Delta V = V\beta\,\Delta T$$
$$= (37\,000\text{ L})(9.50 \times 10^{-4}/\text{C}°)(-23.0\text{ C}°) = -808\text{ L}.$$

Thus, the amount delivered was

$$V_{\text{del}} = V + \Delta V = 37\,000\text{ L} - 808\text{ L}$$
$$= 36\,192\text{ L}. \qquad \text{(Answer)}$$

Note that the thermal contraction of the steel tank does not affect this answer. Question: Who paid for the "missing" diesel fuel?

Touchstone Example 19-11-1

Figure TE19-2 shows the cross section of a wall made of white pine of thickness L_a and brick of thickness L_d ($= 2.0L_a$), sandwiching two layers of unknown material with identical thicknesses and thermal conductivities. The thermal conductivity of the pine is k_a and that of the brick is k_d ($= 5.0k_a$). The face area A of the wall is unknown. Thermal conduction through the wall has reached the steady state; the only known interface temperatures are $T_1 = 25°$C, $T_2 = 20°$C, and $T_5 = -10°$C. What is interface temperature T_4?

SOLUTION: One **Key Idea** here is that temperature T_4 helps determine the rate H_d at which energy is conducted through the brick, as given by Eq. 19-27. However, we lack enough data to solve Eq. 19-27 for T_4. A second **Key Idea** is that because the conduction is steady, the conduction rate H_d through the brick must equal the conduction rate H_a through the pine: From Eq. 19-27 and Fig. TE19-2, we can write

$$H_a = k_a A \frac{T_1 - T_2}{L_a} \quad \text{and} \quad H_d = k_d A \frac{T_4 - T_5}{L_d}.$$

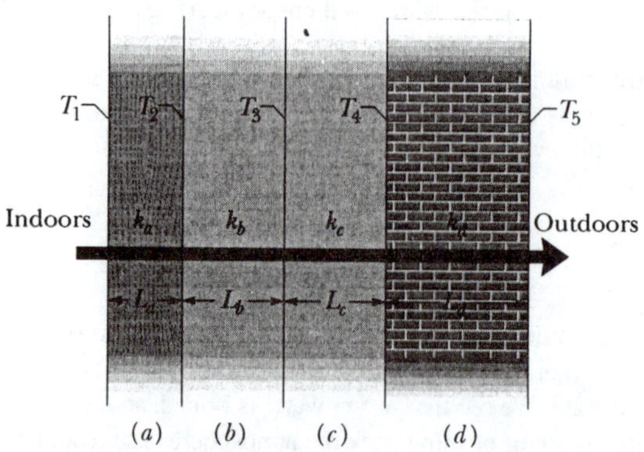

Fig. TE19-2 A wall of four layers through which there is steady-state heat transfer.

Setting $H_a = H_d$ and solving for T_4 yield

$$T_4 = \frac{k_a L_d}{k_d L_a}(T_1 - T_2) + T_5.$$

Letting $L_d = 2.0L_a$ and $k_d = 5.0k_a$, and inserting the known temperatures, we find

$$T_4 = \frac{k_a(2.0L_a)}{(5.0k_a)L_a}(25°\text{C} - 20°\text{C}) + (-10°\text{C})$$
$$= -8.0°\text{C}. \qquad \text{(Answer)}$$

Touchstone Example 19-11-2

When hundreds of Japanese bees form a compact ball around a giant hornet that attempts to invade their hive, they can quickly raise their body temperature from the normal 35°C to 47°C or 48°C. That higher temperature is lethal to the hornet but not to the bees (Fig. TE19-3). Assume the following: 500 bees form a ball of radius $R = 2.0$ cm for a time $t = 20$ min, the primary loss of energy by the ball is by thermal radiation, the ball's surface has emissivity $\varepsilon = 0.80$, and the ball has a uniform temperature. On average, how much additional energy must each bee produce during the 20 min to maintain 47°C?

SOLUTION: The **Key Idea** here is that, because the surface temperature of the bee ball increases after the ball forms, the rate at which energy is radiated by the ball also increases. Thus, the bees lose an additional amount of energy to thermal radiation. We can relate the surface temperature to the rate of radiation (energy per unit time) with Eq. 19-33 ($P_{\text{rad}} = \sigma\varepsilon A T^4$), in which A is the ball's surface area and T is the ball's surface temperature in kelvins. This rate is an energy per unit time; that is,

$$P_{\text{rad}} = \frac{\Delta E}{\Delta t}.$$

Thus, the amount of energy E radiated in time Δt is $\Delta E = P_{\text{rad}}\Delta t$.

At the normal temperature $T_1 = 35°\text{C}$, the radiation rate would be P_{r1} and the amount of energy radiated in time Δt would be $\Delta E_1 = P_{r1}\Delta t$. At the increased temperature $T_2 = 47°\text{C}$, the (greater) radiation rate is P_{r2} and the (greater) amount of energy radiated in time Δt is $\Delta E_2 = P_{r2}\Delta t$. Thus, in maintaining the ball at T_2 for time Δt, the bees must (together) provide an additional energy of $E = E_2 - E_1$.

We can now write

$$E = E_2 - E_1 = P_{r2}\Delta t - P_{r1}\Delta t$$
$$= (\sigma\varepsilon A T_2^4)\Delta t - (\sigma\varepsilon A T_1^4)\Delta t = \sigma\varepsilon A\Delta t(T_2^4 - T_1^4). \qquad \text{(TE 19-6)}$$

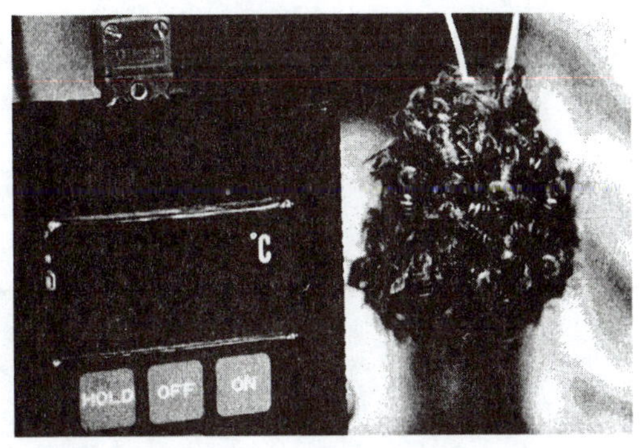

Fig. TE19-3 The bees were unharmed by their increased body temperature, which the hornet could not withstand.

The temperatures here *must* be in kelvins; thus, we write them as

$$T_2 = 47°\text{C} + 273\ \text{C}° = 320\ \text{K}$$

and

$$T_1 = 35°\text{C} + 273\ \text{C}° = 308\ \text{K}.$$

The surface area A of the ball is

$$A = 4\pi R^2 = (4\pi)(0.020\ \text{m})^2 = 5.027 \times 10^{-3}\ \text{m}^2,$$

and the time Δt is 20 min = 1200 s. Substituting these and other known values into Eq. TE19-6, we find

$$E = (5.6703 \times 10^{-8}\ \text{W/m}^2\cdot\text{K}^4)(0.80)(5.027 \times 10^{-3}\ \text{m}^2) \times (1200\ \text{s})[(320\ \text{K})^4 - (308\ \text{K})^4] = 406.8\ \text{J}.$$

Thus, with 500 bees in the ball, each bee must produce an additional energy of

$$\frac{E}{500} = \frac{406.8\ \text{J}}{500} = 0.81\ \text{J}. \qquad \text{(Answer)}$$

20 The Kinetic Theory of Gases

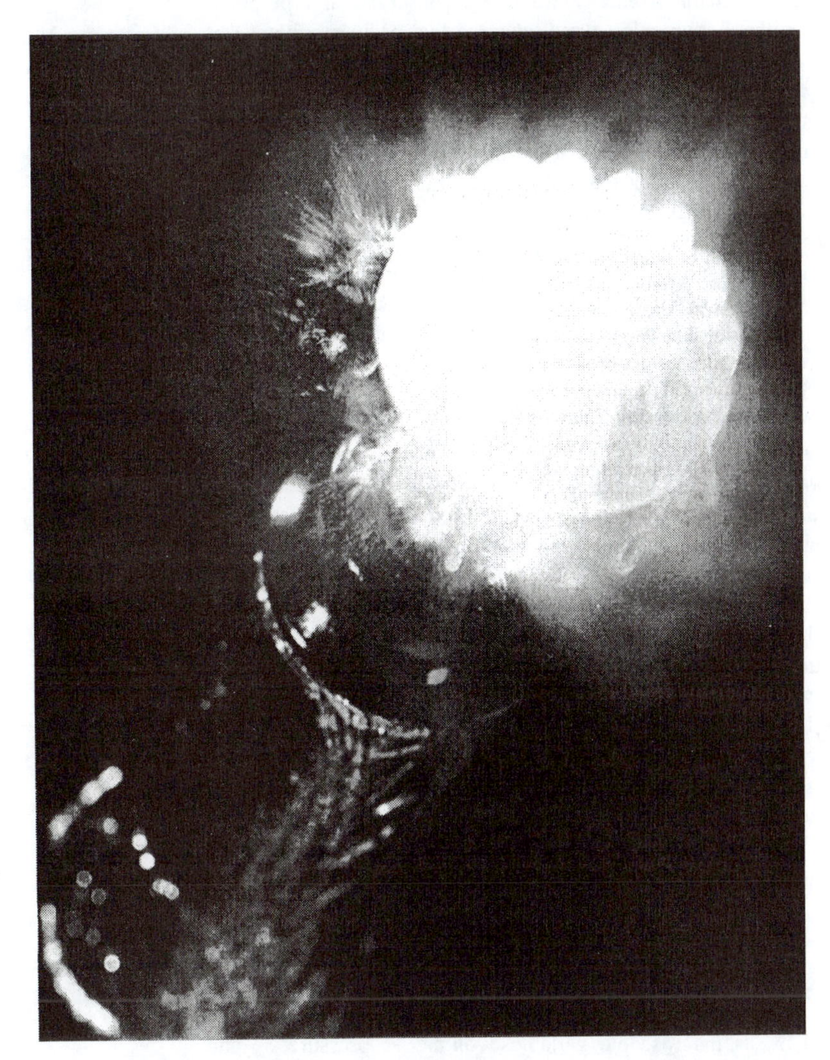

When a container of cold champagne, soda pop, or any other carbonated drink is opened, a slight fog forms around the opening and some of the liquid sprays outward. (In the photograph, the fog is the white cloud that surrounds the stopper, and the spray has formed streaks within the cloud.)

What causes the fog?

The answer is in this chapter.

20-1 Molecules and Thermal Gas Behavior

In our studies of mechanics and thermodynamics we have found a number of strange and interesting results. In mechanics, we saw that moving objects tend to run down and come to a stop. We attributed this to the inevitable presence of friction and drag forces. Without these non-conservative forces mechanical energy would be conserved and perpetual motion would be possible. In thermodynamics, we discovered that ordinary objects, by virtue of their temperature, contain huge quantities of internal energy and that this might be where the "lost" energy resulting from friction forces is hidden. In this chapter, we will learn about some ways that matter can store internal energy. What you are about to learn may be counter-intuitive. Instead of finding that the "natural state" of a system is to lose energy, you will find considerable evidence that the "natural state" of a system is quite the opposite. It is one in which its fundamental parts (atoms and molecules) are traveling every which way—in a state of perpetual motion.

Classical thermodynamics—the subject of the last chapter—has nothing to say about atoms or molecules. Its laws are concerned only with such macroscopic variables as pressure, volume, and temperature. In this chapter we begin an exploration of the atomic and molecular basis of thermodynamics. As is usual in the development of new theories in physics, we start with a simple model. The fact that gases are fluid and compressible is evidence that their molecules are quite small relative to the average spacing between them. If so, we expect that gas molecules are relatively free and independent of one another. For this reason, we believe that the thermal behavior of gases will be easier to understand than that of liquids and solids. Thus, we begin an exploration of the atomic and molecular basis of thermodynamics by developing **the kinetic theory of gases**—a simplified model of gas behavior based on the laws of classical mechanics.

We begin the chapter with a discussion of how the **ideal gas law** characterizes the macroscopic behavior of simple gases. This macroscopic law relates the amount of gas and its pressure, temperature, and volume to each other. Next we consider how kinetic theory, which provides us with a molecular (or microscopic) model of gas behavior, can be used to explain observed macroscopic relationships between gas pressure, volume, and temperature. We then move on to using kinetic theory as an underlying model of the characteristics of an ideal gas. The basic ideas of kinetic theory are that: (1) an ideal gas at a given temperature consists of a collection of tiny particles (atoms or molecules) that are in perpetual motion—colliding with each other and the walls of their container; and (2) the hidden internal energy of the gas is directly proportional to the kinetic energy of its particles.

20-2 The Macroscopic Behavior of Gases

Any gas can be described by its macroscopic variables volume V, pressure P, and temperature T. Simple experiments were performed on low density gases in the seventeenth and eighteenth centuries to relate these variables. Robert Boyle (b. 1627) determined that at a constant temperature the product of pressure and temperature remains constant.

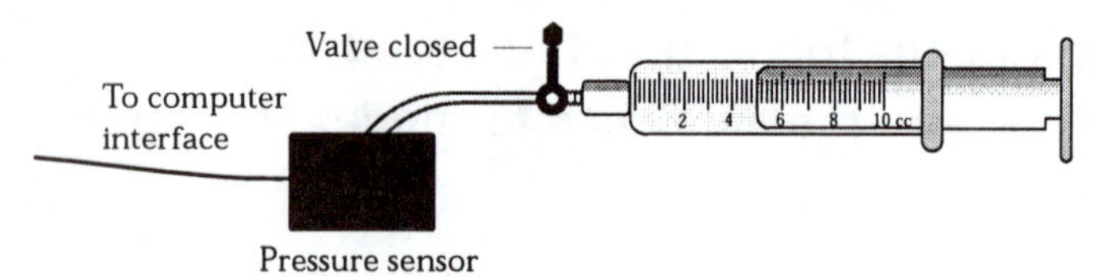

Fig. 20-1: A contemporary setup for determining the relationship between gas pressure and volume using an inexpensive medical syringe and an electronic pressure sensor attached to an MBL system. When temperature is held constant P and V turn out to be inversely proportional to each other so that PV is constant.

French scientists Jacques Charles (b. 1746) and Joseph Gay-Lussac (b. 1778) found that as the Kelvin temperature of a fixed volume of gas is raised its pressure increases proportionally. (See Fig. 19-14*b* and 19-15). Similarly, Charles who was a hot air balloonist, discovered that for a constant pressure (such as atmospheric pressure) the volume of a gas is proportional to its temperature. By combining the results of all three of these experiments we must conclude that there is a proportionality between PV and T:

$$PV \propto T.$$

The Molecular Form of the Ideal Gas Law

If we can find a constant of proportionality between PV and T for a relatively low density gas then we will have formulated a gas law. An examination of the student-generated P vs. T data shown in Fig. 19-15 indicates that the constant of proportionality between the product PV and the variable T (determined by the slopes) of the graphs decreases as the mass of gas confined to the same volume decreases. Similar experiments have shown that the slope of a P vs. T graph will change if the same volume and mass of a different kind of gas is used. This suggests that the constant of proportionality we are looking for must be a function of *both* the mass and type of gas. It was puzzling to early investigators that the slopes of their P vs. T and V vs. T graphs were not just proportional to the mass of the different gases used in the experiments.

The key to finding a constant of proportionality that embodies both gas type and mass was a hypothesis developed in the early nineteenth century by the Italian scientist Amadeo Avogadro (1776-1856). In 1811, Avogadro proposed that equal volumes of any kind of gas at the same pressure would have both the same number of molecules and occupy the same volume. Eventually it was discovered that the constant of proportionality needed for the fledgling gas law was one that is directly proportional to the number of molecules of a gas rather than its mass, so that

$$PV = Nk_BT \quad \text{(molecular ideal gas law)}, \quad (20\text{-}1)$$

where N is the number of molecules of confined gas and k_B is a proportionality constant needed to shift from kelvin to joules, the SI units for the product PV. The experimentally determined value of k_B, is known historically as the Boltzmann constant. Its value is experimentally determined to be

$$k_B = 1.38 \times 10^{-23}\ \text{J/K} \quad \text{(Boltzmann constant)}. \quad (20\text{-}2)$$

Avogadro's Number and the Mole

The problem with the molecular form of the ideal gas law we just presented is that it is hard to count molecules. It is much easier to measure the mass of a sample of gas or its volume at a standard pressure. In this subsection we will define two new quantities—*mole* and *molar mass*. Although these quantities are related to the number of molecules in a gas, they can be measured macroscopically, so it is useful to reformulate the ideal gas law in terms of moles.

Let's start our reformulation of the ideal gas law with definitions of mole and molar mass. In Section 1-7, we presented the SI definition of the *atomic mass unit* in terms of the mass of a carbon-12 atom. In particular, carbon-12 is assigned an atomic mass of exactly 12 u. In a related fashion, the SI definition of the *mole* (or *mol* for short) relates the number of particles in a substance to its macroscopic mass.

➤A **mole** is defined as the amount of any substance that contains the same number of atoms or molecules as there are in *exactly* 12 g of carbon-12.

The results of many different types of experiments, including x-ray diffraction studies in crystals, have revealed that there are a very large number of atoms in 12 grams of carbon-12. The amount of atoms is known as Avogadro's number and is denoted as N_A.

$$N_A = 6.022137 \times 10^{23}\ \text{mol}^{-1} \quad \text{(Avogadro's number)}. \quad (20\text{-}3)$$

Here the symbol mol^{-1} represents the inverse mole or "per mole." Usually we round off the value to three significant figures so that $N_A = 6.02 \times 10^{23}\ \text{mol}^{-1}$.

The number of moles n contained in a sample of any substance is equal to the ratio of the number of molecules N in the sample to the number of molecules N_A in 1 mole of the same substance:

$$n = \frac{N}{N_A}. \qquad (20\text{-}4)$$

(*Caution:* The three symbols in this equation can easily be confused with one another, so you should sort them with their meanings now, before you end in "N-confusion.")

We can easily calculate the mass of one mole of atoms or molecules in any sample, defined as the **molar mass** (denoted as M), by looking in a table of atomic or molecular masses.

Note that if we refer to Appendix F to find the molar mass, M, in grams of a sample of matter, we can determine the number of moles in the sample by determining its mass M_{sam} and using the equation

$$n = \frac{M_{sam}}{M}. \qquad (20\text{-}5)$$

It is puzzling to note that the atomic mass of carbon that is listed in Appendix F is given as 12.01115 u rather than 12.00000 u. This is because a natural sample of carbon does not consist of just carbon-12. Instead it contains a relatively small percentage of carbon-13 which has an extra neutron in its nucleus. Nevertheless, by definition, a mole of pure carbon-12 and a mole of a naturally-occuring mixture of carbon-12 and carbon-13 both contain Avogadro's number of atoms.

The Molar Form of the Ideal Gas Law

We can rewrite the molecular ideal gas law expressed in Eq. 20-1 in an alternative form by using Eq. 20-4, so that

$$PV = Nk_BT = nN_Ak_BT.$$

Since both Avogadro's number and the Boltzmann constant are constants we can replace their product with a new constant R which is called the **universal gas constant** because it has the same value for all ideal gases—namely,

$$R = N_Ak_B = 6.02 \times 10^{23}\,\text{mol}^{-1}(1.38 \times 10^{-23}\ \text{J/K}) = 8.31\ \text{J/mol} \cdot \text{K}. \qquad (20\text{-}6)$$

This allows us to write

$$nR = Nk_B. \qquad (20\text{-}7)$$

Substituting this into Eq. 20-1 gives a second expression for the ideal gas law:

$$PV = nRT \qquad \text{(molar ideal gas law)}, \qquad (20\text{-}8)$$

in which P is the absolute (not gauge) pressure, n is the number of moles of gas present, and T is the temperature in kelvin. Provided the gas density is low, the ideal gas law as represented in either Eq. 20-1 or Eq. 20-8 holds for any single gas or for any mixture of different gases. (For a mixture, n is the total number of moles in the mixture.)

Note the difference between the two expressions for the ideal gas law—Eq. 20-8 involves the number of moles n and Eq. 20-1 involves the number of molecules N. That is, the Boltzmann constant k_B, tells us about individual atoms, whereas the gas constant R, tells us about moles of atoms. Recall that moles are defined via macroscopic measurements that are easily done in the lab—such as 1 mole of carbon has a mass of 12 g. As a result, R is easily measured in the lab. On the other hand, since k_B is about individual atoms, to get to it from a lab measurement we have to count the number of molecules in a mole. This is a decidedly non-trivial task.

You may well ask, "What is an *ideal gas* and what is so 'ideal' about one?" The answer lies in the simplicity of the law (Eqs. 20-1 and 20-8) that describes the macroscopic properties of a gas. Using this law—as you will see—we can deduce many properties of the ideal gas in a simple way. There is no such thing in nature as a truly ideal gas. But *all* gases approach the ideal state at low enough densities—that is, under conditions in which their molecules are far enough apart that they do not interact with one another as much as they do with the walls of their containers. Thus, the two equivalent ideal gas equations allow us to gain useful insights into the behavior of real gases at low densities.

Touchstone Example 20-2-1, at the end of this chapter, illustrates how to use what you learned in this section.

TE

20-3 Work Done by Ideal Gases

Engines are devices that can absorb heat energy and do useful work on their surroundings. As you will see in the next chapter, air which is typically used as a working medium in heat engines behaves like an ideal gas in some circumstances. For this reason engineers are interested in knowing how to calculate the work done by ideal gases. Before we turn our attention to how the action of molecules that make up an ideal gas can be used to explain the ideal gas law, we first consider how to calculate the work done by ideal gases under various conditions.

Work Done by an Ideal Gas at Constant Temperature

Suppose we put an ideal gas in a piston-cylinder arrangement like those in Chapter 19. Suppose also that we allow the gas to expand from an initial volume V_i to a final volume V_f while we keep the temperature T of the gas constant. Such a process, at *constant temperature,* is called an **isothermal expansion** (and the reverse is called an **isothermal compression**).

On a P-V diagram, an *isotherm* is a curve that connects points that have the same temperature. Thus, it is a graph of pressure versus volume for a gas whose temperature T is held constant. For n moles of an ideal gas, it is a graph of the equation

$$P = nRT\frac{1}{V} = (\text{a constant})\ \frac{1}{V}. \tag{20-9}$$

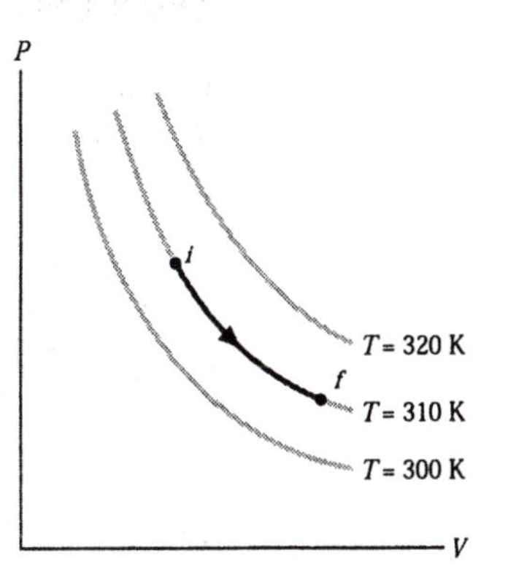

Fig. 20-2: Three isotherms on a P-V diagram. The path shown along the middle isotherm represents an isothermal expansion of a gas from an initial state i to a final state f. The path from f to i along the isotherm would represent the reverse process, an isothermal compression.

Figure 20-2 shows three isotherms, each corresponding to a different (constant) value of T. (Note that the values of T for the isotherms increase upward to the right.) Superimposed on the middle isotherm is the path followed by a gas during an isothermal expansion from state i to state f at a constant temperature of 310 K.

To find the work done by an ideal gas during an isothermal expansion, we start with Eq. 19-11,

$$W = \int_{V_i}^{V_f} P\, dV. \tag{20-10}$$

This is a general expression for the work done during any change in volume of any gas. For an ideal gas, we can use Eq. 20-8 to substitute for P, obtaining

$$W = \int_{V_i}^{V_f} \frac{nRT}{V} dV. \tag{20-11}$$

Because we are considering an isothermal expansion, T is constant and we can move it in front of the integral sign to write

$$W = nRT\int_{V_i}^{V_f} \frac{dV}{V} = nRT\big[\ln\ V\big]_{V_i}^{V_f}. \qquad (20\text{-}12)$$

By evaluating the expression in brackets at the limits and then using the relationship $\ln a - \ln b = \ln(a/b)$, we find that

$$W = nRT\ln\frac{V_f}{V_i} \qquad \text{(ideal gas, isothermal process)}. \; (20\text{-}13)$$

Recall that the symbol ln specifies a *natural* logarithm which has base *e*.

As we often do in science and engineering we have derived a mathematical relationship. Before using this relationship it's a good idea to check our equation to see if it makes sense. We know that an expanding gas does work on its surroundings. If the gas contracts we expect that the surroundings has done work on the gas instead. Is this what Eq. 20-13 tells us? For an expansion, V_f is greater than V_i, so the ratio V_f/V_i in Eq. 20-13 is greater than unity. The natural logarithm of a quantity greater than unity is positive, and so the work W done by an ideal gas during an isothermal expansion is positive, as we expect. For a compression, V_f is less than V_i, so the ratio of volumes in Eq. 20-13 is less than unity. The natural logarithm in that equation—hence the work W—is negative, again as we expect.

Work Done at Constant Volume and at Constant Pressure

Equation 20-13 does not give the work W done by an ideal gas during *every* thermodynamic process. Instead, it gives the work only for a process in which the temperature is held constant. If the temperature varies, then the symbol T in Eq. 20-11 cannot be moved in front of the integral symbol as in Eq. 20-12, and thus we do not end up with Eq. 20-13.

However, we can go back to Eq. 20-10 to find the work W done by an ideal gas (or any other gas) during two more processes—a constant-volume process and a constant-pressure process. If the volume of the gas is constant, then Eq. 20-10 yields

$$W = 0 \qquad \text{(constant-volume process)}. \qquad (20\text{-}14)$$

If, instead, the volume changes while the pressure P of the gas is held constant, then Eq. 20-10 becomes

$$W = P(V_f - V_i) = P\Delta V \qquad \text{(constant-pressure process)}. \qquad (20\text{-}15)$$

READING EXERCISE 20-1: An ideal gas has an initial pressure of 3 pressure units and an initial volume of 4 volume units. The table gives the final pressure and volume of the gas (in those same units) in five processes. Which processes start and end on the same isotherm?

	a	*b*	*c*	*d*	*e*
P	12	6	5	4	1
V	1	2	7	3	12

Touchstone Example 20-3-1, at the end of this chapter, illustrates how to use what you learned in this section.

TE

20-4 Pressure, Temperature, and Molecular Kinetic Energy

In terms of everyday experience, molecules and atoms are hypothetical entities. In just the past forty years or so, scientists have been able to "see" molecules using electron microscopes and field ion microscopes. But long before atoms and molecules could be "seen," nineteenth-century scientists such as James Clerk Maxwell and Ludwig Boltzmann in Europe, and Josiah Willard Gibbs in the United States used these imaginary *microscopic* entities to construct models that made the description and prediction of the *macroscopic* behavior of thermodynamic systems possible.

Is it possible to describe the behavior of an ideal gas that obeys the first law of thermodynamics microscopically as a collection of moving molecules? To answer this question, let's observe the pressure exerted by a hypothetical molecule undergoing perfectly elastic collisions with the walls of a cubical box. By using the laws of mechanics we can derive a mathematical expression for the pressure exerted by just one of the molecules as a function of the volume of the box. Next we can extend our "ideal gas" so it is a low-density collection of molecules. By low-density we mean that the volume occupied by the molecules is negligible compared to the volume of their container. This means that the molecules are far enough apart on the average that attractive interactions between molecules is also negligible. If we then define temperature as being related to the average kinetic energy of the molecules in an ideal gas, we can show that kinetic theory is a powerful construct for explaining both the ideal gas law and the first law of thermodynamics.

We start developing our idealized kinetic theory model by considering N molecules of an ideal gas that are confined in a cubical box of volume V, as in Fig. 20-3. The walls of the box are held at temperature T. How is the pressure P exerted by the gas on the walls related to the speeds of the molecules? Remember from our discussions of fluids in Chapter 15 that pressure is a scalar that is defined as the ratio of the magnitude of force (exerted normal to a surface) and the area of the surface. In the example at hand, a gas confined to a box, the pressure results from the motion of molecules in all directions resulting in collisions between gas molecules and the walls of the box. We ignore (for the time being) collisions of the molecules with one another and consider only elastic collisions with the walls.

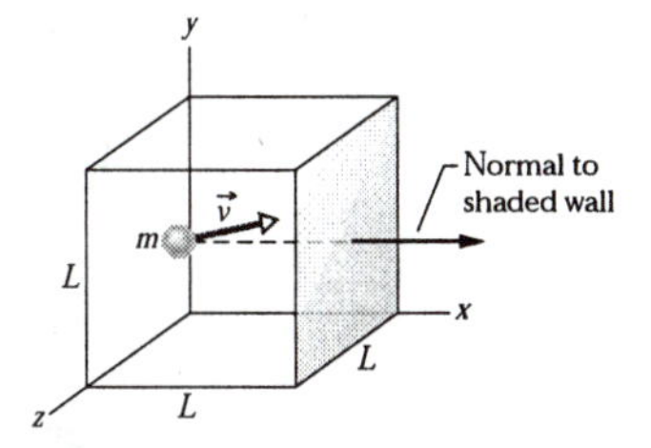

Fig. 20-3: We assume a cubical box of edge L, contains N ideal gas molecules (not shown) that move around perpetually without losing energy. One of the molecules of mass m and velocity $\vec{v}$ is shown heading for a collision with the shaded wall of area L^2. The normal to the shaded wall points in the positive x-direction.

Figure 20-3 shows a typical gas molecule, of mass m and velocity $\vec{v}$, which is about to collide with the shaded wall. Because we assume that any collision of a molecule with a wall is elastic, when this molecule collides with the shaded wall, the only component of its velocity that is changed by the collision is its x-component. That x-component has the same magnitude after collision but its sign is reversed. This means that the only change in the particle's momentum is along the x axis, so

$$(\Delta p_x)_{\text{molecule}} = p_{fx} - p_{ix} = (-mv_x) - (mv_x) = -2mv_x.$$

But the law of conservation of momentum tells us that the momentum change Δp_x that the wall experiences after a molecule collides with it is $+2mv_x$. Remember that in this book $\vec{p}$, p_x, p_y, and p_z denote momentum vectors or vector components while capital P represents pressure. *Be careful not to confuse them.*

The molecule of Fig. 20-3 will hit the shaded wall repeatedly. The time Δt between collisions is the time the molecule takes to travel to the opposite wall and back again (a distance of $2L$) at speed $|\vec{v}_x|$. Thus, Δt is equal to $2L/|\vec{v}_x|$. (Note that this result holds even if the molecule bounces off any of the other walls along the way, because those walls are parallel to x and so cannot change $|\vec{v}_x|$.) Therefore, the average rate at which momentum is delivered to the shaded wall by this single molecule is

$$\frac{(\Delta p_x)_{\text{wall}}}{\Delta t} = \frac{+2m|\vec{v}_x|}{2L/|\vec{v}_x|} = \frac{mv_x^2}{L}.$$

From Newton's Second Law ($\vec{F} = d\vec{p}/dt$), the rate at which momentum is delivered to the wall is the force acting on that wall. To find the total force, we must add up the contributions of all of the N molecules that strike the wall during a short time interval Δt. We will allow for the possibility that all the molecules have different velocities. Then we can divide the magnitude of the total force acting normal to the shaded wall $|F_x|$ by the area of the wall ($= L^2$) to determine the pressure P on that wall. Thus,

$$P=\frac{|F_x|}{L^2}=\frac{mv_{x1}^2/L+mv_{x2}^2/L+\cdots+mv_{xN}^2/L}{L^2}$$
$$=\left(\frac{m}{L^3}\right)(v_{x1}^2+v_{x2}^2+\cdots+v_{xN}^2). \quad (20\text{-}16)$$

Since by definition $(v_x^2)_{\text{avg}}=(v_{x1}^2+v_{x2}^2+\cdots+v_{xN}^2)/N$ we can replace the sum of squares of the velocities in the second parentheses of Eq. 20-16 by $N(v_x^2)_{\text{avg}}$ where $(v_x^2)_{\text{avg}}$ is the average value of the square of the x components of all the speeds. Equation 20-16 for the pressure on the container wall then reduces to

$$P=\frac{Nm}{L^3}(v_x^2)_{\text{avg}}=\frac{Nm}{V}(v_x^2)_{\text{avg}}, \quad (20\text{-}17)$$

since the volume V of the cubical box is just L^3.

It is reasonable to assume molecules are moving at random in three dimensions rather than just in the x direction that we considered initially so that $(v_x^2)_{\text{avg}}=(v_y^2)_{\text{avg}}=(v_z^2)_{\text{avg}}$ and

$$(v^2)_{\text{avg}}=(v_x^2+v_y^2+v_z^2)_{\text{avg}}=(v_x^2)_{\text{avg}}+(v_y^2)_{\text{avg}}+(v_z^2)_{\text{avg}}=3(v_x^2)_{\text{avg}},$$

or

$$(v_x^2)_{\text{avg}}=(v^2)_{\text{avg}}/3.$$

Thus, we can rewrite the expression above as

$$P=\frac{Nm(v^2)_{\text{avg}}}{3V}. \quad (20\text{-}18)$$

The square root of $(v^2)_{\text{avg}}$ is a kind of average speed, called the **root-mean-square speed** of the molecules and symbolized by $|\vec{v}_{\text{rms}}|$. Its name describes it rather well: You *square* each speed, you find the *mean* (that is, the average) of all these squared speeds, and then you take the square *root* of that mean. With $\sqrt{(v^2)_{\text{avg}}}=|\vec{v}_{\text{rms}}|$, we can then write Eq. 20-18 as

$$P=\frac{Nm(v_{\text{rms}}^2)}{3V}. \quad (20\text{-}19)$$

Equation 20-19 is very much in the spirit of kinetic theory. It tells us how the pressure of the gas (a purely macroscopic quantity) depends on the speed of the molecules (a purely microscopic quantity). We can turn Eq. 20-19 around and use it to calculate $|\vec{v}_{\text{rms}}|$ as

$$|\vec{v}_{\text{rms}}|=\sqrt{\frac{3PV}{Nm}}.$$

Combining this with the ideal gas law $(PV=Nk_BT)$ gives us

$$|\vec{v}_{\text{rms}}|=\sqrt{\frac{3k_BT}{m}}, \quad (20\text{-}20)$$

where m is the mass of a single molecule.

Table 20-1 shows some rms speeds calculated from Eq. 20-20. The speeds are surprisingly high. For hydrogen molecules at room temperature (300 K), the rms speed is 1920 m/s or 4300 mi/h—faster than a speeding bullet! On the surface of the Sun, where the temperature is 2×10^6 K, the rms speed of hydrogen molecules would be 82 times greater than at room temperature were it not for the fact that at such high speeds, the molecules cannot survive collisions among themselves. Remember too that the rms speed is only a kind of average speed; many molecules move much faster than this, and some much slower.

TABLE 20-1: Some Molecular Speeds at Room Temperature (T = 300 K)[a]

Gas	Molar Mass $M = mN_A$ (10^{-3} kg/mol)	$\lvert\vec{v}_{rms}\rvert$ (m/s)
Hydrogen (H_2)	2.02	1920
Helium (He)	4.0	1370
Water vapor (H_2O)	18.0	645
Nitrogen (N_2)	28.0	517
Oxygen (O_2)	32.0	483
Carbon dioxide (CO_2)	44.0	412
Sulfur dioxide (SO_2)	64.1	342

[a]For convenience, we often set room temperature = 300 K even though (at 27°C or 81°F) that represents a fairly warm room.

The speed of sound in a gas is closely related to the rms speed of the molecules of that gas. In a sound wave, the disturbance is passed on from molecule to molecule by means of collisions. The wave cannot move any faster than the "average" speed of the molecules. In fact, the speed of sound must be somewhat less than this "average" molecular speed because not all molecules are moving in exactly the same direction as the wave. As examples, at room temperature, the rms speeds of hydrogen and nitrogen molecules are 1920 m/s and 517 m/s, respectively. The speeds of sound in these two gases at this temperature are 1350 m/s and 350 m/s, respectively.

Translational Kinetic Energy

Let's again consider a single molecule of an ideal gas as it moves around in the box of Fig. 20-3, but we now assume that its speed changes when it collides with other molecules. Its translational kinetic energy at any instant is $\frac{1}{2}mv^2$. Its *average* translational kinetic energy over the time that we watch it is

$$K_{avg} = \tfrac{1}{2}(mv^2)_{avg} = \tfrac{1}{2}m(v^2)_{avg} = \tfrac{1}{2}mv_{rms}^2, \qquad (20\text{-}21)$$

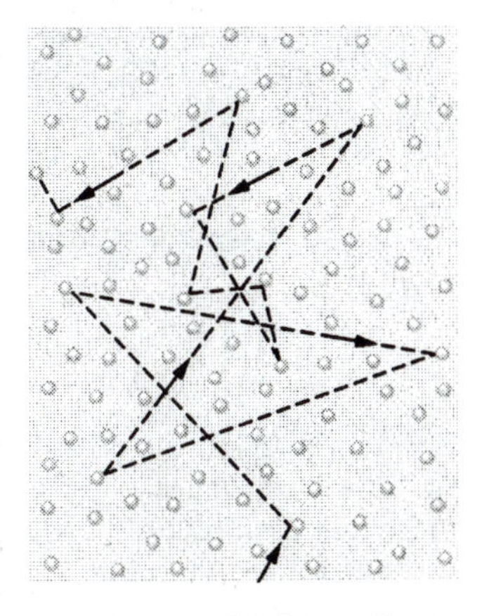

Fig. 20-4: A molecule traveling through a gas, colliding with other gas molecules in its path. Although the other molecules are shown as stationary, we believe they are also moving in a similar fashion.

in which we make the assumption that the average speed of the molecule during our observation is the same as the average speed of all the molecules at any given time. (Provided the total energy of the gas is not changing and we observe our molecule for long enough, this assumption is appropriate.) Substituting for $\lvert\vec{v}_{rms}\rvert$ from Eq. 20-20 leads to

$$K_{avg} = (\tfrac{1}{2}m)\frac{3k_BT}{m}$$

so that

$$K_{avg} = \tfrac{3}{2}k_BT. \qquad (20\text{-}22)$$

This equation tells us something unexpected:

➤At a given temperature T, all ideal gas molecules—no matter what their mass—have the same average translational kinetic energy, namely, $(3/2)k_BT$. When we measure the temperature of a gas, we are also measuring the average translational kinetic energy of its molecules.

READING EXERCISE 20-2: What happens to the average translational kinetic energy of each molecule in a gas when its temperature in kelvin: (a) doubles and (b) is reduced to zero?

READING EXERCISE 20-3: A gas mixture consists of molecules of types 1, 2, and 3, with molecular masses $m_1 > m_2 > m_3$. Rank the three types according to (a) average kinetic energy and (b) rms speed, greatest first.

20-5 Mean Free Path

In considering the motion of molecules, a question often arises: If molecules move so fast, why does it take as long as a minute or so before you can smell perfume when someone opens a bottle across a room? In order to answer this question, we continue to examine the motion of molecules in an ideal gas. Figure 20-4 shows the path of a typical molecule as it moves through the gas, changing both speed and direction abruptly as it collides elastically with other molecules. Between collisions, our typical molecule moves in a straight line at constant speed. Although the figure shows all the other molecules as stationary, they too are moving similarly.

One useful parameter to describe this random motion is the **mean free path** λ of the molecules. As its name implies, λ is the average distance traversed by a molecule between collisions. We expect λ to vary inversely with N/V, the number of molecules per unit volume (or density of molecules). The larger N/V is, the more collisions there should be and the smaller the mean free path. We also expect λ to vary inversely with the size of the molecules, say, with their diameter d. (If the molecules were points, as we have assumed them to be, they would never collide and the mean free path would be infinite.) Thus, the larger the molecules are, the smaller the mean free path. We can even predict that λ should vary (inversely) as the *square* of the molecular diameter because the cross section of a molecule—not its diameter—determines its effective target area.

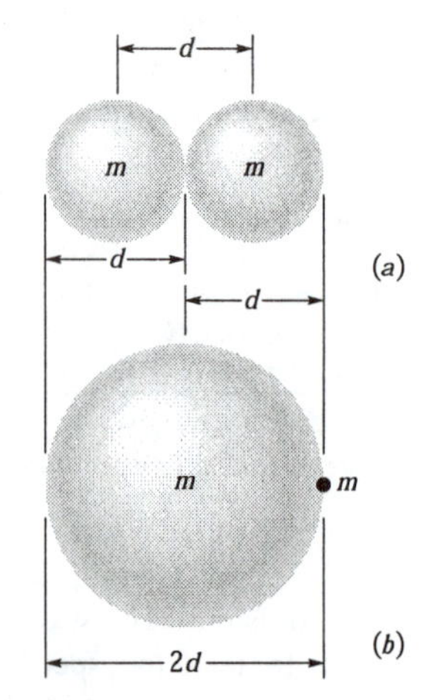

Fig. 20-5: (*a*) A collision occurs when the centers of two molecules come within a distance d of each other, d being the molecular diameter. (*b*) An equivalent but more convenient representation is to think of the moving molecule as having a *radius* d and all other molecules as being points. The condition for a collision is unchanged.

The expression for the mean free path does, in fact, turn out to be

$$\lambda = \frac{1}{\sqrt{2}\pi d^2 \; N/V} \qquad \text{(mean free path).} \qquad (20\text{-}23)$$

To justify Eq. 20-23, we focus attention on a single molecule and assume—as Fig. 20-4 suggests—that our molecule is traveling with a constant speed v and that all the other molecules are at rest. Later, we shall relax this assumption.

We assume further that the molecules are spheres of diameter d. A collision will then take place if the centers of the molecules come within a distance d of each other, as in Fig. 20-5*a*. Another, more helpful way to look at the situation is to consider our single molecule to have a *radius* of d and all the other molecules to be *points,* as in Fig. 20-5*b*. This does not change our criterion for a collision.

As our single molecule zigzags through the gas, it sweeps out a short cylinder of cross-sectional area πd^2 between successive collisions. If we watch this molecule for a time interval Δt, it moves a distance $|\vec{v}|\Delta t$, where $|\vec{v}|$ is its assumed speed. Thus, if we align all the short cylinders swept out in Δt, we form a composite cylinder (Fig. 20-6) of length $|\vec{v}|\Delta t$ and volume $(\pi d^2)(|\vec{v}|\Delta t)$. The number of collisions that occur in time Δt is then equal to the number of (point) molecules that lie within this cylinder.

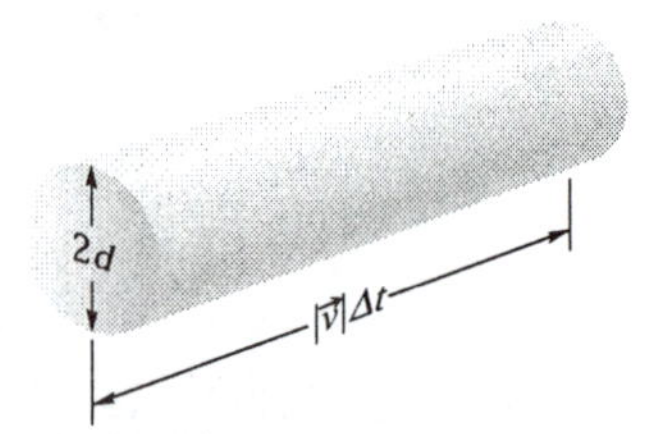

Fig. 20-6: In time Δt the moving molecule effectively sweeps out a cylinder of length $|\vec{v}|\Delta t$ and radius d.

Since N/V is the number of molecules per unit volume, the number of molecules in the cylinder is N/V times the volume of the cylinder, or $(N/V)(\pi d^2|\vec{v}|\Delta t)$. This is also the number of collisions in time Δt. The mean free path is the length of the path (and of the cylinder) divided by this number:

$$\begin{aligned}\lambda &= \frac{\text{length of path}}{\text{number of collisions}} \approx \frac{|\vec{v}|\,\Delta t}{\pi d^2|\vec{v}|\,\Delta t\; N/V} \\ &= \frac{1}{\pi d^2\; N/V}.\end{aligned} \qquad (20\text{-}24)$$

This equation is only approximate because it is based on the assumption that all the molecules except one are at rest. In fact, *all* the molecules are moving; when this is taken properly into account, Eq. 20-23 results. Note that it differs from the (approximate) Eq. 20-24 only by a factor of $1/\sqrt{2}$.

We can even get a glimpse of what is "approximate" about Eq. 20-24. The $|\vec{v}|$ in the numerator and that in the denominator are—strictly—not the same. The $|\vec{v}|$ in the numerator is $|\vec{v}_{\text{avg}}|$, the mean speed of the molecule *relative to the container.* The $|\vec{v}|$ in the denominator is $|\vec{v}_{\text{rel}}|$, the mean speed of our single molecule *relative to the other molecules,* which are moving. It is this latter average speed that determines the number of collisions. A detailed calculation, taking into account the actual speed distribution of the molecules, gives $|\vec{v}_{rel}| = \sqrt{2}\,|\vec{v}_{avg}|$ and thus the factor $\sqrt{2}$.

The mean free path of air molecules at sea level is about 0.1 μm. At an altitude of 100 km, the density of air has dropped to such an extent that the mean free path rises to about 16 cm. At 300 km, the mean free path is about 20 km. A problem faced by those who would study the physics and chemistry of the upper atmosphere in the laboratory is the unavailability of containers large enough to hold gas samples that simulate upper atmospheric conditions. Yet studies of the concentrations of Freon, carbon dioxide, and ozone in the upper atmosphere are of vital public concern.

Recall the question that began this chapter: If molecules move so fast, why does it take as long as a minute or so before you can smell perfume when someone opens a bottle across a room? We now know the answer. In still air, each perfume molecule moves away from the bottle only very slowly because its repeated collisions with other molecules prevent it from moving directly across the room to you.

Touchstone Example 20-5-1, at the end of this chapter, illustrates how to use what you learned in this section.

20-6 The Distribution of Molecular Speeds

The root-mean-square speed $|\vec{v}_{rms}|$ gives us a general idea of molecular speeds in a gas at a given temperature. We often want to know more. For example, what fraction of the molecules have speeds greater than the rms value? Greater than twice the rms value? To answer such questions, we need to know how the possible values of speed are distributed among the molecules. Figure 20-7*a* shows this distribution for oxygen molecules at room temperature (T = 300 K); Fig. 20-7*b* compares it with the distribution at $T = 80\text{ K}$.

In 1852, Scottish physicist James Clerk Maxwell first solved the problem of finding the speed distribution of gas molecules. His result, known as **Maxwell's speed distribution law,** is

$$f(|\vec{v}|) = 4\pi N\left(\frac{m}{2\pi k_B T}\right)^{3/2} v^2 e^{-mv^2/2k_BT}. \qquad (20\text{-}25)$$

Here $|\vec{v}|$ is the molecular speed, T is the gas temperature, m is the mass of a single gas molecule, and k_B is Boltzmann's constant. It is this equation that is plotted in Fig. 20-7*a,b*. The quantity $f(|\vec{v}|)$ in Eq. 20-25 and Fig. 20-7 is a *probability distribution function*: For any speed $|\vec{v}|$, the product $f(|\vec{v}|)d|\vec{v}|$ (a dimensionless quantity) is the fraction of molecules whose speeds lie in the interval of width dv centered on speed $|\vec{v}|$.

As Fig. 20-7*a* shows, this fraction is equal to the area of a strip with height $f(|\vec{v}|)$ and width $d|\vec{v}|$. The total area under the distribution curve corresponds to the fraction of the molecules whose speeds lie between zero and infinity. All molecules fall into this category, so the value of this total area is unity; that is,

$$\int_0^\infty f(|\vec{v}|)\; d|\vec{v}| = 1. \qquad (20\text{-}26)$$

The fraction of molecules with speeds in an interval of, say, $|\vec{v}_1|$ to $|\vec{v}_2|$ is then

$$\text{fraction} = \int_{|\vec{v}_1|}^{|\vec{v}_2|} f(|\vec{v}|)\; d|\vec{v}|. \qquad (20\text{-}27)$$

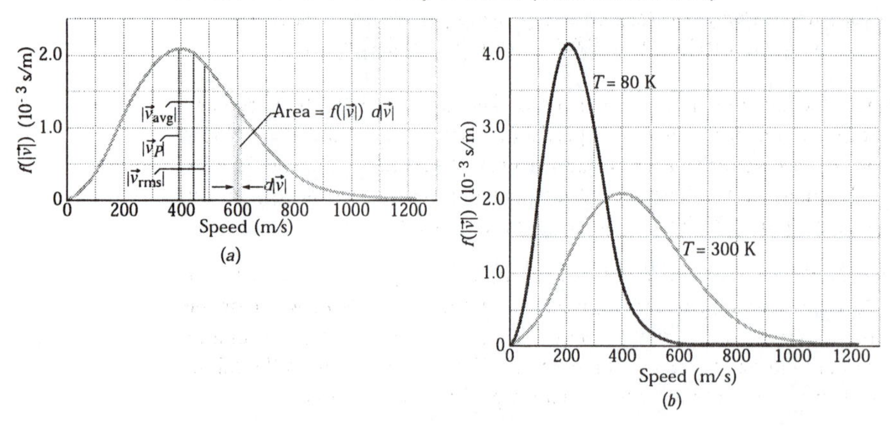

Fig. 20-7: (*a*) The Maxwell speed distribution for oxygen molecules at $T = 300$ K. The three characteristic speeds are marked. (*b*) The curves for 300 K and 80 K. Note that the molecules move more slowly at the lower temperature. Because these are probability distributions, the area under each curve has a numerical value of unity.

Average, RMS, and Most Probable Speeds

In principle, we can find the **average speed** $|\vec{v}_{avg}|$ of the molecules in a gas with the following procedure: We *weight* each value of $|\vec{v}|$ in the distribution; that is, we multiply it by the fraction $f(|\vec{v}|)d|\vec{v}|$ of molecules with speeds in a differential interval $d|\vec{v}|$ centered on $|\vec{v}|$. Then we add up all these values of $|\vec{v}|\,f(|\vec{v}|)\;d|\vec{v}|$. The result is $|\vec{v}_{avg}|$. In practice, we do all this by evaluating

$$|\vec{v}_{avg}| = \int_0^\infty |\vec{v}|\,f(|\vec{v}|)d|\vec{v}|. \qquad (20\text{-}28)$$

Substituting for $f(|\vec{v}|)$ from Eq. 20-25 and using generic integral 20 from the list of integrals in Appendix E, we find

$$|\vec{v}_{avg}| = \sqrt{\frac{8k_BT}{\pi m}} \quad \text{(average speed).} \qquad (20\text{-}29)$$

Similarly, we can find the average of the square of the speeds $(v^2)_{avg}$ with

$$(v^2)_{avg} = \int_0^\infty v^2 f(|\vec{v}|)\;d|\vec{v}|. \qquad (20\text{-}30)$$

Substituting for $f(|\vec{v}|)$ from Eq. 20-26 and using generic integral 16 from the list of integrals in Appendix E, we find

$$(v^2)_{avg} = \frac{3k_BT}{m}. \qquad (20\text{-}31)$$

The square root of $(v^2)_{avg}$ is the **root-mean-square speed** $|\vec{v}_{rms}|$. Thus,

$$|\vec{v}_{rms}| = \sqrt{\frac{3k_BT}{m}} \quad \text{(rms speed),} \qquad (20\text{-}32)$$

which agrees with Eq. 20-20.

The **most probable speed** $|\vec{v}_P|$ is the speed at which $f(|\vec{v}|)$ is maximum (see Fig. 20-7*a*). To calculate $|\vec{v}_P|$, we set $df/d|\vec{v}| = 0$ (the slope of the curve in Fig. 20-7*a* is zero at the maximum of the curve) and then solve for $|\vec{v}|$. Doing so, we find

$$|\vec{v}_P| = \sqrt{\frac{2k_BT}{m}} \quad \text{(most probable speed).} \qquad (20\text{-}33)$$

What is the relationship between the most probable speed, the average speed and the rms speed of a molecule? The relationship is fixed.

➤The most probable speed v_P is always less than the average speed $|\vec{v}_{avg}|$ which in turn is less than the rms speed $|\vec{v}_{rms}|$. More specifically, $|\vec{v}_P| = 0.82|\vec{v}_{rms}|$ and $|\vec{v}_{avg}| = 0.92|\vec{v}_{rms}|$.

This is consistent with the idea that a molecule is more likely to have speed $|\vec{v}_P|$ than any other speed, but some molecules will have speeds that are many times $|\vec{v}_P|$. These molecules lie in the *high-speed tail* of a distribution curve like that in Fig. 20-7*a*. We should be thankful for these few, higher speed molecules because they make possible both rain and sunshine (without which we could not exist). We next see why.

Rain: The speed distribution of water molecules in, say, a pond at summertime temperatures can be represented by a curve similar to that of Fig. 20-7*a*. Most of the molecules do not have nearly enough kinetic energy to escape from the water through its surface. However, small numbers of very fast molecules with speeds far out in the tail of the curve can do so. It is these water molecules that evaporate, making clouds and rain a possibility.

As the fast water molecules leave the surface, carrying energy with them, the temperature of the remaining water is maintained by heat transfer from the surroundings. Other fast molecules—produced in particularly favorable collisions—quickly take the place of those that have left, and the speed distribution is maintained.

Sunshine: Let the distribution curve of Fig. 20-7*a* now refer to protons in the core of the Sun. The Sun's energy is supplied by a nuclear fusion process that starts with the merging of two protons. However, protons repel each other because of their electrical charges, and protons of average speed do not have enough kinetic energy to overcome the repulsion and get close enough to merge. Very fast protons with speeds in the tail of the distribution curve can do so, however, and thus the Sun can shine.

20-7 The Molar Specific Heats of an Ideal Gas

Up to now, we have taken the specific heat of a substance as a quantity to be measured. But now, with the kinetic theory of gases, we know something about the structure of matter and where its energy is stored. With this additional information, we can actually calculate and make predictions about what we expect the specific heats of different kinds of gases to be. If we compare our predictions based on kinetic theory to experimental measurements, we get some good agreement and also some surprises. The surprises are among the first hints that the laws of matter at the atomic level are not just Newton's laws scaled down. In other words, we begin to notice that atoms aren't just little billiard balls but something different from any macroscopic object that we have experience with.

In order to explore this idea, we derive here (from molecular considerations) an expression for the internal energy E_{int} of an ideal gas. In other words, we find an expression for the energy associated with the random motions of the atoms or molecules in the gas. We shall then use that expression to derive the molar specific heats of an ideal gas.

Internal Energy E_{int}

Let us first assume that our ideal gas is a *monatomic gas* (which has individual atoms rather than molecules), such as helium, neon, or argon. Let us also assume that the internal energy E_{int} of our ideal gas is simply the sum of the translational kinetic energies of its atoms.

The average translational kinetic energy of a single atom depends only on the gas temperature and is given by Eq. 20-22 as $K_{avg} = \frac{3}{2}kT$. A sample of n moles of such a gas contains nN_A atoms. The internal energy E_{int} of the sample is then

$$E_{int} = (nN_A)K_{avg} = (nN_A)(\tfrac{3}{2}k_B T). \tag{20-34}$$

Using Eq. 20-6 $(k_B = R/N_A)$, we can rewrite this as

$$E_{int} = \tfrac{3}{2} nRT \qquad \text{(monatomic ideal gas).} \qquad (20\text{-}35)$$

Thus,

➤The internal energy E_{int} of an ideal gas is a function of the gas temperature *only*; it does not depend on any other variable.

With Eq. 20-35 in hand, we are now able to derive an expression for the molar specific heat of an ideal gas. Actually, we shall derive two expressions. One is for the case in which the volume of the gas remains constant as energy is transferred to or from it as heat. The other is for the case in which the pressure of the gas remains constant as energy is transferred to or from it as heat. The symbols for these two molar specific heats are C_V and C_P, respectively. (By convention, the capital letter C is used in both cases, even though C_V and C_P represent types of specific heat and not heat capacities.)

Molar Specific Heat at Constant Volume

Figure 20-8*a* shows n moles of an ideal gas at pressure P and temperature T, confined to a cylinder of fixed volume V. This *initial state i* of the gas is marked on the P-V diagram of Fig. 20-8*b*. Suppose that you add a small amount of energy to the gas as heat Q by slowly turning up the temperature of the thermal reservoir. The gas temperature rises a small amount to $T + \Delta T$, and its pressure rises to $P + \Delta P$, bringing the gas to *final state f*.

In such experiments, we would find that the heat Q is related to the temperature change ΔT by

$$Q = nC_V \, \Delta T \qquad \text{(constant volume),} \qquad (20\text{-}36)$$

where C_V is a constant called the **molar specific heat at constant volume.** Substituting this expression for Q into the first law of thermodynamics as given by Eq. 19-12 ($E_{int} = Q - W$) yields

$$\Delta E_{int} = nC_V \, \Delta T - W. \qquad (20\text{-}37)$$

With the volume held constant, the gas cannot expand and thus cannot do any work. Therefore, $W = 0$, and Eq. 20-37 gives us

$$C_V = \frac{\Delta E_{int}}{n \, \Delta T}. \qquad (20\text{-}38)$$

From Eq. 20-35 we know that $E_{int} = \frac{3}{2} nRT$, so the change in internal energy must be

$$E_{int} = \tfrac{3}{2} nR \quad T. \qquad (20\text{-}39)$$

Substituting this result into Eq. 20-38 yields

$$C_V = \tfrac{3}{2} R = 12.5 \text{ J/mol} \cdot \text{K} \qquad \text{(monatomic gas).} \qquad (20\text{-}40)$$

As Table 20-2 shows, this prediction of the kinetic theory (for ideal gases) agrees very well with experiment for real monatomic gases, the case that we have assumed. The (predicted and) experimental values of C_V for *diatomic gases* (which have molecules with two atoms) and *polyatomic gases* (which have molecules with more than two atoms) are greater than those for monatomic gases. Hence, our theory fails to correctly explain anything but the case of single atom molecules. The reasons for this will be suggested in Section 20-8.

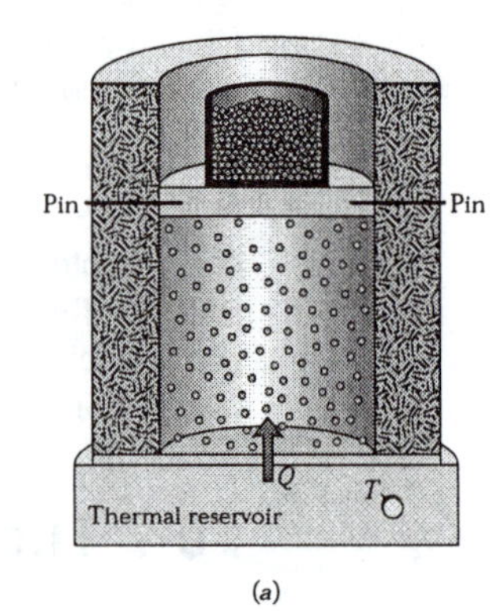

(a)

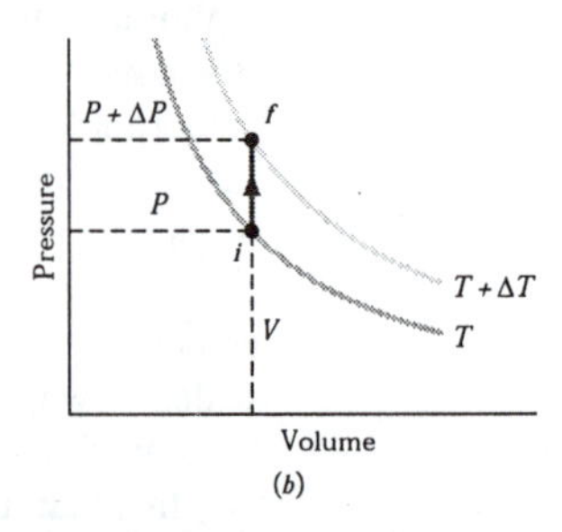

(b)

Fig. 20-8: (*a*) The temperature of an ideal gas is raised from T to $T + \Delta T$ in a constant-volume process. Heat is added, but no work is done. (*b*) The process on a P-V diagram.

TABLE 20-2: Molar Specific Heats

Molecule	Example	C_V (J/mol · K)	
Monatomic	Ideal		$\frac{3}{2}R = 12.5$
	Real	He	12.5
		Ar	12.6
Diatomic	Ideal		$\frac{5}{2}R = 20.8$
	Real	N_2	20.7
		O_2	20.8
Polyatomic	Ideal		$3R = 24.9$
	Real	NH_4	29.0
		CO_2	29.7

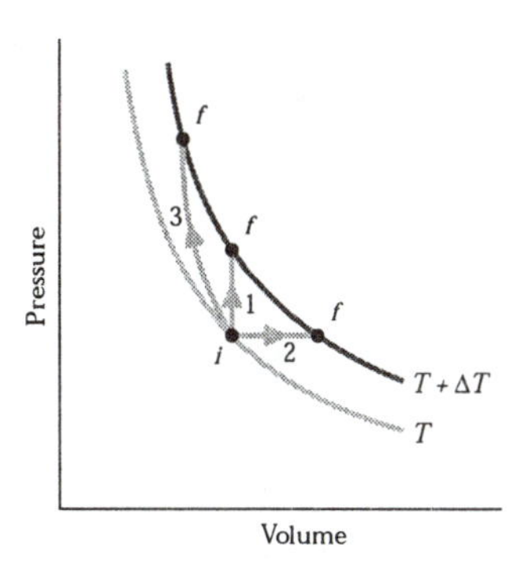

Fig. 20-9: Three paths representing three different processes that take an ideal gas from an initial state i at temperature T to some final state f at temperature $T + \Delta T$. The change ΔE_{int} in the internal energy of the gas is the same for these three processes and for any others that result in the same change of temperature.

We can now generalize Eq. 20-35 for the internal energy of any ideal gas by substituting C_V for $\frac{3}{2}R$; we get

$$E_{\text{int}} = nC_V T \qquad \text{(any ideal gas).} \qquad (20\text{-}41)$$

This equation applies not only to an ideal monatomic gas but also to diatomic and polyatomic ideal gases, provided the appropriate value of C_V is used. Just as with Eq. 20-36, we see that the internal energy of a gas depends on the temperature of the gas but not on its pressure or density.

When an ideal gas that is confined to a container undergoes a temperature change ΔT, then from either Eq. 20-38 or Eq. 20-41 we can write the resulting change in its internal energy as

$$\Delta E_{\text{int}} = nC_V\,\Delta T \qquad \text{(ideal gas, any process).} \qquad (20\text{-}42)$$

This equation tells us:

➤A change in the internal energy E_{int} of a confined ideal gas depends on the change in the gas temperature only; it does *not* depend on what type of process produces the change in the temperature.

As examples, consider the three paths between the two isotherms in the P-V diagram of Fig. 20-9. Path 1 represents a constant-volume process. Path 2 represents a constant-pressure process (that we are about to examine). Path 3 represents a process in which no heat is exchanged with the system's environment (we discuss this in Section 20-11). Although the values of heat Q and work W associated with these three paths differ, as do P_f and V_f, the values of ΔE_{int} associated with the three paths are identical and are all given by Eq. 20-42, because they all involve the same temperature change ΔT. Therefore, no matter what path is actually taken between T and $T + \Delta T$, we can *always* use path 1 and Eq. 20-42 to compute ΔE_{int} easily.

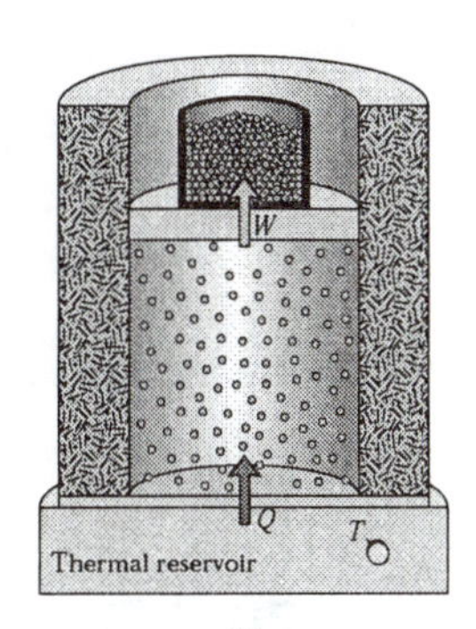

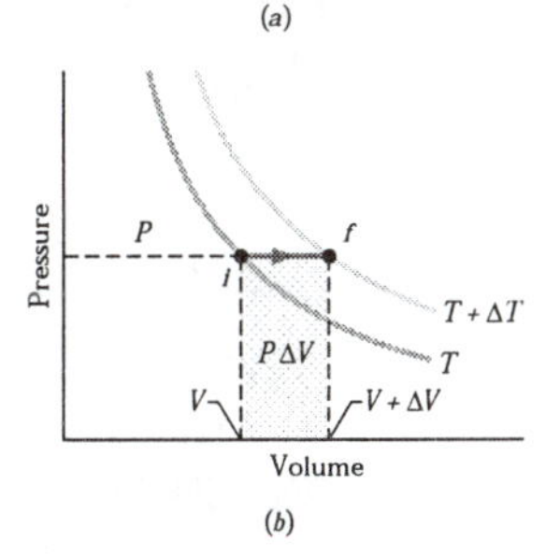

Fig. 20-10: (a) The temperature of an ideal gas is raised from T to $T + \Delta T$ in a constant-pressure process. Heat is added and work is done in lifting the loaded piston. (b) The process on a P-V diagram. The work $P\,\Delta V$ is given by the shaded area.

Molar Specific Heat at Constant Pressure

We now assume that the temperature of the ideal gas is increased by the same small amount ΔT as previously, but that the necessary energy (heat Q) is added with the gas under constant pressure. An experiment for doing this is shown in Fig. 20-10a; the P-V diagram for the process is plotted in Fig. 20-10b. From such experiments we find that the heat Q is related to the temperature change ΔT by

$$Q = nC_P\,\Delta T \qquad \text{(constant pressure),} \qquad (20\text{-}43)$$

where C_P is a constant called the **molar specific heat at constant pressure.** This C_P is *greater* than the molar specific heat at constant volume C_V, because energy must now be supplied not only to raise the temperature of the gas but also for the gas to do work—that is, to lift the weighted piston of Fig. 20-10a.

To relate molar specific heats C_P and C_V, we start with the first law of thermodynamics (Eq. 19-12):

$$\Delta E_{\text{int}} = Q - W. \tag{20-44}$$

We next replace each term in Eq. 20-44. For ΔE_{int}, we substitute from Eq. 20-42. For Q, we substitute from Eq. 20-43. To replace W, we first note that since the pressure remains constant, Eq. 20-15 tells us that $W = P\Delta V$. Then we note that, using the ideal gas equation $(PV = nRT)$, we can write

$$W = P\,\Delta V = nR\,\Delta T. \tag{20-45}$$

Making these substitutions in Eq. 20-44, and then dividing through by $n\,\Delta T$, we find

$$C_V = C_P - R$$

and then

$$C_P = C_V + R. \tag{20-46}$$

This prediction of kinetic theory agrees well with experiment, not only for monatomic gases but for gases in general, as long as their density is low enough so that we may treat them as ideal.

READING EXERCISE 20-4: The figure here shows five paths traversed by a gas on a *P-V* diagram. Rank the paths according to the change in internal energy of the gas, greatest first.

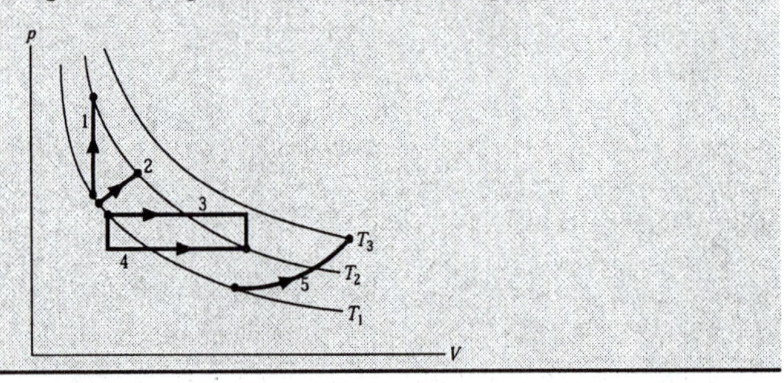

Touchstone Example 20-7-1, at the end of this chapter, illustrates how to use what you learned in this section.

TE

20-8 Degrees of Freedom and Molar Specific Heats

As Table 20-2 shows, the prediction that $C_V = \frac{3}{2}R$ agrees with experiment for monatomic gases but fails for diatomic and polyatomic gases. Let us try to explain the discrepancy by considering the possibility that molecules with more than one atom can store internal energy in forms other than *translational* motion.

Figure 20-11 shows common models of helium (a *monatomic* molecule, containing a single atom), oxygen (a *diatomic* molecule, containing two atoms), and methane (a *polyatomic* molecule). From such models, we would assume that all three types of molecules can have translational motions (say, moving left-right and up-down) and rotational motions (spinning about an axis like a top). In addition, we would assume that the diatomic and polyatomic molecules can have oscillatory motions, with the atoms oscillating slightly toward and away from one another, as if attached to opposite ends of a spring.

To keep account of the various ways in which energy can be stored in a gas, James Clerk Maxwell introduced the theorem of the **equipartition of energy:**

➤Every kind of molecule has a certain number f of *degrees of freedom*, which are independent ways in which the molecule can store energy. Each such degree of freedom has associated with it—on average—an energy of $\frac{1}{2}k_B T$ per molecule (or $\frac{1}{2}RT$ per mole).

Let us apply the theorem to the translational and rotational motions of the molecules in Fig. 20-11. (We discuss oscillatory motion in the next section.) For the translational motion, superimpose an *xyz* coordinate system on any gas. The molecules will, in general, have velocity components along all three axes. Thus, gas molecules of all types have three degrees of translational freedom (three ways to move in translation) and, on average, an associated energy of $3(\frac{1}{2}k_BT)$ per molecule.

For the rotational motion, imagine the origin of our *xyz* coordinate system at the center of each molecule in Fig. 20-11. In a gas, each molecule should be able to rotate with an angular velocity component along each of the three axes, so each gas should have three degrees of rotational freedom and, on average, an additional energy of $3(\frac{1}{2}k_BT)$ per molecule. *However,* experiment shows this is true only for the polyatomic molecules.

A possible solution to this dilemma is that rotations about an axis of symmetry don't count as a degree of freedom. For example, as seen is Fig. 20-11, a single atom molecule is symmetric about all three (mutually perpendicular) axes through the molecule. Hence, according to our proposed solution, these rotations are not additional degrees of freedom. A diatomic molecule is symmetric about only one axis (the axis through the center of both atoms). Accordingly, a diatomic molecule would have two rather than three degrees of freedom associated with rotation of the molecule.

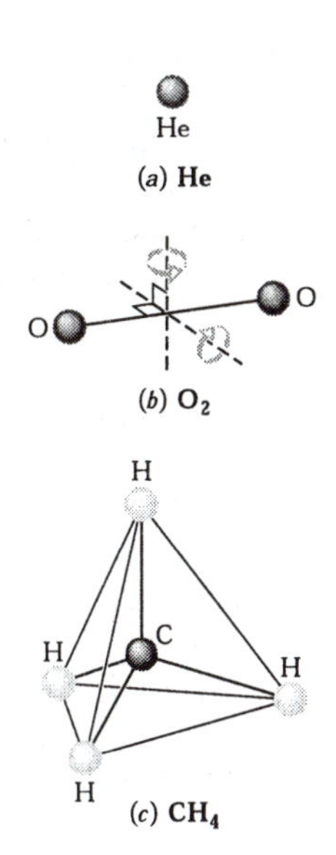

Fig. 20-11: Models of molecules as used in kinetic theory: (*a*) helium, a typical monatomic molecule; (*b*) oxygen, a typical diatomic molecule; and (*c*) methane, a typical polyatomic molecule. The spheres represent atoms, and the lines between them represent bonds. Two rotation axes are shown for the oxygen molecule.

It appears that modifying our theory in this manner brings us more in alignment with the experimental results. However, one should ask what reasoning (other than experimental evidence) supports this modification of the theory. One thing is clear. It is impossible to tell if a molecule is rotating about an axis of symmetry. Unlike a baseball (which has stitches or other marks) molecules have no characteristics that allow us to sense the rotation about an axis of symmetry. Although classical physics gives us no real foundation for ignoring the motion simply because it is indistinguishable from no motion at all, this is what quantum theory would suggest.

So, according to our new model, a monatomic molecule has zero degrees of freedom associated with rotation. A diatomic molecule has two degrees of freedom associated with rotations about the two axes perpendicular to the line connecting the atoms (the axes are shown in Fig. 20-11*b*) but no degree of freedom for rotation about that line itself. Therefore, a diatomic molecule can have a rotational energy of only $2(\frac{1}{2}k_BT)$ per molecule. A polyatomic molecule has a full three degrees of freedom associated with rotational motion.

To extend our analysis of molar specific heats (C_p and C_V, in Section 20-7) to ideal diatomic and polyatomic gases, it is necessary to retrace the derivations of that analysis in detail. First, we replace Eq. 20-35 $(E_{\text{int}} = \frac{3}{2}nRT)$ with $E_{\text{int}} = (f/2)nRT$, where f is the number of degrees of freedom listed in Table 20-3. Doing so leads to the prediction

$$C_V = \left(\frac{f}{2}\right)R = 4.16f \ \ \text{J/mol}\cdot\text{K}, \tag{20-47}$$

which agrees—as it must—with Eq. 20-40 for monatomic gases $(f = 3)$. As Table 20-3 shows, this prediction also agrees with experiment for diatomic gases $(f = 5)$, but it is too low for polyatomic gases. **Note**: The symbol f used here to denote degrees of freedom should not be confused with $f(|\vec{v}|)$ used to describe the velocity distribution function for molecules.

TABLE 20-3: Degrees of Freedom for Various Molecules

		Degrees of Freedom			Predicted Molar Specific Heats	
Molecule	Example	Translational	Rotational	Total (f)	C_V (Eq. 20-47)	$C_p = C_V + R$
Monatomic	He	3	0	3	$\frac{3}{2}R$	$\frac{5}{2}R$
Diatomic	O_2	3	2	5	$\frac{5}{2}R$	$\frac{7}{2}R$
Polyatomic	CH_4	3	3	6	$3R$	$4R$

Touchstone Example 20-8-1, at the end of this chapter, illustrates how to use what you learned in this section.

TE

20-9 A Hint of Quantum Theory

We can improve the agreement of kinetic theory with experiment by including the oscillations of the atoms in a gas of diatomic or polyatomic molecules. For example, the two atoms in the O_2 molecule of Fig. 20-11*b* can oscillate toward and away from each other, with the interconnecting bond acting like a spring. However, experiment shows that such oscillations occur only at relatively high temperatures of the gas—the motion is "turned on" only when the gas molecules have relatively large energies. Rotational motion is also subject to such "turning on," but at a lower temperature.

Figure 20-12 is of help in seeing this turning on of rotational motion and oscillatory motion. The ratio C_V/R for diatomic hydrogen gas (H_2) is plotted there against temperature, with the temperature scale logarithmic to cover several orders of magnitude. Below about 80 K, we find that $C_V/R = 1.5$. This result implies that only the three translational degrees of freedom of hydrogen are involved in the specific heat.

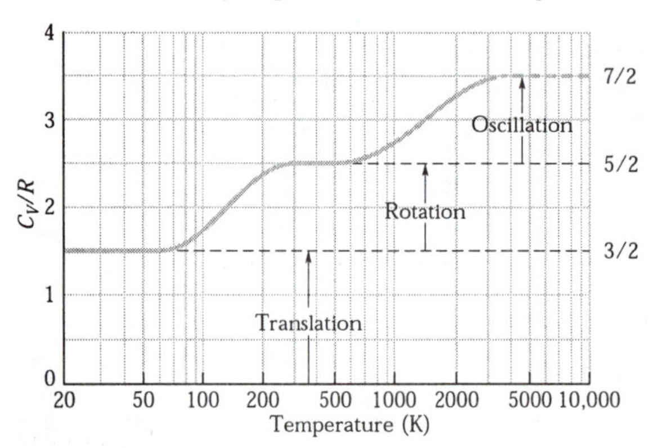

Fig. 20-12: A plot of C_V/R versus temperature for (diatomic) hydrogen gas. Because rotational and oscillatory motions begin at certain energies, only translation is possible at very low temperatures. As the temperature increases, rotational motion can begin. At still higher temperatures, oscillatory motion can begin.

As the temperature increases, the value of C_V/R gradually increases to 2.5, implying that two additional degrees of freedom have become involved. Quantum theory shows that these two degrees of freedom are associated with the rotational motion of the hydrogen molecules and that this motion requires a certain minimum amount of energy. At very low temperatures (below 80 K), the molecules do not have enough energy to rotate. As the temperature increases from 80 K, first a few molecules and then more and more obtain enough energy to rotate, and C_V/R increases, until all of them are rotating and $C_V/R = 2.5$.

Similarly, quantum theory shows that oscillatory motion of the molecules requires a certain (higher) minimum amount of energy. This minimum amount is not met until the molecules reach a temperature of about 1000 K, as shown in Fig. 20-12. As the temperature increases beyond 1000 K, the number of molecules with enough energy to oscillate increases, and C_V/R increases, until all of them are oscillating and $C_V/R = 3.5$. (In Fig. 20-12, the plotted curve stops at 3200 K because at that temperature, the atoms of a hydrogen molecule oscillate so much that they overwhelm their bond, and the molecule then *dissociates* into two separate atoms.)

The observed fact that rotational degrees of freedom are not excited until sufficiently high temperatures are reached implies that rotational kinetic energy is not a continuous function of angular velocity. Instead, a discrete, quantized energy level must be attained before rotation is excited. This discreteness of energy levels is a hallmark of quantum mechanical behavior. It is interesting to note that some of the issues discussed in this chapter are the first examples (with many more to come) that macroscopic properties of matter, which are easily measured in the laboratory, depend critically on (and provide strong evidence for) the quantum theory we will develop later.

The compatibility between microscopic theory and macroscopic observations when coupled with quantum theory and other phenomena in physics and chemistry provided additional support for the theory that matter is composed of atoms and molecules.

20-10 The Adiabatic Expansion of an Ideal Gas

We saw in Section 18-2 that sound waves are propagated through air and other gases as a series of compressions and expansions; these variations in the transmission medium take place so rapidly that there is no time for energy to be transferred from one part of the medium to another as heat. As we saw in Section 19-8, a process for which $Q = 0$ is an *adiabatic process*. We can ensure that $Q = 0$ either by carrying out the process very quickly (as in sound waves) or by doing it (at any rate) in a well-insulated container. Let us see what the kinetic theory has to say about adiabatic processes.

Figure 20-13*a* shows our usual insulated cylinder, now containing an ideal gas and resting on an insulating stand. By removing mass from the piston, we can allow the gas to expand adiabatically. As the volume increases, both the pressure and the temperature drop. We shall prove next that the relation between the pressure and the volume during such an adiabatic process is

$$PV^{\gamma} = \text{a constant} \qquad \text{(adiabatic process)}, \quad (20\text{-}48)$$

in which $\gamma = C_P/C_V$, the ratio of the molar specific heats for the gas. On a *P-V* diagram such as that in Fig. 20-13*b*, the process occurs along a line (called an *adiabat*) that has the equation $P = (\text{a constant})/V^{\gamma}$. Since the gas goes from an initial state *i* to a final state *f*, we can rewrite Eq. 20-48 as

$$P_i V_i^{\gamma} = P_f V_f^{\gamma} \qquad \text{(adiabatic process)}. \quad (20\text{-}49)$$

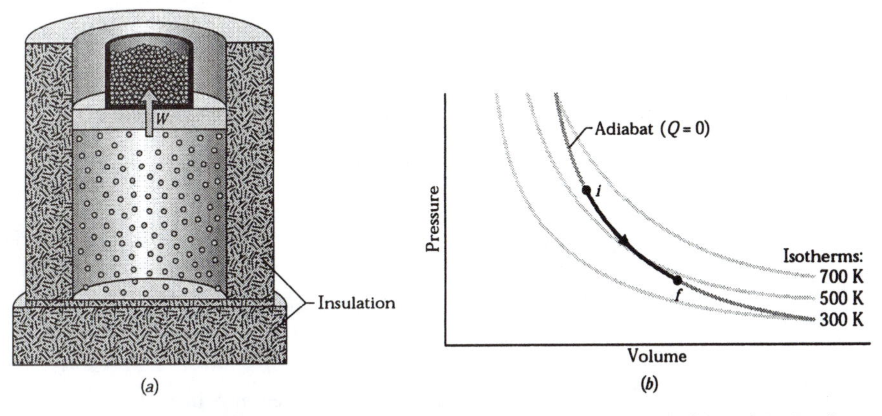

Fig. 20-13: (*a*) The volume of an ideal gas is increased by removing weight from the piston. The process is adiabatic ($Q = 0$). (*b*) The process proceeds from *i* to *f* along an adiabat on a *P-V* diagram.

We can also write an equation for an adiabatic process in terms of *T* and *V*. To do so, we use the ideal gas equation ($PV = nRT$) to eliminate *P* from Eq. 20-48, finding

$$\left(\frac{nRT}{V}\right) V^{\gamma} = \text{a constant}.$$

Because *n* and *R* are constants, we can rewrite this in the alternative form

$$TV^{\gamma-1} = \text{a constant} \qquad \text{(adiabatic process)}, \quad (20\text{-}50)$$

in which the constant is different from that in Eq. 20-48. When the gas goes from an initial state *i* to a final state *f*, we can rewrite Eq. 20-50 as

$$T_i V_i^{\gamma-1} = T_f V_f^{\gamma-1} \qquad \text{(adiabatic process)}. \quad (20\text{-}51)$$

We can now answer the question that opens this chapter. At the top of an unopened carbonated drink, there is a gas of carbon dioxide and water vapor. Because the pressure of the gas is greater than atmospheric pressure, the gas expands out into the atmosphere when the container is opened. Thus, the gas increases its volume, but that means it must do work to push against the atmosphere. Because the expansion is so rapid, it is adiabatic and the only source of energy for the work is the internal energy of the gas. Because the internal energy decreases, the temperature of the gas must also decrease, which can cause the water vapor in the gas to condense into tiny drops, forming the fog. (Note that Eq. 20-51 also tells us that the temperature must decrease during an adiabatic expansion: Since V_f is greater than V_i, then T_f must be less than T_i.)

Proof of Eq. 20-48

Suppose that you remove some shot from the piston of Fig. 20-13*a*, allowing the ideal gas to push the piston and the remaining shot upward and thus to increase the volume by a differential amount dV. Since the volume change is tiny, we may assume that the pressure P of the gas on the piston is constant during the change. This assumption allows us to say that the work dW done by the gas during the volume increase is equal to $P\,dV$. From Eq. 19-13, the first law of thermodynamics can then be written as

$$dE_{\text{int}} = Q - P\,dV. \tag{20-52}$$

Since the gas is thermally insulated (and thus the expansion is adiabatic), we substitute 0 for Q. Then we use Eq. 20-42 to substitute $nC_V\,dT$ for dE_{int}. With these substitutions, and after some rearranging, we have

$$n\,dT = -\left(\frac{P}{C_V}\right)dV. \tag{20-53}$$

Now from the ideal gas law $(PV = nRT)$ we have

$$PdV + V\,dP = nR\,dT. \tag{20-54}$$

Replacing R with its equal, $C_P - C_V$, in Eq. 20-54 yields

$$n\,dT = \frac{P\,dV + V\,dP}{C_p - C_V}. \tag{20-55}$$

Equating Eqs. 20-53 and 20-55 and rearranging then give

$$\frac{dP}{P} + \left(\frac{C_P}{C_V}\right)\frac{dV}{V} = 0.$$

Replacing the ratio of the molar specific heats with γ and integrating (see integral 5 in Appendix E) yield

$$\ln P + \gamma\,\ln V = \text{a constant}.$$

Rewriting the left side as $\ln PV^\gamma$ and then taking the antilog of both sides, we find

$$PV^\gamma = \text{a constant}, \tag{20-56}$$

which is what we set out to prove.

Free Expansions

Recall from Section 19-8 that a free expansion of a gas is an adiabatic process that involves no work done on or by the gas, and no change in the internal energy of the gas. A free expansion is thus quite different from the type of adiabatic process described by Eqs. 20-48 through 20-56, in which work is done and the internal energy changes. Those equations then do *not* apply to a free expansion, even though such an expansion is adiabatic.

Also recall that in a free expansion, a gas is in equilibrium only at its initial and final points; thus, we can plot only those points, but not the expansion itself, on a P-V diagram.

In addition, because $\Delta E_{\text{int}} = 0$, the temperature of the final state must be that of the initial state. Thus, the initial and final points on a *P*-*V* diagram must be on the same isotherm, and instead of Eq. 20-51 we have

$$T_i = T_f \qquad \text{(free expansion).} \qquad (20\text{-}57)$$

If we next assume that the gas is ideal (so that $PV = nRT$), because there is no change in temperature, there can be no change in the product *PV*. Thus, instead of Eq. 20-48 a free expansion involves the relation

$$P_i V_i = P_f V_f \qquad \text{(free expansion).} \qquad (20\text{-}58)$$

Touchstone Example 20-10-1, at the end of this chapter, illustrates how to use what you learned in this section.

TE

Touchstone Example 20-2-1

A cylinder contains 12 L of oxygen at 20°C and 15 atm. The temperature is raised to 35°C, and the volume is reduced to 8.5 L. What is the final pressure of the gas in atmospheres? Assume that the gas is ideal.

SOLUTION: The **Key Idea** here is that, because the gas is ideal, its pressure, volume, temperature, and number of moles are related by the ideal gas law, both in the initial state i and in the final state f (after the changes). Thus, from Eq. 20-8 we can write

$$P_iV_i = nRT_i \quad \text{and} \quad P_fV_f = nRT_f.$$

Dividing the second equation by the first equation and solving for P_f yields

$$P_f = \frac{P_iT_fV_i}{T_iV_f}. \qquad \text{(TE20-1)}$$

Note here that if we converted the given initial and final volumes from liters to the proper units of cubic meters, the multiplying conversion factors would cancel out of Eq. TE 20-1. The same would be true for conversion factors that convert the pressures from atmospheres to the proper pascals. However, to convert the given temperatures to kelvins requires the addition of an amount that would not cancel and thus must be included. Hence, we must write

$$T_i = (273 + 20)\text{ K} = 293\text{ K}$$

and

$$T_f = (273 + 35)\text{ K} = 308\text{ K}.$$

Inserting the given data into Eq. TE20-1 then yields

$$P_f = \frac{(15\text{ atm})(308\text{ K})(12\text{ L})}{(293\text{ K})(8.5\text{ L})} = 22\text{ atm}. \qquad \text{(Answer)}$$

Touchstone Example 20-3-1

One mole of oxygen (assume it to be an ideal gas) expands at a constant temperature T of 310 K from an initial volume V_i of 12 L to a final volume V_f of 19 L. How much work is done by the gas during the expansion?

SOLUTION: The **Key Idea** is this: Generally we find the work by integrating the gas pressure with respect to the gas volume, using Eq. 20-10. However, because the gas here is ideal and the expansion is isothermal, that integration leads to Eq. 20-13. Therefore, we can write

$$W = nRT \ln \frac{V_f}{V_i}$$

$$= (1\text{ mol})(8.31\text{ J/mol}\cdot\text{K})(310\text{ K}) \ln \frac{19\text{ L}}{12\text{ L}}$$

$$= 1180\text{ J}. \qquad \text{(Answer)}$$

The expansion is graphed in the P-V diagram of Fig. TE20-1. The work done by the gas during the expansion is represented by the area beneath the curve *if*.

You can show that if the expansion is now reversed, with the gas undergoing an isothermal compression from 19 L to 12 L, the work done by the gas will be −1180 J. Thus, an external force would have to do 1180 J of work on the gas to compress it.

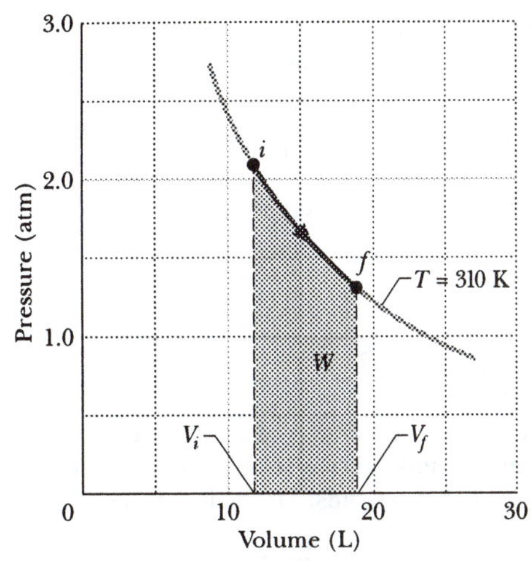

Fig. TE20-1 The shaded area represents the work done by 1 mol of oxygen in expanding from V_i to V_f at a constant temperature T of 310 K.

Touchstone Example 20-5-1

(a) What is the mean free path λ for oxygen molecules at temperature $T = 300$ K and pressure $P = 1.0$ atm? Assume that the molecular diameter is $d = 290$ pm and the gas is ideal.

SOLUTION: The **Key Idea** here is that each oxygen molecule moves among other *moving* oxygen molecules in a zigzag path due to the resulting collisions. Thus, we use Eq. 20-23 for the mean free path, for which we need the number of molecules per unit volume, N/V. Because we assume the gas is ideal, we can use the ideal gas law of Eq. 20-1 ($PV = Nk_BT$) to write $N/V = P/k_BT$. Substituting this into Eq. 20-23, we find

$$\lambda = \frac{1}{\sqrt{2}\pi d^2\, N/V} = \frac{k_B T}{\sqrt{2}\pi d^2\, P}$$

$$= \frac{(1.38 \times 10^{-23}\ \text{J/K})(300\ \text{K})}{\sqrt{2}\pi(2.9 \times 10^{-10}\ \text{m})^2(1.01 \times 10^5\ \text{Pa})}$$

$$= 1.1 \times 10^{-7}\ \text{m}. \qquad \text{(Answer)}$$

This is about 380 molecular diameters.

(b) Assume the average speed of the oxygen molecules is $|\vec{v}| = 450$ m/s. What is the average time Δt between successive collisions for any given molecule? At what rate does the molecule collide; that is, what is the frequency f of its collisions?

SOLUTION: To find the time Δt between collisions, we use this **Key Idea:** Between collisions, the molecule travels, on average, the mean free path λ at speed $|\vec{v}|$. Thus, the average time between collisions is

$$\Delta t = \frac{\text{distance}}{\text{speed}} = \frac{\lambda}{|\vec{v}|} = \frac{1.1 \times 10^{-7}\ \text{m}}{450\ \text{m/s}}$$

$$= 2.44 \times 10^{-10}\ \text{s} \approx 0.24\ \text{ns}. \qquad \text{(Answer)}$$

This tells us that, on average, any given oxygen molecule has less than a nanosecond between collisions.

To find the frequency f of the collisions, we use this **Key Idea:** The average rate or frequency at which the collisions occur is the inverse of the time Δt between collisions. Thus,

$$f = \frac{1}{\Delta t} = \frac{1}{2.44 \times 10^{-10}\ \text{s}} = 4.1 \times 10^{9}\ \text{s}^{-1}. \quad \text{(Answer)}$$

This tells us that, on average, any given oxygen molecule makes about 4 billion collisions per second.

Touchstone Example 20-7-1

A bubble of 5.00 mol of helium is submerged at a certain depth in liquid water when the water (and thus the helium) undergoes a temperature increase ΔT of 20.0 C° at constant pressure. As a result, the bubble expands. The helium is monatomic and ideal.

(a) How much energy is added as heat to the helium during the increase and expansion?

SOLUTION: One **Key Idea** here is that the heat Q is related to the temperature change ΔT by a molar specific heat of the gas. Because the pressure P is held constant during the addition of energy, we use the molar specific heat at constant pressure C_P and Eq. 20-43,

$$Q = nC_P\,\Delta T, \quad \text{(TE20-2)}$$

to find Q. To evaluate C_P we go to Eq. 20-46, which tells us that for any ideal gas, $C_P = C_V + R$. Then from Eq. 20-40, we know that for any *monatomic* gas (like the helium here), $C_V = \frac{3}{2}R$. Thus, Eq. TE20-2 gives us

$$\begin{aligned} Q &= n(C_V + R)\,\Delta T = n(\tfrac{3}{2}R + R)\,\Delta T = n(\tfrac{5}{2}R)\,\Delta T \\ &= (5.00\ \text{mol})(2.5)(8.31\ \text{J/mol}\cdot\text{K})(20.0\ \text{C}^\circ) \\ &= 2077.5\ \text{J} \approx 2080\ \text{J}. \quad \text{(Answer)} \end{aligned}$$

(b) What is the change ΔE_{int} in the internal energy of the helium during the temperature increase?

SOLUTION: Because the bubble expands, this is not a constant-volume process. However, the helium is nonetheless confined (to the bubble). Thus, a **Key Idea** here is that the change ΔE_{int} is the same as *would occur* in a constant-volume process with the same temperature change ΔT. We can easily find the constant-volume change ΔE_{int} with Eq. 20-42:

$$\begin{aligned} \Delta E_{int} &= nC_V\,\Delta T = n(\tfrac{3}{2}R)\,\Delta T \\ &= (5.00\ \text{mol})(1.5)(8.31\ \text{J/mol}\cdot\text{K})(20.0\ \text{C}^\circ) \\ &= 1246.5\ \text{J} \approx 1250\ \text{J}. \quad \text{(Answer)} \end{aligned}$$

(c) How much work W is done by the helium as it expands against the pressure of the surrounding water during the temperature increase?

SOLUTION: One **Key Idea** here is that the work done by *any* gas expanding against the pressure from its environment is given by Eq. 20-10, which tells us to integrate $P\,dV$. When the pressure is constant (as here), we can simplify that to $W = P\,\Delta V$. When the gas is *ideal* (as here), we can use the ideal gas law (Eq. 20-8) to write $P\,\Delta V = nR\,\Delta T$. We end up with

$$\begin{aligned} W &= nR\,\Delta T \\ &= (5.00\ \text{mol})(8.31\ \text{J/mol}\cdot\text{K})(20.0\ \text{C}^\circ) \\ &= 831\ \text{J}. \quad \text{(Answer)} \end{aligned}$$

Because we happen to know Q and ΔE_{int}, we can work this problem another way: The **Key Idea** now is that we can account for the energy changes of the gas with the first law of thermodynamics, writing

$$W = Q - \Delta E_{int} = 2077.5 \text{ J} - 1246.5 \text{ J}$$
$$= 831 \text{ J}. \qquad \text{(Answer)}$$

Note that during the temperature increase, only a portion (1250 J) of the energy (2080 J) that is transferred to the helium as heat goes to increasing the internal energy of the helium and thus the temperature of the helium. The rest (831 J) is transferred out of the helium as work that the helium does during the expansion. If the water were frozen, it would not allow that expansion. Then the same temperature increase of 20.0 C° would require only 1250 J of heat, because no work would be done by the helium.

Touchstone Example 20-8-1

A cabin of volume V is filled with air (which we consider to be an ideal diatomic gas) at an initial low temperature T_1. After you light a wood stove, the air temperature increases to T_2. What is the resulting change ΔE_{int} in the internal energy of the air in the cabin?

SOLUTION: As the air temperature increases, the air pressure P cannot change but must always be equal to the air pressure outside the room. The reason is that, because the room is not air-tight, the air is not confined. As the temperature increases, air molecules leave through various openings and thus the number of moles n of air in the room decreases. Thus, one **Key Idea** here is that we *cannot* use Eq. 20-42 ($\Delta E_{int} = nC_V \Delta T$) to find ΔE_{int}, because it requires constant n.

A second **Key Idea** is that we *can* relate the internal energy E_{int} at any instant to n and the temperature T with Eq. 20-41 ($E_{int} = nC_V T$). From that equation we can then write

$$\Delta E_{int} = \Delta(nC_V T) = C_V \Delta(nT),$$

Next, using Eq. 20-8 ($PV = nRT$), we can replace nT with PV/R, obtaining

$$\Delta E_{int} = C_V \Delta\left(\frac{PV}{R}\right). \qquad \text{(TE20-3)}$$

Now, because P, V, and R are all constants, Eq. TE20-3 yields

$$\Delta E_{int} = 0, \qquad \text{(Answer)}$$

even though the temperature changes.

Why does the cabin feel more comfortable at the higher temperature? There are at least two factors involved: (a) You exchange electromagnetic radiation (thermal radiation) with surfaces inside the room, and (2) you exchange energy with air molecules that collide with you. When the room temperature is increased, (1) the amount of thermal radiation emitted by the surfaces and absorbed by you is increased, and (2) the amount of energy you gain through the collisions of air molecules with you is increased.

Touchstone Example 20-10-1

In Touchstone Example 20-3-1, 1 mol of oxygen (assumed to be an ideal gas) expands isothermally (at 310 K) from an initial volume of 12 L to a final volume of 19 L.

(a) What would be the final temperature if the gas had expanded adiabatically to this same final volume? Oxygen (O_2) is diatomic and here has rotation but not oscillation.

SOLUTION: The **Key Ideas** here are

1. When a gas expands against the pressure of its environment, it must do work.
2. When the process is adiabatic (no energy is transferred as heat), then the energy required for the work can come only from the internal energy of the gas.
3. Because the internal energy decreases, the temperature T must also decrease.

We can relate the initial and final temperatures and volumes with Eq. 20-51:

$$T_i V_i^{\gamma-1} = T_f V_f^{\gamma-1}. \qquad \text{(TE20-4)}$$

Because the molecules are diatomic and have rotation but not oscillation, we can take the molar specific heats from Table 20-3. Thus,

$$\gamma = \frac{C_P}{C_V} = \frac{\frac{7}{2}R}{\frac{5}{2}R} = 1.40.$$

Solving Eq. TE20-4 for T_f and inserting known data then yield

$$T_f = \frac{T_i V_i^{\gamma-1}}{V_f^{\gamma-1}} = \frac{(310\ \text{K})(12\ \text{L})^{1.40-1}}{(19\ \text{L})^{1.40-1}}$$

$$= (310\ \text{K})(\tfrac{12}{19})^{0.40} = 258\ \text{K}. \qquad \text{(Answer)}$$

(b) What would be the final temperature and pressure if, instead, the gas had expanded freely to the new volume, from an initial pressure of 2.0 Pa?

SOLUTION: Here the **Key Idea** is that the temperature does not change in a free expansion:

$$T_f = T_i = 310\ \text{K}. \qquad \text{(Answer)}$$

We find the new pressure using Eq. 20-58, which gives us

$$P_f = P_i \frac{V_i}{V_f} = (2.0\ \text{Pa}) \frac{12\ \text{L}}{19\ \text{L}} = 1.3\ \text{Pa}. \qquad \text{(Answer)}$$

21 Entropy and the Second Law of Thermodynamics

An anonymous graffito on a wall of the Pecan Street Cafe in Austin, Texas, reads: "Time is God's way of keeping things from happening all at once." Time also has direction—some things happen in a certain sequence and could never happen on their own in a reverse sequence. As an example, an accidentally dropped egg splatters in a cup. The reverse process, a splattered egg re-forming into a whole egg and jumping up to an outstretched hand, will never happen on its own—but why not? Why can't that process be reversed, like a videotape run backward?

What in the world gives direction to time?

The answer is in this chapter.

21-1 Some One-Way Processes

Suppose you come indoors on a very cold day and wrap your cold hands around a warm mug of cocoa. Then your hands get warmer and the mug gets cooler. However, it never happens the other way around; that is, your cold hands never get still colder while the warm mug gets still warmer.

The system consisting of your hands and the mug is a *closed system*, one that is isolated from its environment. Here are some other one-way processes that occur in closed systems: (1) A crate sliding over an ordinary surface eventually stops—but you never see an initially stationary crate start to move all by itself. (2) If you drop a glob of putty, it falls to the floor—but an initially motionless glob of putty never leaps spontaneously into the air. (3) If you puncture a helium-filled balloon in a closed room, the helium gas spreads throughout the room—but the individual helium atoms will never clump up again into the shape of the balloon. We say that such one-way processes are **irreversible**, meaning that they cannot be reversed by means of only small changes in their environment.

Many chemical transformations are also irreversible. For example, when methane gas is burned, each methane molecule mixes with an oxygen molecule. Water vapor and carbon dioxide are given off as shown in the following chemical equation:

$$CH_4 + O_2 \rightarrow CO_2 + H_2O.$$

This combustion process is irreversible, since we don't find water and carbon dioxide spontaneously reacting to produce methane and oxygen gas.

The one-way character of such thermodynamic processes is so pervasive that we take it for granted. If these processes were to occur *spontaneously* (on their own) in the "wrong" direction, we would be astonished beyond belief. *Yet none of these "wrong-way" events would violate the law of conservation of energy.* In the cocoa mug example, that law would be obeyed even for a wrong-way transfer of energy as heat between hands and mug. It would be obeyed even if a stationary crate or a stationary glob of putty suddenly were to transfer some of its thermal energy to kinetic energy and begin to move. It would also be obeyed even if the helium atoms released from a balloon were, on their own, to clump together again.

Thus, changes in energy within a closed system do not set the direction of irreversible processes. Rather, that direction is set by another property that we shall discuss in this chapter—the *change in entropy* ΔS of the system. The change in entropy of a system is defined in the next section, but we can here state its central property, often called the *entropy postulate:*

➤If an irreversible process occurs in a *closed* system, the entropy S of the system always increases; it never decreases.

Entropy differs from energy in that it does *not* obey a conservation law. The energy of a closed system is conserved; it always remains constant. For irreversible processes, the *entropy* of a closed system always increases. Because of this property, the change in entropy is sometimes called "the arrow of time." For example, we associate the egg of our opening photograph, breaking irreversibly as it drops into a cup, with the forward direction of time and with an increase in entropy. The backward direction of time (a videotape run backward) would correspond to the broken egg re-forming into a whole egg and rising into the air. This backward process would result in an entropy decrease, so it never happens.

There are two equivalent ways to define the change in entropy of a system: (1) in terms of the system's temperature and the energy it gains or loses as heat, and (2) by counting the ways in which the atoms or molecules that make up the system can be arranged. We use the first approach in the next section, and the second in Section 21-7.

READING EXERCISE 21-1: Consider the irreversible process of dropping a glob of putty on the floor. Describe what energy transformations are taking place that allow us to believe that energy is conserved.

21-2 Change in Entropy

Let's approach this definition of *change in entropy* by looking again at a process that we described in Sections 19-8 and 20-10: the free expansion of an ideal gas. Figure 21-1*a* shows the gas in its initial equilibrium state *i*, confined by a closed stopcock to the left half of a thermally insulated container. If we open the stopcock, the gas rushes to fill the entire container, eventually reaching the final equilibrium state *f* shown in Fig. 21-1*b*. This is an irreversible process; all the molecules of the gas will never return, by themselves, to the left half of the container.

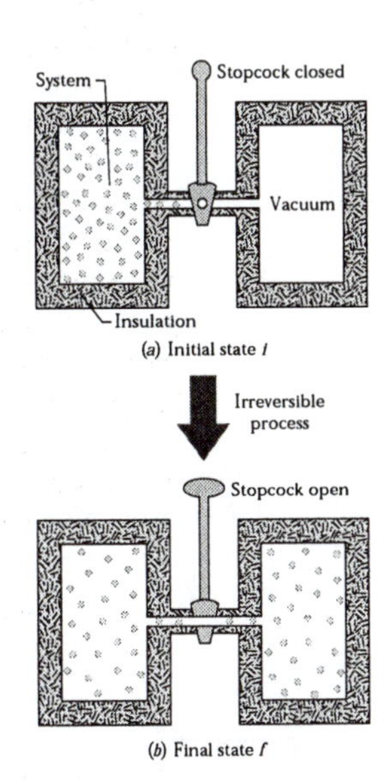

Fig. 21-1: The free expansion of an ideal gas. (*a*) The gas is confined to the left half of an insulated container by a closed stopcock. (*b*) When the stopcock is opened, the gas rushes to fill the entire container. This process is irreversible; that is, it does not occur in reverse, with the gas spontaneously collecting itself in the left half of the container.

The *p-V* plot of the process, in Fig. 21-2, shows the pressure and volume of the gas in its initial state *i* and final state *f*. Pressure and volume are *state properties*, properties that depend only on the state of the gas and not on how it reached that state. Other state properties are temperature and internal energy. We now assume that the gas has still another state property—its entropy. Furthermore, we define the **change in entropy** $S_f - S_i$ of a system during a process that takes the system from an initial state *i* to a final state *f* as

$$\Delta S = S_f - S_i = \int_i^f \frac{dQ}{T} \qquad \text{(change in entropy defined).} \qquad (21\text{-}1)$$

Here Q is the energy transferred as heat to or from the system during the process, and T is the temperature of the system in kelvins. Thus, an entropy change depends not only on the energy transferred as heat but also on the temperature at which the transfer takes place. Because T is always positive, the sign of ΔS is the same as that of Q. We see from Eq. 21-1 that the SI unit for entropy and entropy change is the joule per kelvin.

There is a problem, however, in applying Eq. 21-1 to the free expansion of Fig. 21-1. As the gas rushes to fill the entire container, the pressure, temperature, and volume of the gas fluctuate unpredictably. In other words, they do not have a sequence of well-defined equilibrium values during the intermediate stages of the change from initial equilibrium state *i* to final equilibrium state *f*. Thus, we cannot trace a pressure-volume path for the free expansion on the *P-V* plot of Fig. 21-2 and, more important, we cannot find a relation between Q and T that allows us to integrate as Eq. 21-1 requires.

However, if entropy is truly a state property, the difference in entropy between states *i* and *f* must depend *only on those states* and not at all on the way the system went from one state to the other. Suppose, then, that we replace the irreversible free expansion of Fig. 21-1 with a *reversible* process that connects states *i* and *f*. With a reversible process we can trace a pressure-volume path on a *p-V* plot, and we can find a relation between Q and T that allows us to use Eq. 21-1 to obtain the entropy change.

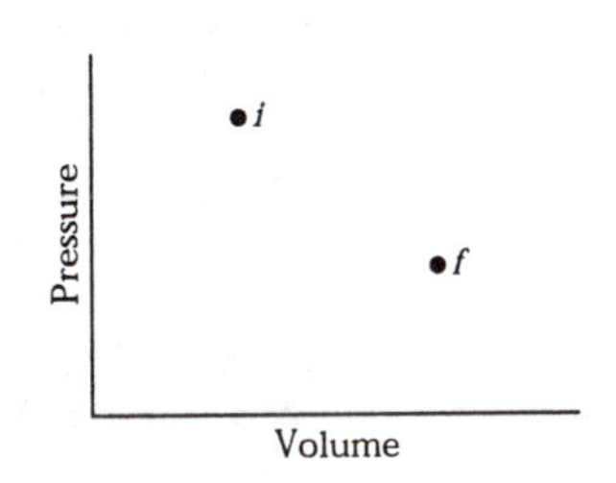

Fig. 21-2: A *P-V* diagram showing the initial state *i* and the final state *f* of the free expansion of Fig. 21-1. The intermediate states of the gas cannot be shown because they are not equilibrium states.

We saw in Section 20-10 that the temperature of an ideal gas does not change during a free expansion: $T_i = T_f = T$. Thus, points *i* and *f* in Fig. 21-2 must be on the same isotherm. A convenient replacement process is then a reversible isothermal expansion from state *i* to state *f*, which actually proceeds *along* that isotherm. Furthermore, because T is constant throughout a reversible isothermal expansion, the integral of Eq. 21-1 is greatly simplified.

Figure 21-3 shows how to produce such a reversible isothermal expansion. We confine the gas to an insulated cylinder that rests on a thermal reservoir maintained at the temperature T. We begin by placing just enough lead shot on the movable piston so that the pressure and volume of the gas are those of the initial state *i* of Fig. 21-1*a*. We then remove shot slowly (piece by piece) until the pressure and volume of the gas are those of the final state *f* of Fig. 21-1*b*. The temperature of the gas does not change because the gas remains in thermal contact with the reservoir throughout the process.

The reversible isothermal expansion of Fig. 21-3 is physically quite different from the irreversible free expansion of Fig. 21-1. However, *both processes have the same initial state and the same final state and thus must have the same change in entropy.* Because we removed the lead shot slowly, the intermediate states of the gas are equilibrium states, so we can plot them on a *p-V* diagram (Fig. 21-4).

To apply Eq. 21-1 to the isothermal expansion, we take the constant temperature T outside the integral, obtaining

$$\Delta S = S_f - S_i = \frac{1}{T}\int_i^f dQ.$$

Because $\int dQ = Q$, where Q is the total energy transferred as heat during the process, we have

$$\Delta S = S_f - S_i = \frac{Q}{T} \qquad \text{(change in entropy, isothermal process).} \qquad (21\text{-}2)$$

To keep the temperature T of the gas constant during the isothermal expansion of Fig. 21-3, heat Q must have been transferred *from* the reservoir to the gas. Thus, Q is positive and the entropy of the gas *increases* during the isothermal process and during the free expansion of Fig. 21-1.

To summarize:

➤To find the entropy change for an irreversible process occurring in a *closed* system, replace that process with any reversible process that connects the same initial and final states. Calculate the entropy change for this reversible process with Eq. 21-1.

When the temperature change ΔT of a system is small relative to the temperature (in kelvins) before and after the process, the entropy change can be approximated as

$$\Delta S = S_f - S_i \approx \frac{Q}{T_{\text{avg}}}, \qquad (21\text{-}3)$$

where T_{avg} is the average temperature of the system in kelvins during the process.

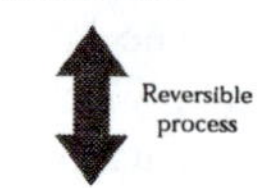

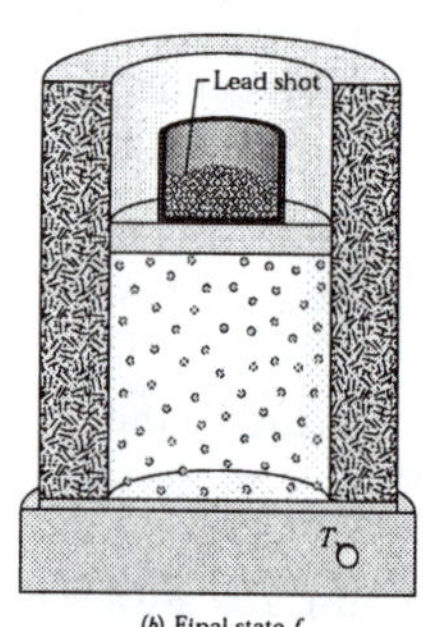

Fig. 21-3: The isothermal expansion of an ideal gas, done in a reversible way. The gas has the same initial state i and same final state f as in the irreversible process of Figs. 21-1 and 21-2.

READING EXERCISE 21-2: Water is heated on a stove. Rank the entropy changes of the water as its temperature rises (a) from 20°C to 30°C, (b) from 30°C to 35°C, and (c) from 80°C to 85°C, greatest first.

Entropy as a State Function

We have assumed that entropy, like pressure, energy, and temperature, is a property of the state of a system and is independent of how that state is reached. That entropy is indeed a *state function* (as state properties are usually called) can only be deduced by experiment. However, we can prove it is a state function for the special and important case in which an ideal gas is taken through a reversible process.

To make the process reversible, it is done slowly in a series of small steps, with the gas in an equilibrium state at the end of each step. For each small step, the energy transferred as heat to or from the gas is dQ, the work done by the gas is dW, and the change in internal energy is dE_{int}. These are related by the first law of thermodynamics in differential form (Eq. 19-13):

$$dE_{\text{int}} = dQ - dW.$$

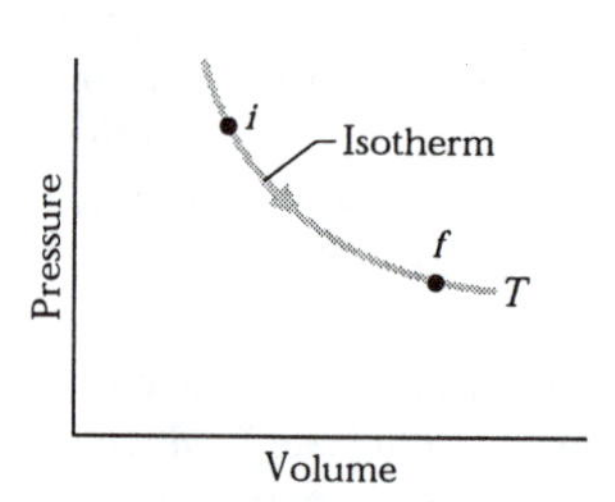

Fig. 21-4: A *P-V* diagram for the reversible isothermal expansion of Fig. 21-3. The intermediate states, which are now equilibrium states, are shown.

Because the steps are reversible, with the gas in equilibrium states, we can use Eq. 19-10 to replace dW with $P\,dV$ and Eq. 20-43 to replace dE_{int} with $nC_V\,dT$. Solving for dQ then leads to

$$dQ = P\,dV + nC_V\,dT.$$

Using the ideal gas law, we replace P in this equation with nRT/V. Then we divide each term in the resulting equation by T, obtaining

$$\frac{dQ}{T} = nR\frac{dV}{V} + nC_V\frac{dT}{T}.$$

Now let us integrate each term of this equation between an arbitrary initial state i and an arbitrary final state f to get

$$\int_i^f \frac{dQ}{T} = \int_i^f nR\frac{dV}{V} + \int_i^f nC_V\frac{dT}{T}.$$

The quantity on the left is the entropy change $\Delta S(= S_f - S_i)$ defined by Eq. 21-1. Substituting this and integrating the quantities on the right yield

$$\Delta S = S_f - S_i = nR\ln\frac{V_f}{V_i} + nC_V\ln\frac{T_f}{T_i}. \tag{21-4}$$

Note that we did not have to specify a particular reversible process when we integrated. Therefore, the integration must hold for all reversible processes that take the gas from state i to state f. Thus, the change in entropy ΔS between the initial and final states of an ideal gas depends only on properties of the initial state (V_i and T_i) and properties of the final states (V_f and T_f); ΔS does not depend on how the gas changes between the two states.

Touchstone Examples 21-2-1 and 21-2-2, at the end of this chapter, illustrate how to use what you learned in this section.

TE

21-3 The Second Law of Thermodynamics

Here is a puzzle. We saw in Touchstone Example 21-2-1 that if we cause the reversible process of Fig. 21-3 to proceed from (a) to (b) in that figure, the change in entropy of the gas—which we take as our system—is positive. However, because the process is reversible, we can just as easily make it proceed from (b) to (a), simply by slowly adding lead shot to the piston of Fig. 21-3b until the original volume of the gas is restored. In this reverse process, energy must be extracted as heat *from the gas* to keep its temperature from rising. Hence Q is negative and so, from Eq. 21-2, the entropy of the gas must decrease.

Doesn't this decrease in the entropy of the gas violate the entropy postulate of Section 21-1, which states that entropy always increases? No, because that postulate holds only for *irreversible* processes occurring in closed systems. The procedure suggested here does not meet these requirements. The process is *not* irreversible and (because energy is transferred as heat from the gas to the reservoir) the system—which is the gas alone—is not closed.

However, if we include the reservoir, along with the gas, as part of the system, then we do have a closed system. Let's check the change in entropy of the enlarged system *gas + reservoir* for the process that takes it from (b) to (a) in Fig. 21-3. During this reversible process, energy is transferred as heat from the gas to the reservoir—that is, from one part of the enlarged system to another. Let $|Q|$ represent the absolute value (or magnitude) of this heat. With Eq. 21-2, we can then calculate separately the entropy changes for the gas (which loses $|Q|$) and the reservoir (which gains $|Q|$). We get

$$\Delta S_{\text{gas}} = -\frac{|Q|}{T}$$

and

$$\Delta S_{\text{res}} = +\frac{|Q|}{T}$$

The entropy change of the closed system is the sum of these two quantities, *which is zero.*

With this result, we can modify the entropy postulate of Section 21-1 to include both reversible and irreversible processes:

➤If a process occurs in a *closed* system, the entropy of the system increases for irreversible processes and remains constant for reversible processes. It never decreases.

Although entropy may decrease in part of a closed system, there will always be an equal or larger entropy increase in another part of the system, so that the entropy of the system as a whole never decreases. This fact is one form of the **second law of thermodynamics** and can be written as

$$\Delta S \geq 0 \quad \text{(second law of thermodynamics)}, \qquad (21\text{-}5)$$

where the greater-than sign applies to irreversible processes, and the equals sign to reversible processes. Equation 21-5 applies only to closed systems.

In the real world almost all processes are irreversible to some extent because of friction, turbulence, and other factors, so the entropy of real closed systems undergoing real processes always increases. Processes in which the system's entropy remains constant are always idealizations.

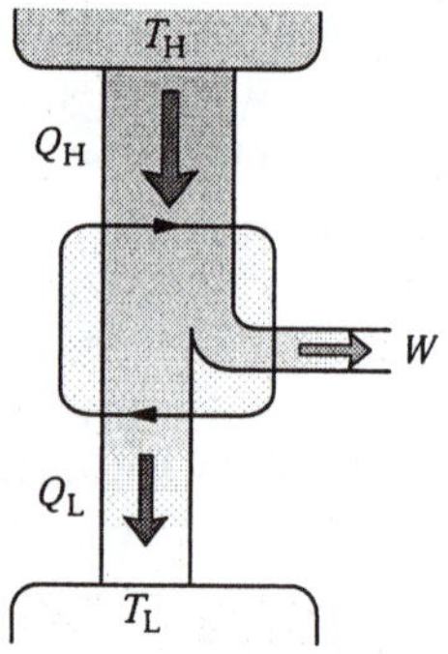

Fig. 21-5: The elements of an engine. The two black arrowheads on the central loop suggest the working substance operating in a cycle, as if on a p-V plot. Energy $|Q_H|$ is transferred from the high-temperature reservoir at temperature T_H to the working substance. Energy $|Q_L|$ is transferred as heat from the working substance to the low-temperature reservoir at temperature T_L. Work W is done by the engine (actually by the working substance) on something in the environment.

21-4 Entropy in the Real World: Engines

A **heat engine**, or more simply, an **engine**, is a device that extracts energy from its environment in the form of heat and does useful work. At the heart of every engine is a *working substance*. In a steam engine, the working substance is water, in both its vapor and its liquid form. In an automobile engine the working substance is a gasoline-air mixture. If an engine is to do work on a sustained basis, the working substance must operate in a *cycle*; that is, the working substance must pass through a closed series of thermodynamic processes, called *strokes*, returning again and again to each state in its cycle. Let us see what the laws of thermodynamics can tell us about the operation of engines.

A Carnot Engine

We have seen that we can learn much about real gases by analyzing an ideal gas, which obeys the simple law $PV = nRT$. This is a useful plan because, although an ideal gas does not exist, any real gas approaches ideal behavior as closely as you wish if its density is low enough. In much the same spirit we choose to study real engines by analyzing the behavior of an **ideal engine.**

➤In an ideal engine, all processes are reversible and no wasteful energy transfers occur due to, say, friction and turbulence.

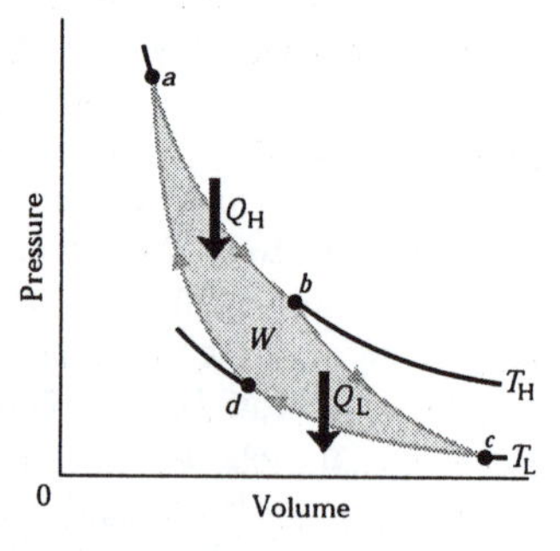

Fig. 21-6: A pressure-volume plot of the cycle followed by the working substance of the Carnot engine in Fig. 21-5. The cycle consists of two isotherms (*ab* and *cd*) and two adiabatic processes (*bc* and *da*). the shaded area enclosed by the cycle is equal to the work W per cycle done by the Carnot engine.

We shall focus on a particular ideal engine called a **Carnot engine** after the French scientist and engineer N. L. Sadi Carnot (pronounced "car-no"), who first proposed the engine's concept in 1824. This ideal engine turns out to be the best (in principle) at using energy as heat to do useful work. Surprisingly, Carnot was able to analyze the performance of this engine before the first law of thermodynamics and the concept of entropy had been discovered.

Figure 21-5 shows schematically the operation of a Carnot engine. During each cycle of the engine, the working substance absorbs heat $|Q_H|$ from a thermal reservoir at constant temperature T_H and discharges heat $|Q_L|$ to a second thermal reservoir at a constant lower temperature T_L.

Figure 21-6 shows a P-V plot of the *Carnot cycle*—the cycle followed by the working substance. As indicated by the arrows, the cycle is traversed in the clockwise direction. Imagine the working substance to be a gas, confined to an insulating cylinder with a weighted, movable piston. The cylinder may be placed at will on either of the two thermal reservoirs, as in Fig. 21-3, or on an insulating slab. Figure 21-6 shows that, if we place the cylinder in contact with the high-temperature reservoir at temperature T_H, heat $|Q_H|$ is transferred *to* the working substance *from* this reservoir as the gas undergoes an isothermal *expansion* from volume V_a to volume V_b. Similarly, with the working substance in contact with the low-temperature reservoir at temperature T_L, heat $|Q_L|$ is transferred *from* the working substance *to* this reservoir, as the gas undergoes an isothermal *compression* from volume V_c to volume V_d.

In the engine of Fig. 21-5, we assume that heat transfers to or from the working substance can take place *only* during the isothermal processes *ab* and *cd* of Fig. 21-6. Therefore, processes *bc* and *da* in that figure, which connect the two isotherms at temperatures T_H and T_L, must be (reversible) adiabatic processes; that is, they must be processes in which no heat is transferred. To ensure this, during processes *bc* and *da* the cylinder is placed on an insulating slab as the volume of the working substance is changed.

During the consecutive processes *ab* and *bc* of Fig. 21-6, the working substance is expanding and thus doing positive work as it raises the weighted piston. This work is represented in Fig. 21-6 by the area under curve *abc*. During the consecutive processes *cd* and *da*, the working substance is being compressed, which means that it is doing negative work on its environment or, equivalently, that its environment is doing work on it as the loaded piston descends. This work is represented by the area under curve *cda*. The *net work per cycle*, which is represented by W in Figs. 21-5 and 21-6, is the difference between these two areas, a positive quantity equal to the area enclosed by cycle *abcda* in Fig. 21-6. This work W is performed on some outside object. The engine might, for example, be used to lift a weight.

Equation 21-1 $(\Delta S = \int dQ/T)$ tells us that any energy transfer as heat must involve a change in entropy. To illustrate the entropy changes for a Carnot engine, we can plot the Carnot cycle on a temperature-entropy (*T-S*) diagram as in Fig. 21-7. The lettered points *a*, *b*, *c*, and *d* in Fig. 21-7 correspond to the lettered points in the *p-V* diagram in Fig. 21-6. The two horizontal lines in Fig. 21-7 correspond to the two isothermal processes of the Carnot cycle (because the temperature is constant). Process *ab* is the isothermal expansion of the cycle. As the working substance (reversibly) absorbs heat $|Q_H|$ at constant temperature T_H during that process, its entropy increases. Similarly, during the isothermal compression *cd*, the working substance (reversibly) loses heat $|Q_L|$ at constant temperature T_L, and its entropy decreases.

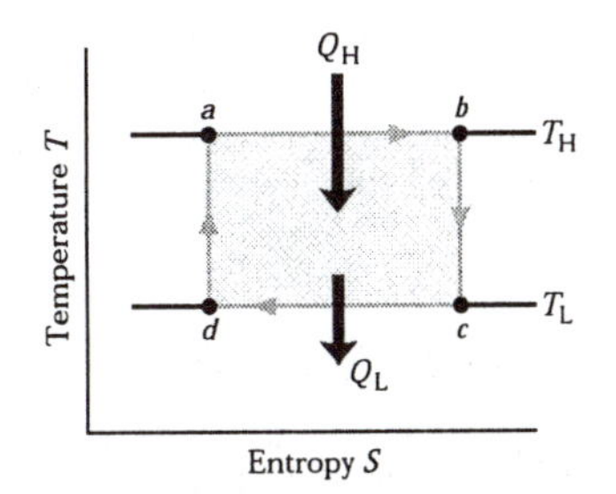

Fig. 21-7: The Carnot cycle of Fig. 21-6 plotted on a temperature-entropy diagram. During processes *ab* and *cd* the temperature remains constant. During processes *bc* and *da* the entropy remains constant.

The two vertical lines in Fig. 21-7 correspond to the two adiabatic processes of the Carnot cycle. Because no energy is transferred as heat during the two processes, the entropy of the working substance is constant during them.

The Work: To calculate the net work done by a Carnot engine during a cycle, let us apply the first law of thermodynamics $(\Delta E_{\text{int}} = Q - W)$, to the working substance of a Carnot engine. That substance must return again and again to any arbitrarily selected state in that cycle. Thus, if X represents any state property of the working substance, such as pressure, temperature, volume, internal energy, or entropy, we must have $\Delta X = 0$ for every cycle. It follows that $\Delta E_{\text{int}} = 0$ for a complete cycle of the working substance. Recalling that Q in Eq. 19-13 is the *net* heat transfer per cycle and W is the net work, we can write the first law of thermodynamics for the Carnot cycle as

$$W = |Q_H| - |Q_L|. \tag{21-6}$$

Entropy Changes: In a Carnot engine, there are *two* (and only two) reversible energy transfers as heat, and thus two changes in entropy—one at temperature T_H and one at T_L. The net entropy change per cycle is then

$$\Delta S = \Delta S_H + \Delta S_L = \frac{|Q_H|}{T_H} - \frac{|Q_L|}{T_L}. \tag{21-7}$$

Here ΔS_H is positive because energy $|Q_H|$ is *added to* the working substance as heat (an increase in entropy) and ΔS_L is negative because energy $|Q_L|$ is removed from the working substance as heat (a decrease in entropy). Because entropy is a state function, we must have $\Delta S = 0$ for a complete cycle. Putting $\Delta S = 0$ in Eq. 21-7 requires that

$$\frac{|Q_H|}{T_H} = \frac{|Q_L|}{T_L}. \tag{21-8}$$

Note that, because $T_H > T_L$, we must have $|Q_H| > |Q_L|$; that is, more energy is extracted as heat from the high-temperature reservoir than is delivered to the low-temperature reservoir.

We shall now use Eqs. 21-6 and 21-8 to derive an expression for the efficiency of a Carnot engine.

Efficiency of a Carnot Engine

The purpose of any engine is to transform as much of the extracted heat Q_H into work as possible. We measure its success in doing so by its **thermal efficiency** ε, defined as the work the engine does per cycle ("energy we get") divided by the heat energy it absorbs per cycle ("energy we pay for"):

$$\varepsilon = \frac{\text{energy we get}}{\text{energy we pay for}} = \frac{|W|}{|Q_H|} \quad \text{(efficiency, any engine).} \tag{21-9}$$

For a Carnot engine we can substitute for W from Eq. 21-6 to write

$$\varepsilon_C = \frac{|Q_H| - |Q_L|}{|Q_H|} = 1 - \frac{|Q_L|}{|Q_H|}. \tag{21-10}$$

Using Eq. 21-8 we can write this as

$$\varepsilon_C = 1 - \frac{T_L}{T_H} \quad \text{(efficiency, Carnot engine),} \tag{21-11}$$

where the temperatures T_L and T_H are in kelvins. Because $T_L < T_H$, the Carnot engine necessarily has a thermal efficiency less than unity—that is, less than 100%. This is indicated in Fig. 21-5, which shows that only part of the energy extracted as heat from the high-temperature reservoir is available to do work, and the rest is delivered to the low-temperature reservoir. We will show in Section 21-6 that no real engine can have a thermal efficiency greater than that calculated from Eq. 21-11.

Inventors continually try to improve engine efficiency by reducing the energy $|Q_L|$ that is "thrown away" during each cycle. The inventor's dream is to produce the *perfect engine*, diagrammed in Fig. 21-8, in which $|Q_L|$ is reduced to zero and $|Q_H|$ is converted completely into work. Such an engine on an ocean liner, for example, could extract energy as heat from the water and use it to drive the propellers, with no fuel cost. An automobile, fitted with such an engine, could extract energy as heat from the surrounding air and use it to drive the car, again with no fuel cost. Alas, a perfect engine is only a dream: Inspection of Eq. 21-11 shows that we can achieve 100% engine efficiency (that is, $\varepsilon = 1$) only if $T_L = 0$ or $T_H \to \infty$, requirements that are impossible to meet. Instead, decades of practical engineering experience have led to the following alternative version of the second law of thermodynamics:

➤No series of processes is possible whose sole result is the transfer of energy as heat from a thermal reservoir and the complete conversion of this energy to work.

In short, *there are no perfect engines.*

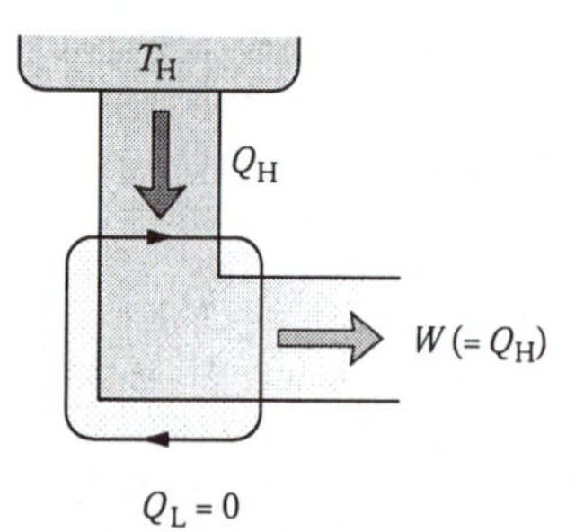

Fig. 21-8: The elements of a perfect engine—that is, one that converts heat Q_H from a high-temperature reservoir directly to work W with 100% efficiency.

Fig. 21-9: The North Anna nuclear power plant near Charlottesville, Virginia, which generates electrical energy at the rate of 900 MW. At the same time, by design, it discards energy into the nearby river at the rate of 2100 MW. This plant—and all others like it—throws away more energy than it delivers in useful form. It is a real counterpart to the ideal engine of Fig. 21-5.

To summarize: The thermal efficiency given by Eq. 21-11 applies only to Carnot engines. Real engines, in which the processes that form the engine cycle are not reversible, have lower efficiencies. If your car were powered by a Carnot engine, it would have an efficiency of about 55% according to Eq. 21-11; its actual efficiency is probably about 25%. A nuclear power plant (Fig. 21-9), taken in its entirety, is an engine. It extracts heat from a reactor core, does work by means of a turbine, and discharges heat to a nearby river. If the power plant operated as a Carnot engine, its efficiency would be about 40%; its actual efficiency is about 30%. In designing engines of any type, there is simply no way to beat the efficiency limitation imposed by Eq. 21-11.

Stirling Engine

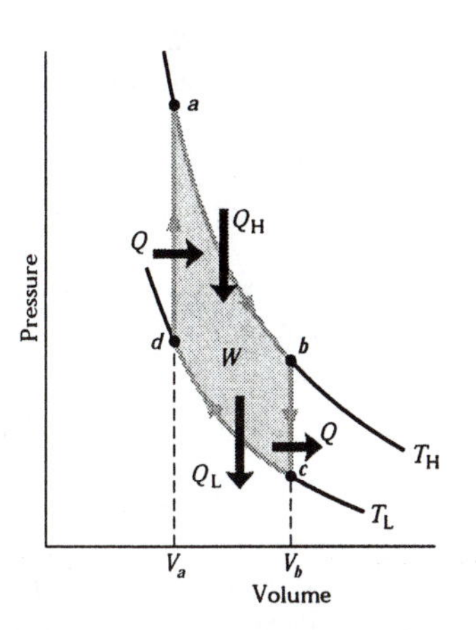

Fig. 21-10: A *P-V* plot for the working substance of an ideal Stirling engine, assumed for convenience to be an ideal gas.

Equation 21-11 does not apply to all ideal engines, but only to engines that can be represented as in Fig. 21-5—that is, to Carnot engines. For example, Fig. 21-10 shows the operating cycle of an ideal **Stirling engine**. Comparison with the Carnot cycle of Fig. 21-6 shows that each engine has isothermal heat transfers at temperatures T_H and T_L. However, the two isotherms of the Stirling engine cycle of Fig. 21-10 are connected, not by adiabatic processes as for the Carnot engine, but by constant-volume processes. To increase the temperature of a gas at constant volume reversibly from T_L to T_H (as in process *da* of Fig. 21-10) requires a heat transfer to the working substance from a thermal reservoir whose temperature can be varied smoothly between those limits. Note that reversible heat transfers (and corresponding entropy changes) occur in all four of the processes that form the cycle of a Stirling engine, not just two processes as in a Carnot engine. Thus, the derivation that led to Eq. 21-11 does not apply to an ideal Stirling engine. More important, the efficiency of an ideal Stirling engine is lower than that of a Carnot engine operating between the same two temperatures. Real Stirling engines have even lower efficiencies.

The Stirling engine was developed in 1816 by Robert Stirling. This engine, long neglected, is now being developed for use in automobiles and spacecraft. A Stirling engine delivering 5000 hp (3.7 MW) has been built.

READING EXERCISE 21-3: Three Carnot engines operate between reservoir temperatures of (a) 400 and 500 K, (b) 600 and 800 K, and (c) 400 and 600 K. Rank the engines according to their thermal efficiencies, greatest first.

Touchstone Examples 21-4-1 and 21-4-2, at the end of this chapter, illustrate how to use what you learned in this section.

TE

21-5 Entropy in the Real World: Refrigerators

A **refrigerator** is a device that uses work to transfer thermal energy from a low-temperature reservoir to a high-temperature reservoir as it continuously repeats a set series of thermodynamic processes. In a household refrigerator, for example, work is done by an electrical compressor to transfer thermal energy from the food storage compartment (a low-temperature reservoir) to the room (a high-temperature reservoir).

Air conditioners and heat pumps are also refrigerators. The differences are only in the nature of the high- and low-temperature reservoirs. For an air conditioner, the low-temperature reservoir is the room that is to be cooled, and the high-temperature reservoir is the (presumably warmer) outdoors. A heat pump is an air conditioner that can be operated in reverse to heat a room; the room is the high-temperature reservoir and heat is transferred to it from the (presumably cooler) outdoors.

Let us consider an *ideal refrigerator:*

➤In an ideal refrigerator, all processes are reversible and no wasteful energy transfers occur due to, say, friction and turbulence.

Figure 21-11 shows the basic elements of an ideal refrigerator that operates in the reverse of the Carnot engine of Fig. 21-5. In other words, all the energy transfers, as either heat or work, are reversed from those of a Carnot engine. We can call such an ideal refrigerator a **Carnot refrigerator.**

The designer of a refrigerator would like to extract as much heat $|Q_L|$ as possible from the low-temperature reservoir (what we want) for the least amount of work W (what we pay for). A measure of the efficiency of a refrigerator, then, is

$$K = \frac{\text{what we want}}{\text{what we pay for}} = \frac{|Q_L|}{|W|} \quad \text{(coefficient of performance, any refrigerator),} \quad (21\text{-}12)$$

where K is called the *coefficient of performance*. For a Carnot refrigerator, the first law of thermodynamics gives $|W| = |Q_H| - |Q_L|$, where $|Q_H|$ is the magnitude of the energy transferred as heat to the high-temperature reservoir. Equation 21-12 then becomes

$$K_C = \frac{|Q_L|}{|Q_H| - |Q_L|}. \quad (21\text{-}13)$$

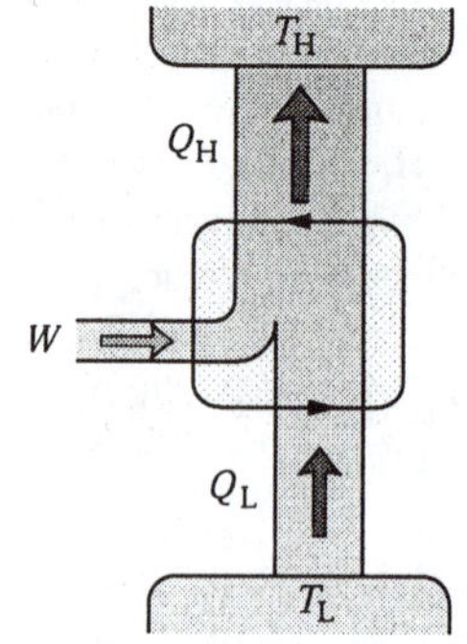

Fig. 21-11: The elements of a refrigerator. The two black arrowheads on the central loop suggest the working substance operating in a cycle, as if on a P-V plot. Energy Q_L is transferred as heat to the working substance from the low-temperature reservoir. Energy Q_H is transferred as heat to the high-temperature reservoir from the working substance. Work W is done on the refrigerator (on the working substance) by something in the environment.

Because a Carnot refrigerator is a Carnot engine operating in reverse, we can combine Eq. 21-8 with Eq. 21-13; after some algebra we find

$$K_C = \frac{T_L}{T_H - T_L} \quad \text{(coefficient of performance, Carnot refrigerator).} \quad (21\text{-}14)$$

For typical room air conditioners, $K \approx 2.5$. For household refrigerators, $K \approx 5$. Perversely, the value of K is higher the closer the temperatures of the two reservoirs are to each other. That is why heat pumps are more effective in temperate climates than in climates where the outside temperature varies widely.

It would be nice to own a refrigerator that did not require some input of work—that is, one that would run without being plugged in. Figure 21-12 represents another "inventor's dream," a *perfect refrigerator* that transfers heat Q from a cold reservoir to a warm reservoir without the need for work. Because the unit operates in cycles, the entropy of the working substance does not change during a complete cycle. The entropies of the two reservoirs, however, do change: The entropy change for the cold reservoir is $-|Q|/T_L$, and that for the warm reservoir is $+|Q|/T_H$. Thus, the net entropy change for the entire system is

$$\Delta S = -\frac{|Q|}{T_L} + \frac{|Q|}{T_H}.$$

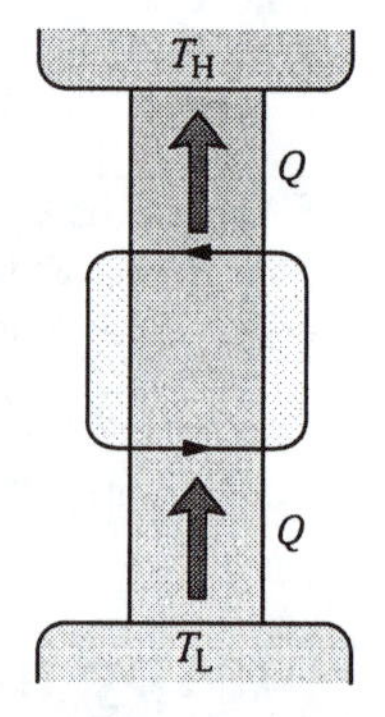

Fig. 21-12: The elements of a perfect refrigerator—that is, one that transfers energy from a low-temperature reservoir to a high-temperature reservoir without any input of work.

Because $T_H > T_L$, the right side of this equation is negative and thus the net change in entropy per cycle for the closed system *refrigerator + reservoirs* is also negative. Because such a decrease in entropy violates the second law of thermodynamics (Eq. 21-5), a perfect refrigerator does not exist. (If you want your refrigerator to operate, you must plug it in.)

This result leads us to another (equivalent) formulation of the second law of thermodynamics:

➤No series of processes is possible whose sole result is the transfer of energy as heat from a reservoir at a given temperature to a reservoir at a higher temperature.

In short, *there are no perfect refrigerators.*

READING EXERCISE 21-4: You wish to increase the coefficient of performance of an ideal refrigerator. You can do so by (a) running the cold chamber at a slightly higher temperature, (b) running the cold chamber at a slightly lower temperature, (c) moving the unit to a slightly warmer room, or (d) moving it to a slightly cooler room. The temperature changes are to be the same in all four cases. List the changes according to the resulting coefficients of performance, greatest first.

21-6 The Efficiencies of Real Engines

Let ε_C be the efficiency of a Carnot engine operating between two temperatures. In this section we prove that no real engine operating between those temperatures can have an efficiency greater than ε_C. If it could, the engine would violate the second law of thermodynamics.

Let us assume that an inventor, working in her garage, has constructed an engine X, which she claims has an efficiency ε_X that is greater than ε_C:

$$\varepsilon_X > \varepsilon_C \qquad \text{(a claim).} \qquad (21\text{-}15)$$

Let us couple engine X to a Carnot refrigerator, as in Fig. 21-13*a*. We adjust the strokes of the Carnot refrigerator so that the work it requires per cycle is just equal to that provided by engine X. Thus, no (external) work is performed on or by the combination *engine + refrigerator* of Fig. 21-13*a*, which we take as our system.

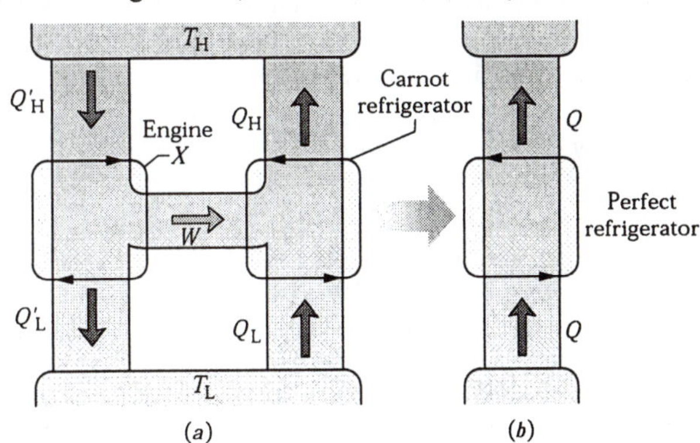

Fig. 21-13: (*a*) Engine X drives a Carnot refrigerator. (*b*) If, as claimed, engine X is more efficient than a Carnot engine, then the combination shown in (*a*) is equivalent to the perfect refrigerator shown here. This violates the second law of thermodynamics, so we conclude that engine X cannot be more efficient than a Carnot engine.

If Eq. 21-15 is true, from the definition of efficiency (Eq. 21-9), we must have

$$\frac{|W|}{|Q'_H|} > \frac{|W|}{|Q_H|},$$

where the prime refers to engine X and the right side of the inequality is the efficiency of the Carnot refrigerator when it operates as an engine. This inequality requires that

$$|Q_H| > |Q'_H|. \qquad (21\text{-}16)$$

Because the work done by engine X is equal to the work done on the Carnot refrigerator, we have, from the first law of thermodynamics (see Eq. 21-6),

$$|Q_H| - |Q_L| = |Q'_H| - |Q'_L|,$$

which we can write as

$$|Q_H| - |Q'_H| = |Q_L| - |Q'_L| = Q. \qquad (21\text{-}17)$$

Because of Eq. 21-16, the quantity Q in Eq. 21-17 must be positive.

Comparison of Eq. 21-17 with Fig. 21-13 shows that the net effect of engine X and the Carnot refrigerator, working in combination, is to transfer energy Q as heat from a low-temperature reservoir to a high-temperature reservoir without the requirement of work. Thus, the combination acts like the perfect refrigerator of Fig. 21-12, whose existence is a violation of the second law of thermodynamics.

Something must be wrong with one or more of our assumptions, and it can only be Eq. 21-15. We conclude that *no real engine can have an efficiency greater than that of a Carnot engine when both engines work between the same two temperatures.* At most, it can have an efficiency equal to that of a Carnot engine. In that case, engine X *is* a Carnot engine.

21-7 A Statistical View of Entropy

In Chapter 20 we saw that the macroscopic properties of gases can be explained in terms of their microscopic, or molecular, behavior. For one example, recall that we were able to account for the pressure exerted by a gas on the walls of its container in terms of the momentum transferred to those walls by rebounding gas molecules. Such explanations are part of a study called **statistical mechanics.**

Here we shall focus our attention on a single problem, involving the distribution of gas molecules between the two halves of an insulated box. This problem is reasonably simple to analyze, and it allows us to use statistical mechanics to calculate the entropy change for the free expansion of an ideal gas. You will see in Touchstone Example 21-7-2 that statistical mechanics leads to the same entropy change we obtain in Touchstone Example 21-2-2 using thermodynamics.

Figure 21-14 shows a box that contains six identical (and thus indistinguishable) molecules of a gas. At any instant, a given molecule will be in either the left or the right half of the box; because the two halves have equal volumes, the molecule has the same likelihood, or probability, of being in either half.

Table 21-1 shows four of the seven possible *configurations* of the six molecules, each configuration labeled with Roman numerals. For example, in configuration I, all six molecules are in the left half of the box $(n_1 = 6)$, and none are in the right half $(n_2 = 0)$. The three configurations not shown are V with a (2, 4) split, VI with a (1, 5) split, and VII with a (0, 6) split. In configuration II, five molecules are in one half of the box, leaving one molecule in the other half. We see that, in general, a given configuration can be achieved in a number of different ways. We call these different arrangements of the molecules *microstates*. Let us see how to calculate the number of microstates that correspond to a given configuration.

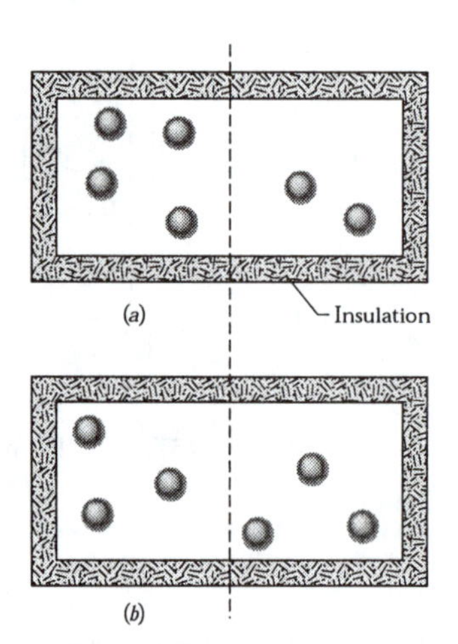

Fig. 21-14: An insulated box contains six gas molecules. Each molecule has the same probability of being in the left half of the box as in the right half. The arrangement in (a) corresponds to configuration III in Table 21-1, and that in (b) corresponds to configuration IV.

Table 21-1: Six Molecules in a Box

Configuration Label	n_1	n_2	Multiplicity W (number of microstates)	Calculation of W (Eq. 21-18)	Entropy 10^{-23} J/K (Eq. 21-19)
I	6	0	1	$6!/(6!0!) = 1$	0
II	5	1	6	$6!/(5!\ 1!) = 6$	2.47
III	4	2	15	$6!/(4!2!) = 15$	3.74
IV	3	3	20	$6!/(3!3!) = 20$	4.13
Total number of microstates = 64					

Suppose we have N molecules, distributed with n_1 molecules in one half of the box and n_2 in the other. (Thus $n_1 + n_2 = N$.) Let us imagine that we distribute the molecules "by hand," one at a time. If $N = 6$, we can select the first molecule in six independent ways; that is, we can pick any one of the six molecules. We can pick the second molecule in five ways, by picking any one of the remaining five molecules; and so on. The total number of ways in which we can select all six molecules is the product of these independent ways, or $6 \times 5 \times 4 \times 3 \times 2 \times 1 = 720$. In mathematical shorthand we write this product as $6! = 720$, where 6! is pronounced "six factorial." Your hand calculator can probably calculate factorials. For later use you will need to know that $0! = 1$. (Check this on your calculator.)

However, because the molecules are indistinguishable, these 720 arrangements are not all different. In the case that $n_1 = 4$ and $n_2 = 2$ (which is configuration III in Table 21-1), for example, the order in which you put four molecules in one half of the box does not matter, because after you have put all four in, there is no way that you can tell the order in which you did so. The number of ways in which you can order the four molecules is 4! or 24. Similarly, the number of ways in which you can order two molecules for the other half of the box is simply 2! or 2. To get the number of different arrangements that lead to the 4, 2 split of configuration III, we must divide 720 by 24 and also by 2. We call the resulting quantity, which is the number of microstates that correspond to a given configuration, the *multiplicity* W of that configuration. Thus, for configuration III,

$$W_{\text{III}} = \frac{6!}{4!\,2!} = \frac{720}{24 \times 2} = 15.$$

Thus, Table 21-1 tells us there are 15 independent microstates that correspond to configuration III. Note that, as the table also tells us, the total number of microstates for six molecules distributed over four configurations is 42.

Extrapolating from six molecules to the general case of N molecules, we have

$$W = \frac{N!}{n_1!\,n_2!} \quad \text{(multiplicity of configuration).} \qquad (21\text{-}18)$$

You should verify that Eq. 21-18 gives the multiplicities for all the configurations listed in Table 21-1.

The basic assumption of statistical mechanics is:

➤All microstates are equally probable.

In other words, if we were to take a great many snapshots of the six molecules as they jostle around in the box of Fig. 21-14 and then count the number of times each microstate occurred, we would find that all 42 microstates will occur equally often. In other words, the system will spend, on average, the same amount of time in each of the 42 microstates listed in Table 21-1.

Because the microstates are equally probable, but different configurations have different numbers of microstates, the configurations are *not* equally probable. In Table 21-1 configuration IV, with 20 microstates, is the *most probable configuration*, with a probability of $20/64 = 0.313$. This means that the system is in configuration IV 31.3% of the time. Configurations I and VII, in which all the molecules are in one half of the box, are the least probable, each with a probability of $1/64 = 0.016$ or 1.6%. It is not surprising that the most probable configuration is the one in which the molecules are evenly divided between the two halves of the box, because that is what we expect at thermal equilibrium. However, it *is* surprising that there is any probability, however small, of finding all six molecules clustered in half of the box, with the other half empty. In Touchstone Example 21-7-1 we show that this state can occur because six molecules is an extremely small number.

For large values of N there are extremely large numbers of microstates, but nearly all the microstates belong to the configuration in which the molecules are divided equally between the two halves of the box, as Fig. 21-15 indicates. Even though the measured temperature and pressure of the gas remain constant, the gas is churning away endlessly as its molecules "visit" all probable microstates with equal probability. However, because so few microstates lie outside the very narrow central configuration peak of Fig. 21-15, we might as well assume that the gas molecules are always divided equally between the two halves of the box. As we shall see, this is the configuration with the greatest entropy.

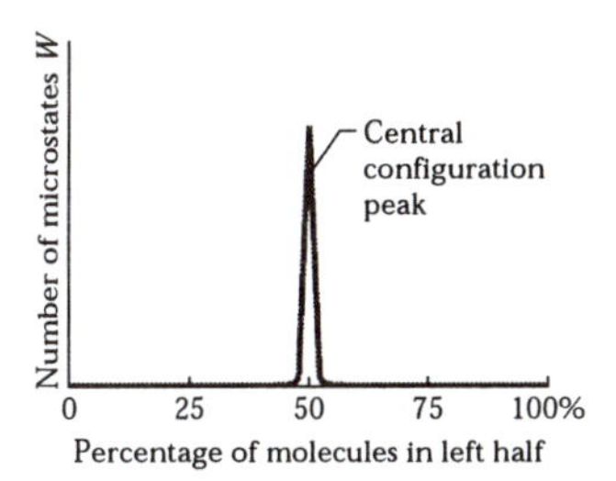

Fig. 21-15: For a large number of molecules in a box, a plot of the number of microstates that require various percentages of the molecules to be in the left half of the box. Nearly all the microstates correspond to an approximately equal sharing of the molecules between the two halves of the box; those microstates form the central configuration peak on the plot. For $N \approx 10^{22}$, the central configuration peak is much too narrow to be drawn on this plot.

Probability and Entropy

In 1877, Austrian physicist Ludwig Boltzmann (the Boltzmann of Boltzmann's constant k) derived a relationship between the entropy S of a configuration of a gas and the multiplicity W of that configuration. That relationship is

$$S = k \ln W \qquad \text{(Boltzmann's entropy equation).} \qquad (21\text{-}19)$$

This famous formula is engraved on Boltzmann's tombstone.

It is natural that S and W should be related by a logarithmic function. The total entropy of two systems is the *sum* of their separate entropies. The probability of occurrence of two independent systems is the *product* of their separate probabilities. Because $\ln ab = \ln a + \ln b$, the logarithm seems the logical way to connect these quantities.

Table 21-1 displays the entropies of the configurations of the six-molecule system of Fig. 21-14, computed using Eq. 21-19. Configuration IV, which has the greatest multiplicity, also has the greatest entropy.

When you use Eq. 21-18 to calculate W, your calculator may signal "OVERFLOW" if you try to find the factorial of a number greater than a few hundred. Fortunately, there is a very good approximation, known as **Stirling's approximation**, not for $N!$ but for $\ln N!$, which as it happens is exactly what is needed in Eq. 21-19. Stirling's approximation is

$$\ln N! \approx N(\ln N) - N \qquad \text{(Stirling's approximation).} \qquad (21\text{-}20)$$

The Stirling of this approximation is not the Stirling of the Stirling engine.

READING EXERCISE 21-5: A box contains one mole of a gas. Consider two configurations: (a) each half of the box contains half the molecules, and (b) each third of the box contains one-third of the molecules. Which configuration has more microstates?

Touchstone Examples 21-7-1 and 21-7-2, at the end of this chapter, illustrate how to use what you learned in this section.

TE

Touchstone Example 21-2-1

One mole of nitrogen gas is confined to the left side of the container of Fig. 21-1*a*. You open the stopcock and the volume of the gas doubles. What is the entropy change of the gas for this irreversible process? Treat the gas as ideal.

SOLUTION We need two **Key Ideas** here. One is that we can determine the entropy change for the irreversible process by calculating it for a reversible process that provides the same change in volume. The other is that the temperature of the gas does not change in the free expansion. Thus, the reversible process should be an isothermal expansion—namely, the one of Figs. 21-3 and 21-4.

From Table 20-4, the energy Q added as heat to the gas as it expands isothermally at temperature T from an initial volume V_i to a final volume V_f is

$$Q = nRT \ln \frac{V_f}{V_i},$$

in which n is the number of moles of gas present. From Eq. 21-2 the entropy change for this reversible process is

$$\Delta S_{\text{rev}} = \frac{Q}{T} = \frac{nRT \ln(V_f/V_i)}{T} = nR \ln \frac{V_f}{V_i}.$$

Substituting $n = 1.00$ mol and $V_f/V_i = 2$, we find

$$\Delta S_{\text{rev}} = nR \ln \frac{V_f}{V_i} = (1.00 \text{ mol})(8.31 \text{ J/mol} \cdot \text{K})(\ln 2)$$
$$= +5.76 \text{ J/K}.$$

Thus, the entropy change for the free expansion (and for all other processes that connect the initial and final states shown in Fig. 21-2) is

$$\Delta S_{\text{irrev}} = \Delta S_{\text{rev}} = +5.76 \text{ J/K}. \quad \text{(Answer)}$$

ΔS is positive, so the entropy increases, in accordance with the entropy postulate of Section 21-1.

Touchstone Example 21-2-2

Figure TE21-1*a* shows two identical copper blocks of mass $m = 1.5$ kg: block L at temperature $T_{iL} = 60°\text{C}$ and block R at temperature $T_{iR} = 20°\text{C}$. The blocks are in a thermally insulated box and are separated by an insulating shutter. When we lift the shutter, the

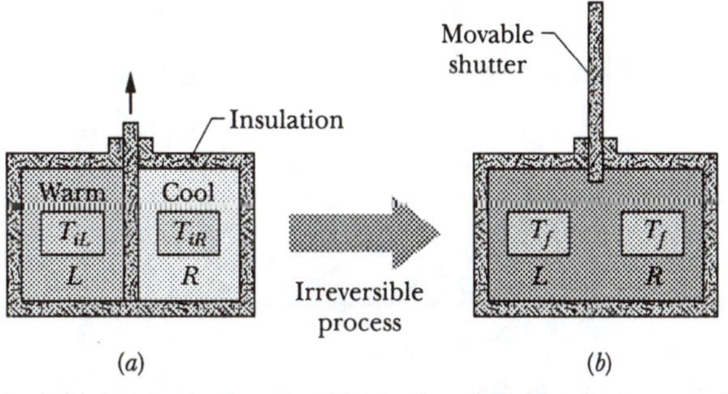

Fig. TE21-1 (*a*) In the initial state, two copper blocks L and R, identical except for their temperatures, are in an insulating box and are separated by an insulating shutter. (*b*) When the shutter is removed, the blocks exchange energy as heat and come to a final state, both with the same temperature T_f.

blocks eventually come to the equilibrium temperature $T_f = 40°C$ (Fig. TE21-1*b*). What is the net entropy change of the two-block system during this irreversible process? The specific heat of copper is 386 J/kg · K.

SOLUTION: The **Key Idea** here is that to calculate the entropy change, we must find a reversible process that takes the system from the initial state of Fig. TE21-1*a* to the final state of Fig. TE21-1*b*. We can calculate the net entropy change ΔS_{rev} of the reversible process using Eq. 21-1, and then the entropy change for the irreversible process is equal to ΔS_{rev}. For such a reversible process we need a thermal reservoir whose temperature can be changed slowly (say, by turning a knob). We then take the blocks through the following two steps, illustrated in Fig. TE21-2.

Step 1. With the reservoir's temperature set at 60°C, put block L on the reservoir. (Since block and reservoir are at the same temperature, they are already in thermal equilibrium.) Then slowly lower the temperature of the reservoir and the block to 40°C. As the block's temperature changes by each increment dT during this process, energy dQ is transferred as heat *from* the block to the reservoir. Using Eq. 19-14, we can write this transferred energy as $dQ = mc\,dT$, where c is the specific heat of copper. According to Eq. 21-1, the entropy change ΔS_L of block L during the full temperature change from initial temperature T_{iL} (= 60°C = 333 K) to final temperature T_f (= 40°C = 313 K) is

$$\Delta S_L = \int_i^f \frac{dQ}{T} = \int_{T_{iL}}^{T_f} \frac{mc\,dT}{T} = mc \int_{T_{iL}}^{T_f} \frac{dT}{T}$$
$$= mc \ln \frac{T_f}{T_{iL}}.$$

Inserting the given data yields

$$\Delta S_L = (1.5 \text{ kg})(386 \text{ J/kg}\cdot\text{K}) \ln \frac{313 \text{ K}}{333 \text{ K}}$$
$$= -35.86 \text{ J/K}.$$

Step 2. With the reservoir's temperature now set at 20°C, put block R on the reservoir. Then slowly raise the temperature of the reservoir and the block to 40°C. With the same reasoning used to find ΔS_L, you can show that the entropy change ΔS_R of block R during this process is

$$\Delta S_R = (1.5 \text{ kg})(386 \text{ J/kg}\cdot\text{K}) \ln \frac{313 \text{ K}}{293 \text{ K}}$$
$$= +38.23 \text{ J/K}.$$

The net entropy change ΔS_{rev} of the two-block system undergoing this two-step reversible process is then

$$\Delta S_{rev} = \Delta S_L + \Delta S_R$$
$$= -35.86 \text{ J/K} + 38.23 \text{ J/K} = 2.4 \text{ J/K}.$$

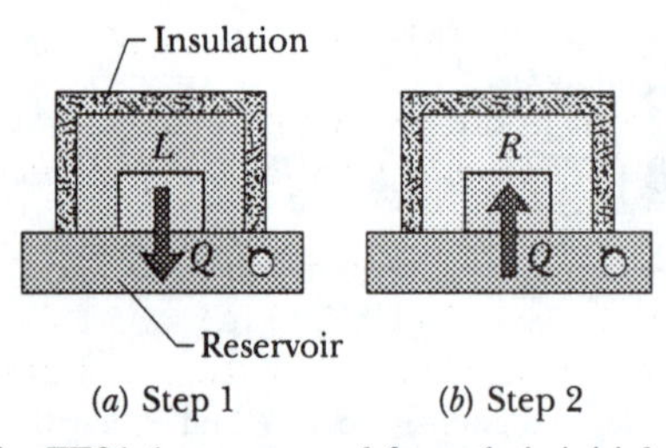

(*a*) Step 1 (*b*) Step 2

Fig. TE21-2 The blocks of Fig. TE21-1 can proceed from their initial state to their final state in a reversible way if we use a reservoir with a controllable temperature (*a*) to extract heat reversibly from block L and (*b*) to add heat reversibly to block R.

Thus, the net entropy change ΔS_{irrev} for the two-block system undergoing the actual irreversible process is

$$\Delta S_{irrev} = \Delta S_{rev} = 2.4 \text{ J/K}. \qquad \text{(Answer)}$$

This result is positive, in accordance with the entropy postulate of Section 21-1.

Touchstone Example 21-4-1

Imagine a Carnot engine that operates between the temperatures $T_H = 850$ K and $T_L = 300$ K. The engine performs 1200 J of work each cycle, which takes 0.25 s.

(a) What is the efficiency of this engine?

SOLUTION: The **Key Idea** here is that the efficiency ε of a Carnot engine depends only on the ratio T_L/T_H of the temperatures (in kelvins) of the thermal reservoirs to which it is connected. Thus, from Eq. 21-11, we have

$$\varepsilon = 1 - \frac{T_L}{T_H} = 1 - \frac{300 \text{ K}}{850 \text{ K}} = 0.647 \approx 65\%. \qquad \text{(Answer)}$$

(b) What is the average power of this engine?

SOLUTION: Here the **Key Idea** is that the average power P of an engine is the ratio of the work W it does per cycle to the time t that each cycle takes. For this Carnot engine, we find

$$P = \frac{W}{t} = \frac{1200 \text{ J}}{0.25 \text{ s}} = 4800 \text{ W} = 4.8 \text{ kW}. \qquad \text{(Answer)}$$

(c) How much energy $|Q_H|$ is extracted as heat from the high-temperature reservoir every cycle?

SOLUTION: Now the **Key Idea** is that for any engine, including a Carnot engine, the efficiency ε is the ratio of the work W that is done per cycle to the energy $|Q_H|$ that is extracted as heat from the high-temperature reservoir per cycle ($\varepsilon = W/|Q_H|$). Thus,

$$|Q_H| = \frac{W}{\varepsilon} = \frac{1200 \text{ J}}{0.647} = 1855 \text{ J}. \qquad \text{(Answer)}$$

(d) How much energy $|Q_L|$ is delivered as heat to the low-temperature reservoir every cycle?

SOLUTION: The **Key Idea** here is that for a Carnot engine, the work W done per cycle is equal to the difference in the energy transfers as heat: $|Q_H| - |Q_L|$, as in Eq. 21-6. Thus, we have

$$\begin{aligned} |Q_L| &= |Q_H| - W \\ &= 1855 \text{ J} - 1200 \text{ J} = 655 \text{ J}. \end{aligned} \qquad \text{(Answer)}$$

(e) What is the entropy change of the working substance for the energy transfer to it from the high-temperature reservoir? From it to the low-temperature reservoir?

SOLUTION: The **Key Idea** here is that the entropy change ΔS during a transfer of energy Q as heat at constant temperature T is given by Eq. 21-2 ($\Delta S = Q/T$). Thus, for the *positive* transfer of energy Q_H from the high-temperature reservoir at T_H, the change in the entropy of the working substance is

$$\Delta S_H = \frac{Q_H}{T_H} = \frac{1855 \text{ J}}{850 \text{ K}} = +2.18 \text{ J/K}. \qquad \text{(Answer)}$$

Similarly, for the *negative* transfer of energy Q_L to the low-temperature reservoir at T_L, we have

$$\Delta S_L = \frac{Q_L}{T_L} = \frac{-655 \text{ L}}{300 \text{ K}} = -2.18 \text{ J/K}. \qquad \text{(Answer)}$$

Note that the net entropy change of the working substance for one cycle is zero, as we discussed in deriving Eq. 21-8.

Touchstone Example 21-4-2

An inventor claims to have constructed an engine that has an efficiency of 75% when operated between the boiling and freezing points of water. Is this possible?

SOLUTION: The **Key Idea** here is that the efficiency of a real engine (with its irreversible processes and wasteful energy transfers) must be less than the efficiency of a Carnot engine operating between the same two temperatures. From Eq. 21-11, we find that the efficiency of a Carnot engine operating between the boiling and freezing points of water is

$$\varepsilon = 1 - \frac{T_L}{T_H} = 1 - \frac{(0 + 273)\text{ K}}{(100 + 273)\text{ K}} = 0.268 \approx 27\%.$$

Thus, the claimed efficiency of 75% for a real engine operating between the given temperatures is impossible.

Touchstone Example 21-7-1

Suppose that there are 100 indistinguishable molecules in the box of Fig. 21-14. How many microstates are associated with the configuration $n_1 = 50$ and $n_2 = 50$? How many are associated with the configuration $n_1 = 100$ and $n_2 = 0$? Interpret the results in terms of the relative probabilities of the two configurations.

SOLUTION: The **Key Idea** here is that the multiplicity W of a configuration of indistinguishable molecules in a closed box is the number of independent microstates with that configuration, as given by Eq. 21-18. For the (n_1, n_2) configuration (50, 50), that equation yields

$$\begin{aligned} W &= \frac{N!}{n_1!\, n_2!} = \frac{100!}{50!\, 50!} \\ &= \frac{9.33 \times 10^{157}}{(3.04 \times 10^{64})(3.04 \times 10^{64})} \\ &= 1.01 \times 10^{29}. \end{aligned} \qquad \text{(Answer)}$$

Similarly, for the configuration of (100, 0), we have

$$W = \frac{N!}{n_1!\, n_2!} = \frac{100!}{100!\, 0!} = \frac{1}{0!} = \frac{1}{1} = 1. \qquad \text{(Answer)}$$

Thus, a 50–50 distribution is more likely than a 100–0 distribution by the enormous factor of about 1×10^{29}. If you could count, at one per nanosecond, the number of microstates that correspond to the 50–50 distribution, it would take you about 3×10^{12} years, which is about 750 times longer than the age of the universe. Even 100 molecules is *still* a very small number. Imagine what these calculated probabilities would be like for a mole of molecules, say about $N = 10^{24}$. Thus, you need never worry about suddenly finding all the air molecules clustering in one corner of your room!

Touchstone Example 21-7-2

In Touchstone Example 21-2-1 we showed that when n moles of an ideal gas doubles its volume in a free expansion, the entropy increase from the initial state i to the final state f is $S_f - S_i = nR \ln 2$. Derive this result with statistical mechanics.

SOLUTION: One **Key Idea** here is that we can relate the entropy S of any given configuration of the molecules in the gas to the multiplicity W of microstates for that configuration, using Eq. 21-19 ($S = k \ln W$). We are interested in two configurations: the final configuration f (with the molecules occupying the full volume of their container in Fig. 21-1b) and the initial configuration i (with the molecules occupying the left half of the container).

A second **Key Idea** is that, because the molecules are in a closed container, we can calculate the multiplicity W of their microstates with Eq. 21-18. Here we have N molecules in the n moles of the gas. Initially, with the molecules all in the left half of the container, their (n_1, n_2) configuration is $(N, 0)$. Then, Eq. 21-18 gives their multiplicity as

$$W_i = \frac{N!}{N!\,0!} = 1.$$

Finally, with the molecules spread through the full volume, their (n_1, n_2) configuration is $(N/2, N/2)$. Then, Eq. 21-18 gives their multiplicity as

$$W_f = \frac{N!}{(N/2)!\,(N/2)!}.$$

From Eq. 21-19, the initial and final entropies are

$$S_i = k \ln W_i = k \ln 1 = 0$$

and

$$S_f = k \ln W_f = k \ln(N!) - 2k \ln[(N/2)!]. \qquad \text{(TE21-1)}$$

In writing Eq. TE21-1, we have used the relation

$$\ln \frac{a}{b^2} = \ln a - 2 \ln b.$$

Now, applying Eq. 21-201 to evaluate Eq. TE21-1, we find that

$$\begin{aligned} S_f &= k \ln(N!) - 2k \ln[(N/2)!] \\ &= k[N(\ln N) - N] - 2k[(N/2)\ln(N/2) - (N/2)] \\ &= k[N(\ln N) - N - N\ln(N/2) + N] \\ &= k[N(\ln N) - N(\ln N - \ln 2)] = Nk \ln 2. \end{aligned} \qquad \text{(TE21-2)}$$

From Section 20-3 we can substitute nR for Nk, where R is the universal gas constant. Equation TE21-2 then becomes

$$S_f = nR \ln 2.$$

The change in entropy from the initial state to the final is thus

$$\begin{aligned} S_f - S_i &= nR \ln 2 - 0 \\ &= nR \ln 2, \end{aligned} \qquad \text{(Answer)}$$

which is what we set out to show. In Touchstone Example 21-2-1 we calculated this entropy increase for a free expansion with thermodynamics by finding an equivalent reversible process and calculating the entropy change for *that* process in terms of temperature and heat transfer. In this sample problem, we calculate the same increase in entropy with statistical mechanics using the fact that the system consists of molecules.

Exercises & Problems

Chapter 13

SEC. 13-3 The Center of Gravity

1E. A physics Brady Bunch, whose weights in newtons are indicated in Fig. 13-11, is balanced on a seesaw. What is the number of the person who causes the largest torque, about the rotation axis at *fulcrum f*, directed (a) out of the page and (b) into the page?

Fig. 13-11 Exercise 1.

2E. The leaning Tower of Pisa (Fig. 13-12) is 55 m high and 7.0 m in diameter. The top of the tower is displaced 4.5 m from the vertical. Treat the tower as a uniform, circular cylinder. (a) What additional displacement, measured at the top, would bring the tower to the verge of toppling? (b) What angle would the tower then make with the vertical?

Fig. 13-12 Exercise 2.

3E. A particle is acted on by forces given, in newtons, by $\vec{F}_1 = 10\hat{\mathrm{i}} - 4\hat{\mathrm{j}}$ and $\vec{F}_2 = 17\hat{\mathrm{i}} + 2\hat{\mathrm{j}}$. (a) What force $\vec{F}_3$ balances these forces? (b) What direction does $\vec{F}_3$ have relative to the x axis?

4E. A bow is drawn at its midpoint until the tension in the string is equal to the force exerted by the archer. What is the angle between the two halves of the string.

5E. A rope of negligible mass is stretched horizontally between two supports that are 3.44 m apart. When an object of weight 3160 N is hung at the center of the rope, the rope is observed to sag by 35.0 cm. What is the tension in the rope?

6E. A scaffold of mass 60 kg and length 5.0 m is supported in a horizontal position by a vertical cable at each end. A window washer of mass 80 kg stands at a point 1.5 m from one end. What is the tension in (a) the nearer cable and (b) the farther cable?

7E. In Fig. 13-13, a uniform sphere of mass m and radius r is held in place by a massless rope attached to a frictionless wall a distance L above the center of the sphere. Find (a) the tension in the rope and (b) the force on the sphere from the wall.

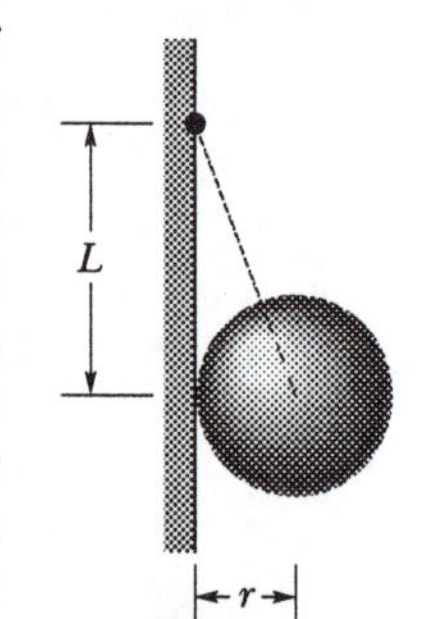

Fig. 13-13 Exercise 7.

8E. An automobile with a mass of 1360 kg has 3.05 m between the front and rear axles. Its center of gravity is located 1.78 m behind the front axle. With the automobile on level ground, determine the magnitude of the force from the ground on (a) each front wheel (assuming equal forces on the front wheels) and (b) each rear wheel (assuming equal forces on the rear wheels).

9E. A diver of weight 580 N stands at the end of a 4.5 m diving board of negligible mass (Fig. 13-14). The board is attached to two pedestals 1.5 m apart. What are the magnitude and direction of the force on the board from (a) the left pedestal and (b) the right pedestal? (c) Which pedestal is being stretched, and (d) which compressed?

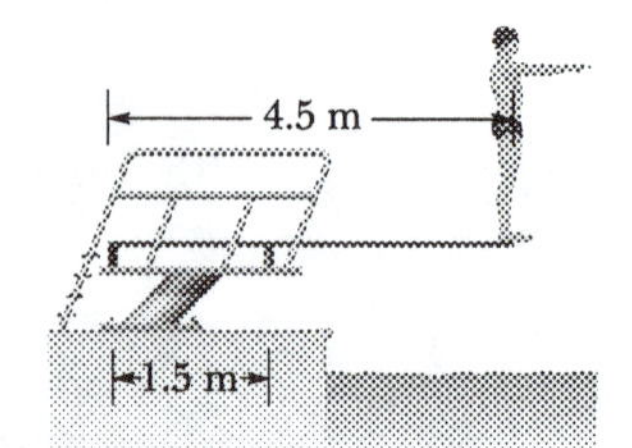

Fig. 13-14 Exercise 9.

10E. In Fig. 13-15, a man is trying to get his car out of mud on the shoulder of a road. He ties one end of a rope tightly around the front bumper and the other end tightly around a utility pole 18 m away. He then pushes sideways on the rope at its midpoint with a force of 550 N, displacing the center of the rope 0.30 m from its previous position, and the car barely moves. What is the magnitude of the force on the car from the rope? (The rope stretches somewhat.)

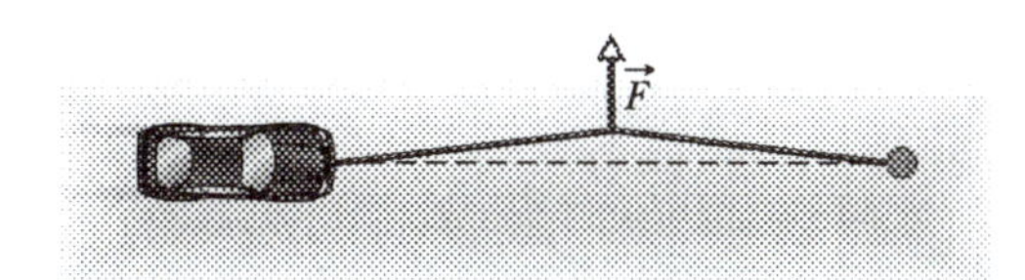

Fig. 13-15 Exercise 10.

11E. A meter stick balances horizontally on a knife-edge at the 50.0 cm mark. With two 5.0 g coins stacked over the 12.0 cm mark, the stick is found to balance at the 45.5 cm mark. What is the mass of the meter stick?

12E. A uniform cubical crate is 0.750 m on each side and weighs 500 N. It rests on a floor with one edge against a very small, fixed obstruction. At what least height above the floor must a horizontal force of magnitude 350 N be applied to the crate to tip it?

13E. A 75 kg window cleaner uses a 10 kg ladder that is 5.0 m long. He places one end on the ground 2.5 m from a wall, rests the upper end against a cracked window, and climbs the ladder. He is 3.0 m up along the ladder when the window breaks. Neglecting friction between the ladder and window and assuming that the base of the ladder does not slip, find (a) the magnitude of the force on the window from the ladder just before the window breaks and (b) the magnitude and direction of the force on the ladder from the ground just before the window breaks.

14E. Figure 13-16 shows the anatomical structures in the lower leg and foot that are involved in standing tiptoe with the heel raised off the floor so the foot effectively contacts the floor at only one point, shown as P in the figure. Calculate, in terms of a person's weight W, the forces on the foot from (a) the calf muscle (at A) and (b) the lower-leg bones (at B) when the person stands tiptoe on one foot. Assume that $a = 5.0$ cm and $b = 15$ cm.

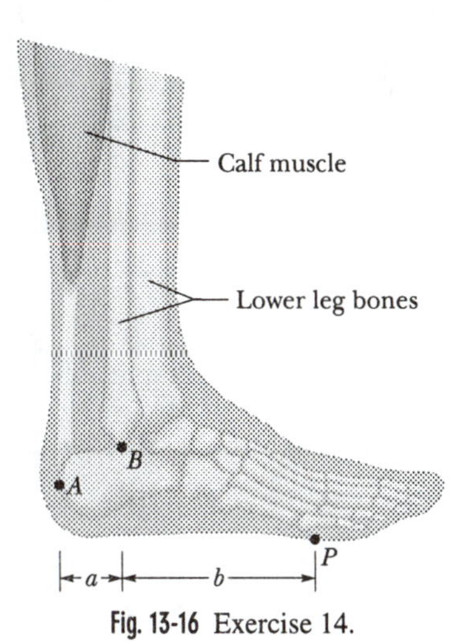

Fig. 13-16 Exercise 14.

15P. In Fig. 13-17, an 817 kg construction bucket is suspended by a cable A that is attached at O to two other cables B and C, making angles of 51.0° and 66.0° with the horizontal. Find the tensions in (a) cable A, (b) cable B, and (c) cable C. (*Hint:* To avoid solving two equations in two unknowns, position the axes as shown in the figure.)

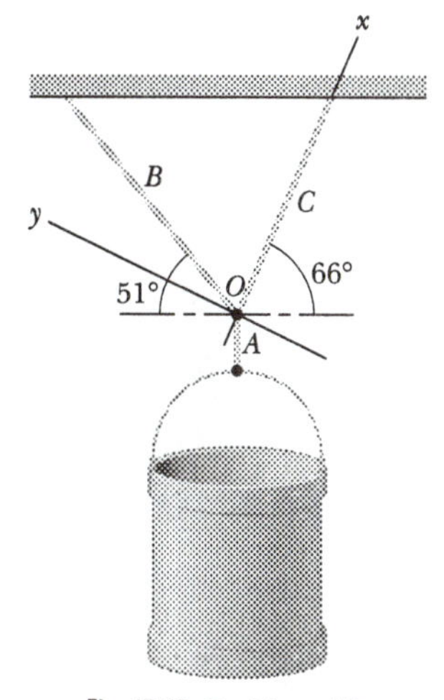

Fig. 13-17 Problem 15.

16P. The system in Fig. 13-18 is in equilibrium, with the string in the center exactly horizontal. Find (a) tension T_1, (b) tension T_2, (c) tension T_3, and (d) angle θ.

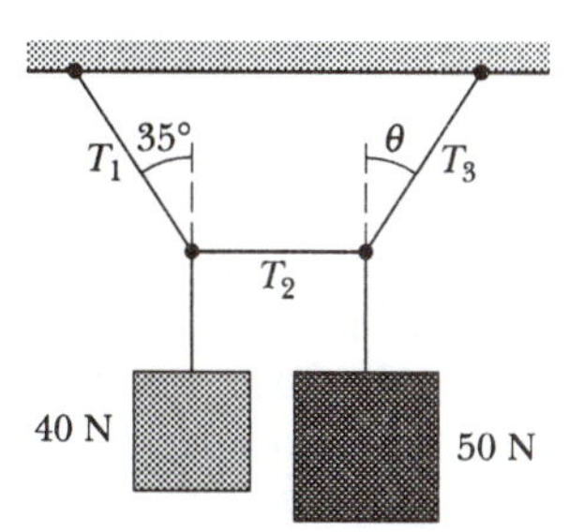

Fig. 13-18 Problem 16.

17P. The force $\vec{F}$ in Fig. 13-19 keeps the 6.40 kg block and the pulleys in equilibrium. The pulleys have negligible mass and friction. Calculate the tension T in the upper cable. (*Hint:* When a cable wraps halfway around a pulley as here, the magnitude of its net force on the pulley is twice the tension in the cable.)

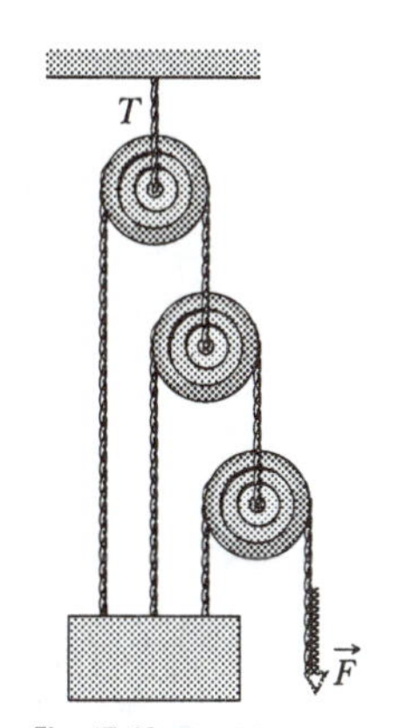

Fig. 13-19 Problem 17.

18P. A 15 kg block is being lifted by the pulley system shown in Fig. 13-20. The upper arm is vertical, whereas the forearm makes an angle of 30° with the horizontal. What are the forces on the forearm from (a) the triceps muscle and (b) the upper-arm bone (the humerus)? The forearm and hand together have a mass of 2.0 kg with a center of mass 15 cm (measured along the arm) from the point where the forearm and upper-arm bones are in contact. The triceps muscle pulls vertically upward at a point 2.5 cm behind that contact point.

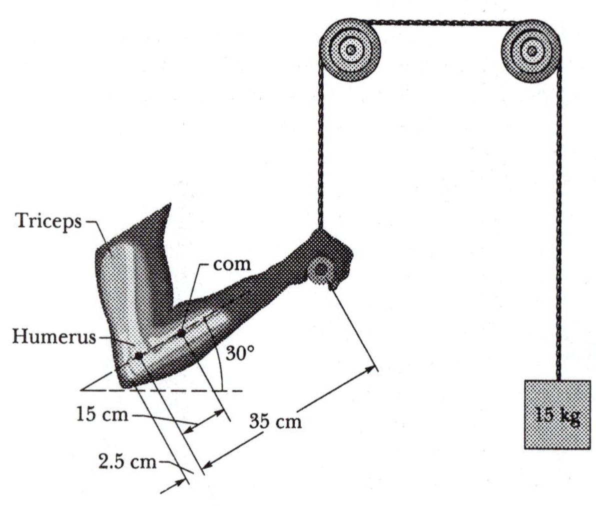

Fig. 13-20 Problem 18.

19P. Forces $\vec{F}_1$, $\vec{F}_2$, and $\vec{F}_3$ act on the structure of Fig. 13-21, shown in an overhead view. We wish to put the structure in equilibrium by applying a fourth force, at a point such as P. The fourth force has vector components $\vec{F}_h$ and $\vec{F}_v$. We are given that $a = 2.0$ m, $b = 3.0$ m, $c = 1.0$ m, $|\vec{F}_1| = 20$ N, $|\vec{F}_2| = 10$ N, and $|\vec{F}_3| = 5.0$ N. Find (a) $|\vec{F}_h|$, (b) $|\vec{F}_v|$, and (c) d.

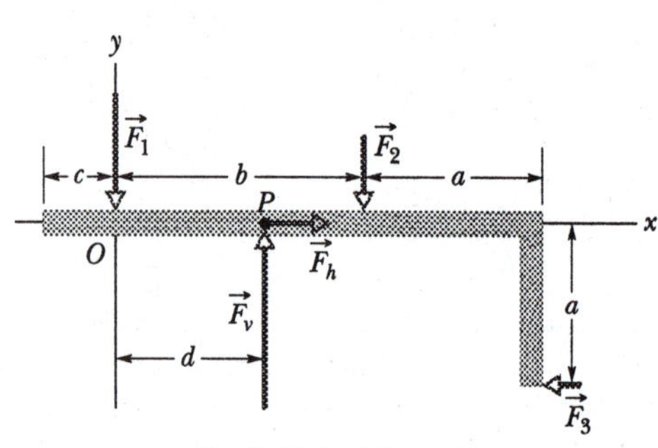

Fig. 13-21 Problem 19.

20P. In Fig. 13-22, a 50.0 kg uniform square sign, 2.00 m on a side, is hung from a 3.00 m horizontal rod of negligible mass. A cable is attached to the end of the rod and to a point on the wall 4.00 m above the point where the rod is hinged to the wall. (a) What is the tension in the cable? What are the magnitudes and directions of the (b) horizontal and (c) vertical components of the force on the rod from the wall?

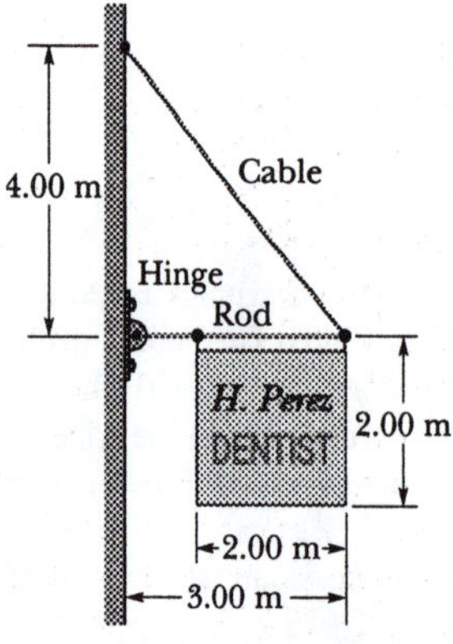

Fig. 13-22 Problem 20.

21P. In Fig. 13-23, what magnitude of force $\vec{F}$ applied horizontally at the axle of the wheel is necessary to raise the wheel over an obstacle of height h? The wheel's radius is r and its mass is m.

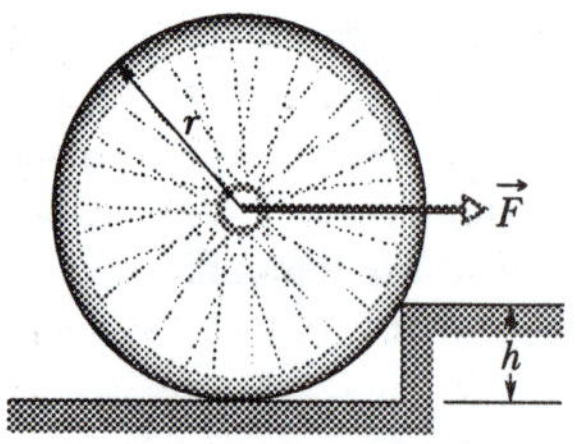

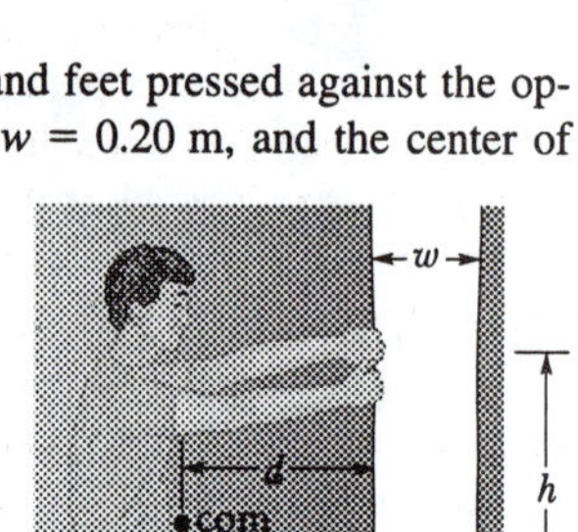

Fig. 13-23 Problem 21.

22P. In Fig. 13-24, a 55 kg rock climber is in a lie-back climb along a fissure, with hands pulling on one side of the fissure and feet pressed against the opposite side. The fissure has width $w = 0.20$ m, and the center of mass of the climber is a horizontal distance $d = 0.40$ m from the fissure. The coefficient of static friction between hands and rock is $\mu_1 = 0.40$, and between boots and rock it is $\mu_2 = 1.2$. (a) What is the least horizontal pull by the hands and push by the feet that will keep the climber stable? (b) For the horizontal pull of (a), what must be the vertical distance h between hands and feet? (c) If the climber encounters wet rock, so that μ_1 and μ_2 are reduced, what happens to the answers to (a) and (b), respectively?

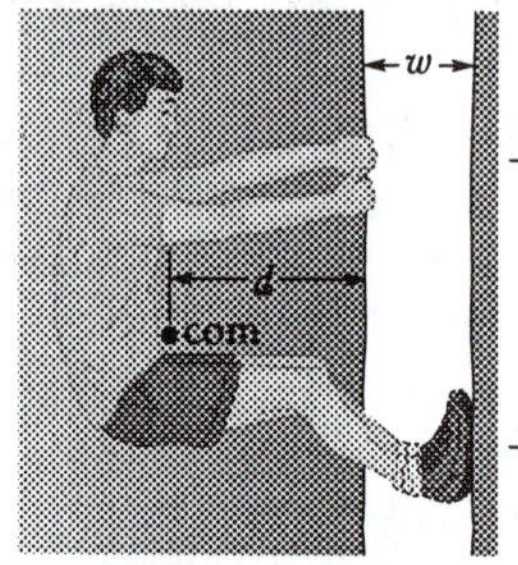

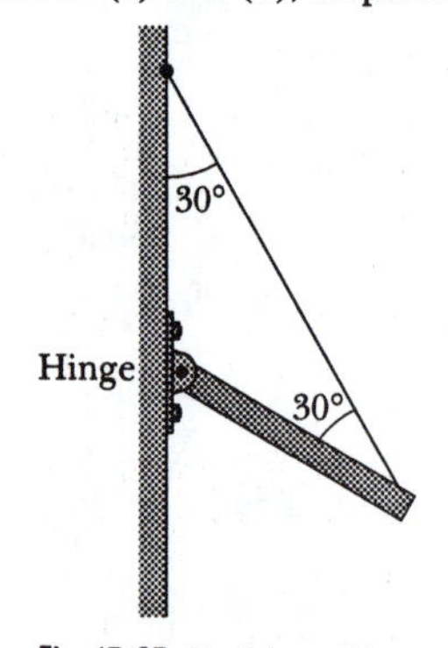

Fig. 13-24 Problem 22.

23P. In Fig. 13-25, one end of a uniform beam that weighs 222 N is attached to a wall with a hinge. The other end is supported by a wire. (a) Find the tension in the wire. What are the (b) horizontal and (c) vertical components of the force of the hinge on the beam?

Fig. 13-25 Problem 23.

24P. Four bricks of length L, identical and uniform, are stacked on top of one another (Fig. 13-26) in such a way that part of each extends beyond the one beneath. Find, in terms of L, the maximum values of (a) a_1, (b) a_2, (c) a_3, (d) a_4, and (e) h, such that the stack is in equilibrium.

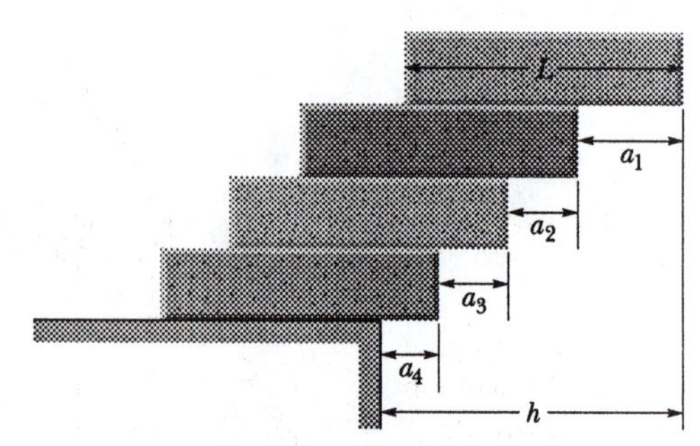

Fig. 13-26 Problem 24.

25P. The system in Fig. 13-27 is in equilibrium. A concrete block of mass 225 kg hangs from the end of the uniform strut whose mass is 45.0 kg. Find (a) the tension T in the cable and the (b) horizontal and (c) vertical force components on the strut from the hinge.

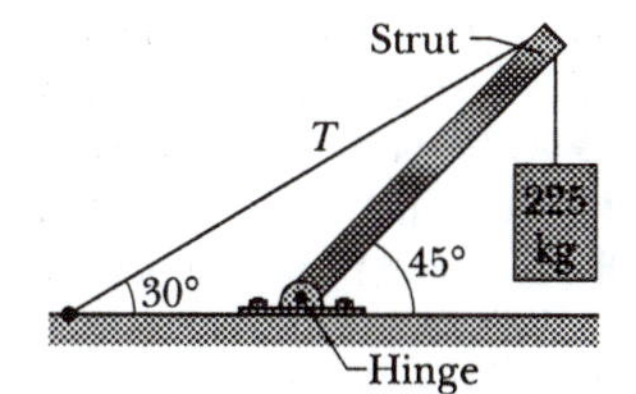

Fig. 13-27 Problem 25.

26P. A door 2.1 m high and 0.91 m wide has a mass of 27 kg. A hinge 0.30 m from the top and another 0.30 m from the bottom each support half the door's mass. Assume that the center of gravity is at the geometrical center of the door, and determine the (a) vertical and (b) horizontal components of the force from each hinge on the door.

27P. A nonuniform bar is suspended at rest in a horizontal position by two massless cords as shown in Fig. 13-28. One cord makes the angle $\theta = 36.9°$ with the vertical; the other makes the angle $\phi = 53.1°$ with the vertical. If the length L of the bar is 6.10 m, compute the distance x from the left-hand end of the bar to its center of mass.

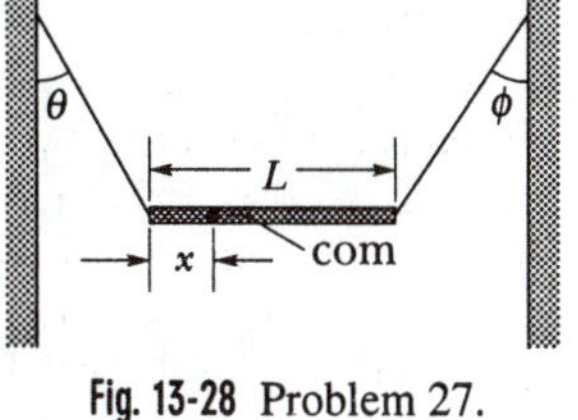

Fig. 13-28 Problem 27.

28P. In Fig. 13-29, a thin horizontal bar AB of negligible weight and length L is hinged to a vertical wall at A and supported at B by a thin wire BC that makes an angle θ with the horizontal. A load of weight W can be moved anywhere along the bar; its position is defined by the distance x from the wall to its center of mass. As a function of x, find (a) the tension in the wire, and the (b) horizontal and (c) vertical components of the force on the bar from the hinge at A.

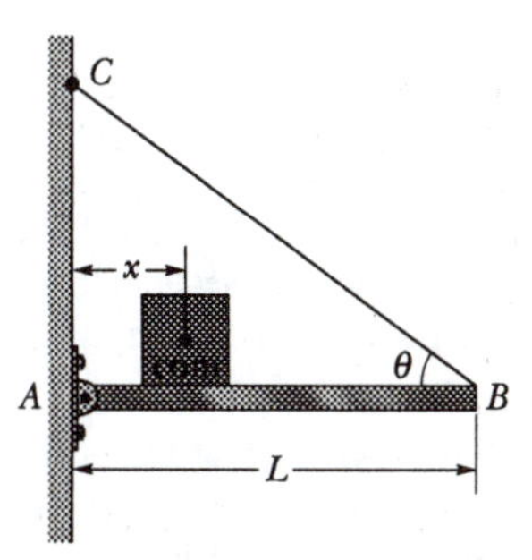

Fig. 13-29 Problems 28 and 30.

29P. In Fig. 13-30, a uniform plank, with a length L of 6.10 m and a weight of 445 N, rests on the ground and against a frictionless roller at the top of a wall of height $h = 3.05$ m. The plank remains in equilibrium for any value of $\theta \geq 70°$ but slips if $\theta < 70°$. Find the coefficient of static friction between the plank and the ground.

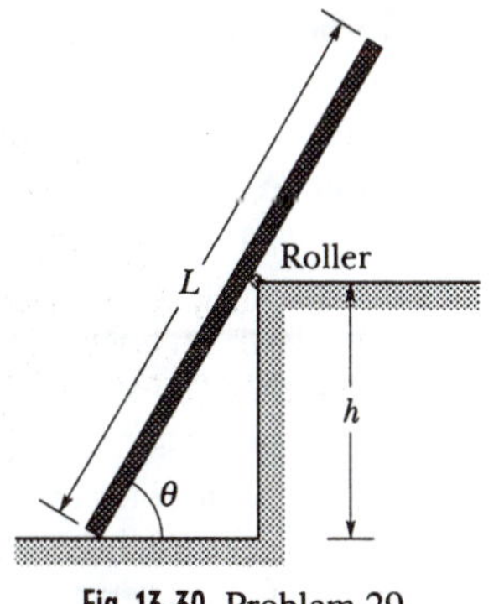

Fig. 13-30 Problem 29.

30P. In Fig. 13-29, suppose the length L of the uniform bar is 3.0 m and its weight is 200 N. Also, let the load's weight $W = 300$ N and the angle $\theta = 30°$. The wire can withstand a maximum tension of 500 N. (a) What is the maximum possible distance x before the wire breaks? With the load placed at this maximum x, what are the (b) horizontal and (c) vertical components of the force on the bar from the hinge at A?

31P. For the stepladder shown in Fig. 13-31, sides AC and CE are each 2.44 m long and hinged at C. Bar BD is a tie-rod 0.762 m long, halfway up. A man weighing 854 N climbs 1.80 m along the ladder. Assuming that the floor is frictionless and neglecting the mass of the ladder, find (a) the tension in the tie-rod and the magnitudes of the forces on the ladder from the floor at (b) A and (c) E. (*Hint:* It will help to isolate parts of the ladder in applying the equilibrium conditions.)

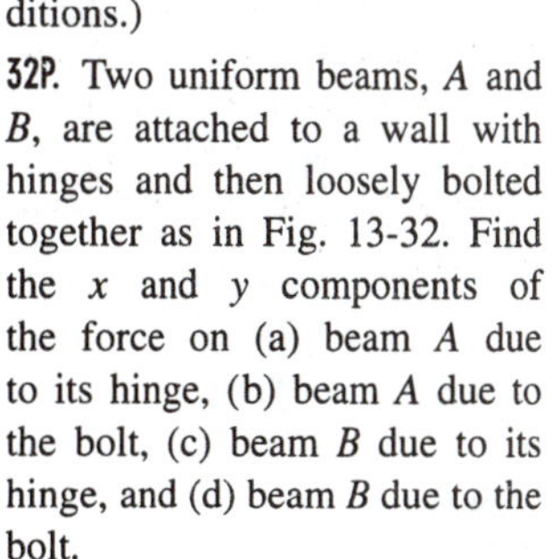

Fig. 13-31 Problem 31.

32P. Two uniform beams, A and B, are attached to a wall with hinges and then loosely bolted together as in Fig. 13-32. Find the x and y components of the force on (a) beam A due to its hinge, (b) beam A due to the bolt, (c) beam B due to its hinge, and (d) beam B due to the bolt.

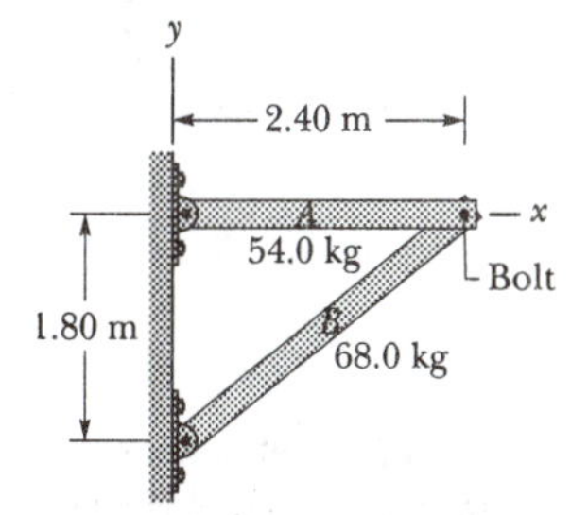

Fig. 13-32 Problem 32.

33P. A cubical box is filled with sand and weighs 890 N. We wish to "roll" the box by pushing horizontally on one of the upper edges. (a) What minimum force is required? (b) What minimum coefficient of static friction between box and floor is required? (c) Is there a more efficient way to roll the box? If so, find the smallest possible force that would have to be applied directly to the box to roll it. (*Hint:* At the onset of tipping, where is the normal force located?)

34P. Four bricks of length L, identical and uniform, are stacked on a table in two ways, as shown in Fig. 13-33 (compare with Problem 24). We seek to maximize the overhang distance h in both arrangements. Find the optimum distances a_1, a_2, b_1, and b_2, and calculate h for the two arrangements. (See "The Amateur Scientist," *Scientific American,* June 1985, pp. 133–134, for a discussion and an even better version of arrangement (b).)

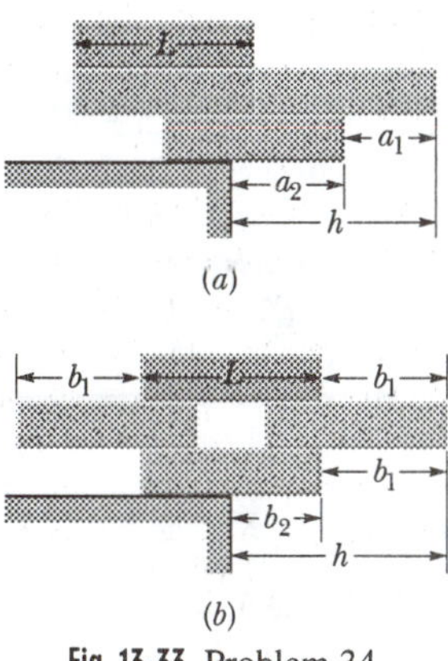

Fig. 13-33 Problem 34.

35P. A crate, in the form of a cube with edge lengths of 1.2 m, contains a piece of machinery; the center of mass of the crate and its contents is located 0.30 m above the crate's geometrical center. The crate rests on a ramp that makes an angle θ with the horizontal. As θ is increased from zero, an angle will be reached at which the crate will either start to slide down the ramp or tip over. Which event will occur (a) when the coefficient of static friction between ramp and crate is 0.60 and (b) when it is 0.70? In each case, give the angle at which the event occurs. (*Hint:* At the onset of tipping, where is the normal force located?)

SEC. 13-5 Elasticity

36E. Figure 13-34 shows the stress–strain curve for quartzite. What are (a) the Young's modulus and (b) the approximate yield strength for this material?

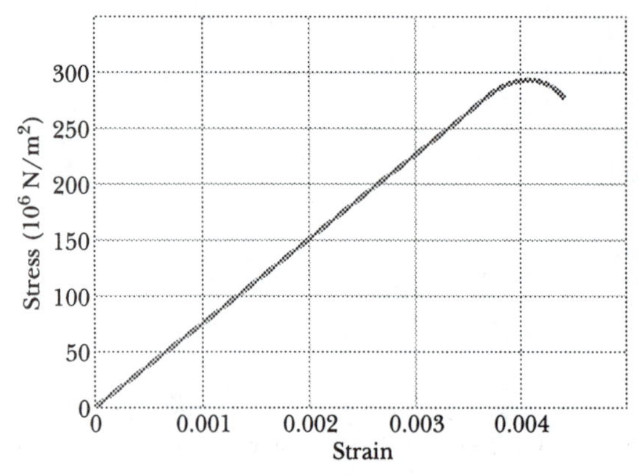

Fig. 13-34 Exercise 36.

37E. A horizontal aluminum rod 4.8 cm in diameter projects 5.3 cm from a wall. A 1200 kg object is suspended from the end of the rod. The shear modulus of aluminum is 3.0×10^{10} N/m^2. Neglecting the rod's mass, find (a) the shear stress on the rod and (b) the vertical deflection of the end of the rod.

38P. In Fig. 13-35, a lead brick rests horizontally on cylinders A and B. The areas of the top faces of the cylinders are related by $A_A = 2A_B$; the Young's moduli of the cylinders are related by $E_A = 2E_B$. The cylinders had identical lengths before the brick was placed on them. What fraction of the brick's mass is supported (a) by cylinder A and (b) by cylinder B? The horizontal distances between the center of mass of the brick and the centerlines of the cylinders are d_A for cylinder A and d_B for cylinder B. (c) What is the ratio d_A/d_B?

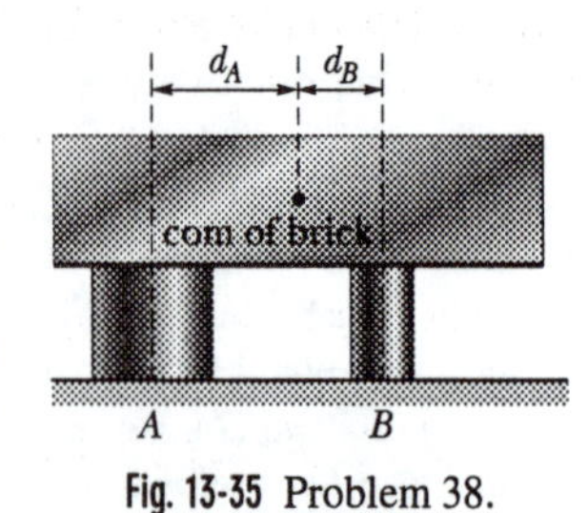

Fig. 13-35 Problem 38.

39P. In Fig. 13-36, a 103 kg uniform log hangs by two steel wires, A and B, both of radius 1.20 mm. Initially, wire A was 2.50 m long and 2.00 mm shorter than wire B. The log is now horizontal. What are the magnitudes of the forces on it from (a) wire A and (b) wire B? (c) What is the ratio d_A/d_B?

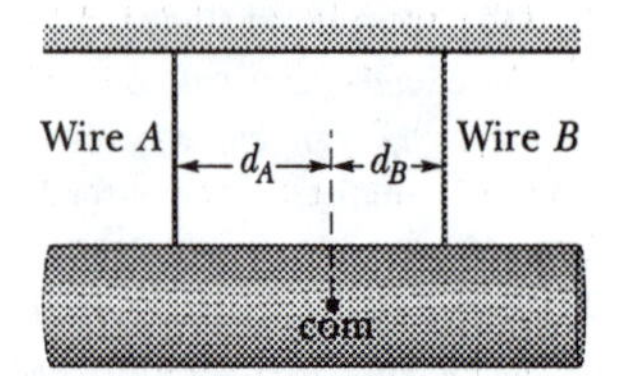

Fig. 13-36 Problem 39.

40P. A tunnel 150 m long, 7.2 m high, and 5.8 m wide (with a flat roof) is to be constructed 60 m beneath the ground. (See Fig. 13-37.) The tunnel roof is to be supported entirely by square steel columns, each with a cross-sectional area of 960 cm^2. The density of the ground material is 2.8 g/cm^3. (a) What is the total mass of the material that the columns must support? (b) How many columns are needed to keep the compressive stress on each column at one-half its ultimate strength?

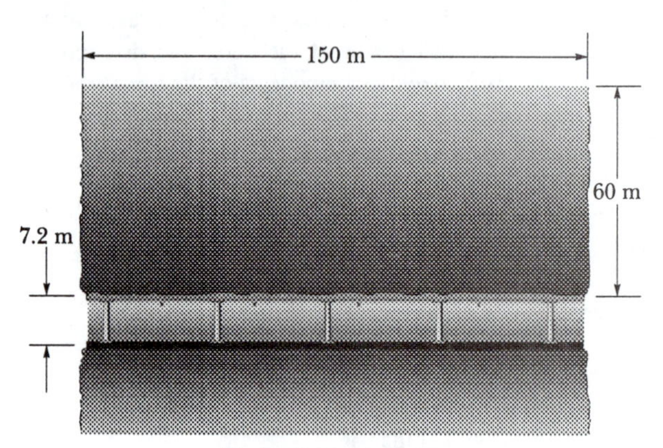

Fig. 13-37 Problem 40.

Chapter 14

SEC. 14-2 Newton's Law of Gravitation

1E. What must the separation be between a 5.2 kg particle and a 2.4 kg particle for their gravitational attraction to have a magnitude of 2.3×10^{-12} N?

2E. Some believe that the positions of the planets at the time of birth influence the newborn. Others deride this belief and claim that the gravitational force exerted on a baby by the obstetrician is greater than that exerted by the planets. To check this claim, calculate and compare the magnitude of the gravitational force exerted on a 3 kg baby (a) by a 70 kg obstetrician who is 1 m away and roughly approximated as a point mass, (b) by the massive planet Jupiter ($m = 2 \times 10^{27}$ kg) at its closest approach to Earth ($= 6 \times 10^{11}$ m), and (c) by Jupiter at its greatest distance from Earth ($= 9 \times 10^{11}$ m). (d) Is the claim correct?

3E. One of the *Echo* satellites consisted of an inflated spherical aluminum balloon 30 m in diameter and of mass 20 kg. Suppose a meteor having a mass of 7.0 kg passes within 3.0 m of the surface of the satellite. What is the magnitude of the gravitational force on the meteor from the satellite at the closest approach?

4E. The Sun and Earth each exert a gravitational force on the Moon. What is the ratio $|\vec{F}_{\text{Sun}}/\vec{F}_{\text{Earth}}|$ of the magnitudes of these two forces? (The average Sun–Moon distance is equal to the Sun–Earth distance.)

5E. A mass M is split into two parts, m and $M - m$, which are then separated by a certain distance. What ratio m/M maximizes the magnitude of the gravitational force between the parts?

SEC. 14-3 Gravitation and the Principle of Superposition

6E. A spaceship is on a straight-line path between Earth and its moon. At what distance from Earth is the net gravitational force on the spaceship zero?

7E. How far from Earth must a space probe be along a line toward the Sun so that the Sun's gravitational pull on the probe balances Earth's pull?

8P. Three 5.0 kg spheres are located in the xy plane as shown in Fig. 14-18. What is the magnitude of the net gravitational force on the sphere at the origin due to the other two spheres?

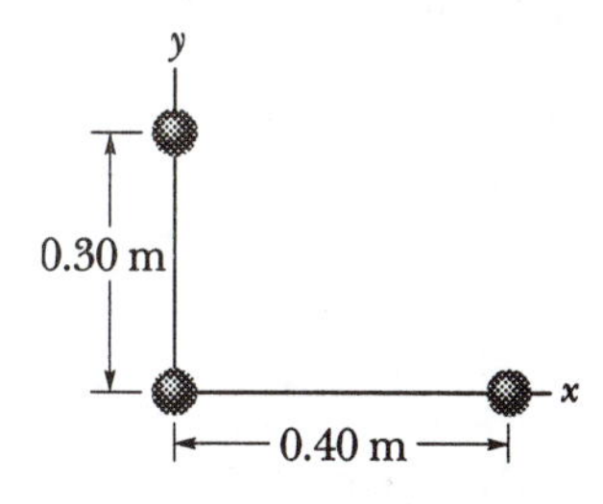

Fig. 14-18 Problem 8.

9P. In Fig. 14-19a, four spheres form the corners of a square whose side is 2.0 cm long. What are the magnitude and direction of the net gravitational force from them on a central sphere with mass $m_5 = 250$ kg?

10P. In Fig. 14-19b, two spheres of mass m and a third sphere of mass M form an equilateral triangle, and a fourth sphere of mass m_4 is at the center of the triangle. The net gravitational force on that central sphere from the three other spheres is zero. (a) What is M in terms of m? (b) If we double the value of m_4, what then is the magnitude of the net gravitational force on the central sphere?

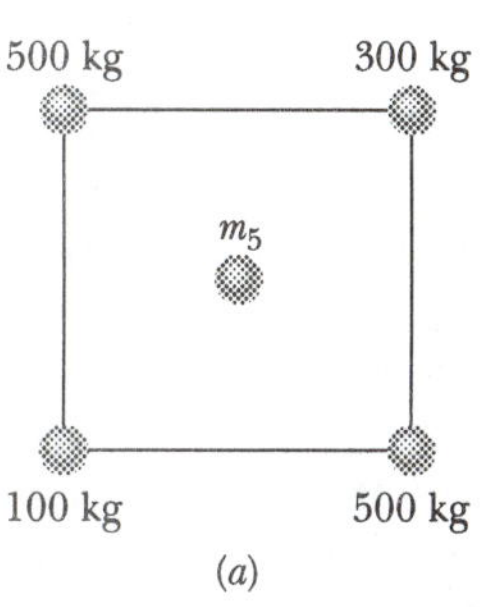

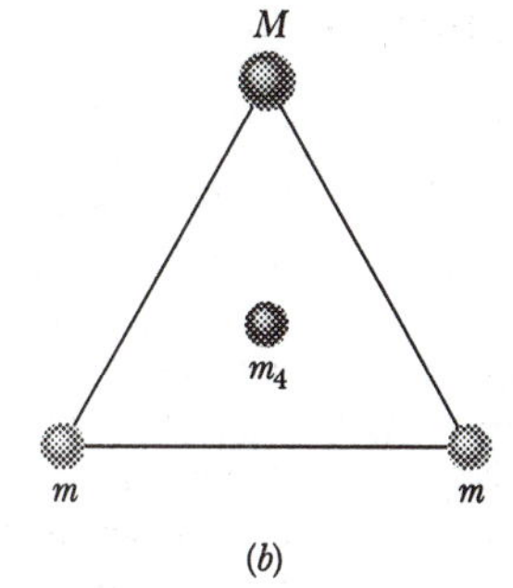

Fig. 14-19 Problems 9 and 10.

11P. The masses and coordinates of three spheres are as follows: 20 kg, $x = 0.50$ m, $y = 1.0$ m; 40 kg, $x = -1.0$ m, $y = -1.0$ m; 60 kg, $x = 0$ m, $y = -0.50$ m. What is the magnitude of the gravitational force on a 20 kg sphere located at the origin due to the other spheres?

12P. Four uniform spheres, with masses $m_A = 400$ kg, $m_B = 350$ kg, $m_C = 2000$ kg, and $m_D = 500$ kg, have (x, y) coordinates of (0, 50 cm), (0, 0), (-80 cm, 0), and (40 cm, 0), respectively. What is the net gravitational force on sphere B due to the other spheres?

13P. Figure 14-20 shows a spherical hollow inside a lead sphere of radius R; the surface of the hollow passes through the center of the sphere and "touches" the right side of the sphere. The mass of the sphere before hollowing was M. With what gravitational force does the hollowed-out lead sphere attract a small sphere of mass m that lies at a distance d from the center of the lead sphere, on the straight line connecting the centers of the spheres and of the hollow?

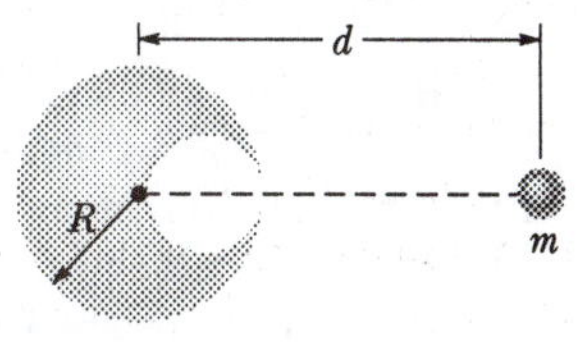

Fig. 14-20 Problem 13.

SEC. 14-4 Gravitation Near Earth's Surface

14E. You weigh 530 N at sidewalk level outside the World Trade Center in New York City. Suppose that you ride from this level to the top of one of its 410 m towers. Ignoring Earth's rotation, how much less would you weigh there (because you are slightly farther from the center of Earth)?

15E. At which altitude above Earth's surface would the gravitational acceleration be 4.9 m/s^2?

16E. (a) What will an object weigh on the Moon's surface if it weighs 100 N on Earth's surface? (b) How many Earth radii must this same object be from the center of Earth if it is to weigh the same as it does on the Moon?

17P. The fastest possible rate of rotation of a planet is that for which the gravitational force on material at the equator just barely provides the centripetal force needed for the rotation. (Why?) (a) Show that the corresponding shortest period of rotation is

$$T = \sqrt{\frac{3\pi}{G\rho}},$$

where ρ is the uniform density of the spherical planet. (b) Calculate the rotation period assuming a density of 3.0 g/cm^3, typical of many planets, satellites, and asteroids. No astronomical object has ever been found to be spinning with a period shorter than that determined by this analysis.

18P. One model for a certain planet has a core of radius R and mass M surrounded by an outer shell of inner radius R, outer radius $2R$, and mass $4M$. If $M = 4.1 \times 10^{24}$ kg and $R = 6.0 \times 10^6$ m, what is the gravitational acceleration of a particle at points (a) R and (b) $3R$ from the center of the planet?

19P. A body is suspended from a spring scale in a ship sailing along the equator with speed $|\vec{v}|$. (a) Show that the scale reading will be very close to $W_0(1 \pm 2|\vec{\omega}||\vec{v}|/g)$, where $|\vec{\omega}|$ is the angular speed of Earth and W_0 is the scale reading when the ship is at rest. (b) Explain the $\pm$ sign.

20P. Certain neutron stars (extremely dense stars) are believed to be rotating at about 1 rev/s. If such a star has a radius of 20 km, what must be its minimum mass so that material on its surface remains in place during the rapid rotation?

SEC. 14-5 Gravitation Inside Earth

21E. Two concentric shells of uniform density having masses M_1 and M_2 are situated as shown in Fig. 14-21. Find the magnitude of the net gravitational force on a particle of mass m, due to the shells, when the particle is located at (a) point A, at distance $r = a$ from the center, (b) point B at $r = b$, and

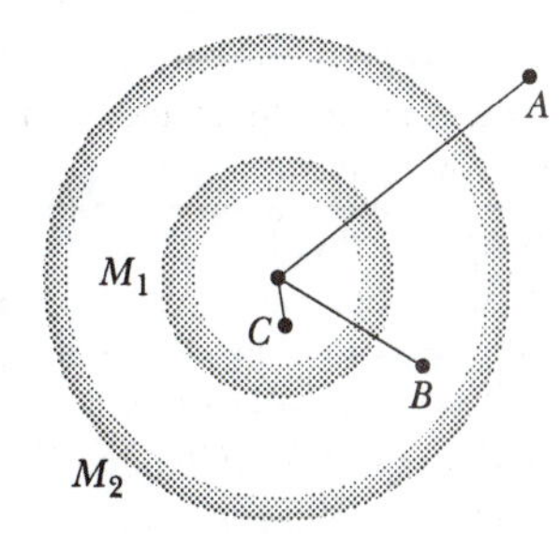

Fig. 14-21 Exercise 21.

(c) point C at $r = c$. The distance r is measured from the center of the shells.

22P. A solid sphere of uniform density has a mass of 1.0×10^4 kg and a radius of 1.0 m. What is the magnitude of the gravitational force due to the sphere on a particle of mass m located at a distance of (a) 1.5 m and (b) 0.50 m from the center of the sphere? (c) Write a general expression for the magnitude of the gravitational force on the particle at a distance $r \leq 1.0$ m from the center of the sphere.

23P. A uniform solid sphere of radius R produces a gravitational acceleration of a_g on its surface. At what two distances from the center of the sphere is the gravitational acceleration $|\vec{a}_g|3$? (*Hint:* Consider distances both inside and outside the sphere.)

24P. Figure 14-22 shows, not to scale, a cross section through the interior of Earth. Rather than being uniform throughout, Earth is divided into three zones: an outer *crust,* a *mantle,* and an inner *core.* The dimensions of these zones and the masses contained within them are shown on the figure. Earth has a total mass of 5.98×10^{24} kg and a radius of 6370 km. Ignore rotation and assume that Earth is spherical. (a) Calculate the magnitude of $|\vec{a}_g|$ at the surface. (b) Suppose that a bore hole (the *Mohole*) is driven to the crust–mantle interface at a depth of 25 km; what would be the value of $|\vec{a}_g|$ at the bottom of the hole? (c) Suppose that Earth were a uniform sphere with the same total mass and size. What would be the value of $|\vec{a}_g|$ at a depth of 25 km? (Precise measurements of $|\vec{a}_g|$ are sensitive probes of the interior structure of Earth, although results can be clouded by local density variations.)

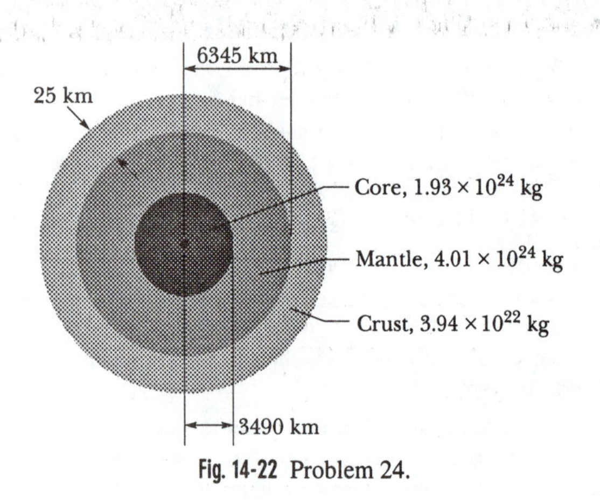

Fig. 14-22 Problem 24.

SEC. 14-6 Gravitational Potential Energy

25E. (a) What is the gravitational potential energy of the two-particle system in Exercise 1? If you triple the separation between the particles, how much work is done (b) by the gravitational force between the particles and (c) by you?

26E. (a) In Problem 12, remove sphere A and calculate the gravitational potential energy of the remaining three-particle system. (b) If A is then put back in place, is the potential energy of the four-particle system more or less than that of the system in (a)? (c) In (a), is the work done by you to remove A positive or negative? (d) In (b), is the work done by you to replace A positive or negative?

27E. In Problem 5, what ratio m/M gives the least gravitational potential energy for the system?

28E. The mean diameters of Mars and Earth are 6.9×10^3 km and 1.3×10^4 km, respectively. The mass of Mars is 0.11 times Earth's mass. (a) What is the ratio of the mean density of Mars to that of Earth? (b) What is the value of the gravitational acceleration on Mars? (c) What is the escape speed on Mars?

29E. Calculate the amount of energy required to escape from (a) Earth's moon and (b) Jupiter relative to that required to escape from Earth.

30P. The three spheres in Fig. 14-23, with masses $m_A = 800$ g, $m_B = 100$ g, and $m_C = 200$ g, have their centers on a common line, with $L = 12$ cm and $d = 4.0$ cm. You move sphere B along the line until its center-to-center separation from C is $d = 4.0$ cm. How much work is done on sphere B (a) by you and (b) by the net gravitational force on B due to spheres A and C?

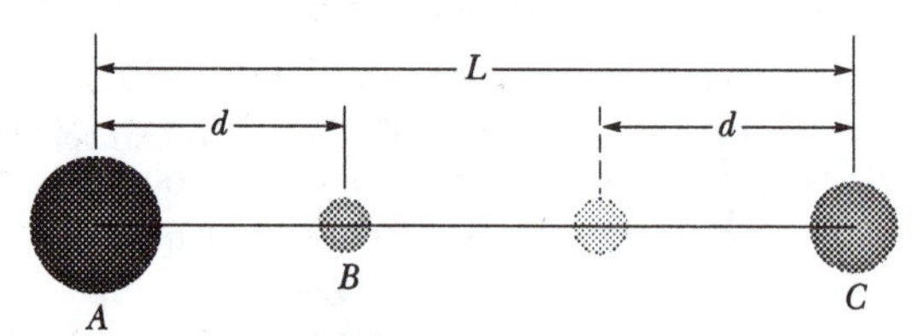

Fig. 14-23 Problem 30.

31P. Zero, a hypothetical planet, has a mass of 5.0×10^{23} kg, a radius of 3.0×10^6 m, and no atmosphere. A 10 kg space probe is to be launched vertically from its surface. (a) If the probe is launched with an initial energy of 5.0×10^7 J, what will be its kinetic energy when it is 4.0×10^6 m from the center of Zero? (b) If the probe is to achieve a maximum distance of 8.0×10^6 m from the center of Zero, with what initial kinetic energy must it be launched from the surface of Zero?

32P. A rocket is accelerated to speed $|\vec{v}| = 2\sqrt{gR_e}$ near Earth's surface (where Earth's radius is R_e), and it then coasts upward. (a) Show that it will escape from Earth. (b) Show that very far from Earth its speed will be $|\vec{v}| = \sqrt{2gR_e}$.

33P. Planet Roton, with a mass of 7.0×10^{24} kg and a radius of 1600 km, gravitationally attracts a meteorite that is initially at rest relative to the planet, at a great enough distance to take as infinite. The meteorite falls toward the planet. Assuming the planet is airless, find the speed of the meteorite when it reaches the planet's surface.

34P. (a) What is the escape speed on a spherical asteroid whose radius is 500 km and whose gravitational acceleration at the surface is 3.0 m/s^2? (b) How far from the surface will a particle go if it leaves the asteroid's surface with a radial speed of 1000 m/s? (c) With what speed will an object hit the asteroid if it is dropped from 1000 km above the surface?

35P. A 150.0 kg rocket moving radially outward from Earth has a speed of 3.70 km/s when its engine shuts off 200 km above Earth's surface. (a) Assuming negligible air drag, find the rocket's kinetic energy when the rocket is 1000 km above Earth's surface. (b) What maximum height above the surface is reached by the rocket?

36P. Two neutron stars are separated by a distance of 10^{10} m. They each have a mass of 10^{30} kg and a radius of 10^5 m. They are initially at rest with respect to each other. As measured from that rest frame, how fast are they moving when (a) their separation has decreased to one-half its initial value and (b) they are about to collide?

37P. In deep space, sphere A of mass 20 kg is located at the origin of an x axis and sphere B of mass 10 kg is located on the axis at $x = 0.80$ m. Sphere B is released from rest while sphere A is held at the origin. (a) What is the gravitational potential energy of the two-sphere system as B is released? (b) What is the kinetic energy of B when it has moved 0.20 m toward A?

38P. A projectile is fired vertically from Earth's surface with an initial speed of 10 km/s. Neglecting air drag, how far above the surface of Earth will it go?

SEC. 14-7 Planets and Satellites: Kepler's Laws

39E. The mean distance of Mars from the Sun is 1.52 times that of Earth from the Sun. From Kepler's law of periods, calculate the number of years required for Mars to make one revolution about the Sun; compare your answer with the value given in Appendix C.

40E. The Martian satellite Phobos travels in an approximately circular orbit of radius 9.4×10^6 m with a period of 7 h 39 min. Calculate the mass of Mars from this information.

41E. Determine the mass of Earth from the period T (27.3 days) and the radius r (3.82×10^5 km) of the Moon's orbit about Earth. Assume the Moon orbits the center of Earth rather than the center of mass of the Earth–Moon system.

42E. Our Sun, with mass 2.0×10^{30} kg, revolves about the center of the Milky Way galaxy, which is 2.2×10^{20} m away, once every 2.5×10^8 years. Assuming that each of the stars in the galaxy has a mass equal to that of our Sun, that the stars are distributed uniformly in a sphere about the galactic center, and that our Sun is essentially at the edge of that sphere, estimate roughly the number of stars in the galaxy.

43E. A satellite is placed in a circular orbit about Earth with a radius equal to one-half the radius of the Moon's orbit. What is its period of revolution in lunar months? (A lunar month is the period of revolution of the Moon.)

44E. (a) What linear speed must an Earth satellite have to be in a circular orbit at an altitude of 160 km? (b) What is the period of revolution?

45E. The Sun's center is at one focus of Earth's orbit. How far from this focus is the other focus, (a) in meters and (b) in terms of the solar radius, 6.96×10^8 m? The eccentricity of Earth's orbit is 0.0167, and the semimajor axis is 1.50×10^{11} m.

46E. A satellite, moving in an elliptical orbit, is 360 km above Earth's surface at its farthest point and 180 km above at its closest point. Calculate (a) the semimajor axis and (b) the eccentricity of the orbit. (*Hint:* See Touchstone Example 14-7-1.)

47E. A satellite hovers over a certain spot on the equator of (rotating) Earth. What is the altitude of its orbit (called a *geosynchronous orbit*)?

48E. A comet that was seen in April 574 by Chinese astronomers on a day known by them as the Woo Woo day was spotted again in May 1994. Assume the time between observations is the period of the Woo Woo day comet and take its eccentricity as 0.11. What are (a) the semimajor axis of the comet's orbit and (b) its greatest distance from the Sun in terms of the mean orbital radius R_P of Pluto?

49E. In 1993 the spacecraft *Galileo* sent home an image (Fig. 14-24) of asteroid 243 Ida and a tiny orbiting moon (now known as Dactyl), the first confirmed example of an asteroid–moon system. In the image, the moon, which is 1.5 km wide, is 100 km from the center of the asteroid, which is 55 km long. The shape of the moon's orbit is not well known; assume it is circular with a period of 27 h. (a) What is the mass of the asteroid? (b) The volume of the asteroid, measured from the *Galileo* images, is 14 100 km^3. What is the density of the asteroid?

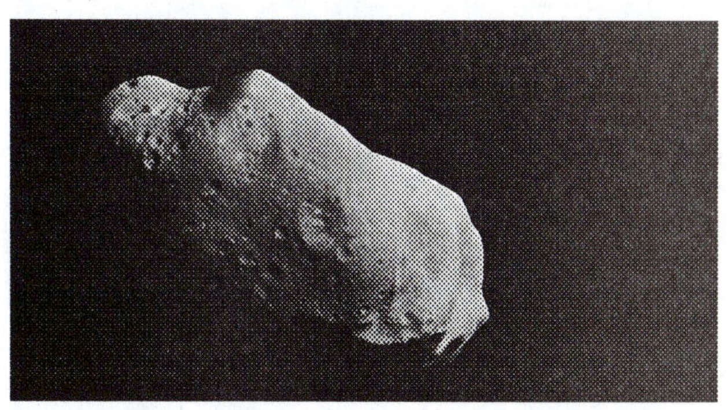

Fig. 14-24 Exercise 49. An image from the spacecraft *Galileo* shows a tiny moon orbiting asteroid 243 Ida.

50P. In 1610, Galileo used his telescope to discover four prominent moons around Jupiter. Their mean orbital radii a and periods T are as follows:

Name	a (10^8 m)	T (days)
Io	4.22	1.77
Europa	6.71	3.55
Ganymede	10.7	7.16
Callisto	18.8	16.7

(a) Plot log a (y axis) against log T (x axis) and show that you get a straight line. (b) Measure the slope of the line and compare it with the value that you expect from Kepler's third law. (c) Find the mass of Jupiter from the intercept of this line with the y axis.

51P. A 20 kg satellite has a circular orbit with a period of 2.4 h and a radius of 8.0×10^6 m around a planet of unknown mass. If the magnitude of the gravitational acceleration on the surface of the planet is 8.0 m/s^2, what is the radius of the planet?

52P. In a certain binary-star system, each star has the same mass as our Sun, and they revolve about their center of mass. The distance between them is the same as the distance between Earth and the Sun. What is their period of revolution in years?

53P. A certain triple-star system consists of two stars, each of mass m, revolving about a central star of mass M in the same circular orbit of radius r (Fig. 14-25). The two stars are always at opposite ends of a diameter of the circular orbit. Derive an expression for the period of revolution of the stars.

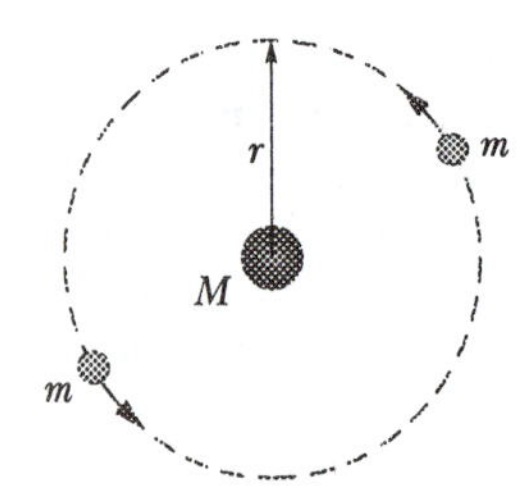

Fig. 14-25 Problem 53.

54P*. Three identical stars of mass M are located at the vertices of an equilateral triangle with side L. At what speed must they move if they all revolve under the influence of one another's gravitational force in a circular orbit circumscribing the triangle while still preserving the equilateral triangle?

SEC. 14-8 Satellites: Orbits and Energy

55E. Consider two satellites, A and B, both of mass m, moving in the same circular orbit of radius r around Earth, of mass M_E, but in opposite senses of rotation and therefore on a collision course (see Fig. 14-26). (a) In terms of G, M_E, m, and r, find the total mechanical energy $E_A + E_B$ of the two-satellite-plus-Earth system before collision. (b) If the collision is completely inelastic so that the wreckage remains as one piece of tangled material (mass = $2m$), find the total mechanical energy immediately after collision. (c) Describe the subsequent motion of the wreckage.

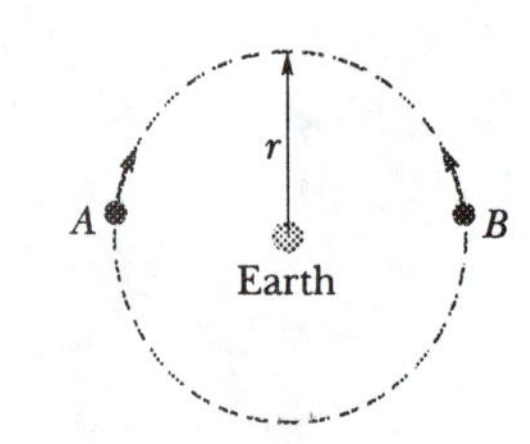

Fig. 14-26 Exercise 55.

56E. An asteroid, whose mass is 2.0×10^{-4} times the mass of Earth, revolves in a circular orbit around the Sun at a distance that is twice Earth's distance from the Sun. (a) Calculate the period of revolution of the asteroid in years. (b) What is the ratio of the kinetic energy of the asteroid to that of Earth?

57P. Two Earth satellites, A and B, each of mass m, are to be launched into circular orbits about Earth's center. Satellite A is to orbit at an altitude of 6370 km. Satellite B is to orbit at an altitude of 19 110 km. The radius of Earth R_E is 6370 km. (a) What is the ratio of the potential energy of satellite B to that of satellite A, in orbit? (b) What is the ratio of the kinetic energy of satellite B to that of satellite A, in orbit? (c) Which satellite has the greater total energy if each has a mass of 14.6 kg? By how much?

58P. Show that if an object is in an elliptical orbit with semimajor axis a about a planet of mass M, then its distance r from the planet and speed $|\vec{v}|$ are related by

$$v^2 = GM\left(\frac{2}{r} - \frac{1}{a}\right).$$

(*Hint:* Use the law of conservation of mechanical energy and Eq. 14-30.)

59P. Use the result of Problem 58 and data contained in Touchstone Example 14-7-1 to calculate (a) the speed $|\vec{v}_p|$ of comet Halley at perihelion and (b) its speed $|\vec{v}_a|$ at aphelion. (c) Using the law of conservation of angular momentum relative to the Sun, find the ratio of the comet's perihelion distance R_p to its aphelion distance R_a in terms of $|\vec{v}_p|$ and $|\vec{v}_a|$.

60P. (a) Does it take more energy to get a satellite up to 1500 km above Earth than to put it in circular orbit once it is there? (Take Earth's radius to be 6370 km.) (b) What about 3185 km? (c) What about 4500 km?

61P. One way to attack a satellite in Earth orbit is to launch a swarm of pellets in the same orbit as the satellite but in the opposite direction. Suppose a satellite in a circular orbit 500 km above Earth's surface collides with a pellet having mass 4.0 g. (a) What is the kinetic energy of the pellet in the reference frame of the satellite just before the collision? (b) What is the ratio of this kinetic energy to the kinetic energy of a 4.0 g bullet from a modern army rifle with a muzzle speed of 950 m/s?

62P. What are (a) the speed and (b) the period of a 220 kg satellite in an approximately circular orbit 640 km above the surface of Earth? Suppose the satellite loses mechanical energy at the average rate of 1.4×10^5 J per orbital revolution. Adopting the reasonable approximation that the satellite's orbit becomes a "circle of slowly diminishing radius," determine the satellite's (c) altitude, (d) speed, and (e) period at the end of its 1500th revolution. (f) What is the magnitude of the average retarding force on the satellite? Is angular momentum around Earth's center conserved for (g) the satellite and (h) the satellite–Earth system?

SEC. 14-9 Einstein and Gravitation

63E. In Fig. 14-15*b*, the scale on which the 60 kg physicist stands reads 220 N. How long will the cantaloupe take to reach the floor if the physicist drops it from rest (relative to himself), 2.1 m from the floor?

Chapter 15

SEC. 15-3 Pressure and Density

1E. Find the pressure increase in the fluid in a syringe when a nurse applies a force of 42 N to the syringe's circular piston, which has a radius of 1.1 cm.

2E. Three liquids that will not mix are poured into a cylindrical container. The volumes and densities of the liquids are 0.50 L, 2.6 g/cm^3; 0.25 L, 1.0 g/cm^3; and 0.40 L, 0.80 g/cm^3. What is the force on the bottom of the container due to these liquids? One liter = 1 L = 1000 cm^3. (Ignore the contribution due to the atmosphere.)

3E. An office window has dimensions 3.4 m by 2.1 m. As a result of the passage of a storm, the outside air pressure drops to 0.96 atm, but inside the pressure is held at 1.0 atm. What net force pushes out on the window?

4E. You inflate the front tires on your car to 28 psi. Later, you measure your blood pressure, obtaining a reading of 120/80, the readings being in mm Hg. In metric countries (which is to say, most of the world), these pressures are customarily reported in kilopascals (kPa). In kilopascals, what are (a) your tire pressure and (b) your blood pressure?

5E. A fish maintains its depth in fresh water by adjusting the air content of porous bone or air sacs to make its average density the same as that of the water. Suppose that with its air sacs collapsed, a fish has a density of 1.08 g/cm^3. To what fraction of its expanded body volume must the fish inflate the air sacs to reduce its density to that of water?

6P. An airtight container having a lid with negligible mass and an area of 77 cm^2 is partially evacuated. If a 480 N force is required to pull the lid off the container and the atmospheric pressure is 1.0×10^5 Pa, what is the air pressure in the container before it is opened?

7P. In 1654 Otto von Guericke, inventor of the air pump, gave a demonstration before the noblemen of the Holy Roman Empire in which two teams of eight horses could not pull apart two evacuated brass hemispheres. (a) Assuming that the hemispheres have thin walls, so that R in Fig. 15-28 may be considered both the inside and outside radius, show that the force $\vec{F}$ required to pull apart the hemispheres has magnitude $|\vec{F}| = \pi R^2 \Delta P$, where ΔP is the difference between the pressures outside and inside the sphere. (b) Taking R as 30 cm, the inside pressure as 0.10 atm, and the outside pressure as 1.00 atm, find the force magnitude the teams of horses would have had to exert to pull apart the hemispheres. (c) Explain why one team of horses could have proved the point just as well if the hemispheres were attached to a sturdy wall.

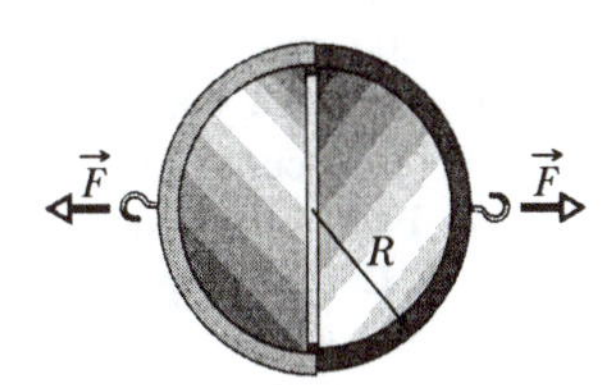

Fig. 15-28 Problem 7.

SEC. 15-4 Gravitational Forces and Fluids at Rest

8E. Calculate the hydrostatic difference in blood pressure between the brain and the foot in a person of height 1.83 m. The density of blood is 1.06×10^3 kg/m^3.

9E. The sewage outlet of a house constructed on a slope is 8.2 m below street level. If the sewer is 2.1 m below street level, find the minimum pressure difference that must be created by the sewage pump to transfer waste of average density 900 kg/m^3 from outlet to sewer.

10E. Figure 15-29 displays the *phase diagram* of carbon, showing the ranges of temperature and pressure in which carbon will crystallize either as diamond or graphite. What is the minimum depth at which diamonds can form if the temperature at that depth is 1000°C and the rocks there have density 3.1 g/cm^3? Assume that, as in a fluid, the pressure at any level is due to the gravitational force on the material lying above that level.

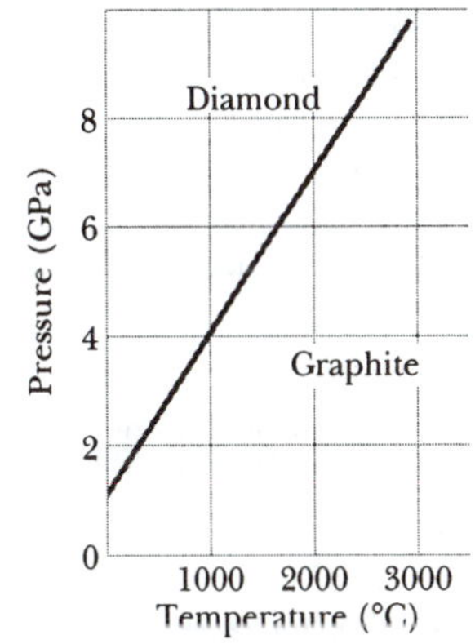

Fig. 15-29 Exercise 10.

11E. A swimming pool has the dimensions 24 m $\times$ 9.0 m $\times$ 2.5 m. When it is filled with water, what is the force (resulting from the water alone) on (a) the bottom, (b) each short side, and (c) each long side? (d) If you are concerned with the possibility that the concrete walls and floor will collapse, is it appropriate to take the atmospheric pressure into account? Why?

12E. (a) Assuming the density of seawater is 1.03 g/cm^3, find the total weight of water on top of a nuclear submarine at a depth of 200 m if its (horizontal cross-sectional) hull area is 3000 m^2. (b) In atmospheres, what water pressure would a diver experience at this depth? Do you think that occupants of a damaged submarine at this depth could escape without special equipment?

13E. Crew members attempt to escape from a damaged submarine 100 m below the surface. What force must be applied to a pop-out hatch, which is 1.2 m by 0.60 m, to push it out at that depth? Assume that the density of the ocean water is 1025 kg/m^3.

14E. A cylindrical barrel has a narrow tube fixed to the top, as shown (with dimensions) in Fig. 15-30. The vessel is filled with water to the top of the tube. Calculate the ratio of the hydrostatic force on the bottom of the barrel to the gravitational force on the water contained inside the barrel. Why is that ratio not equal to one? (You need not consider the atmospheric pressure.)

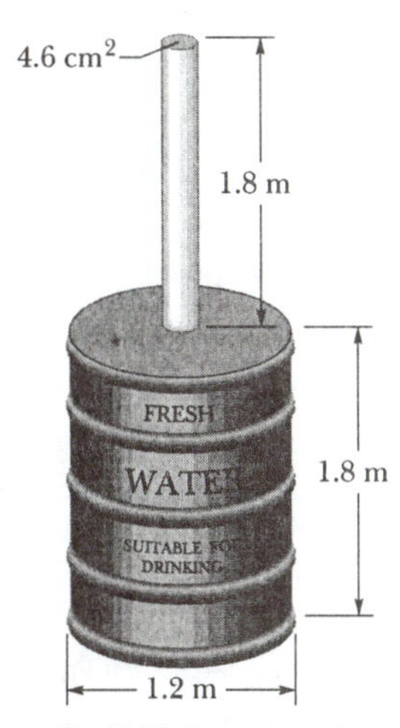

Fig. 15-30 Exercise 14.

15P. Two identical cylindrical vessels with their bases at the same level each contain a liquid of density ρ. The area of each base is A, but in one vessel the liquid height is h_1, and in the other it is h_2. Find the work done by the gravitational force in equalizing the levels when the two vessels are connected.

16P. In analyzing certain geological features, it is often appropriate to assume that the pressure at some horizontal *level of compensation*, deep inside Earth, is the same over a large region and is equal to the pressure due to the gravitational force on the overlying material. Thus, the pressure on the level of compensation is given by the fluid pressure formula. This model requires, for one thing, that mountains have *roots* of continental rock extending into the denser mantle (Fig. 15-31). Consider a mountain 6.0 km high. The continental rocks have a density of 2.9 g/cm^3, and beneath the continent the mantle has a density of 3.3 g/cm^3. Calculate the depth D of the root. (*Hint:* Set the pressure at points a and b equal; the depth y of the level of compensation will cancel out.)

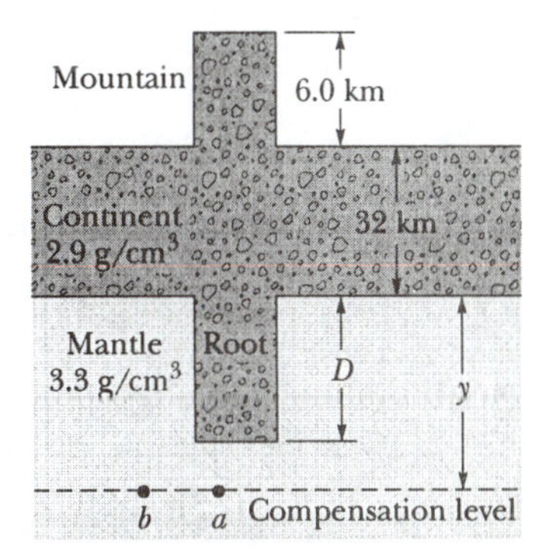

Fig. 15-31 Problem 16.

17P. Figure 15-32 shows the juncture of ocean and continent. Find the depth h of the ocean using the level-of-compensation technique presented in Problem 16.

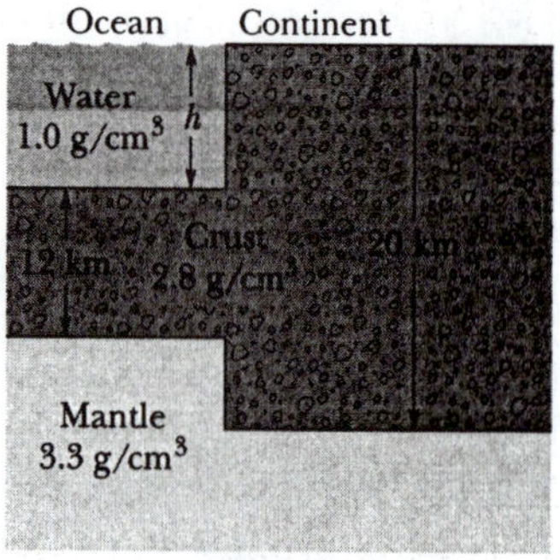

Fig. 15-32 Problem 17.

18P. The L-shaped tank shown in Fig. 15-33 is filled with water and is open at the top. If $d = 5.0$ m, what are (a) the force on face A and (b) the force on face B due to the water?

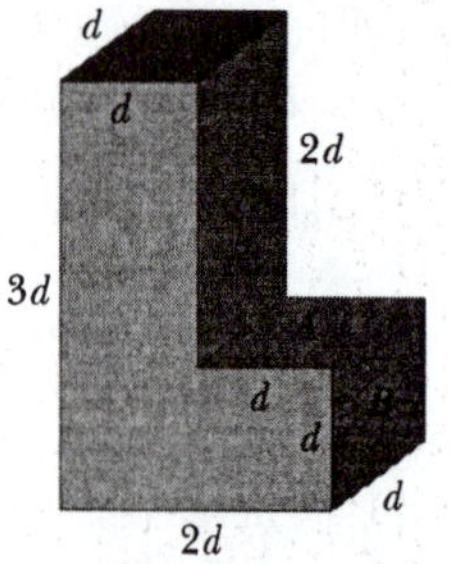

Fig. 15-33 Problem 18.

19P. Water stands at a depth D behind the vertical upstream face of a dam, as shown in Fig. 15-34. Let W be the width of the dam. Find (a) the net horizontal force on the dam from the gauge pressure of the water and (b) the net torque due to that force (and thus gauge pressure) about a line through O parallel to the width of the dam. (c) Find the moment arm of the net horizontal force about the line through O.

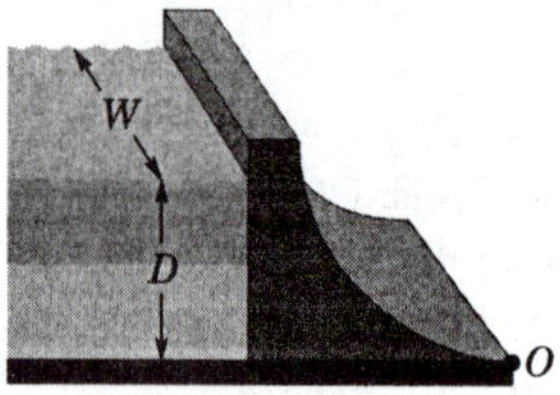

Fig. 15-34 Problem 19.

SEC. 15-5 Measuring Pressure

20E. To suck lemonade of density 1000 kg/m^3 up a straw to a maximum height of 4.0 cm, what minimum gauge pressure (in atmospheres) must you produce in your lungs?

21E. What would be the height of the atmosphere if the air density (a) were uniform and (b) decreased linearly to zero with height? Assume that at sea level the air pressure is 1.0 atm and the air density is 1.3 kg/m^3.

SEC. 15-6 Pascal's Principle

22E. A piston of small cross-sectional area a is used in a hydraulic press to exert a small force $\vec{f}$ on the enclosed liquid. A connecting pipe leads to a larger piston of cross-sectional area A (Fig. 15-35). (a) What force magnitude $|\vec{F}|$ will the larger piston sustain without moving? (b) If the small piston has a diameter of 3.80 cm and the large piston one of 53.0 cm, what force magnitude on the small piston will balance a 20.0 kN force on the large piston?

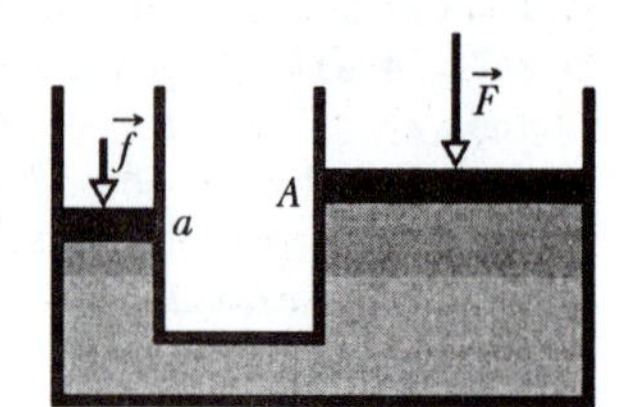

Fig. 15-35 Exercises 22 and 23.

23E. In the hydraulic press of Exercise 22, through what distance must the large piston be moved to raise the small piston a distance of 0.85 m?

SEC. 15-7 Archimedes' Principle

24E. A boat floating in fresh water displaces water weighing 35.6 kN. (a) What is the weight of the water that this boat would displace if it were floating in salt water with a density of 1.10×10^3 kg/m^3? (b) Would the volume of the displaced water change? If so, by how much?

25E. An iron anchor of density 7870 kg/m^3 appears 200 N lighter in water than in air. (a) What is the volume of the anchor? (b) How much does it weigh in air?

26E. In Fig. 15-36, a cubical object of dimensions $L = 0.600$ m on a side and with a mass of 450 kg is suspended by a rope in an open tank of liquid of density 1030 kg/m^3. (a) Find the magnitude of the total downward force on the top of the object from the liquid and the atmosphere, assuming that atmospheric pressure is 1.00 atm. (b) Find the magnitude of the total upward force on the bottom of the object. (c) Find the tension in the rope. (d) Calculate the magnitude of the buoyant force on the object using Archimedes' principle. What relation exists among all these quantities?

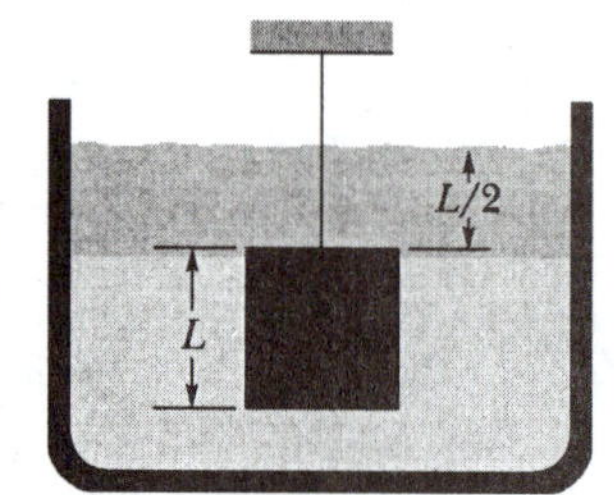

Fig. 15-36 Exercise 26.

27E. A block of wood floats in fresh water with two-thirds of its volume submerged. In oil the block floats with 0.90 of its volume submerged. Find the density of (a) the wood and (b) the oil.

28E. A blimp is cruising slowly at low altitude, filled as usual with helium gas. Its maximum useful payload, including crew and cargo, is 1280 kg. The volume of the helium-filled interior space is 5000 m^3. The density of helium gas is 0.16 kg/m^3, and the density of hydrogen is 0.081 kg/m^3. How much more payload could the blimp carry if you replaced the helium with hydrogen? (Why not do it?)

29P. A hollow sphere of inner radius 8.0 cm and outer radius 9.0 cm floats half-submerged in a liquid of density 800 kg/m^3. (a) What is the mass of the sphere? (b) Calculate the density of the material of which the sphere is made.

30P. About one-third of the body of a person floating in the Dead Sea will be above the water line. Assuming that the human body density is 0.98 g/cm^3, find the density of the water in the Dead Sea. (Why is it so much greater than 1.0 g/cm^3?)

31P. A hollow spherical iron shell floats almost completely submerged in water. The outer diameter is 60.0 cm, and the density of iron is 7.87 g/cm^3. Find the inner diameter.

32P. A block of wood has a mass of 3.67 kg and a density of 600 kg/m^3. It is to be loaded with lead so that it will float in water with 0.90 of its volume submerged. What mass of lead is needed (a) if the lead is attached to the top of the wood and (b) if the lead is attached to the bottom of the wood? The density of lead is 1.13×10^4 kg/m^3.

33P. An iron casting containing a number of cavities weighs 6000 N in air and 4000 N in water. What is the total volume of all the cavities in the casting? The density of iron (that is, a sample with no cavities) is 7.87 g/cm^3.

34P. Assume the density of brass weights to be 8.0 g/cm^3 and that of air to be 0.0012 g/cm^3. What percent error arises from neglecting the buoyancy of air in weighing an object of mass m and density ρ on a beam balance, as in Fig. 6-2?

35P. (a) What is the minimum area of the top surface of a slab of ice 0.30 m thick floating on fresh water that will hold up an automobile of mass 1100 kg? (b) Does it matter where the car is placed on the block of ice?

36P. Three children, each of weight 356 N, make a log raft by lashing together logs of diameter 0.30 m and length 1.80 m. How many logs will be needed to keep them afloat in fresh water? Take the density of the logs to be 800 kg/m^3.

37P. A metal rod of length 80 cm and mass 1.6 kg has a uniform cross-sectional area of 6.0 cm^2. Due to a nonuniform density, the center of mass of the rod is 20 cm from one end of the rod. The rod is suspended in a horizontal position in water by ropes attached to both ends (Fig. 15-37). (a) What is the tension in the rope closer to the center of mass? (b) What is the tension in the rope farther from the center of mass? (*Hint:* The buoyancy force on the rod effectively acts at the rod's center.)

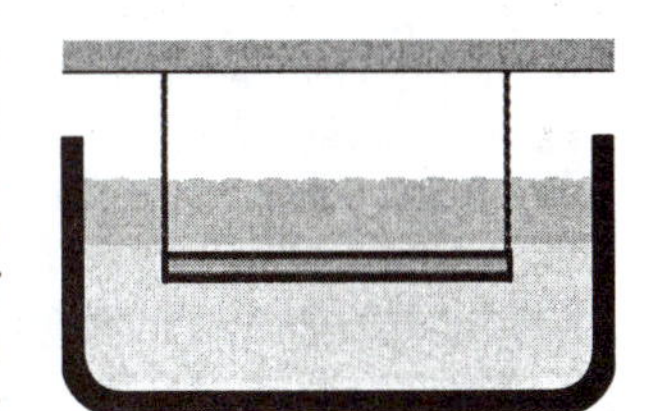

Fig. 15-37 Problem 37.

38P. A car has a total mass of 1800 kg. The volume of air space in the passenger compartment is 5.00 m^3. The volume of the motor and front wheels is 0.750 m^3, and the volume of the rear wheels, gas tank, and trunk is 0.800 m^3; water cannot enter these areas. The car is parked on a hill; the handbrake cable snaps and the car rolls down the hill into a lake (Fig. 15-38). (a) At first, no water enters the passenger compartment. How much of the car, in cubic meters, is below the water surface with the car floating as shown? (b) As water slowly enters, the car sinks. How many cubic meters of water are in the car as it disappears below the water surface? (The car, with a heavy load in the trunk, remains horizontal.)

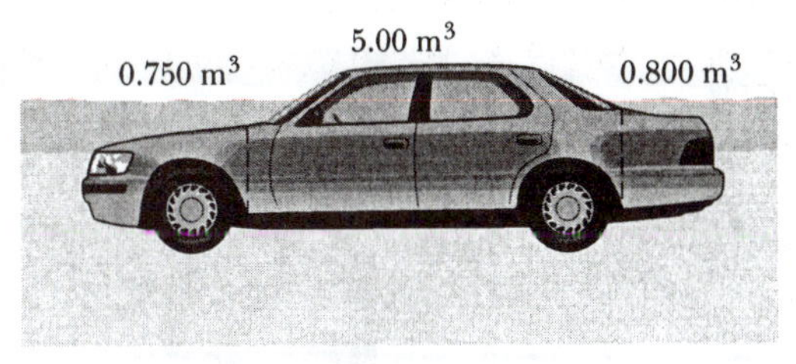

Fig. 15-38 Problem 38.

SEC. 15-9 The Equation of Continuity

39E. A garden hose with an internal diameter of 1.9 cm is connected to a (stationary) lawn sprinkler that consists merely of an enclosure with 24 holes, each 0.13 cm in diameter. If the water in the hose has a speed of 0.91 m/s, at what speed does it leave the sprinkler holes?

40E. Figure 15-39 shows the merging of two streams to form a river. One stream has a width of 8.2 m, depth of 3.4 m, and current speed of 2.3 m/s. The other stream is 6.8 m wide and 3.2 m deep, and flows at 2.6 m/s. The width of the river is 10.5 m, and the current speed is 2.9 m/s. What is its depth?

Fig. 15-39 Problem 40.

41P. Water is pumped steadily out of a flooded basement at a speed of 5.0 m/s through a uniform hose of radius 1.0 cm. The hose passes out through a window 3.0 m above the waterline. What is the power of the pump?

42E. The water flowing through a 1.9 cm (inside diameter) pipe flows out through three 1.3 cm pipes. (a) If the flow rates in the three smaller pipes are 26, 19, and 11 L/min, what is the flow rate in the 1.9 cm pipe? (b) What is the ratio of the speed of water in the 1.9 cm pipe to that in the pipe carrying 26 L/min?

SEC. 15-11 Bernoulli's Equation

43E. Water is moving with a speed of 5.0 m/s through a pipe with a cross-sectional area of 4.0 cm^2. The water gradually descends 10 m as the pipe increases in area to 8.0 cm^2. (a) What is the speed at the lower level? (b) If the pressure at the upper level is 1.5×10^5 Pa, what is the pressure at the lower level?

44E. Models of torpedoes are sometimes tested in a horizontal pipe of flowing water, much as a wind tunnel is used to test model airplanes. Consider a circular pipe of internal diameter 25.0 cm and a torpedo model, aligned along the axis of the pipe, with a diameter of 5.00 cm. The model is to be tested with water flowing past it at 2.50 m/s. (a) With what speed must the water flow in the part of the pipe that is unconstricted by the model? (b) What will the pressure difference be between the constricted and unconstricted parts of the pipe?

45E. A water pipe having a 2.5 cm inside diameter carries water into the basement of a house at a speed of 0.90 m/s and a pressure of 170 kPa. If the pipe tapers to 1.2 cm and rises to the second floor 7.6 m above the input point, what are (a) the speed and (b) the water pressure at the second floor?

46E. A water intake at a pump storage reservoir (Fig. 15-40) has a cross-sectional area of 0.74 m^2. The water flows in at a speed of 0.40 m/s. At the generator building 180 m below the intake point, the cross-sectional area is smaller than at the intake and the water flows out at 9.5 m/s. What is the difference in pressure, in megapascals, between inlet and outlet?

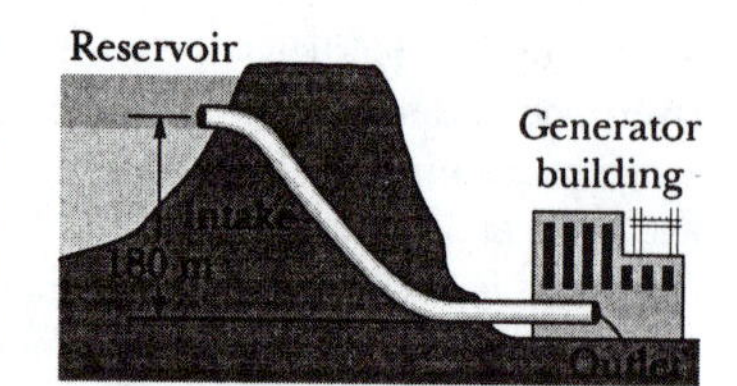

Fig. 15-40 Exercise 46.

47E. A tank of large area is filled with water to a depth $D = 0.30$ m. A hole of cross-sectional area $A = 6.5\ \text{cm}^2$ in the bottom of the tank allows water to drain out. (a) What is the rate at which water flows out, in cubic meters per second? (b) At what distance below the bottom of the tank is the cross-sectional area of the stream equal to one-half the area of the hole?

48E. Air flows over the top of an airplane wing of area A with speed $|\vec{v}_t|$ and past the underside of the wing (also of area A) with speed $|\vec{v}_u|$. Show that in this simplified situation Bernoulli's equation predicts that the magnitude $|\vec{L}|$ of the upward lift force on the wing will be

$$|\vec{L}| = \tfrac{1}{2}\rho A(v_t^2 - v_u^2),$$

where ρ is the density of the air.

49E. If the speed of flow past the lower surface of an airplane wing is 110 m/s, what speed of flow over the upper surface will give a pressure difference of 900 Pa between upper and lower surfaces? Take the density of air to be $1.30 \times 10^{-3}\ \text{g/cm}^3$, and see Exercise 48.

50E. Suppose that two tanks, 1 and 2, each with a large opening at the top, contain different liquids. A small hole is made in the side of each tank at the same depth d below the liquid surface, but the hole in tank 1 has half the cross-sectional area of the hole in tank 2. (a) What is the ratio ρ_1/ρ_2 of the densities of the liquids if the mass flow rate is the same for the two holes? (b) What is the ratio of the volume flow rates from the two tanks? (c) To what height above the hole in the second tank should liquid be added or drained to equalize the volume flow rates?

51P. In Fig. 15-41, water flows through a horizontal pipe, and then out into the atmosphere at a speed of 15 m/s. The diameters of the left and right sections of the pipe are 5.0 cm and 3.0 cm, respectively. (a) What volume of water flows into the atmosphere during a 10 min period? In the left section of the pipe, what are (b) the speed $|\vec{v}_2|$, and (c) the gauge pressure?

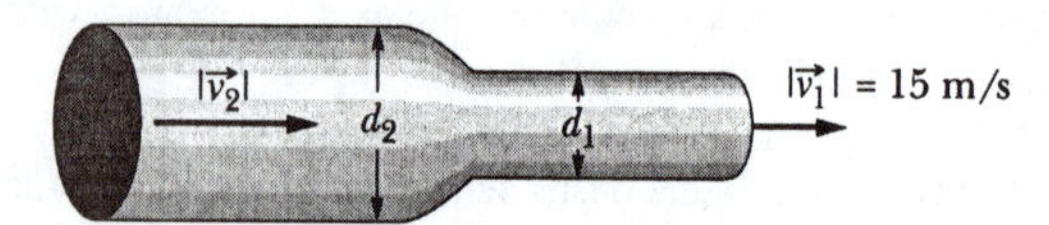

Fig. 15-41 Problem 51.

52P. An opening of area 0.25 cm^2 in an otherwise closed beverage keg is 50 cm below the level of the liquid (of density 1.0 g/cm^3) in the keg. What is the speed of the liquid flowing through the opening if the gauge pressure in the air space above the liquid is (a) zero and (b) 0.40 atm?

53P. The fresh water behind a reservoir dam is 15 m deep. A horizontal pipe 4.0 cm in diameter passes through the dam 6.0 m below the water surface, as shown in Fig. 15-42. A plug secures the pipe opening. (a) Find the magnitude of the frictional force between plug and pipe wall. (b) The plug is removed. What volume of water flows out of the pipe in 3.0 h?

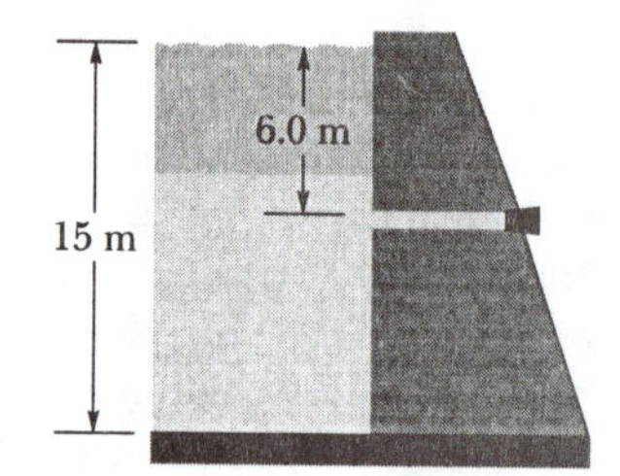

Fig. 15-42 Problem 53.

54P. A tank is filled with water to a height H. A hole is punched in one of the walls at a depth h below the water surface (Fig. 15-43). (a) Show that the distance x from the base of the tank to the point at which the resulting stream strikes the floor is given by $x = 2\sqrt{h(H-h)}$. (b) Could a hole be punched at another depth to produce a second stream that would have the same range? If so, at what depth? (c) At what depth should the hole be placed to make the emerging stream strike the ground at the maximum distance from the base of the tank?

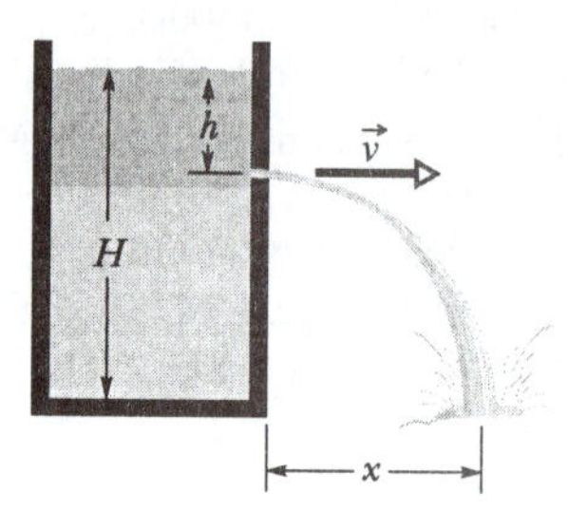

Fig. 15-43 Problem 54.

55P. A *venturi meter* is used to measure the flow speed of a fluid in a pipe. The meter is connected between two sections of the pipe (Fig. 15-44); the cross-sectional area A of the entrance and exit of the meter matches the pipe's cross-sectional area. Between the entrance and exit, the fluid flows from the pipe with speed $|\vec{V}|$ and then through a narrow "throat" of cross-sectional area a with speed $|\vec{v}|$. A manometer connects the wider portion of the meter to the narrower portion. The change in the fluid's speed is accompanied by a change ΔP in the fluid's pressure, which causes a height difference h of the liquid in the two arms of the manometer. (Here ΔP means pressure in the throat minus pressure in the pipe.) (a) By applying Bernoulli's equation and the equation of continuity to points 1 and 2 in Fig. 15-44, show that

$$|\vec{V}| = \sqrt{\frac{2a^2\,\Delta P}{\rho(a^2 - A^2)}},$$

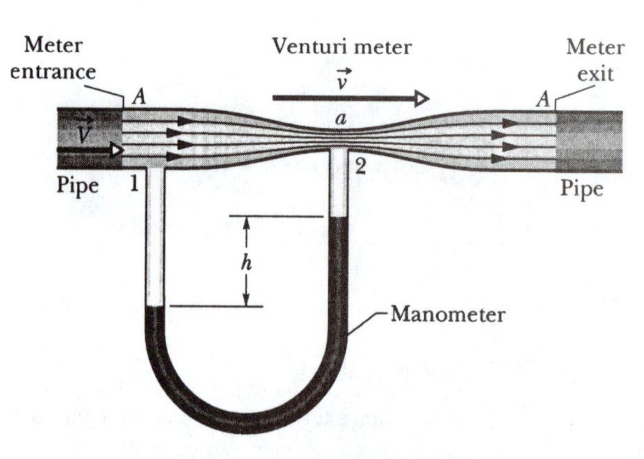

Fig. 15-44 Problems 55 and 56.

where ρ is the density of the fluid. (b) Suppose that the fluid is fresh water, that the cross-sectional areas are 64 cm^2 in the pipe and 32 cm^2 in the throat, and that the pressure is 55 kPa in the pipe and 41 kPa in the throat. What is the rate of water flow in cubic meters per second?

56P. Consider the venturi tube of Problem 55 and Fig. 15-44 without the manometer. Let A equal $5a$. Suppose that the pressure P_1 at A is 2.0 atm. Compute the values of (a) $|\vec{V}|$ at A and (b) $|\vec{v}|$ at a that would make the pressure P_2 at a equal to zero. (c) Compute the corresponding volume flow rate if the diameter at A is 5.0 cm. The phenomenon that occurs at a when P_2 falls to nearly zero is known as cavitation. The water vaporizes into small bubbles.

57P. A pivot tube (Fig. 15-45) is used to determine the airspeed of an airplane. It consists of an outer tube with a number of small holes B (four are shown) that allow air into the tube; that tube is connected to one arm of a U-tube. The other arm of the U-tube is connected to hole A at the front end of the device, which points in the direction the plane is headed. At A the air becomes stagnant so that $|\vec{v}_A| = 0$. At B, however, the speed of the air presumably equals the airspeed $|\vec{v}|$ of the aircraft. (a) Use Bernoulii's equation to show that

$$|\vec{v}| = \sqrt{\frac{2\rho g h}{\rho_{air}}},$$

where ρ is the density of the liquid in the U-tube and h is the difference in the fluid levels in that tube. (b) Suppose that the tube contains alcohol and indicates a level difference h of 26.0 cm. What is the plane's speed relative to the air? The density of the air is 1.03 kg/m^3 and that of alcohol is 810 kg/m^3.

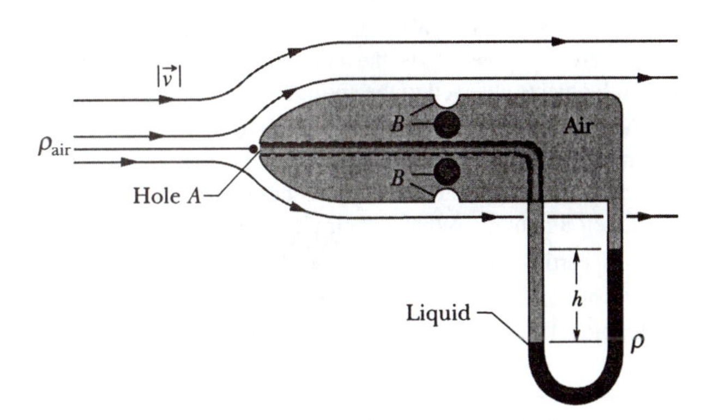

Fig. 15-45 Problems 57 and 58.

58P. A pivot tube (see Problem 57) on a high-altitude aircraft measures a differential pressure of 180 Pa. What is the airspeed if the density of the air is 0.031 kg/m^3?

Chapter 16

SEC. 16-4 Velocity and Acceleration for Simple Harmonic Motion

1E. An object undergoing simple harmonic motion takes 0.25 s to travel from one point of zero velocity to the next such point. The distance between those points is 36 cm. Calculate (a) the period, (b) the frequency, and (c) the amplitude of the motion.

2E. An oscillating block–spring system takes 0.75 s to begin repeating its motion. Find (a) the period, (b) the frequency in hertz, and (c) the angular frequency in radians per second.

3E. An oscillator consists of a block of mass 0.500 kg connected to a spring. When set into oscillation with amplitude 35.0 cm, the oscillator repeats its motion every 0.500 s. Find (a) the period, (b) the frequency, (c) the angular frequency, (d) the spring constant, (e) the maximum speed, and (f) the magnitude of the maximum force on the block from the spring.

4E. What is the maximum acceleration of a platform that oscillates with an amplitude of 2.20 cm at a frequency of 6.60 Hz?

5E. A loudspeaker produces a musical sound by means of the oscillation of a diaphragm. If the amplitude of oscillation is limited to 1.0×10^{-3} mm, what frequencies will result in the magnitude of the diaphragm's acceleration exceeding g?

6E. The scale of a spring balance that reads from 0 to 15.0 kg is 12.0 cm long. A package suspended from the balance is found to oscillate vertically with a frequency of 2.00 Hz. (a) What is the spring constant? (b) How much does the package weigh?

7E. A particle with a mass of 1.00×10^{-20} kg is oscillating with simple harmonic motion with a period of 1.00×10^{-5} s and a maximum speed of 1.00×10^3 m/s. Calculate (a) the angular frequency and (b) the maximum displacement of the particle.

8E. A small body of mass 0.12 kg is undergoing simple harmonic motion of amplitude 8.5 cm and period 0.20 s. (a) What is the magnitude of the maximum force acting on it? (b) If the oscillations are produced by a spring, what is the spring constant?

9E. In an electric shaver, the blade moves back and forth over a distance of 2.0 mm in simple harmonic motion, with frequency 120 Hz. Find (a) the amplitude, (b) the maximum blade speed, and (c) the magnitude of the maximum blade acceleration.

10E. A loudspeaker diaphragm is oscillating in simple harmonic motion with a frequency of 440 Hz and a maximum displacement of 0.75 mm. What are (a) the angular frequency, (b) the maximum speed, and (c) the magnitude of the maximum acceleration?

11E. An automobile can be considered to be mounted on four identical springs as far as vertical oscillations are concerned. The springs of a certain car are adjusted so that the oscillations have a frequency of 3.00 Hz. (a) What is the spring constant of each spring if the mass of the car is 1450 kg and the mass is evenly distributed over the springs? (b) What will be the oscillation frequency if five passengers, averaging 73.0 kg each, ride in the car? (Again, consider an even distribution of mass.)

12E. A body oscillates with simple harmonic motion according to the equation

$$x = (6.0 \text{ m}) \cos[(3\pi \text{ rad/s})t + \pi/3 \text{ rad}].$$

At $t = 2.0$ s, what are (a) the displacement, (b) the velocity, (c) the acceleration, and (d) the phase of the motion? Also, what are (e) the frequency and (f) the period of the motion?

13E. The piston in the cylinder head of a locomotive has a stroke (twice the amplitude) of 0.76 m. If the piston moves with simple harmonic motion with an angular frequency of 180 rev/min, what is its maximum speed?

14P. Figure 16-25 shows an astronaut on a body-mass measuring device (BMMD). Designed for use on orbiting space vehicles, its purpose is to allow astronauts to measure their mass in the "weightless" conditions in Earth orbit. The BMMD is a spring-mounted chair; an astronaut measures his or her period of oscillation in the chair; the mass follows from the formula for the period of an oscillating block–spring system. (a) If M is the mass of the astronaut and m the effective mass of that part of the BMMD that also oscillates, show that

$$M = (k/4\pi^2)T^2 - m,$$

where T is the period of oscillation and k is the spring constant. (b) The spring constant was $k = 605.6$ N/m for the BMMD on Skylab Mission Two; the period of oscillation of the empty chair was 0.90149 s. Calculate the effective mass of the chair. (c) With an astronaut in the chair, the period of oscillation became 2.08832 s. Calculate the mass of the astronaut.

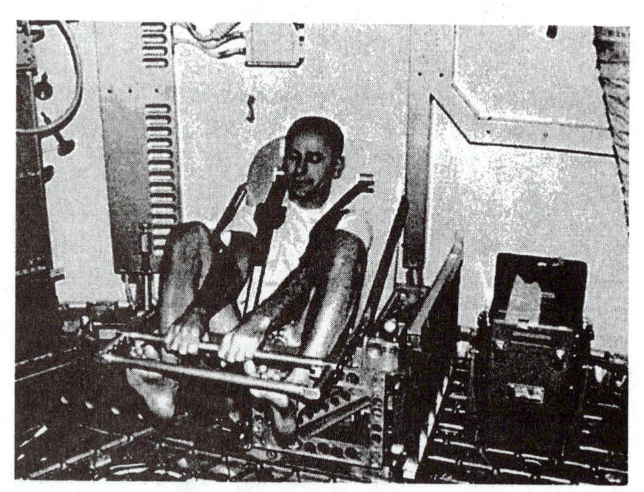

Fig. 16-25 Problem 14.

15P. At a certain harbor, the tides cause the ocean surface to rise and fall a distance d (from highest level to lowest level) in simple harmonic motion, with a period of 12.5 h. How long does it take for the water to fall a distance $d/4$ from its highest level?

16P. In Fig. 16-26, two blocks ($m = 1.0$ kg and $M = 10$ kg) and a spring ($k = 200$ N/m) are arranged on a horizontal, frictionless surface. The coefficient of static friction between the two blocks is 0.40. What amplitude of simple harmonic motion of the spring–blocks system puts the smaller block on the verge of slipping over the larger block?

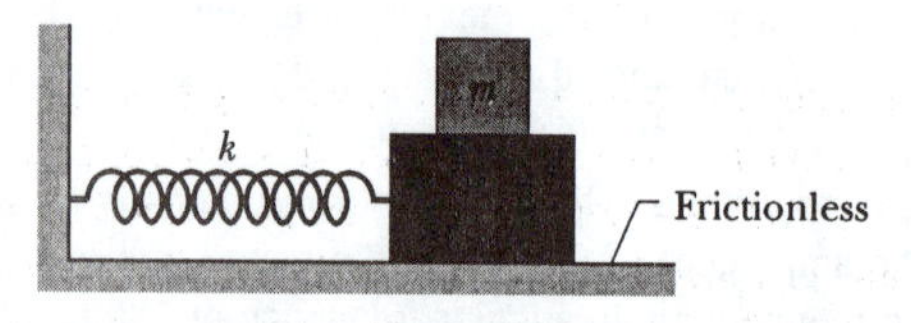

Fig. 16-26 Problem 16.

17P. A block is on a horizontal surface (a shake table) that is moving back and forth horizontally with simple harmonic motion of frequency 2.0 Hz. The coefficient of static friction between block and surface is 0.50. How great can the amplitude of the SHM be if the block is not to slip along the surface?

18P. A block rides on a piston that is moving vertically with simple harmonic motion. (a) If the SHM has period 1.0 s, at what amplitude of motion will the block and piston separate? (b) If the piston has an amplitude of 5.0 cm, what is the maximum frequency for which the block and piston will be in contact continuously?

19P. An oscillator consists of a block attached to a spring ($k = 400$ N/m). At some time t, the position (measured from the system's equilibrium location), velocity, and acceleration of the block are $x = 0.100$ m, $v = -13.6$ m/s, and $a = -123$ m/s^2. Calculate (a) the frequency of oscillation, (b) the mass of the block, and (c) the amplitude of the motion.

20P. A simple harmonic oscillator consists of a block of mass 2.00 kg attached to a spring of spring constant 100 N/m. When $t = 1.00$ s, the position and velocity of the block are $x = 0.129$ m and $v = 3.415$ m/s. (a) What is the amplitude of the oscillations? What were the (b) position and (c) velocity of the block at $t = 0$ s?

21P. A massless spring hangs from the ceiling with a small object attached to its lower end. The object is initially held at rest in a position y_i such that the spring is at its rest length. The object is then released from y_i and oscillates up and down, with its lowest position being 10 cm below y_i. (a) What is the frequency of the oscillation? (b) What is the speed of the object when it is 8.0 cm below the initial position? (c) An object of mass 300 g is attached to the first object, after which the system oscillates with half the original frequency. What is the mass of the first object? (d) Relative to y_i, where is the new equilibrium (rest) position with both objects attached to the spring?

22P. Two particles execute simple harmonic motion of the same amplitude and frequency along close parallel lines. They pass each other moving in opposite directions each time their displacement is half their amplitude. What is their phase difference?

23P. Two particles oscillate in simple harmonic motion along a common straight-line segment of length A. Each particle has a period of 1.5 s, but they differ in phase by $\pi/6$ rad. (a) How far apart are they (in terms of A) 0.50 s after the lagging particle leaves one end of the path? (b) Are they then moving in the same direction, toward each other, or away from each other?

24P. In Fig. 16-27, two identical springs of spring constant k are attached to a block of mass m and to fixed supports. Show that the block's frequency of oscillation on the frictionless surface is

$$f = \frac{1}{2\pi}\sqrt{\frac{2k}{m}}.$$

25P. Suppose that the two springs in Fig. 16-27 have different spring constants k_1 and k_2. Show that the frequency f of oscillation of the block is then given by

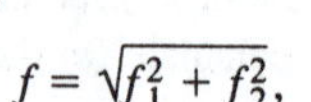

$$f = \sqrt{f_1^2 + f_2^2},$$

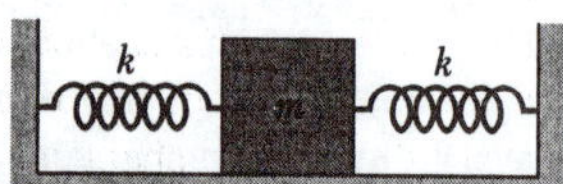

Fig. 16-27 Problems 24 and 25.

where f_1 and f_2 are the frequencies at which the block would oscillate if connected only to spring 1 or only to spring 2.

26P. The end of one of the prongs of a tuning fork that executes simple harmonic motion of frequency 1000 Hz has an amplitude of 0.40 mm. Find (a) the magnitude of the maximum acceleration and (b) the maximum speed of the end of the prong. Find (c) the magnitude of the acceleration and (d) the speed of the end of the prong when the end has a displacement of 0.20 mm.

27P. In Fig. 16-28, two springs are joined and connected to a block of mass m. The surface is frictionless. If the springs both have spring constant k, show that

$$f = \frac{1}{2\pi}\sqrt{\frac{k}{2m}}$$

gives the block's frequency of oscillation.

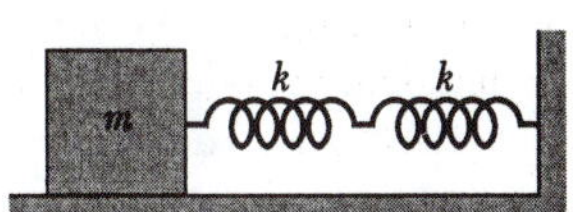

Fig. 16-28 Problem 27.

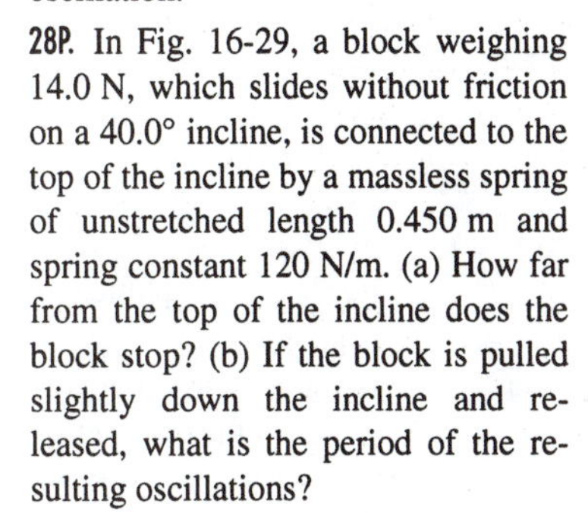

28P. In Fig. 16-29, a block weighing 14.0 N, which slides without friction on a 40.0° incline, is connected to the top of the incline by a massless spring of unstretched length 0.450 m and spring constant 120 N/m. (a) How far from the top of the incline does the block stop? (b) If the block is pulled slightly down the incline and released, what is the period of the resulting oscillations?

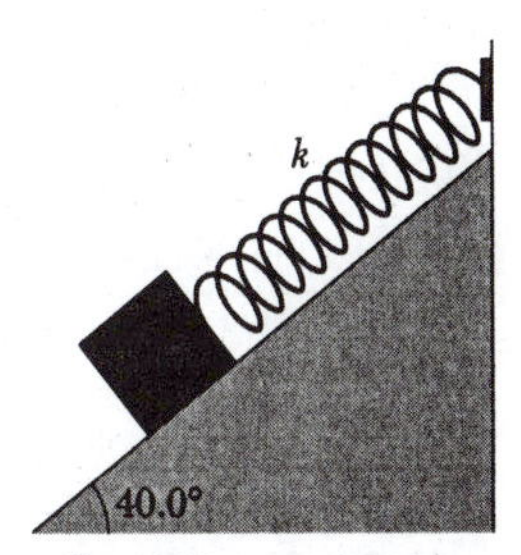

Fig. 16-29 Problem 28.

29P. A uniform spring with unstretched length L and spring constant k is cut into two pieces of unstretched lengths L_1 and L_2, with $L_1 = nL_2$. What are the corresponding spring constants (a) k_1 and (b) k_2 in terms of n and k? If a block is attached to the original spring, as in Fig. 16-9, it oscillates with frequency f. If the spring is replaced with the piece L_1 or L_2, the corresponding frequency is f_1 or f_2. Find (c) f_1 and (d) f_2 in terms of f.

30P. In Fig. 16-30, three 10 000 kg ore cars are held at rest on a 30° incline on a mine railway using a cable that is parallel to the incline. The cable stretches 15 cm just before the coupling between the two lower cars breaks, detaching the lowest car. Assuming that the cable obeys Hooke's law, find (a) the frequency and (b) the amplitude of the resulting oscillations of the remaining two cars.

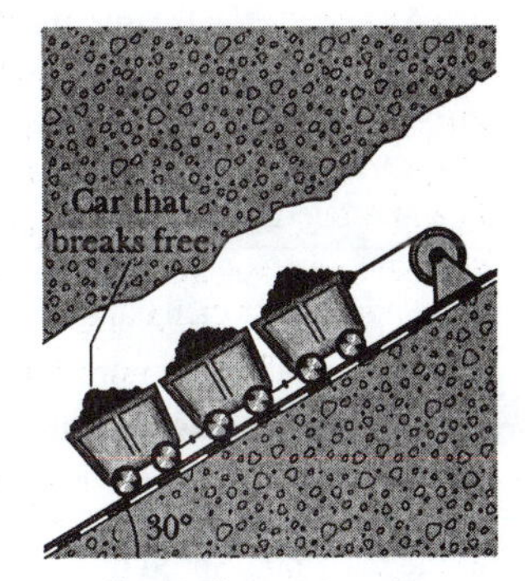

Fig. 16-30 Problem 30.

31E. A flat uniform circular disk has a mass of 3.00 kg and a radius of 70.0 cm. It is suspended in a horizontal plane by a vertical wire attached to its center. If the disk is rotated 2.50 rad about the wire, a torque of 0.0600 N · m is required to maintain that orientation. Calculate (a) the rotational inertia of the disk about the wire, (b) the torsion constant, and (c) the angular frequency of this torsion pendulum when it is set oscillating.

32P. The balance wheel of a watch oscillates with an angular amplitude of π rad and a period of 0.500 s. Find (a) the maximum angular speed of the wheel, (b) the angular speed of the wheel when its displacement is $\pi/2$ rad, and (c) the magnitude of the angular acceleration of the wheel when its displacement is $\pi/4$ rad.

SEC. 16-5 Gravitational Pendulums

33E. In Fig. 16-31, a 2500 kg demolition ball swings from the end of a crane. The length of the swinging segment of cable is 17 m. (a) Find the period of the swinging, assuming that the system can be treated as a simple pendulum. (b) Does the period depend on the ball's mass?

Fig. 16-31 Exercise 33.

34E. What is the length of a simple pendulum that marks seconds by completing a full swing from left to right and then back again every 2.0 s?

35E. A performer seated on a trapeze is swinging back and forth with a period of 8.85 s. If she stands up, thus raising the center of mass of the *trapeze + performer* system by 35.0 cm, what will be the new period of the system? Treat *trapeze + performer* as a simple pendulum.

36E. A physical pendulum consists of a meter stick that is pivoted at a small hole drilled through the stick a distance d from the 50 cm mark. The period of oscillation is 2.5 s. Find d.

37E. In Fig. 16-32, a physical pendulum consists of a uniform solid disk (of mass M and radius R) supported in a vertical plane by a pivot located a distance d from the center of the disk. The disk is displaced by a small angle and released. Find an expression for the period of the resulting simple harmonic motion.

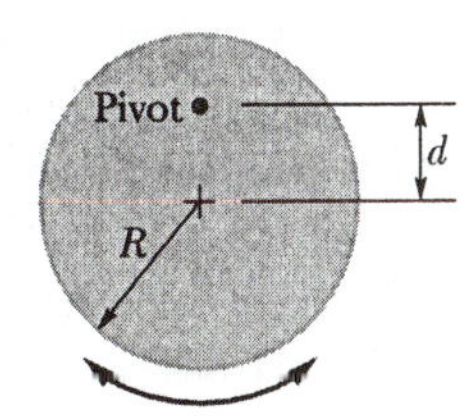

Fig. 16-32 Exercise 37.

38E. A pendulum is formed by pivoting a long thin rod of length L and mass m about a point on the rod that is a distance d above the center of the rod. (a) Find the period of this pendulum in terms of d, L, m, and g, assuming small-amplitude swinging. What happens to the period if (b) d is decreased, (c) L is increased, or (d) m is increased?

39E. A uniform circular disk whose radius R is 12.5 cm is suspended as a physical pendulum from a point on its rim. (a) What is its period? (b) At what radial distance $r < R$ is there a pivot point that gives the same period?

40E. The pendulum in Fig. 16-33 consists of a uniform disk with radius 10.0 cm and mass 500 g attached to a uniform rod with length 500 mm and mass 270 g. (a) Calculate the rotational inertia of the pendulum about the pivot point. (b) What is the distance between the pivot point and the center of mass of the pendulum? (c) Calculate the period of oscillation.

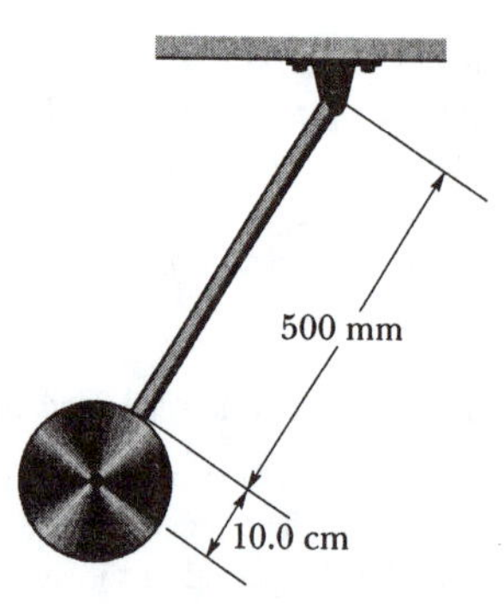

Fig. 16-33 Exercise 40.

41P. A stick with length L oscillates as a physical pendulum, pivoted about point O in Fig. 16-34. (a) Derive an expression for the period of the pendulum in terms of L and x, the distance from the pivot point to the center of mass of the pendulum. (b) For what value of x/L is the period a minimum? (c) Show that if $L =$ 1.00 m and $g = 9.80\ \text{m/s}^2$, this minimum is 1.53 s.

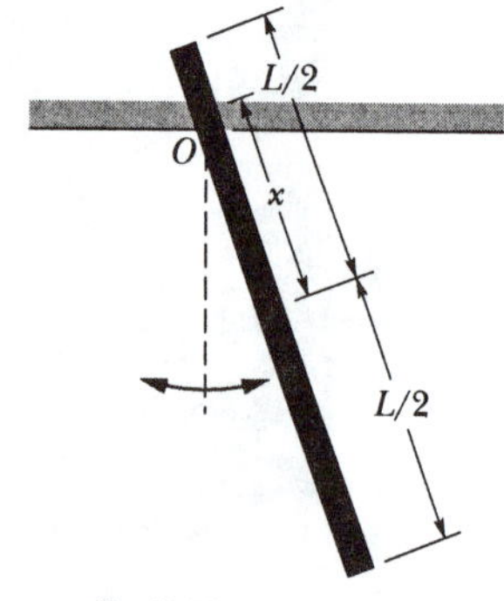

Fig. 16-34 Problem 41.

42P. In the overhead view of Fig. 16-35, a long uniform rod of length L and mass m is free to rotate in a horizontal plane about a vertical axis through its center. A spring with force constant k is connected horizontally between one end of the rod and a fixed wall. When the rod is in equilibrium, it is parallel to the wall. What is the period of the small oscillations that result when the rod is rotated slightly and released?

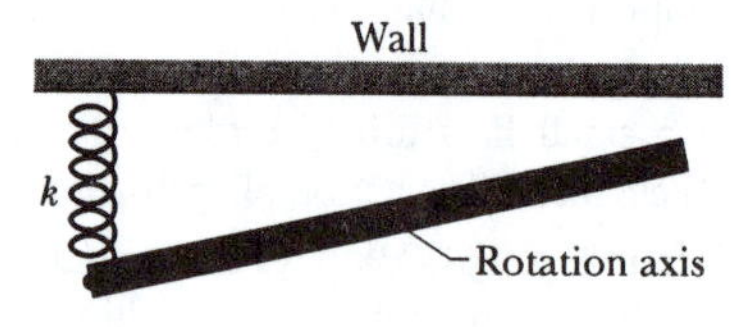

Fig. 16-35 Problem 42.

43P. A simple pendulum of length L and mass m is suspended in a car that is traveling with constant speed $|\vec{v}|$ around a circle of radius R. If the pendulum undergoes small oscillations in a radial direction about its equilibrium position, what will be its frequency of oscillation?

44P. What is the frequency of a simple pendulum 2.0 m long (a) in a room, (b) in an elevator accelerating upward at a rate of 2.0 m/s^2, and (c) in free fall?

45P. For a simple pendulum, find the angular amplitude θ_m at which the restoring torque required for simple harmonic motion deviates from the actual restoring torque by 1.0%. (See "Trigonometric Expansions" in Appendix E.)

46P. The bob on a simple pendulum of length R moves in an arc of a circle. (a) By considering that the radial acceleration of the bob as it moves through its equilibrium position is that for uniform circular motion (v^2/R), show that the tension in the string at that position is $mg(1 + \theta_m^2)$ if the angular amplitude θ_m is small. (See "Trigonometric Expansions" in Appendix E.) (b) Is the tension at other positions of the bob greater, smaller, or the same?

47P. A wheel is free to rotate about its fixed axle. A spring is attached to one of its spokes a distance r from the axle, as shown in Fig. 16-36. (a) Assuming that the wheel is a hoop of mass m and radius R, obtain the angular frequency of small oscillations of this system in terms of m, R, r, and the spring constant k. How does the result change if (b) $r = R$ and (c) $r = 0$?

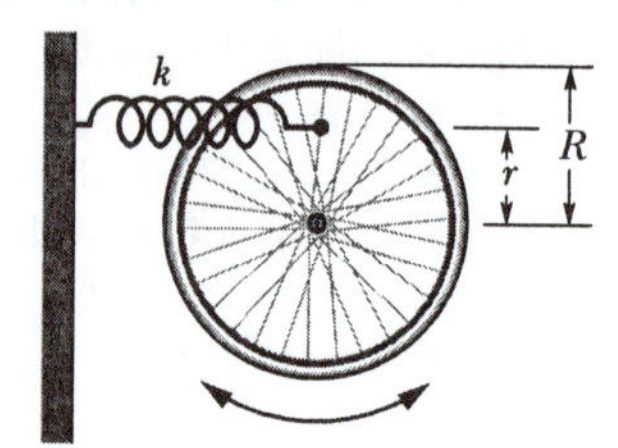

Fig. 16-36 Problem 47.

SEC. 16-6 Energy in Simple Harmonic Motion

48E. Find the mechanical energy of a block–spring system having a spring constant of 1.3 N/cm and an oscillation amplitude of 2.4 cm.

49E. An oscillating block–spring system has a mechanical energy of 1.00 J, an amplitude of 10.0 cm, and a maximum speed of 1.20 m/s. Find (a) the spring constant, (b) the mass of the block and (c) the frequency of oscillation.

50E. A 5.00 kg object on a horizontal frictionless surface is attached to a spring with spring constant 1000 N/m. The object is displaced from equilibrium 50.0 cm horizontally and given an initial velocity of 10.0 m/s back toward the equilibrium position. (a) What is the frequency of the motion? What are (b) the initial potential energy of the block–spring system, (c) the initial kinetic energy, and (d) the amplitude of the oscillation?

51E. A (hypothetical) large slingshot is stretched 1.50 m to launch a 130 g projectile with speed sufficient to escape from Earth (11.2 km/s). Assume the elastic bands of the slingshot obey Hooke's law. (a) What is the spring constant of the device, if all the elastic potential energy is converted to kinetic energy? (b) Assume that an average person can exert a force of 220 N. How many people are required to stretch the elastic bands?

52E. A vertical spring stretches 9.6 cm when a 1.3 kg block is hung from its end. (a) Calculate the spring constant. This block is then displaced an additional 5.0 cm downward and released from rest. Find (b) the period, (c) the frequency, (d) the amplitude, and (e) the maximum speed of the resulting SHM.

53E. A block of mass M, at rest on a horizontal frictionless table, is attached to a rigid support by a spring of constant k. A bullet of mass m and velocity $\vec{v}$ strikes the block as shown in Fig. 16-37. The bullet is embedded in the block. Determine (a) the speed of the block immediately after the collision and (b) the amplitude of the resulting simple harmonic motion.

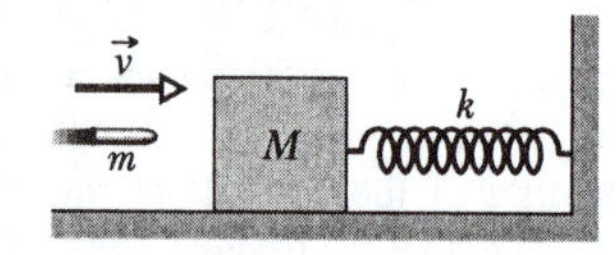

Fig. 16-37 Exercise 53.

54E. When the displacement in SHM is one-half the amplitude x_m, what fraction of the total energy is (a) kinetic energy and (b) potential energy? (c) At what displacement, in terms of the amplitude, is the energy of the system half kinetic energy and half potential energy?

55P. A 10 kg particle is undergoing simple harmonic motion with an amplitude of 2.0×10^{-3} m and a maximum acceleration of magnitude 8.0×10^{-3} m/s^2. The phase constant is $-\pi/3$ rad. (a) Write an equation for the force on the particle as a function of time. (b) What is the period of the motion? (c) What is the maximum speed of the particle? (d) What is the total mechanical energy of this simple harmonic oscillator?

56P*. A 4.0 kg block is suspended from a spring with a spring constant of 500 N/m. A 50 g bullet is fired into the block from directly below with a speed of 150 m/s and becomes embedded in the block. (a) Find the amplitude of the resulting simple harmonic motion. (b) What fraction of the original kinetic energy of the bullet is transferred to mechanical energy of the harmonic oscillator?

SEC. 16-7 Damped Simple Harmonic Motion

57E. In Touchstone Example 16-7-1, what is the ratio of the amplitude of the damped oscillations to the initial amplitude when 20 full oscillations have elapsed?

58E. The amplitude of a lightly damped oscillator decreases by 3.0% during each cycle. What fraction of the mechanical energy of the oscillator is lost in each full oscillation?

59E. For the system shown in Fig. 16-22, the block has a mass of 1.50 kg and the spring constant is 8.00 N/m. The damping force is given by $-b(dx/dt)$, where $b = 230$ g/s. Suppose that the block is initially pulled down a distance 12.0 cm and released. (a) Calculate the time required for the amplitude of the resulting oscillations to fall to one-third of its initial value. (b) How many oscillations are made by the block in this time?

60P. Assume that you are examining the oscillation characteristics of the suspension system of a 2000 kg automobile. The suspension "sags" 10 cm when the entire automobile is placed on it. Also, the amplitude of oscillation decreases by 50% during one complete oscillation. Estimate the values of (a) the spring constant k and (b) the damping constant b for the spring and shock absorber system of one wheel, assuming each wheel supports 500 kg.

SEC. 16-8 Forced Oscillations and Resonance

61E. For Eq. 16-37, suppose the amplitude x_m is given by

$$x_m = \frac{F_m}{[m^2(\omega_d^2 - \omega^2)^2 + b^2\omega_d^2]^{1/2}},$$

where F_m is the (constant) amplitude of the external oscillating force exerted on the spring by the rigid support in Fig. 16-22. At resonance, what are (a) the amplitude and (b) the velocity amplitude of the oscillating object?

62P. A 1000 kg car carrying four 82 kg people travels over a rough "washboard" dirt road with corrugations 4.0 m apart, which cause the car to bounce on its spring suspension. The car bounces with maximum amplitude when its speed is 16 km/h. The car now stops, and the four people get out. By how much does the car body rise on its suspension owing to this decrease in mass?

Chapter 17

SEC. 17-5 The "Wave" Speed of a Traveling Distortion

1E. A wave has an angular frequency of 110 rad/s and a wavelength of 1.80 m. Calculate (a) the angular wave number and (b) the speed of the wave.

2E. The speed of electromagnetic waves (which include visible light, radio, and x rays) in vacuum is 3.0×10^8 m/s. (a) Wavelengths of visible light waves range from about 400 nm in the violet to about 700 nm in the red. What is the range of frequencies of these waves? (b) The range of frequencies for shortwave radio (for example, FM radio and VHF television) is 1.5 to 300 MHz. What is the corresponding wavelength range? (c) X ray wavelengths range from about 5.0 nm to about 1.0×10^{-2} nm. What is the frequency range for x rays?

3E. A sinusoidal wave travels along a string. The time for a particular point to move from maximum displacement to zero is 0.170 s. What are the (a) period and (b) frequency? (c) The wavelength is 1.40 m; what is the wave speed?

4E. Write the equation for a sinusoidal wave traveling in the negative direction along an x axis and having an amplitude of 0.010 m, a frequency of 550 Hz, and a speed of 330 m/s.

5E. Show that

$$y = y_m \sin k(x - vt), \qquad y = y_m \sin 2\pi\left(\frac{x}{\lambda} - ft\right),$$

$$y = y_m \sin \omega\left(\frac{x}{v} - t\right), \qquad y = y_m \sin 2\pi\left(\frac{x}{\lambda} - \frac{t}{T}\right)$$

are all equivalent to $y = y_m \sin(kx - \omega t)$.

6P. The equation of a transverse wave traveling along a very long string is $y = 6.0 \sin(0.020\pi x + 4.0\pi t)$, where x and y are expressed in centimeters and t is in seconds. Determine (a) the amplitude, (b) the wavelength, (c) the frequency, (d) the speed, (e) the direction of propagation of the wave, and (f) the maximum transverse speed of a particle in the string. (g) What is the transverse displacement at $x = 3.5$ cm when $t = 0.26$ s?

7P. (a) Write an equation describing a sinusoidal transverse wave traveling on a cord in the $+x$ direction with a wavelength of 10 cm, a frequency of 400 Hz, and an amplitude of 2.0 cm. (b) What is the maximum speed of a point on the cord? (c) What is the speed of the wave?

8P. A transverse sinusoidal wave of wavelength 20 cm is moving along a string in the positive x direction. The transverse displacement of the string particle at $x = 0$ as a function of time is shown in Fig. 17-24. (a) Make a rough sketch of one wavelength of the wave (the por-

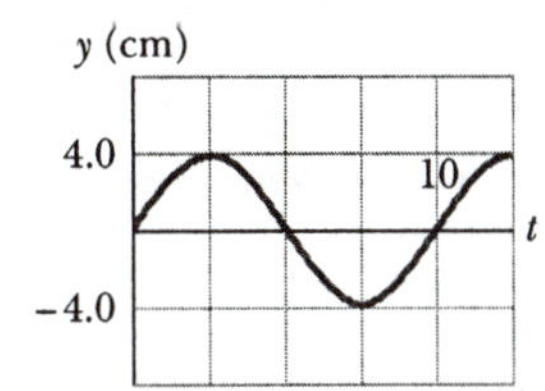

Fig. 17-24 Problem 8.

tion between $x = 0$ and $x = 20$ cm) at time $t = 0$. (b) What is the speed of the wave? (c) Write the equation for the wave with all the constants evaluated. (d) What is the transverse velocity of the particle at $x = 0$ at $t = 5.0$ s?

9P. A sinusoidal wave of frequency 500 Hz has a speed of 350 m/s. (a) How far apart are two points that differ in phase by $\pi/3$ rad? (b) What is the phase difference between two displacements at a certain point at times 1.00 ms apart?

SEC. 17-6 Wave Speed on a Stretched String

10E. The heaviest and lightest strings on a certain violin have linear densities of 3.0 and 0.29 g/m. What is the ratio of the diameter of the heaviest string to that of the lightest string, assuming that the strings are of the same material?

11E. What is the speed of a transverse wave in a rope of length 2.00 m and mass 60.0 g under a tension of 500 N?

12E. The tension in a wire clamped at both ends is doubled without appreciably changing the wire's length between the clamps. What is the ratio of the new to the old wave speed for transverse waves traveling along this wire?

13E. The linear density of a string is 1.6×10^{-4} kg/m. A transverse wave on the string is described by the equation

$$y = (0.021 \text{ m}) \sin[(2.0 \text{ m}^{-1})x + (30 \text{ s}^{-1})t].$$

What is (a) the wave speed and (b) the tension in the string?

14E. The equation of a transverse wave on a string is

$$y = (2.0 \text{ mm}) \sin[(20 \text{ m}^{-1})x - (600 \text{ s}^{-1})t].$$

The tension in the string is 15 N. (a) What is the wave speed? (b) Find the linear density of this string in grams per meter.

15P. A stretched string has a mass per unit length of 5.0 g/cm and a tension of 10 N. A sinusoidal wave on this string has an amplitude of 0.12 mm and a frequency of 100 Hz and is traveling in the negative direction of x. Write an equation for this wave.

16P. What is the fastest transverse wave that can be sent along a steel wire? For safety reasons, the maximum tensile stress to which steel wires should be subjected is 7.0×10^8 N/m^2. The density of steel is 7800 kg/m^3. Show that your answer does not depend on the diameter of the wire.

17P. A sinusoidal transverse wave of amplitude y_m and wavelength λ travels on a stretched cord. (a) Find the ratio of the maximum particle speed (the speed with which a single particle in the cord moves transverse to the wave) to the wave speed. (b) If a wave having a certain wavelength and amplitude is sent along a cord, would this speed ratio depend on the material of which the cord is made, such as wire or nylon?

18P. A sinusoidal wave is traveling on a string with speed 40 cm/s. The displacement of the particles of the string at $x = 10$ cm is found to vary with time according to the equation $y = (5.0 \text{ cm}) \sin[1.0 - (4.0 \text{ s}^{-1})t]$. The linear density of the string is 4.0 g/cm. What are (a) the frequency and (b) the wavelength of the wave? (c) Write the general equation giving the transverse displacement of the particles of the string as a function of position and time. (d) Calculate the tension in the string.

19P. A sinusoidal transverse wave is traveling along a string in the negative direction of an x axis. Figure 17-25 shows a plot of the displacement as a function of position at time $t = 0$; the y intercept is 4.0 cm. The string tension is 3.6 N, and its linear density is 25 g/m. Find (a) the amplitude, (b) the wavelength, (c) the wave speed, and (d) the period of the wave. (e) Find the maximum transverse speed of a particle in the string. (f) Write an equation describing the traveling wave.

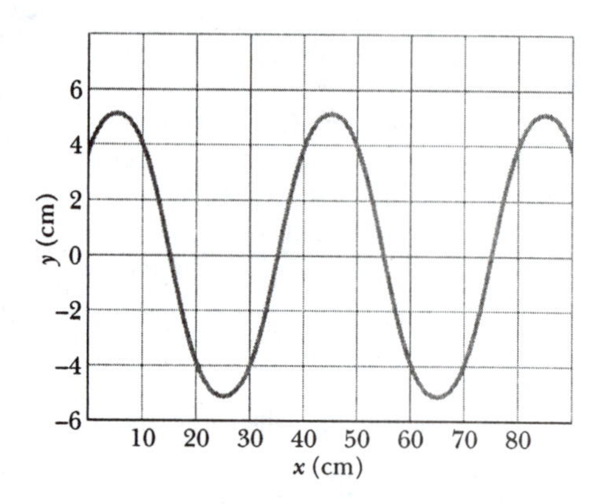

Fig. 17-25 Problem 19.

20P. In Fig. 17-26*a*, string 1 has a linear density of 3.00 g/m, and string 2 has a linear density of 5.00 g/m. They are under tension owing to the hanging block of mass $M = 500$ g. Calculate the wave speed on (a) string 1 and (b) string 2. (*Hint:* When a string loops halfway around a pulley, it pulls on the pulley with a net force that is twice the tension in the string.) Next the block is divided into two blocks (with $M_1 + M_2 = M$) and the apparatus is rearranged as shown in Fig. 17-26*b*. Find (c) M_1 and (d) M_2 such that the wave speeds in the two strings are equal.

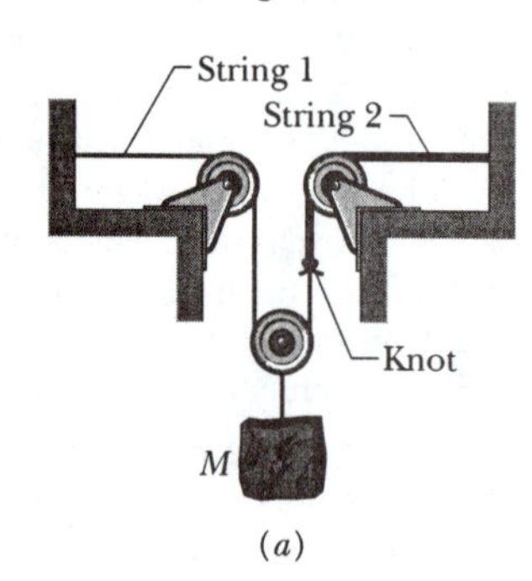

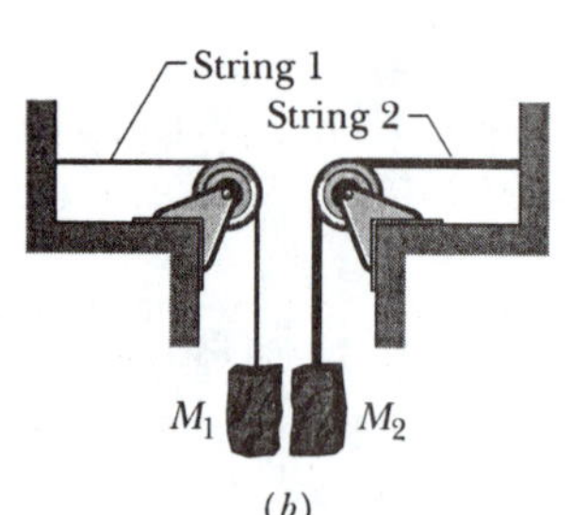

Fig. 17-26 Problem 20.

21P. A wire 10.0 m long and having a mass of 100 g is stretched under a tension of 250 N. If two pulses, separated in time by 30.0 ms, are generated, one at each end of the wire, where will the pulses first meet?

22P. The type of rubber band used inside some baseballs and golf balls obeys Hooke's law over a wide range of elongation of the band. A segment of this material has an unstretched length ℓ and a mass m. When a force $\vec{F}$ is applied, the band stretches an additional length $\Delta\ell$. (a) What is the speed (in terms of m, $\Delta\ell$, and the spring constant k) of transverse waves on this stretched rubber band? (b) Using your answer to (a), show that the time required for a transverse pulse to travel the length of the rubber band is proportional to $1/\sqrt{\Delta\ell}$ if $\Delta\ell \ll \ell$ and is constant if $\Delta\ell \gg \ell$.

23P*. A uniform rope of mass m and length L hangs from a ceiling. (a) Show that the speed of a transverse wave on the rope is a function of y, the distance from the lower end, and is given by $|\vec{v}_w| = \sqrt{gy}$. (b) Show that the time a transverse wave takes to travel the length of the rope is given by $t = 2\sqrt{L/g}$.

SEC. 17-7 Energy and Power Transported by a Traveling Wave in a String

24E. A string along which waves can travel is 2.70 m long and has a mass of 260 g. The tension in the string is 36.0 N. What must be the frequency of traveling waves of amplitude 7.70 mm for the average power to be 85.0 W?

SEC. 17-9 Interference of Waves

25E. What phase difference between two otherwise identical traveling waves, moving in the same direction along a stretched string, will result in the combined wave having an amplitude 1.50 times that of the common amplitude of the two combining waves? Express your answer in (a) degrees, (b) radians, and (c) wavelengths.

26E. Two identical traveling waves, moving in the same direction, are out of phase by $\pi/2$ rad. What is the amplitude of the resultant wave in terms of the common amplitude y_m of the two combining waves?

27P. Two sinusoidal waves, identical except for phase, travel in the same direction along a string and interfere to produce a resultant wave given by $y'(x,t) = (3.0 \text{ mm}) \sin(20 \text{ m}^{-1}) - (4.0 \text{ s}^{-1}) + 0.820 \text{ rad})$, with x in meters and t in seconds. What are (a) the wavelength λ of the two waves, (b) the phase difference between them, and (c) their amplitude y_m?

SEC. 17-11 Standing Waves and Resonance

28E. A string under tension τ_i oscillates in the third harmonic at frequency f_3, and the waves on the string have wavelength λ_3. If the tension is increased to $\tau_f = 4\tau_i$ and the string is again made to oscillate in the third harmonic, what then are (a) the frequency of oscillation in terms of f_3 and (b) the wavelength of the waves in terms of λ_3?

29E. A nylon guitar string has a linear density of 7.2 g/m and is under a tension of 150 N. The fixed supports are 90 cm apart. The string is oscillating in the standing wave pattern shown in Fig. 17-27. Calculate the (a) speed, (b) wavelength, and (c) frequency of the traveling waves whose superposition gives this standing wave.

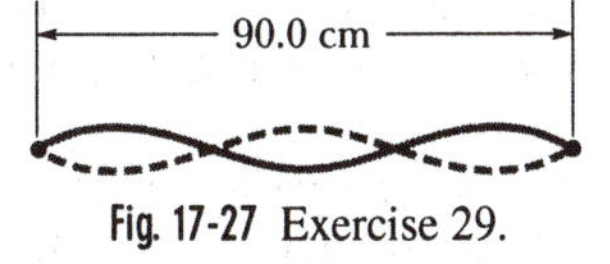

Fig. 17-27 Exercise 29.

30E. Two sinusoidal waves with identical wavelengths and amplitudes travel in opposite directions along a string with a speed of 10 cm/s. If the time interval between instants when the string is flat is 0.50 s, what is the wavelength of the waves?

31E. A string fixed at both ends is 8.40 m long and has a mass of 0.120 kg. It is subjected to a tension of 96.0 N and set oscillating. (a) What is the speed of the waves on the string? (b) What is the longest possible wavelength for a standing wave? (c) Give the frequency of that wave.

32E. A 125 cm length of string has a mass of 2.00 g. It is stretched with a tension of 7.00 N between fixed supports. (a) What is the wave speed for this string? (b) What is the lowest resonant frequency of this string?

33E. What are the three lowest frequencies for standing waves on a wire 10.0 m long having a mass of 100 g, which is stretched under a tension of 250 N?

34E. String A is stretched between two clamps separated by distance L. String B, with the same linear density and under the same tension as string A, is stretched between two clamps separated by distance $4L$. Consider the first eight harmonics of string B. Which, if any, has a resonant frequency that matches a resonant frequency of string A?

35P. A string that is stretched between fixed supports separated by 75.0 cm has resonant frequencies of 420 and 315 Hz, with no intermediate resonant frequencies. What are (a) the lowest resonant frequency and (b) the wave speed?

36P. In Fig. 17-28, two pulses travel along a string in opposite directions. The wave speed $|\vec{v}|$ is 2.0 m/s and the pulses are 6.0 cm apart at $t = 0$. (a) Sketch the wave patterns when t is equal to 5.0, 10, 15, 20, and 25 ms. (b) In what form (or type) is the energy of the pulses at $t = 15$ ms?

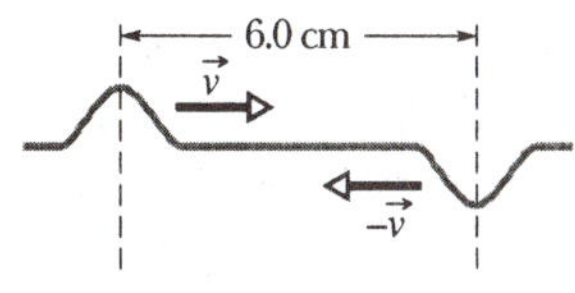

Fig. 17-28 Problem 36.

37P. A string oscillates according to the equation

$$y' = (0.50 \text{ cm}) \sin\left[\left(\frac{\pi}{3} \text{ cm}^{-1}\right)x\right] \cos[(40\pi \text{ s}^{-1})t].$$

What are (a) the amplitude and (b) the speed of the two waves (identical except for direction of travel) whose superposition gives this oscillation? (c) What is the distance between nodes? (d) What is the speed of a particle of the string at the position $x = 1.5$ cm when $t = \frac{9}{8}$ s?

38P. A standing wave results from the sum of two transverse traveling waves given by

$$y_1 = (0.050 \text{ m}) \cos(\pi \text{ m}^{-1})x - (4\pi \text{ s}^{-1})t$$

and

$$y_2 = (0.050 \text{ m}) \cos(\pi \text{ m}^{-1})x + (4\pi \text{ s}^{-1})t.$$

(a) What is the smallest positive value of x that corresponds to a node? (b) At what times during the interval $0 \leq t \leq 0.50$ s will the particle at $x = 0$ have zero velocity?

39P. A string 3.0 m long is oscillating as a three-loop standing wave with an amplitude of 1.0 cm. The wave speed is 100 m/s. (a) What is the frequency? (b) Write equations for two waves that, when combined, will result in this standing wave.

40P. In an experiment on standing waves, a string 90 cm long is attached to the prong of an electrically driven tuning fork that oscillates perpendicular to the length of the string at a frequency of 60 Hz. The mass of the string is 0.044 kg. What tension must the string be under (weights are attached to the other end) if it is to oscillate in four loops?

41P. Oscillation of a 600 Hz tuning fork sets up standing waves in a string clamped at both ends. The wave speed for the string is 400 m/s. The standing wave has four loops and an amplitude of 2.0 mm. (a) What is the length of the string? (b) Write an equation for the displacement of the string as a function of position and time.

42P. A rope, under a tension of 200 N and fixed at both ends, oscillates in a second-harmonic standing wave pattern. The displacement of the rope is given by

$$y = (0.10\text{ m})(\sin\left(\left(\frac{\pi}{2}\text{ m}^{-1}\right)x\right)\sin((12\pi\text{ s}^{-1})t),$$

where $x = 0$ at one end of the rope, x is in meters, and t is in seconds. What are (a) the length of the rope, (b) the speed of the waves on the rope, and (c) the mass of the rope? (d) If the rope oscillates in a third-harmonic standing wave pattern, what will be the period of oscillation?

43P. A generator at one end of a very long string creates a wave given by

$$y = (6.0\text{ cm})\cos\frac{\pi}{2}[(2.0\text{ m}^{-1})x + (8.0\text{ s}^{-1})t],$$

and one at the other end creates the wave

$$y = (6.0\text{ cm})\cos\frac{\pi}{2}[(2.0\text{ m}^{-1})x - (8.0\text{ s}^{-1})t].$$

Calculate (a) the frequency, (b) the wavelength, and (c) the speed of each wave. At what x values are (d) the nodes and (e) the antinodes?

44P. A standing wave pattern on a string is described by

$$y(x, t) = (0.040\text{ m})\sin((5\pi\text{ m}^{-1})x)\cos((40\pi\text{ s}^{-1})t),$$

(a) Determine the location of all nodes for $0 \leq x \leq 0.40$ m. (b) What is the period of the oscillatory motion of any (nonnode) point on the string? What are (c) the speed and (d) the amplitude of the two traveling waves that interfere to produce this wave? (e) At what times for $0 \leq t \leq 0.050$ s will all the points on the string have zero transverse velocity?

45P. Show that the maximum kinetic energy in each loop of a standing wave produced by two traveling waves of identical amplitudes is $2\pi^2\mu y_m^2 f|\vec{v}_w|$.

46P. For a certain transverse standing wave on a long string, an antinode is at $x = 0$ and a node is at $x = 0.10$ m. The displacement $y(t)$ of the string particle at $x = 0$ is shown in Fig. 17-29. When $t = 0.50$ s, what are the displacements of the string particles at (a) $x = 0.20$ m and (b) $x = 0.30$ m? At $x = 0.20$ m, what are the transverse velocities of the string particles at (c) $t = 0.50$ s and (d) $t = 1.0$ s? (e) Sketch the standing wave at $t = 0.50$ s for the range $x = 0$ to $x = 0.40$ m.

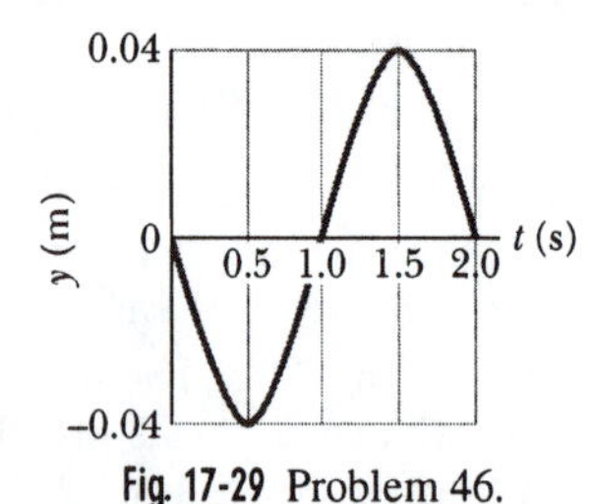

Fig. 17-29 Problem 46.

47P. In Fig. 17-30, an aluminum wire, of length $L_1 = 60.0$ cm, cross-sectional area 1.00×10^{-2} cm^2, and density 2.60 g/cm^3, is joined to a steel wire, of density 7.80 g/cm^3 and the same cross-sectional area. The compound wire, loaded with a block of mass $m = 10.0$ kg, is arranged so that the distance L_2 from the joint to the supporting pulley is 86.6 cm. Transverse waves are set up in the wire by using an external source of variable frequency; a node is located at the pulley. (a) Find the lowest frequency of excitation for which standing waves are observed such that the joint in the wire is one of the nodes. (b) How many nodes are observed at this frequency?

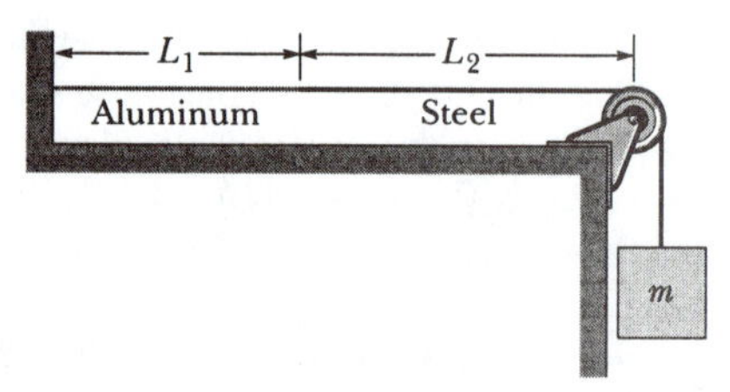

Fig. 17-30 Problem 47.

SEC. 17-12 Phasors

48E. Determine the amplitude of the resultant wave when two sinusoidal string waves having the same frequency and traveling in the same direction on the same string are combined, if their amplitudes are 3.0 cm and 4.0 cm and they have phase constants of 0 and $\pi/2$ rad, respectively.

49P. Two sinusoidal waves of the same period, with amplitudes of 5.0 and 7.0 mm, travel in the same direction along a stretched string; they produce a resultant wave with an amplitude of 9.0 mm. The phase constant of the 5.0 mm wave is 0. What is the phase constant of the 7.0 mm wave?

50P. Three sinusoidal waves of the same frequency travel along a string in the positive direction of an x axis. Their amplitudes are y_1, $y_1/2$, and $y_1/3$, and their phase constants are 0, $\pi/2$, and π, respectively. What are (a) the amplitude and (b) the phase constant of the resultant wave? (c) Plot the wave form of the resultant wave at $t = 0$, and discuss its behavior as t increases.

Chapter 18

Where needed in the problems, use

$$\text{speed of sound in air} = 343\text{ m/s}$$

and

$$\text{density of air} = 1.21\text{ kg/m}^3$$

unless otherwise specified.

SEC. 18-2 The Speed of Sound

1E. Devise a rule for finding your distance in kilometers from a lightning flash by counting the seconds from the time you see the flash until you hear the thunder. Assume that the sound travels to you along a straight line.

2E. You are at a large outdoor concert, seated 300 m from the speaker system. The concert is also being broadcast live via satellite (at the speed of light, 3.0×10^8 m/s). Consider a listener 5000 km away who receives the broadcast. Who hears the music first, you or the listener, and by what time difference?

3E. Two spectators at a soccer game in Montjuic Stadium see, and a moment later hear, the ball being kicked on the playing field. The time delay for one spectator is 0.23 s and for the other 0.12 s.

Sight lines from the two spectators to the player kicking the ball meet at an angle of 90°. (a) How far is each spectator from the player? (b) How far are the spectators from each other?

4E. A column of soldiers, marching at 120 paces per minute, keep in step with the beat of a drummer at the head of the column. It is observed that the soldiers in the rear end of the column are striding forward with the left foot when the drummer is advancing with the right. What is the approximate length of the column?

5P. Earthquakes generate sound waves inside Earth. Unlike a gas, Earth can experience both transverse (S) and longitudinal (P) sound waves. Typically, the speed of S waves is about 4.5 km/s, and that of P waves 8.0 km/s. A seismograph records P and S waves from an earthquake. The first P waves arrive 3.0 min before the first S waves (Fig. 18-21). Assuming the waves travel in a straight line, how far away does the earthquake occur?

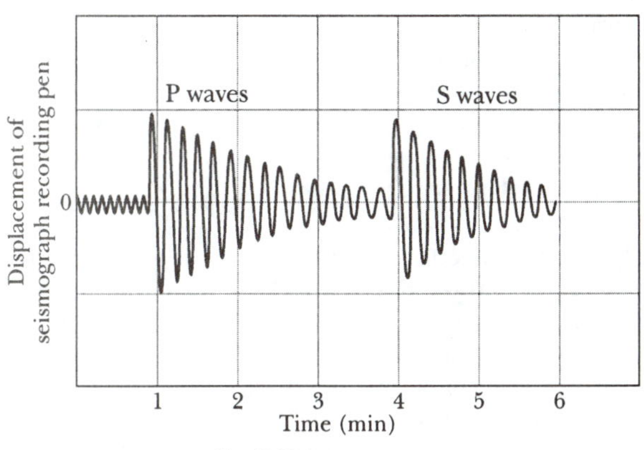

Fig. 18-21 Problem 5.

6P. The speed of sound in a certain metal is $|\vec{V}|$. One end of a long pipe of that metal of length L is struck a hard blow. A listener at the other end hears two sounds, one from the wave that travels along the pipe and the other from the wave that travels through the air. (a) If $|\vec{v}|$ is the speed of sound in air, what time interval Δt elapses between the arrivals of the two sounds? (b) Suppose that $\Delta t = 1.00$ s and the metal is steel. Find the length L.

7P. A stone is dropped into a well. The sound of the splash is heard 3.00 s later. What is the depth of the well?

8E. The audible frequency range for normal hearing is from about 20 Hz to 20 kHz. What are the wavelengths of sound waves at these frequencies?

9E. Diagnostic ultrasound of frequency 4.50 MHz is used to examine tumors in soft tissue. (a) What is the wavelength in air of such a sound wave? (b) If the speed of sound in tissue is 1500 m/s, what is the wavelength of this wave in tissue?

10P. The pressure in a traveling sound wave is given by the equation

$$\Delta P = (1.50 \text{ Pa}) \sin \pi[(0.900 \text{ m}^{-1})x - (315 \text{ s}^{-1})t].$$

Find (a) the pressure amplitude, (b) the frequency, (c) the wavelength, and (d) the speed of the wave.

SEC. 18-3 Interference

11P. Two point sources of sound waves of identical wavelength λ and amplitude are separated by distance $D = 2.0\lambda$. The sources are in phase. (a) How many points of maximum signal (that is, maximum constructive interference) lie along a large circle around the sources? (b) How many points of minimum signal (destructive interference) lie around the circle?

12P. In Fig. 18-22, two loudspeakers, separated by a distance of 2.00 m, are in phase. Assume the amplitudes of the sound from the speakers are approximately the same at the position of a listener, who is 3.75 m directly in front of one of the speakers. (a) For what frequencies in the audible range (20 Hz to 20 kHz) does the listener hear a minimum signal? (b) For what frequencies is the signal a maximum?

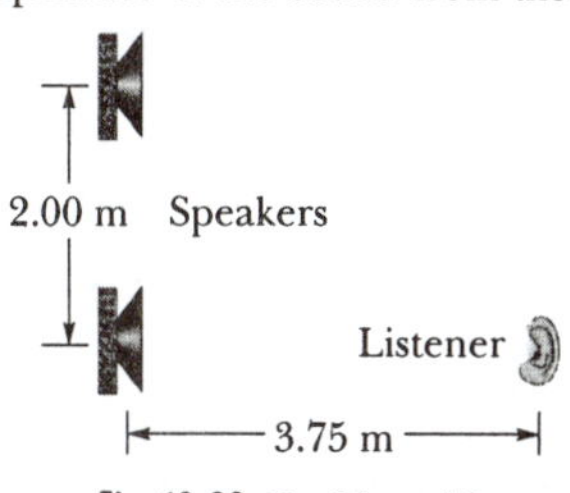

Fig. 18-22 Problem 12.

13P. Two sound waves, from two different sources with the same frequency, 540 Hz, travel in the same direction at 330 m/s. The sources are in phase. What is the phase difference of the waves at a point that is 4.40 m from one source and 4.00 m from the other?

14P. Two loudspeakers are located 3.35 m apart on an outdoor stage. A listener is 18.3 m from one and 19.5 m from the other. During the sound check, a signal generator drives the two speakers in phase with the same amplitude and frequency. The transmitted frequency is swept through the audible range (20 Hz to 20 kHz). (a) What are the three lowest frequencies at which the listener will hear a minimum signal because of destructive interference? (b) What are the three lowest frequencies at which the listener will hear a maximum signal?

15P. In Fig. 18-23, sound with a 40.0 cm wavelength travels rightward from a source and through a tube that consists of a straight portion and a half-circle. Part of the sound wave travels through the half-circle and then rejoins the rest of the wave, which goes directly through the straight portion. This rejoining results in interference. What is the smallest radius r that results in an intensity minimum at the detector?

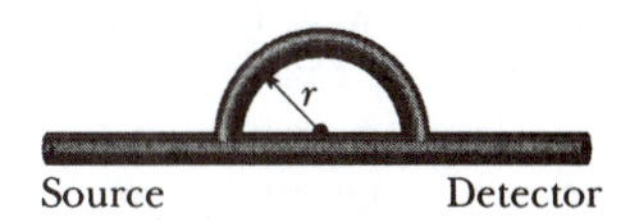

Fig. 18-23 Problem 15.

SEC. 18-4 Intensity and Sound Level

16E. A source emits sound waves isotropically. The intensity of the waves 2.50 m from the source is 1.91×10^{-4} W/m^2. Assuming that the energy of the waves is conserved, find the power of the source.

17E. A 1.0 W point source emits sound waves isotropically. Assuming that the energy of the waves is conserved, find the intensity (a) 1.0 m from the source and (b) 2.5 m from the source.

18E. Two sounds differ in sound level by 1.00 dB. What is the ratio of the greater intensity to the smaller intensity?

19E. A certain sound source is increased in sound level by 30 dB. By what multiple is (a) its intensity increased and (b) its pressure amplitude increased?

20E. The source of a sound wave has a power of 1.00 μW. If it is a point source, (a) what is the intensity 3.00 m away and (b) what is the sound level in decibels at that distance?

21E. (a) If two sound waves, one in air and one in (fresh) water, are equal in intensity, what is the ratio of the pressure amplitude of the wave in water to that of the wave in air? Assume the water and the air are at 20°C. (See Table 15-2.) (b) If the pressure amplitudes are equal instead, what is the ratio of the intensities of the waves?

22P. Assume that a noisy freight train on a straight track emits a cylindrical, expanding sound wave, and that the air absorbs no energy. How does the amplitude ΔP_m of the wave depend on the perpendicular distance r from the source?

23P. Find the ratios (greater to smaller) of (a) the intensities, and (b) the pressure amplitudes for two sounds whose sound levels differ by 37 dB.

24P. A point source emits 30.0 W of sound isotropically. A small microphone intercepts the sound in an area of 0.750 cm^2, 200 m from the source. Calculate (a) the sound intensity there and (b) the power intercepted by the microphone.

25P*. Figure 18-24 shows an air-filled, acoustic interferometer, used to demonstrate the interference of sound waves. Sound source S is an oscillating diaphragm; D is a sound detector, such as the ear or a microphone. Path SBD can be varied in length, but path SAD is fixed. At D, the sound wave coming along path SBD interferes with that coming along path SAD. In one demonstration, the sound intensity at D has a minimum value of 100 units at one position of the movable arm and continuously climbs to a maximum value of 900 units when that arm is shifted by 1.65 cm. Find (a) the frequency of the sound emitted by the source and (b) the ratio of the amplitude at D of the SAD wave to that of the SBD wave. (c) How can it happen that these waves have different amplitudes, considering that they originate at the same source?

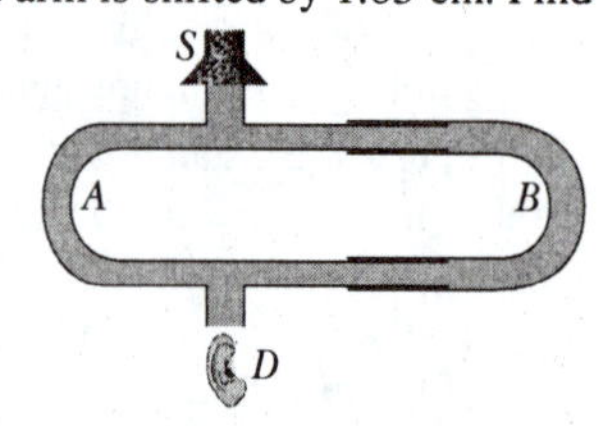

Fig. 18-24 Problem 25.

SEC. 18-5 Sources of Musical Sound

26E. A violin string 15.0 cm long and fixed at both ends oscillates in its $n = 1$ mode. The speed of waves on the string is 250 m/s, and the speed of sound in air is 348 m/s. What are (a) the frequency and (b) the wavelength of the emitted sound wave?

27E. Organ pipe A, with both ends open, had a fundamental frequency of 300 Hz. The third harmonic of organ pipe B, with one end open, has the same frequency as the second harmonic of pipe A. How long are (a) pipe A and (b) pipe B?

28E. The water level in a vertical glass tube 1.00 m long can be adjusted to any position in the tube. A tuning fork vibrating at 686 Hz is held just over the open top end of the tube, to set up a standing wave of sound in the air-filled top portion of the tube. (That air-filled top portion acts as a tube with one end closed and the other end open.) At what positions of the water level is there resonance?

29E. (a) Find the speed of waves on a violin string of mass 800 mg and length 22.0 cm if the fundamental frequency is 920 Hz. (b) What is the tension in the string? For the fundamental, what is the wavelength of (c) the waves on the string and (d) the sound waves emitted by the string?

30P. A certain violin string is 30 cm long between its fixed ends and has a mass of 2.0 g. The "open" string (no applied finger) sounds an A note (440 Hz). (a) To play a C note (523 Hz), how far down the string must one place a finger? (b) What is the ratio of the wavelength of the string waves required for an A note to that required for a C note? (c) What is the ratio of the wavelength of the sound wave for an A note to that for a C note?

31P. In Fig. 18-25, S is a small loudspeaker driven by an audio oscillator and amplifier, adjustable in frequency from 1000 to 2000 Hz only. Tube D is a piece of cylindrical sheet-metal pipe 45.7 cm long and open at both ends. (a) If the speed of sound in air is 344 m/s at the existing temperature, at what frequencies will resonance occur in the pipe when the frequency emitted by the speaker is varied from 1000 Hz to 2000 Hz? (b) Sketch the standing wave (using the style of Fig. 18-10b) for each resonant frequency.

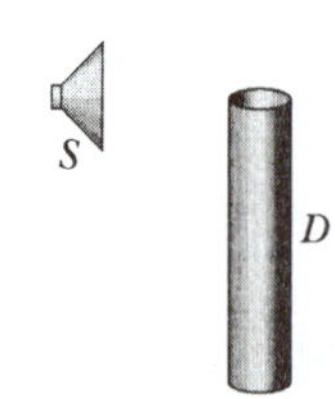

Fig. 18-25 Problem 31.

32P. A string on a cello has length L, for which the fundamental frequency is f. (a) By what length l must the string be shortened by fingering to change the fundamental frequency to rf? (b) What is l if $L = 0.80$ m and $r = 1.2$? (c) For $r = 1.2$, what is the ratio of the wavelength of the new sound wave emitted by the string to that of the wave emitted before fingering?

33P. A well with vertical sides and water at the bottom resonates at 7.00 Hz and at no lower frequency. (The air-filled portion of the well acts as a tube with one closed end and one open end.) The air in the well has a density of 1.10 kg/m^3 and a bulk modulus of 1.33×10^5 Pa. How far down in the well is the water surface?

34P. A tube 1.20 m long is closed at one end. A stretched wire is placed near the open end. The wire is 0.330 m long and has a mass of 9.60 g. It is fixed at both ends and oscillates in its fundamental mode. By resonance, it sets the air column in the tube into oscillation at that column's fundamental frequency. Find (a) that frequency and (b) the tension in the wire.

35P. The period of a pulsating variable star may be estimated by considering the star to be executing *radial* longitudinal pulsations in the fundamental standing wave mode; that is, the star's radius varies periodically with time, with a displacement antinode at the star's surface. (a) Would you expect the center of the star to be a displacement node or antinode? (b) By analogy with a pipe with one open end, show that the period of pulsation T is given by

$$T = \frac{4R}{|\vec{v}|},$$

where R is the equilibrium radius of the star and $|\vec{v}|$ is the average sound speed in the material of the star. (c) Typical white dwarf stars are composed of material with a bulk modulus of 1.33×10^{22} Pa and a density of 10^{10} kg/m^3. They have radii equal to

9.0×10^{-3} solar radius. What is the approximate pulsation period of a white dwarf?

36P. Pipe A, which is 1.2 m long and open at both ends, oscillates at its third lowest harmonic frequency. It is filled with air for which the speed of sound is 343 m/s. Pipe B, which is closed at one end, oscillates at its second lowest harmonic frequency. The frequencies of pipes A and B happen to match. (a) If an x axis extends along the interior of pipe A, with $x = 0$ at one end, where along the axis are the displacement nodes? (b) How long is pipe B? (c) What is the lowest harmonic frequency of pipe A?

37P. A violin string 30.0 cm long with linear density 0.650 g/m is placed near a loudspeaker that is fed by an audio oscillator of variable frequency. It is found that the string is set into oscillation only at the frequencies 880 and 1320 Hz as the frequency of the oscillator is varied over the range 500–1500 Hz. What is the tension in the string?

SEC. 18-6 Beats

38E. The A string of a violin is a little too tightly stretched. Four beats per second are heard when the string is sounded together with a tuning fork that is oscillating accurately at concert A (440 Hz). What is the period of the violin string oscillation?

39E. A tuning fork of unknown frequency makes three beats per second with a standard fork of frequency 384 Hz. The beat frequency decreases when a small piece of wax is put on a prong of the first fork. What is the frequency of this fork?

40P. You have five tuning forks that oscillate at close but different frequencies. What are the (a) maximum and (b) minimum number of different beat frequencies you can produce by sounding the forks two at a time, depending on how the frequencies differ?

41P. Two identical piano wires have a fundamental frequency of 600 Hz when kept under the same tension. What fractional increase in the tension of one wire will lead to the occurrence of 6 beats/s when both wires oscillate simultaneously?

SEC. 18-7 The Doppler Effect

42E. Trooper B is chasing speeder A along a straight stretch of road. Both are moving at a speed of 160 km/h. Trooper B, failing to catch up, sounds his siren again. Take the speed of sound in air to be 343 m/s and the frequency of the source to be 500 Hz. What is the Doppler shift in the frequency heard by speeder A?

43E. The 16 000 Hz whine of the turbines in the jet engines of an aircraft moving with speed 200 m/s is heard at what frequency by the pilot of a second craft trying to overtake the first at a speed of 250 m/s?

44E. An ambulance with a siren emitting a whine at 1600 Hz overtakes and passes a cyclist pedaling a bike at 2.44 m/s. After being passed, the cyclist hears a frequency of 1590 Hz. How fast is the ambulance moving?

45P. A whistle of frequency 540 Hz moves in a circle of radius 60.0 cm at an angular speed of 15.0 rad/s. What are (a) the lowest and (b) the highest frequencies heard by a listener a long distance away, at rest with respect to the center of the circle?

46P. A stationary motion detector sends sound waves of frequency 0.150 MHz toward a truck approaching at a speed of 45.0 m/s. What is the frequency of the waves reflected back to the detector?

47P. A French submarine and a U.S. submarine move toward each other during maneuvers in motionless water in the North Atlantic (Fig. 18-26). The French sub moves at 50.0 km/h, and the U.S. sub at 70.0 km/h. The French sub sends out a sonar signal (sound wave in water) at 1000 Hz. Sonar waves travel at 5470 km/h. (a) What is the signal's frequency as detected by the U.S. sub? (b) What frequency is detected by the French sub in the signal reflected back to it by the U.S. sub?

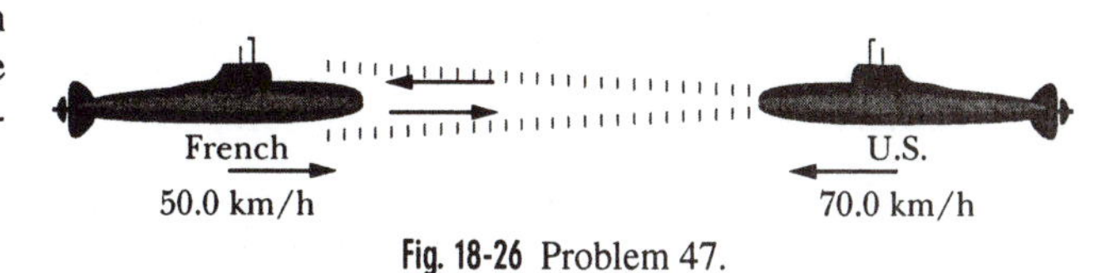

Fig. 18-26 Problem 47.

48P. A sound source A and a reflecting surface B move directly toward each other. Relative to the air, the speed of source A is 29.9 m/s, the speed of surface B is 65.8 m/s, and the speed of sound is 329 m/s. The source emits waves at frequency 1200 Hz as measured in the source frame. In the reflector frame, what are (a) the frequency and (b) the wavelength of the arriving sound waves? In the source frame, what are (c) the frequency and (d) the wavelength of the sound waves reflected back to the source?

49P. An acoustic burglar alarm consists of a source emitting waves of frequency 28.0 kHz. What is the beat frequency between the source waves and the waves reflected from an intruder walking at an average speed of 0.950 m/s directly away from the alarm?

50P. A bat is flitting about in a cave, navigating via ultrasonic bleeps. Assume that the sound emission frequency of the bat is 39 000 Hz. During one fast swoop directly toward a flat wall surface, the bat is moving at 0.025 times the speed of sound in air. What frequency does the bat hear reflected off the wall?

51P. A girl is sitting near the open window of a train that is moving at a velocity of 10.00 m/s to the east. The girl's uncle stands near the tracks and watches the train move away. The locomotive whistle emits sound at frequency 500.0 Hz. The air is still. (a) What frequency does the uncle hear? (b) What frequency does the girl hear? A wind begins to blow from the east at 10.00 m/s. (c) What frequency does the uncle now hear? (d) What frequency does the girl now hear?

52P. A 2000 Hz siren and a civil defense official are both at rest with respect to the ground. What frequency does the official hear if the wind is blowing at 12 m/s (a) from source to official and (b) from official to source?

53P. Two trains are traveling toward each other at 30.5 m/s relative to the ground. One train is blowing a whistle at 500 Hz. (a) What frequency is heard on the other train in still air? (b) What frequency is heard on the other train if the wind is blowing at 30.5 m/s toward the whistle and away from the listener? (c) What frequency is heard if the wind direction is reversed?

SEC. 18-8 Supersonic Speeds; Shock Waves

54E. A bullet is fired with a speed of 685 m/s. Find the angle made by the shock cone with the line of motion of the bullet.

55P. A jet plane passes over you at a height of 5000 m and a speed of Mach 1.5. (a) Find the Mach cone angle. (b) How long after the jet passes directly overhead does the shock wave reach you? Use 331 m/s for the speed of sound.

56P. A plane flies at 1.25 times the speed of sound. Its sonic boom reaches a man on the ground 1.00 min after the plane passes directly overhead. What is the altitude of the plane? Assume the speed of sound to be 330 m/s.

Chapter 19

SEC. 19-2 An Introduction to Thermometers and Temperature Scales

1E. At what temperature is the Fahrenheit scale reading equal to (a) twice that of the Celsius and (b) half that of the Celsius?

2E. (a) In 1964, the temperature in the Siberian village of Oymyakon reached −71°C. What temperature is this on the Fahrenheit scale? (b) The highest officially recorded temperature in the continental United States was 134°F in Death Valley, California. What is this temperature on the Celsius scale?

SEC. 19-4 Heating, Cooling, and Temperature

3P. It is an everyday observation that hot and cold objects cool down or warm up to the temperature of their surroundings. If the temperature difference ΔT between an object and its surroundings ($\Delta T = T_{obj} - T_{sur}$) is not too great, the rate of cooling or warming of the object is proportional, approximately, to this temperature difference; that is,

$$\frac{d\,\Delta T}{dt} = -A(\Delta T),$$

where A is a constant. (The minus sign appears because ΔT decreases with time if ΔT is positive and increases if ΔT is negative.) This is known as *Newton's law of cooling.* (a) On what factors does A depend? What are its dimensions? (b) If at some instant $t = 0$ the temperature difference is ΔT_0, show that it is

$$\Delta T = \Delta T_0 e^{-At}$$

at a later time t.

4P. The heater of a house breaks down one day when the outside temperature is 7.0°C. As a result, the inside temperature drops from 22°C to 18°C in 1.0 h. The owner fixes the heater and adds insulation to the house. Now she finds that, on a similar day, the house takes twice as long to drop from 22°C to 18°C when the heater is not operating. What is the ratio of the new value of constant A in Newton's law of cooling (see Problem 3) to the previous value?

SEC. 19-5 Heat Energy Transfer to Solids and Liquids

5E. A certain diet doctor encourages people to diet by drinking ice water. His theory is that the body must burn off enough fat to raise the temperature of the water from 0.00°C to the body temperature of 37.0°C. How many liters of ice water would have to be consumed to burn off 454 g (about 1 lb) of fat, assuming that this much fat burning requires 3500 Cal be transferred to the ice water? Why is it not advisable to follow this diet? (One liter = 10^3 cm^3. The density of water is 1.00 g/cm^3.)

6E. A certain substance has a mass per mole of 50 g/mol. When 314 J is added as heat energy to a 30.0 g sample, the sample's temperature rises from 25.0°C to 45.0°C. What are (a) the specific heat and (b) the molar specific heat of this substance? (c) How many moles are present?

7E. How much water remains unfrozen after 50.2 kJ is transferred as heat from 260 g of liquid water initially at its freezing point?

8E. Calculate the minimum amount of energy, in joules, required to completely melt 130 g of silver initially at 15.0°C.

9E. A room is lighted by four 100 W incandescent lightbulbs. (The power of 100 W is the rate at which a bulb converts electrical energy to heat and the energy of visible light.) Assuming that 90% of the energy is converted to heat, how much heat does the room receive in 1.00 h?

10E. An energetic athlete can use up all the energy from a diet of 4000 Cal/day. If he were to use up this energy at a steady rate, how would his rate of energy use compare with the power of a 100 W bulb? (The power of 100 W is the rate at which the bulb converts electrical energy to heat and the energy of visible light.)

11E. How many grams of butter, which has a usable energy content of 6.0 Cal/g (= 6000 cal/g), would be equivalent to the change in gravitational potential energy of a 73.0 kg man who ascends from sea level to the top of Mt. Everest, at elevation 8.84 km? Assume that the average value of g is 9.80 m/s^2.

12E. A power of 0.400 hp is required for 2.00 min to drill a hole in a 1.60-lb copper block. (a) If the full power is the rate at which thermal energy is generated, how much is generated in Btu? (b) What is the rise in temperature of the copper if the copper absorbs 75.0% of this energy? (Use the energy conversion 1 ft · lb = 1.285×10^{-3} Btu.)

13E. One way to keep the contents of a garage from becoming too cold on a night when a severe subfreezing temperature is forecast is to put a tub of water in the garage. If the mass of the water is 125 kg and its initial temperature is 20°C, (a) how much energy must the water transfer to its surroundings in order to freeze completely and (b) what is the lowest possible temperature of the water and its surroundings until that happens?

14E. A small electric immersion heater is used to heat 100 g of water for a cup of instant coffee. The heater is labeled "200 watts," which means that it converts electrical energy to thermal energy at this rate. Calculate the time required to bring all this water from 23°C to 100°C, ignoring any heat losses.

15P. A 150 g copper bowl contains 220 g of water, both at 20.0°C. A very hot 300 g copper cylinder is dropped into the water, causing the water to boil, with 5.00 g being converted to steam. The final temperature of the system is 100°C. Neglect energy transfers with the environment. (a) How much energy (in calories) is transferred to the water as heat? (b) How much to the bowl? (d) What is the original temperature of the cylinder?

16P. A chef, on finding his stove out of order, decides to boil the water for his wife's coffee by shaking it in a thermos flask. Suppose that he uses tap water at 15°C and that the water falls 30 cm each shake, the chef making 30 shakes each minute. Neglecting any loss of thermal energy by the flask, how long must he shake the flask until the water reaches 100°C?

17P. *Nonmetric version:* How long does a 2.0×10^5 Btu/h water heater take to raise the temperature of 40 gal of water from 70°F to 100°F? *Metric version:* How long does a 59 kW water heater take to raise the temperature of 150 L of water from 21°C to 38°C?

18P. Ethyl alcohol has a boiling point of 78°C, a freezing point of −114°C, a heat of vaporization of 879 kJ/kg, a heat of fusion of 109 kJ/kg, and a specific heat of 2.43 kJ/kg · C°. How much energy must be removed from 0.510 kg of ethyl alcohol that is initially a gas at 78°C so that it becomes a solid at −114°C?

19P. A 1500 kg Buick moving at 90 km/h brakes to a stop, at uniform deceleration and without skidding, over a distance of 80 m. At what average rate is mechanical energy transferred to thermal energy in the brake system?

20P. In a solar water heater, energy from the Sun is gathered by water that circulates through tubes in a rooftop collector. The solar radiation enters the collector through a transparent cover and warms the water in the tubes; this water is pumped into a holding tank. Assume that the efficiency of the overall system is 20% (that is, 80% of the incident solar energy is lost from the system). What collector area is necessary to raise the temperature of 200 L of water in the tank from 20°C to 40°C in 1.0 h when the intensity of incident sunlight is 700 W/m^2?

21P. What mass of steam at 100°C must be mixed with 150 g of ice at its melting point, in a thermally insulated container, to produce liquid water at 50°C?

22P. A person makes a quantity of iced tea by mixing 500 g of hot tea (essentially water) with an equal mass of ice at its melting point. If the initial hot tea is at a temperature of (a) 90°C and (b) 70°C, what are the temperature and mass of the remaining ice when the tea and ice reach a common temperature? Neglect energy transfers with the environment.

23P. (a) Two 50 g ice cubes are dropped into 200 g of water in a thermally insulated container. If the water is initially at 25°C, and the ice comes directly from a freezer at −15°C, what is the final temperature of the drink when the drink reaches thermal equilibrium? (b) What is the final temperature if only one ice cube is used?

24P. An insulated Thermos contains 130 cm^3 of hot coffee, at a temperature of 80.0°C. You put in a 12.0 g ice cube at its melting point to cool the coffee. By how many degrees has your coffee cooled once the ice has melted? Treat the coffee as though it were pure water and neglect energy transfers with the environment.

SEC. 19-8 Some Special Cases of the First Law of Thermodynamics

25E. Consider that 200 J of work is done on a system and 70.0 cal is extracted from the system as heat. In the sense of the first law of thermodynamics, what are the values (including algebraic signs) of (a) W, (b) Q, and (c) ΔE_{int}?

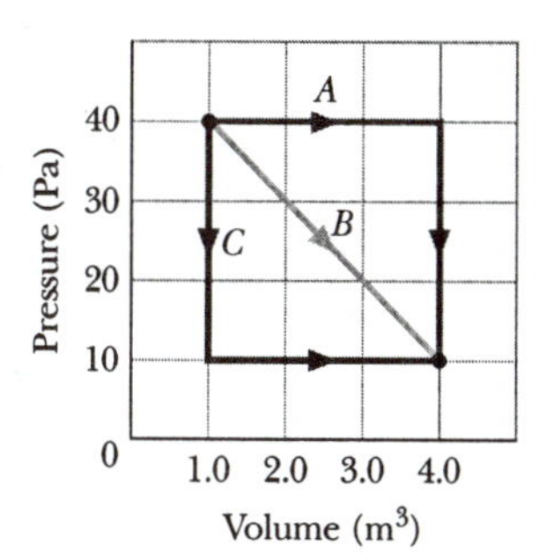

Fig. 19-26 Exercise 26.

26E. A sample of gas expands from 1.0 m^3 to 4.0 m^3 while its pressure decreases from 40 Pa to 10 Pa. How much work is done by the gas if its pressure changes with volume via each of the three paths shown in the P-V diagram in Fig. 19-26?

27E. A thermodynamic system is taken from an initial state A to another state B and back again to A, via state C, as shown by path $ABCA$ in the P-V diagram of Fig. 19-27a. (a) Complete the table in Fig. 19-27b by filling in either + or − for the sign of each thermodynamic quantity associated with each step of the cycle. (b) Calculate the numerical value of the work done by the system for the complete cycle $ABCA$.

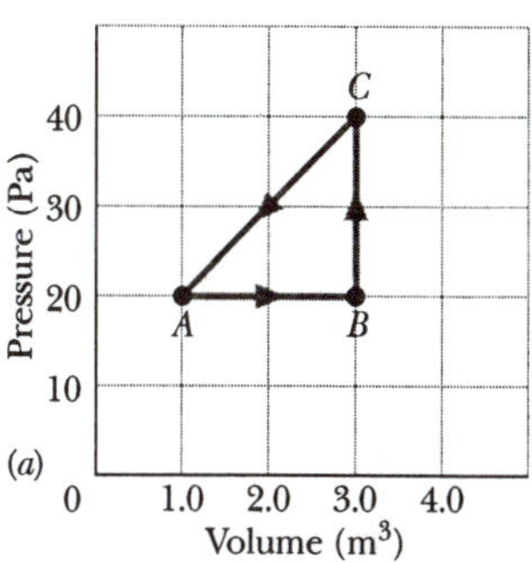

(b)

	Q	W	ΔE_{int}
$A \rightarrow B$			+
$B \rightarrow C$	+		
$C \rightarrow A$			

Fig. 19-27 Exercise 27.

28E. Gas within a closed chamber undergoes the cycle shown in the P-V diagram of Fig. 19-28. Calculate the net energy added to the system as heat during one complete cycle.

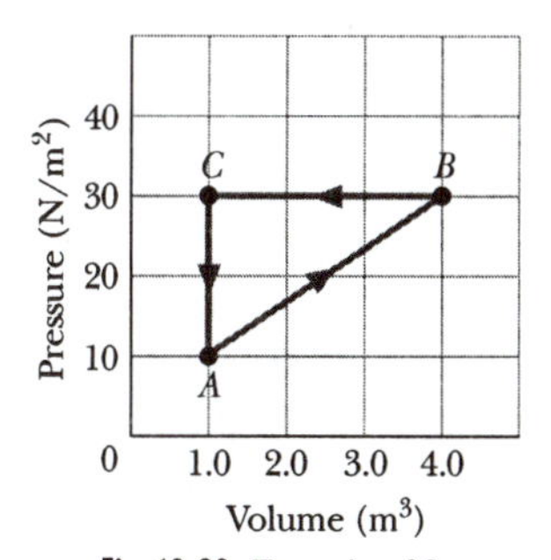

Fig. 19-28 Exercise 28.

29E. Gas within a chamber passes through the cycle shown in Fig. 19-29. Determine the energy transferred by the system as heat during process CA if the energy added as heat Q_{AB} during process AB is 20.0 J, no energy is transferred as heat during process BC, and the net work done during the cycle is 15.0 J.

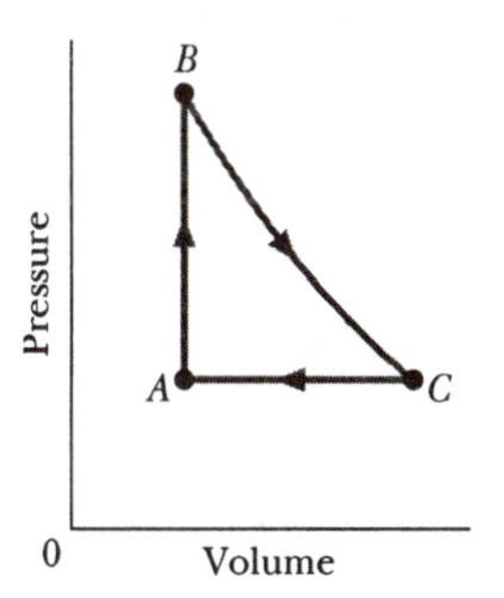

Fig. 19-29 Exercise 29.

30P. When a system is taken from state i to state f along path iaf in Fig. 19-30. Q = 50 cal and W = 20 cal. Along path ibf, Q = 36 cal. (a) What is W along path ibf? (b) If W = −13 cal for the return path fi, what is Q for this path? (c) Take $E_{int,i}$ = 10 cal. What is $E_{int,f}$? (d) If $E_{int,b}$ = 22 cal, what are the values of Q for path ib and path bf?

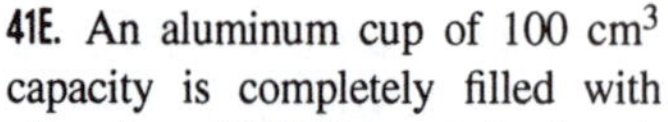

Fig. 19-30 Problem 30.

SEC. 19-9 More on Temperature Measurement

31E. Suppose the temperature of a gas at the boiling point of water is 373.15 K. What then is the limiting value of the ratio of the pressure of the gas at that boiling point to its pressure at the triple point of water? (Assume the volume of the gas is the same at both temperatures.)

32E. A particular gas thermometer is constructed of two gas-containing bulbs, each of which is put into a water bath, as shown in Fig. 19-31. The pressure difference between the two bulbs is measured by a mercury manometer as shown. Appropriate reservoirs, not shown in the diagram, maintain constant gas volume in the two bulbs. There is no difference in pressure when both baths are at the triple point of water. The pressure difference is 120 torr when one bath is at the triple point and the other is at the boiling point of water. It is 90.0 torr when one bath is at the triple point and the other is at an unknown temperature to be measured. What is the unknown temperature?

Fig. 19-31 Exercise 32.

33E. At what temperature do the following pairs of scales read the same, if ever: (a) Fahrenheit and Celsius (verify the listing in Table 19-1), (b) Fahrenheit and Kelvin, and (c) Celsius and Kelvin?

SEC. 19-10 Thermal Expansion (Optional)

34E. An aluminum flagpole is 33 m high. By how much does its length increase as the temperature increases by 15 C°?

35E. The Pyrex glass mirror in the telescope at the Mt. Palomar Observatory has a diameter of 200 in. The temperature ranges from −10°C to 50°C on Mt. Palomar. In micrometers, what is the maximum change in the diameter of the mirror, assuming that the glass can freely expand and contract?

36E. An aluminum-alloy rod has a length of 10.000 cm at 20.000°C and a length of 10.015 cm at the boiling point of water. (a) What is the length of the rod at the freezing point of water? (b) What is the temperature if the length of the rod is 10.009 cm?

37E. A circular hole in an aluminum plate is 2.725 cm in diameter at 0.000°C. What is its diameter when the temperature of the plate is raised to 100.0°C?

38E. What is the volume of a lead ball at 30°C if the ball's volume at 60°C is 50 cm^3?

39E. Find the change in volume of an aluminum sphere with an initial radius of 10 cm when the sphere is heated from 0.0°C to 100°C.

40E. The area A of a rectangular plate is ab. Its coefficient of linear expansion is α. After a temperature rise ΔT, side a is longer by Δa and side b is longer by Δb (Fig. 19-32). Show that if the small quantity $(\Delta a\, \Delta b)/ab$ is neglected, then $\Delta A = 2\alpha A\, \Delta T$.

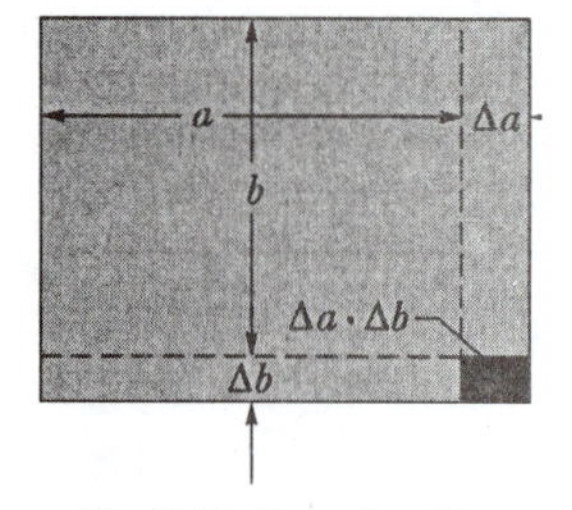

Fig. 19-32 Exercise 40.

41E. An aluminum cup of 100 cm^3 capacity is completely filled with glycerin at 22°C. How much glycerin, if any, will spill out of the cup if the temperature of both the cup and glycerin is increased to 28°C? (The coefficient of volume expansion of glycerin is 5.1×10^{-4}/C°.)

42P. At 20°C, a rod is exactly 20.05 cm long on a steel ruler. Both the rod and the ruler are placed in an oven at 270°C, where the rod now measures 20.11 cm on the same ruler. What is the coefficient of thermal expansion for the material of which the rod is made?

43P. A steel rod is 3.000 cm in diameter at 25°C. A brass ring has an interior diameter of 2.992 cm at 25°C. At what common temperature will the ring just slide onto the rod?

44P. When the temperature of a metal cylinder is raised from 0.0°C to 100°C, its length increases by 0.23%. (a) Find the percent change in density. (b) What is the metal? Use Table 19-5.

45P. Show that when the temperature of a liquid in a barometer changes by ΔT and the pressure is constant, the liquid's height h changes by $\Delta h = \beta h\, \Delta T$, where β is the coefficient of volume expansion. Neglect the expansion of the glass tube.

46P. When the temperature of a copper coin is raised by 100 C°, its diameter increases by 0.18%. To two significant figures, give the percent increase in (a) the area of a face, (b) the thickness, (c) the volume, and (d) the mass of the coin. (e) Calculate the coefficient of linear expansion of the coin.

47P. A pendulum clock with a pendulum made of brass is designed to keep accurate time at 20°C. If the clock operates at 0.0°C, what is the magnitude of its error, in seconds per hour, and does the clock run fast or slow?

48P. In a certain experiment, a small radioactive source must move at selected, extremely slow speeds. This motion is accomplished by fastening the source to one end of an aluminum rod and heating the central section of the rod in a controlled way. If the effective heated section of the rod in Fig. 19-33 is 2.00 cm, at what constant rate must the temperature of the rod be changed if the source is to move at a constant speed of 100 nm/s?

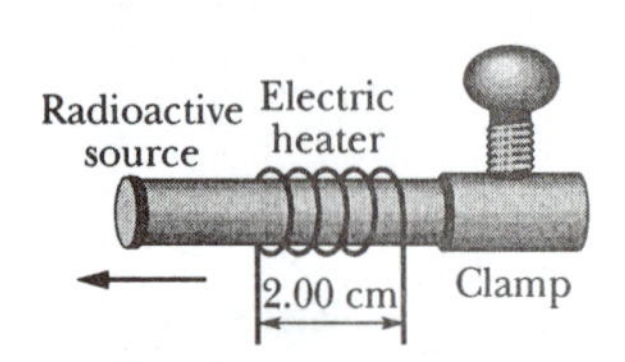

Fig. 19-33 Problem 48.

49P. As a result of a temperature rise of 32 C°, a bar with a crack at its center buckles upward (Fig. 19-34). If the fixed distance L_0 is 3.77 m and the coefficient of linear expansion of the bar is 25×10^{-6}/C°, find the rise x of the center.

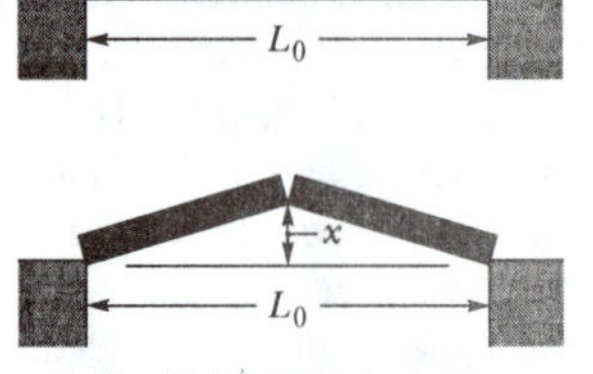

Fig. 19-34 Problem 49.

50P. A 20.0 g copper ring has a diameter of 2.54000 cm at its temperature of 0.000°C. An aluminum sphere has a diameter of 2.54508 cm at its temperature of 100.0°C. The sphere is placed on top of the ring (Fig. 19-35), and the two are allowed to come to thermal equilibrium, with no heat lost to the surroundings. The sphere just passes through the ring at the equilibrium temperature. What is the mass of the sphere?

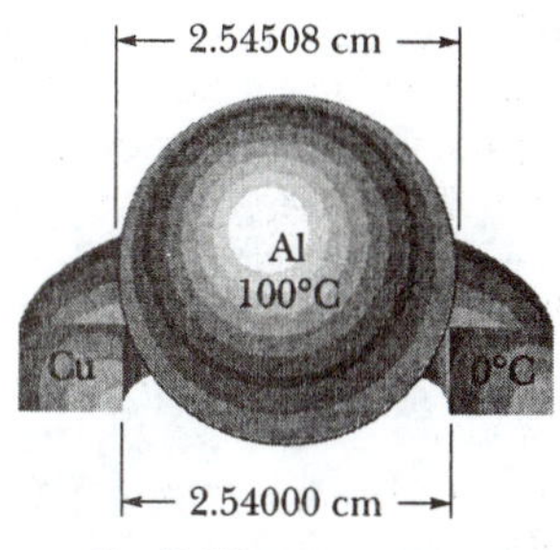

Fig. 19-35 Problem 50.

SEC. 19-11 More on Heat Energy Transfer Mechanisms (Optional)

51E. The average rate at which energy is conducted outward through the ground surface in North America is 54.0 mW/m^2, and the average thermal conductivity of the near-surface rocks is 2.50 W/m · K. Assuming a surface temperature of 10.0°C, find the temperature at a depth of 35.0 km (near the base of the crust). Ignore the heat generated by the presence of radioactive elements.

52E. The ceiling of a single-family dwelling in a cold climate should have an R-value of 30. To give such insulation, how thick would a layer of (a) polyurethane foam and (b) silver have to be?

53E. (a) Calculate the rate at which body heat is conducted through the clothing of a skier in a steady-state process, given the following data: the body surface area is 1.8 m^2 and the clothing is 1.0 cm thick; the skin surface temperature is 33°C and the outer surface of the clothing is at 1.0°C; the thermal conductivity of the clothing is 0.040 W/m · K. (b) How would the answer to (a) change if, after a fall, the skier's clothes became soaked with water of thermal conductivity 0.60 W/m · K?

54E. Consider the slab shown in Fig. 19-23. Suppose that $L =$ 25.0 cm, $A =$ 90.0 cm^2, and the material is copper. If $T_H =$ 125°C, $T_C =$ 10.0°C, and a steady state is reached, find the conduction rate through the slab.

55E. If you were to walk briefly in space without a spacesuit while far from the Sun (as an astronaut does in the movie *2001*), you would feel the cold of space—while you radiated energy, you would absorb almost none from your environment. (a) At what rate would you lose energy? (b) How much energy would you lose in 30 s? Assume that your emissivity is 0.90, and estimate other data needed in the calculations.

56E. A cylindrical copper rod of length 1.2 m and cross-sectional area 4.8 cm^2 is insulated to prevent heat loss through its surface. The ends are maintained at a temperature difference of 100 C° by having one end in a water–ice mixture and the other in boiling water and steam. (a) Find the rate at which energy is conducted along the rod. (b) Find the rate at which ice melts at the cold end.

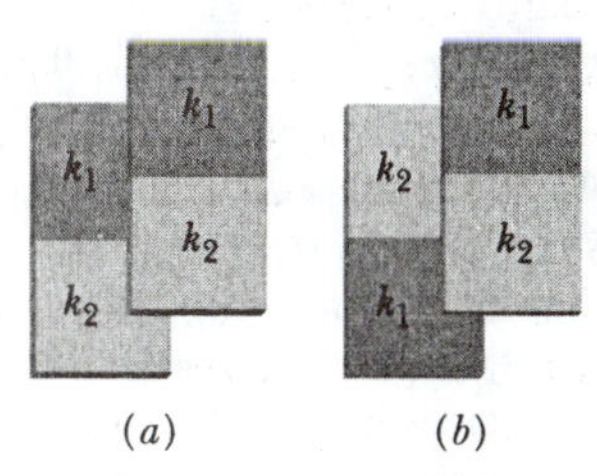

Fig. 19-36 Exercise 57.

57E. Four square pieces of insulation of two different materials, all with the same thickness and area A, are available to cover an opening of area $2A$. This can be done in either of the two ways shown in Fig. 19-36. Which arrangement, (*a*) or (*b*), gives the lower energy flow if $k_2 \neq k_1$?

58P. Two identical rectangular rods of metal are welded end to end as shown in Fig. 19-37*a*, and 10 J is conducted (in a steady-state process) through the rods as heat in 2.0 min. How long would it take for 10 J to be conducted through the rods if they were welded together as shown in Fig. 19-37*b*?

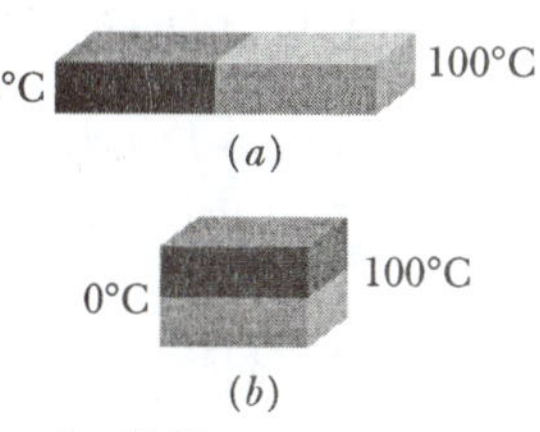

Fig. 19-37 Problem 58.

59P. A sphere of radius 0.500 m, temperature 27.0°C, and emissivity 0.850 is located in an environment of temperature 77.0°C. At what rate does the sphere (a) emit and (b) absorb thermal radiation? (c) What is the sphere's net rate of energy exchange?

60P. (a) What is the rate of energy loss in watts per square meter through a glass window 3.0 mm thick if the outside temperature is −20°F and the inside temperature is +72°F? (b) A storm window having the same thickness of glass is installed parallel to the first window, with an air gap of 7.5 cm between the two windows. What now is the rate of energy loss if conduction is the only important energy-loss mechanism?

61P. Figure 19-38 shows (in cross section) a wall that consists of four layers. The thermal conductivities are $k_1 = 0.060$ W/m · K, $k_3 = 0.040$ W/m · K, and $k_4 = 0.12$ W/m · K (k_2 is not known). The layer thicknesses are $L_1 = 1.5$ cm, $L_3 = 2.8$ cm, and $L_4 =$ 3.5 cm (L_2 is not known). Energy transfer through the wall is steady. What is the temperature of the interface indicated?

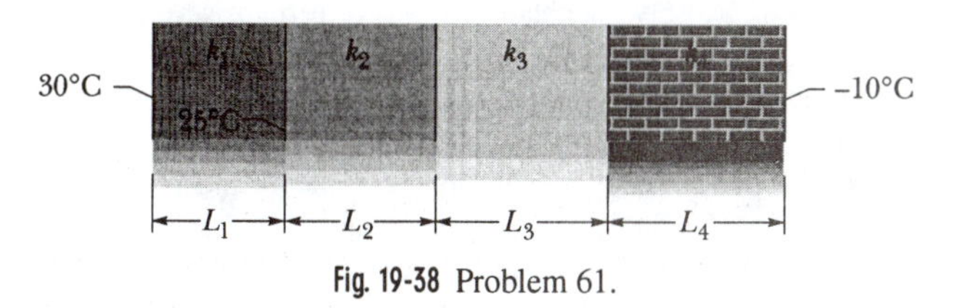

Fig. 19-38 Problem 61.

62P. A tank of water has been outdoors in cold weather, and a slab of ice 5.0 cm thick has formed on its surface (Fig. 19-39). The air above the ice is at −10°C. Calculate the rate of formation of ice (in centimeters per hour) on the ice slab. Take the thermal conductivity and density of ice to be 0.0040 cal/s · cm · C° and 0.92 g/cm^3. Assume that energy is not transferred through the walls or bottom of the tank.

Fig. 19-39 Problem 62.

63P. Ice has formed on a shallow pond and a steady state has been reached, with the air above the ice at −5.0°C and the bottom of the pond at 4.0°C. If the total depth of *ice + water* is 1.4 m, how thick is the ice? (Assume that the thermal conductivities of ice and water are 0.40 and 0.12 cal/m · C° · s, respectively.)

64P. You can join the semi-secret "300 F" club at the Amundsen–Scott South Pole Station only when the outside temperature is below −70°C. On such a day, you first bask in a hot sauna and then run outside wearing only your shoes. (This is, of course, extremely dangerous, but the rite is effectively a protest against the constant danger of the winter cold at the south pole.)

Assume that when you step out of the sauna, your skin temperature is 102°F and the walls, ceiling, and floor of the sauna room have a temperature of 30°C. Estimate your surface area, and take your skin emissivity to be 0.80. (a) What is the approximate net rate P_{net} at which you lose energy via thermal radiation exchanges with the room? Next, assume that when you are outside half your surface area exchanges thermal radiation with the sky at a temperature of −25°C and the other half exchanges thermal radiation with the snow and ground at a temperature of −80°C. What is the approximate net rate at which you lose energy via thermal radiation exchanges with (b) the sky and (c) the snow and ground?

65P. Emperor penguins (Fig. 19-40), those large penguins that resemble stuffy English butlers, breed and hatch their young even during severe Antarctic winters. Once an egg is laid, the father balances the egg on his feet to prevent the egg from freezing. He must do this for the full incubation period of 105 to 115 days, during which he cannot eat because his food is in the water. He can survive this long without food only if he can reduce his consumption of internal energy significantly. If he is alone, he consumes that energy too quickly to stay warm, and eventually abandons the egg in order to eat. To protect themselves and each other from the cold so as to reduce the consumption of internal energy, penguin fathers huddle closely together, in groups of perhaps several thousand. In addition to providing other benefits, the huddling reduces the rate at which the penguins thermally radiate energy to their surroundings.

Assume that a penguin father is a circular cylinder with top surface area a, height h, surface temperature T, and emissivity ε. (a) Find an expression for the rate P_i at which an individual father would radiate energy to the environment from his top surface and his side surface were he alone with his egg.

If N identical fathers were well apart from one another, the total rate of energy loss via radiation would be NP_i. Suppose, instead, that they huddle closely to form a *huddled cylinder* with top surface area Na and height h. (b) Find an expression for the rate P_h at which energy is radiated by the top surface and the side surface of the huddled cylinder.

(c) Assuming $a = 0.34$ m^2 and $h = 1.1$ m and using the expressions you obtained for P_i and P_h, graph the ratio P_h/NP_i versus N_h. Of course, the penguins know nothing about algebra or graphing, but their instinctive huddling reduces this ratio so that more of their eggs survive to the hatching stage. From the graphs (as you will see, you probably need more than one version), approximate how many penguins must huddle so that P_h/NP_i is reduced to (d) 0.5, (e) 0.4, (f) 0.3, (g) 0.2, and (h) 0.15. (i) For the assumed data, what is the lower limiting value for P_h/NP_i?

Fig. 19-40 Problem 65.

Chapter 20

SEC. 20-2 The Macroscopic Behavior of Gases

1E. Find the mass in kilograms of 7.50×10^{24} atoms of arsenic, which has a molar mass of 74.9 g/mol.

2E. Gold has a molar mass of 197 g/mol. (a) How many moles of gold are in a 2.50 g sample of pure gold? (b) How many atoms are in the sample?

3P. If the water molecules in 1.00 g of water were distributed uniformly over the surface of Earth, how many such molecules would there be on 1.00 cm^2 of the surface?

4P. A distinguished scientist has written: "There are enough molecules in the ink that makes one letter of this sentence to provide not only one for every inhabitant of Earth, but one for every creature if each star of our galaxy had a planet as populous as Earth." Check this statement. Assume the ink sample (molar mass = 18 g/mol) to have a mass of 1 μg, the population of Earth to be 5×10^9, and the number of stars in our galaxy to be 10^{11}.

5E. Compute (a) the number of moles and (b) the number of molecules in 1.00 cm^3 of an ideal gas at a pressure of 100 Pa and a temperature of 220 K.

6E. The best laboratory vacuum has a pressure of about 1.00×10^{-18} atm, or 1.01×10^{-13} Pa. How many gas molecules are there per cubic centimeter in such a vacuum at 293 K?

7E. Oxygen gas having a volume of 1000 cm^3 at 40.0°C and 1.01×10^5 Pa expands until its volume is 1500 cm^3 and its pressure is 1.06×10^5 Pa. Find (a) the number of moles of oxygen present and (b) the final temperature of the sample.

8E. An automobile tire has a volume of 1.64×10^{-2} m^3 and contains air at a gauge pressure (pressure above atmospheric pressure) of 165 kPa when the temperature is 0.00°C. What is the gauge pressure of the air in the tires when its temperature rises to 27.0°C and its volume increases to 1.67×10^{-2} m^3? Assume atmospheric pressure is 1.0×10^5 Pa.

9E. A quantity of ideal gas at 10.0°C and 100 kPa occupies a volume of 2.50 m^3. (a) How many moles of the gas are present? (b) If the pressure is now raised to 300 kPa and the temperature is raised to 30.0°C, how much volume does the gas occupy? Assume no leaks.

SEC. 20-3 Work Done by Ideal Gases

10E. Calculate the work done by an external agent during an isothermal compression of 1.00 mol of oxygen from a volume of 22.4 L at 0°C and 1.00 atm pressure to 16.8 L.

11P. Pressure P, volume V, and temperature T for a certain material are related by

$$P = \frac{AT - BT^2}{V},$$

where A and B are constants. Find an expression for the work done by the material if the temperature changes from T_1 to T_2 while the pressure remains constant.

12P. A container encloses two ideal gases. Two moles of the first gas are present, with molar mass M_1. The second gas has molar mass $M_2 = 3M_1$, and 0.5 mol of this gas is present. What fraction of the total pressure on the container wall is attributable to the second gas? (The kinetic theory explanation of pressure leads to the experimentally discovered law of partial pressures for a mixture of gases that do not react chemically: *The total pressure exerted by the mixture is equal to the sum of the pressures that the several gases would exert separately if each were to occupy the vessel alone.*)

13P. Air that initially occupies 0.14 m^3 at a gauge pressure of 103.0 kPa is expanded isothermally to a pressure of 101.3 kPa and then cooled at constant pressure until it reaches its initial volume. Compute the work done by the air. (Gauge pressure is the difference between the actual pressure and atmospheric pressure.)

14P. A sample of an ideal gas is taken through the cyclic process *abca* shown in Fig. 20-14; at point a, $T = 200$ K. (a) How many moles of gas are in the sample? What are (b) the temperature of the gas at point b, (c) the temperature of the gas at point c, and (d) the net energy added to the gas as heat during the cycle?

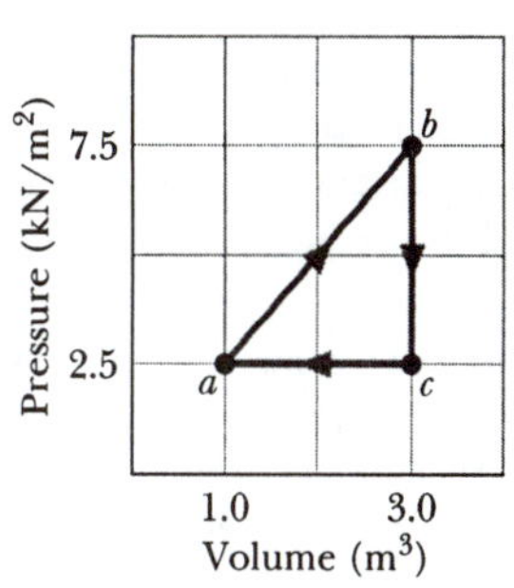

Fig. 20-14 Problem 14.

15P. An air bubble of 20 cm^3 volume is at the bottom of a lake 40 m deep where the temperature is 4.0°C. The bubble rises to the surface, which is at a temperature of 20°C. Take the temperature of the bubble's air to be the same as that of the surrounding water. Just as the bubble reaches the surface, what is its volume?

16P. A pipe of length $L = 25.0$ m that is open at one end contains air at atmospheric pressure. It is thrust vertically into a freshwater lake until the water rises halfway up in the pipe, as shown in Fig. 20-15. What is the depth h of the lower end of the pipe? Assume that the temperature is the same everywhere and does not change.

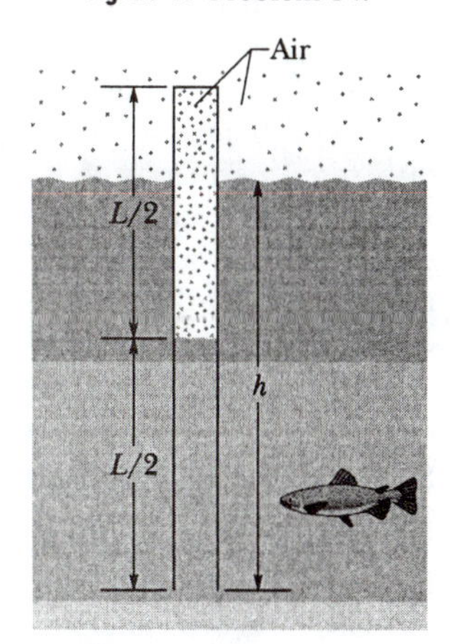

Fig. 20-15 Problem 16.

17P. Container A in Fig. 20-16 holds an ideal gas at a pressure of 5.0×10^5 Pa and a temperature of 300 K. It is connected by a thin tube (and a closed valve) to container B, with four times the volume of A. Container B holds the same ideal gas at a pressure of 1.0×10^5 Pa and a temperature of 400 K. The valve is opened to allow the pressures to equalize, but the temperature of each container is kept constant at its initial value. What then is the pressure in the two containers?

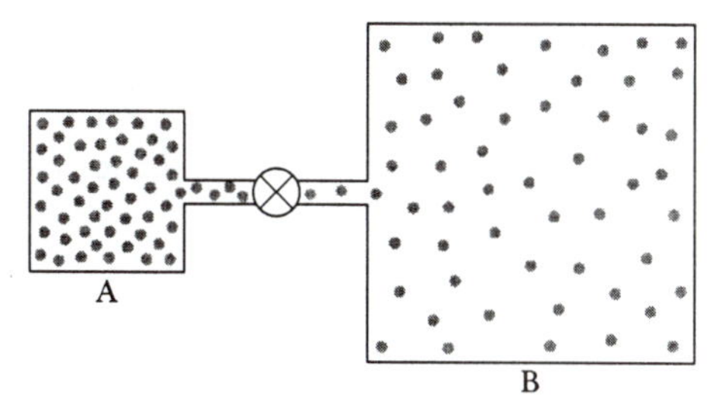

Fig. 20-16 Problem 17.

SEC. 20-4 Pressure, Temperature, and Molecular Kinetic Energy

18E. Calculate the rms speed of helium atoms at 1000 K. See Appendix F for the molar mass of helium atoms.

19E. The lowest possible temperature in outer space is 2.7 K. What is the root-mean-square speed of hydrogen molecules at this temperature? (The molar mass of hydrogen molecules (H_2) is given in Table 20-1.)

20E. Find the rms speed of argon atoms at 313 K. See Appendix F for the molar mass of argon atoms.

21E. The temperature and pressure in the Sun's atmosphere are 2.00×10^6 K and 0.0300 Pa. Calculate the rms speed of free electrons (mass $= 9.11 \times 10^{-31}$ kg) there, assuming they are an ideal gas.

22E. (a) Compute the root-mean-square speed of a nitrogen molecule at 20.0°C. The molar mass of nitrogen molecules (N_2) is given in Table 20-1. At what temperatures will the root-mean-square speed be (b) half that value and (c) twice that value?

23P. A beam of hydrogen molecules (H_2) is directed toward a wall, at an angle of 55° with the normal to the wall. Each molecule in the beam has a speed of 1.0 km/s and a mass of 3.3×10^{-24} g. The beam strikes the wall over an area of 2.0 cm^2, at the rate of 10^{23} molecules per second. What is the beam's pressure on the wall?

24P. At 273 K and 1.00×10^{-2} atm, the density of a gas is 1.24×10^{-5} g/cm^3. (a) Find $|\vec{v}_{rms}|$ for the gas molecules. (b) Find the molar mass of the gas and identify the gas. (*Hint:* The gas is listed in Table 20-1.)

25E. What is the average translational kinetic energy of nitrogen molecules at 1600 K?

26E. Determine the average value of the translational kinetic energy of the molecules of an ideal gas at (a) 0.00°C and (b) 100°C.

What is the translational kinetic energy per mole of an ideal gas at (c) 0.00°C and (d) 100°C?

27P. Water standing in the open at 32.0°C evaporates because of the escape of some of the surface molecules. The heat of vaporization (539 cal/g) is approximately equal to εn, where ε is the average energy of the escaping molecules and n is the number of molecules per gram. (a) Find ε. (b) What is the ratio of ε to the average kinetic energy of H_2O molecules, assuming the latter is related to temperature in the same way as it is for gases?

28P. Show that the ideal gas equation, Eq. 20-8, can be written in the alternative form $P = \rho RT/M$, where ρ is the mass density of the gas and M is the molar mass.

29P. *Avogadro's law* states that under the same conditions of temperature and pressure, equal volumes of gas contain equal numbers of molecules. Is this law equivalent to the ideal gas law? Explain.

SEC. 20-5 Mean Free Path

30E. The mean free path of nitrogen molecules at 0.0°C and 1.0 atm is 0.80×10^{-5} cm. At this temperature and pressure there are 2.7×10^{19} molecules/cm^3. What is the molecular diameter?

31E. At 2500 km above Earth's surface, the density of the atmosphere is about 1 molecule/cm^3. (a) What mean free path is predicted by Eq. 20-23 and (b) what is its significance under these conditions? Assume a molecular diameter of 2.0×10^{-8} cm.

32E. At what frequency would the wavelength of sound in air be equal to the mean free path of oxygen molecules at 1.0 atm pressure and 0.00°C? Take the diameter of an oxygen molecule to be 3.0×10^{-8} cm.

33E. What is the mean free path for 15 spherical jelly beans in a bag that is vigorously shaken? The volume of the bag is 1.0 L, and the diameter of a jelly bean is 1.0 cm. (Consider bean–bean collisions, not bean–bag collisions.)

34P. At 20°C and 750 torr pressure, the mean free paths for argon gas (Ar) and nitrogen gas (N_2) are $\lambda_{Ar} = 9.9 \times 10^{-6}$ cm and $\lambda_{N_2} = 27.5 \times 10^{-6}$ cm. (a) Find the ratio of the effective diameter of argon to that of nitrogen. What is the mean free path of argon at (b) 20°C and 150 torr, and (c) −40°C and 750 torr?

35P. In a certain particle accelerator, protons travel around a circular path of diameter 23.0 m in an evacuated chamber, whose residual gas is at 295 K and 1.00×10^{-6} torr pressure. (a) Calculate the number of gas molecules per cubic centimeter at this pressure. (b) What is the mean free path of the gas molecules if the molecular diameter is 2.00×10^{-8} cm?

SEC. 20-6 The Distribution of Molecular Speeds

36E. Twenty-one particles have speeds as follows (N_i represents the number of particles that have speed $|\vec{v}_i|$):

N_i	2	4	6	8	2
$|\vec{v}_i|$ (cm/s)	1.0	2.0	3.0	4.0	5.0

(a) Computer their average speed $|\vec{v}_{avg}|$. (b) Compute their root-mean-square speed $|\vec{v}_{rms}|$. (c) Of the five speeds shown, which is the most probable speed $|\vec{v}_P|$?

37E. The speeds of 10 molecules are 2.0, 3.0, 4.0, . . . , 11 km/s. (a) What is their average speed? (b) What is their root-mean-square speed?

38E. (a) Ten particles are moving with the following speeds: four at 200 m/s, two at 500 m/s, and four at 600 m/s. Calculate their average and root-mean-square speeds. Is $|\vec{v}_{rms}| > |\vec{v}_{avg}|$? (b) Make up your own speed distribution for the 10 particles and show that $|\vec{v}_{rms}| \geq |\vec{v}_{avg}|$ for your distribution. (c) Under what condition (if any) does $|\vec{v}_{rms}| = |\vec{v}_{avg}|$?

39P. (a) Compute the temperatures at which the rms speed for (a) molecular hydrogen and (b) molecular oxygen is equal to the speed of escape from Earth. (c) Do the same for the speed of escape from the Moon, assuming the gravitational acceleration on its surface to be $0.16g$. (d) The temperature high in Earth's upper atmosphere is about 1000 K. Would you expect to find much hydrogen there? Much oxygen? Explain.

40P. It is found that the most probable speed of molecules in a gas when it has (uniform) temperature T_2 is the same as the rms speed of the molecules in this gas when it has (uniform) temperature T_1. Calculate T_2/T_1.

41P. A molecule of hydrogen (diameter 1.0×10^{-8} cm), traveling with the rms speed, escapes from a furnace (T = 4000 K) into a chamber containing atoms of *cold* argon (diameter 3.0×10^{-8} cm) at a density of 4.0×10^{19} atoms/cm^3. (a) What is the speed of the hydrogen molecule? (b) If the H_2 molecule collides with an argon atom, what is the closest their centers can be, considering each as spherical? (c) What is the initial number of collisions per second experienced by the hydrogen molecule? (*Hint:* Assume that the cold argon atoms are stationary. Then the mean free path of the hydrogen molecule is given by Eq. 20-24 and not Eq. 20-23.)

42P. Two containers are at the same temperature. The first contains gas with pressure P_1, molecular mass m_1, and root-mean-square speed $|\vec{v}_{rms1}|$. The second contains gas with pressure $2P_1$, molecular mass m_2, and average speed $|\vec{v}_{avg2}| = 2|\vec{v}_{rms1}|$. Find the mass ratio m_1/m_2.

43P. Figure 20-17 shows a hypothetical speed distribution for a sample of N gas particles (note that $f(v) = 0$ for $|\vec{v}| > 2|\vec{v}_0|$). (a) Express a in terms of N and $|\vec{v}_0|$. (b) How many of the particles have speeds between $1.5|\vec{v}_0|$ and $2.0|\vec{v}_0|$? (c) Express the average speed of the particles in terms of $|\vec{v}_0|$. (d) Find v_{rms}.

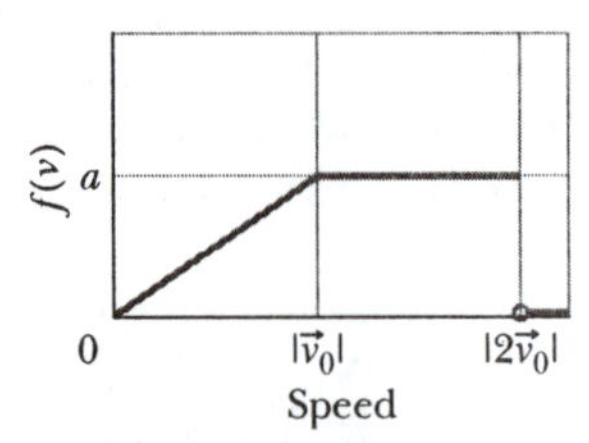

Fig. 20-17 Problem 43.

SEC. 20-7 The Molar Specific Heats of an Ideal Gas

44E. What is the internal energy of 1.0 mol of an ideal monatomic gas at 273 K?

45E. One mole of an ideal gas undergoes an isothermal expansion. Find the energy added to the gas as heat in terms of the initial and final volumes and the temperature. (*Hint:* Use the first law of thermodynamics.)

46P. When 20.9 J was added as heat to a particular ideal gas, the volume of the gas changed from 50.0 cm^3 to 100 cm^3 while the pressure remained constant at 1.00 atm. (a) By how much did the internal energy of the gas change? If the quantity of gas present is 2.00×10^{-3} mol, find the molar specific heat of the gas at (b) constant pressure and (c) constant volume.

47P. A container holds a mixture of three nonreacting gases: n_1 moles of the first gas with molar specific heat at constant volume C_1, and so on. Find the molar specific heat at constant volume of the mixture, in terms of the molar specific heats and quantities of the separate gases.

48P. One mole of an ideal diatomic gas goes from a to c along the diagonal path in Fig. 20-18. During the transition, (a) what is the change in internal energy of the gas, and (b) how much energy is added to the gas as heat? (c) How much heat energy is required if the gas goes from a to c along the indirect path abc?

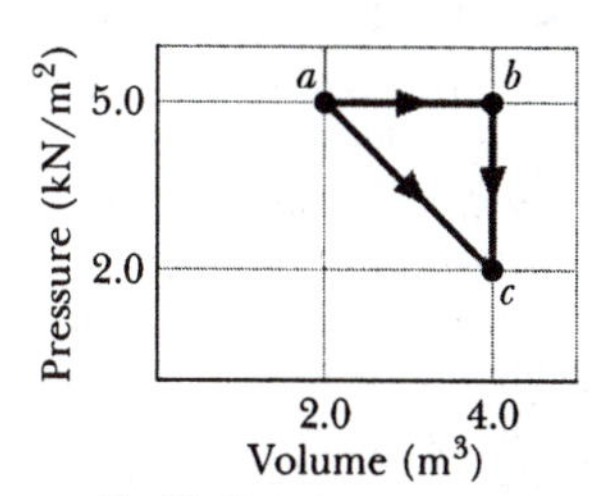

Fig. 20-18 Problem 48.

49P. The mass of a gas molecule can be computed from its specific heat at constant volume c_V. Take $c_V = 0.075$ cal/g · C° for argon and calculate (a) the mass of an argon atom and (b) the molar mass of argon.

SEC. 20-8 Degrees of Freedom and Molar Specific Heats

50E. We give 70 J as heat to a diatomic gas, which then expands at constant pressure. The gas molecules rotate but do not oscillate. By how much does the internal energy of the gas increase?

51E. One mole of oxygen (O_2) is heated at constant pressure starting at 0°C. How much energy must be added to the gas as heat to double its volume? (The molecules rotate but do not oscillate.)

52E. Suppose 12.0 g of oxygen (O_2) is heated at constant atmospheric pressure from 25.0°C to 125°C. (a) How many moles of oxygen are present? (See Table 20-1 for the molar mass.) (b) How much energy is transferred to the oxygen as heat? (The molecules rotate but do not oscillate.) (c) What fraction of the heat is used to raise the internal energy of the oxygen?

53P. Suppose 4.00 mol of an ideal diatomic gas, with molecular rotation but not oscillation, experienced a temperature increase of 60.0 K under constant-pressure conditions. (a) How much energy was transferred to the gas as heat? (b) How much did the internal energy of the gas increase? (c) How much work was done by the gas? (d) How much did the translational kinetic energy of the gas increase?

SEC. 20-10 The Adiabatic Expansion of an Ideal Gas

54E. (a) One liter of a gas with $\gamma = 1.3$ is at 273 K and 1.0 atm pressure. It is suddenly compressed adiabatically to half its original volume. Find its final pressure and temperature. (b) The gas is now cooled back to 273 K at constant pressure. What is its final volume?

55E. A certain gas occupies a volume of 4.3 L at a pressure of 1.2 atm and a temperature of 310 K. It is compressed adiabatically to a volume of 0.76 L. Determine (a) the final pressure and (b) the final temperature, assuming the gas to be an ideal gas for which $\gamma = 1.4$.

56E. We know that for an adiabatic process PV^γ = a constant. Evaluate "a constant" for an adiabatic process involving exactly 2.0 mol of an ideal gas passing through the state having exactly $P = 1.0$ atm and $T = 300$ K. Assume a diatomic gas whose molecules have rotation but not oscillation.

57E. Let n moles of an ideal gas expand adiabatically from an initial temperature T_1 to a final temperature T_2. Prove that the work done by the gas is $nC_V(T_1 - T_2)$, where C_V is the molar specific heat at constant volume. (*Hint:* Use the first law of thermodynamics.)

58E. For adiabatic processes in an ideal gas, show that (a) the bulk modulus is given by

$$B = -V\frac{dP}{dV} = \gamma P,$$

and therefore (b) the speed of sound in the gas is

$$|\vec{v}_s| = \sqrt{\frac{P}{\rho}} = \sqrt{\frac{\gamma RT}{M}}.$$

See Eqs. 18-1 and 18-2.

59E. Air at 0.000°C and 1.00 atm pressure has a density of 1.29×10^{-3} g/cm^3, and the speed of sound in air is 331 m/s at that temperature. Use those data to compute the ratio γ of the molar specific heats of air. (*Hint:* See Exercise 58.)

60P. (a) An ideal gas initially at pressure P_0 undergoes a free expansion until its volume is 3.00 times its initial volume. What then is its pressure? (b) The gas is next slowly and adiabatically compressed back to its original volume. The pressure after compression is $(3.00)^{1/3}$ P_0. Is the gas monatomic, diatomic, or polyatomic? (c) How does the average kinetic energy per molecule in this final state compare with that in the initial state?

61P. One mole of an ideal monatomic gas traverses the cycle of Fig. 20-19. Process 1 → 2 occurs at constant volume, process 2 → 3 is adiabatic, and process 3 → 1 occurs at constant pressure. (a) Compute the heat Q, the change in internal energy ΔE_{int}, and the work done W, for each of the three processes and for the cycle as a whole. (b) The initial pressure at point 1 is 1.00 atm. Find the pressure and the volume at points 2 and 3. Use 1.00 atm = 1.013×10^5 Pa and $R = 8.314$ J/mol · K.

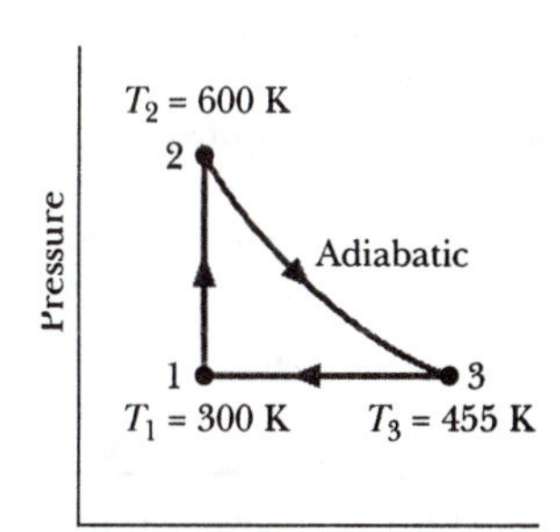

Fig. 20-19 Problem 61.

Chapter 21

SEC. 21-2 Change in Entropy

1E. A 2.50 mol sample of an ideal gas expands reversibly and isothermally at 360 K until its volume is doubled. What is the increase in entropy of the gas?

2E. How much heat is required for a reversible isothermal expansion of an ideal gas at 132°C if the entropy of the gas increases by 46.0 J/K?

3E. Four moles of an ideal gas undergo a reversible isothermal expansion from volume V_1 to volume $V_2 = 2V_1$ at temperature $T = 400$ K. Find (a) the work done by the gas and (b) the entropy change of the gas. (c) If the expansion is reversible and adiabatic instead of isothermal, what is the entropy change of the gas?

4E. An ideal gas undergoes a reversible isothermal expansion at 77.0°C, increasing its volume from 1.30 L to 3.40 L. The entropy change of the gas is 22.0 J/K. How many moles of gas are present?

5E. Find (a) the energy absorbed as heat and (b) the change in entropy of a 2.00 kg block of copper whose temperature is increased reversibly from 25°C to 100°C. The specific heat of copper is 386 J/kg · K.

6E. An ideal monatomic gas at initial temperature T_0 (in kelvins) expands from initial volume V_0 to volume $2V_0$ by each of the five processes indicated in the T-V diagram of Fig. 21-16. In which process is the expansion (a) isothermal, (b) isobaric (constant pressure), and (c) adiabatic? Explain your answers. (d) In which processes does the entropy of the gas decrease?

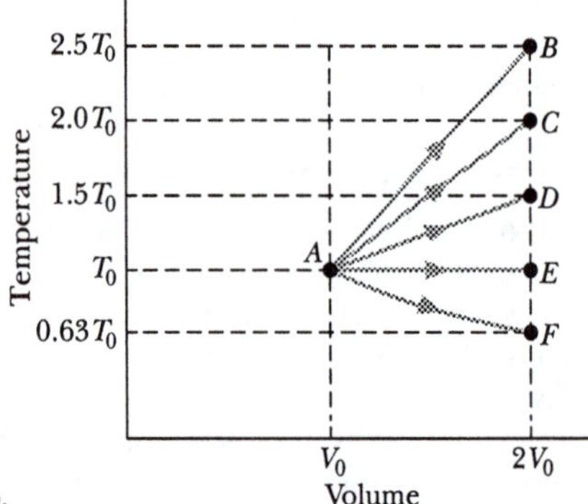

Fig. 21-16 Exercise 6.

7E. (a) What is the entropy change of a 12.0 g ice cube that melts completely in a bucket of water whose temperature is just above the freezing point of water? (b) What is the entropy change of a 5.00 g spoonful of water that evaporates completely on a hot plate whose temperature is slightly above the boiling point of water?

8P. A 2.0 mol sample of an ideal monatomic gas undergoes the reversible process shown in Fig. 21-17. (a) How much energy is absorbed as heat by the gas? (b) What is the change in the internal energy of the gas? (c) How much work is done by the gas?

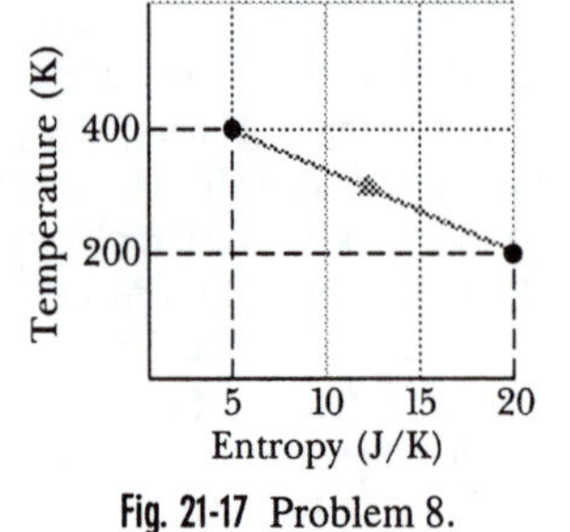

Fig. 21-17 Problem 8.

9P. In an experiment, 200 g of aluminum (with a specific heat of 900 J/kg · K) at 100°C is mixed with 50.0 g of water at 20.0°C, with the mixture thermally isolated. (a) What is the equilibrium temperature? What are the entropy changes of (b) the aluminum, (c) the water, and (d) the aluminum–water system?

10P. In the irreversible process of Fig. TE21-1, let the initial temperatures of identical blocks L and R be 305.5 and 294.5 K, respectively, and let 215 J be the energy transfer between the blocks required to reach equilibrium. Then for the reversible processes of Fig. TE21-2, what are the entropy changes of (a) block L, (b) its reservoir, (c) block R, (d) its reservoir, (e) the two-block system, and (f) the system of the two blocks and the two reservoirs?

11P. Use the reversible apparatus of Fig. TE21-2 to show that, if the process of Fig. TE21-1 happened in reverse, the entropy of the system would decrease, a violation of the second law of thermodynamics.

12P. An ideal diatomic gas, whose molecules are rotating but not oscillating, is taken through the cycle in Fig. 21-18. Determine for all three processes, in terms of P_1, V_1, T_1, and R: (a) P_2, P_3, and T_3 and (b) W, Q, ΔE_{int}, and ΔS per mole.

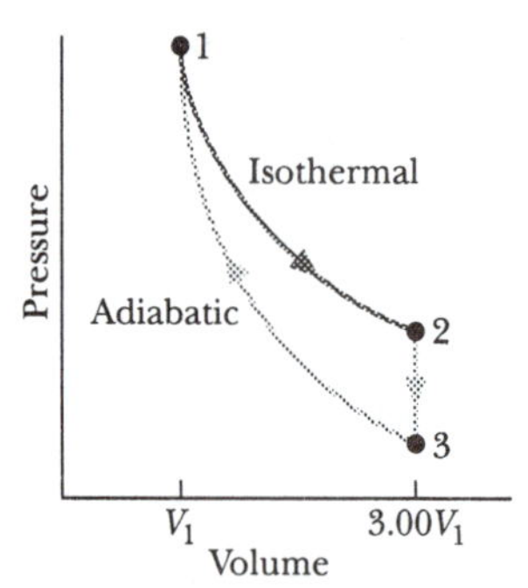

Fig. 21-18 Problem 12.

13P. A 50.0 g block of copper whose temperature is 400 K is placed in an insulating box with a 100 g block of lead whose temperature is 200 K. (a) What is the equilibrium temperature of the two-block system? (b) What is the change in the internal energy of the two-block system between the initial state and the equilibrium state? (c) What is the change in the entropy of the two-block system? (See Table 19-3.)

14P. One mole of a monatomic ideal gas is taken from an initial pressure P and volume V to a final pressure $2P$ and volume $2V$ by two different processes: (I) It expands isothermally until its volume is doubled, and then its pressure is increased at constant volume to the final pressure. (II) It is compressed isothermally until its pressure is doubled, and then its volume is increased at constant pressure to the final volume. (a) Show the path of each process on a P-V diagram. For each process calculate, in terms of P and V, (b) the energy absorbed by the gas as heat in each part of the process, (c) the work done by the gas in each part of the process, (d) the change in internal energy of the gas, $E_{int,f} - E_{int,i}$, and (e) the change in entropy of the gas, $S_f - S_i$.

15P. A 10 g ice cube at −10°C is placed in a lake whose temperature is 15°C. Calculate the change in entropy of the cube–lake system as the ice cube comes to thermal equilibrium with the lake. The specific heat of ice is 2220 J/kg · K. (*Hint:* Will the ice cube affect the temperature of the lake?)

16P. An 8.0 g ice cube at −10°C is put into a Thermos flask containing 100 cm^3 of water at 20°C. By how much has the entropy of the cube–water system changed when a final equilibrium state is reached? The specific heat of ice is 2220 J/kg · K.

17P. A mixture of 1773 g of water and 227 g of ice is in an initial equilibrium state at 0.00°C. The mixture is then, in a reversible process, brought to a second equilibrium state where the water–ice ratio, by mass, is 1:1 at 0.00°C. (a) Calculate the entropy change of the system during this process. (The heat of fusion for water is 333 kJ/kg.) (b) The system is then returned to the initial equilibrium state in an irreversible process (say, by using a Bunsen burner). Calculate the entropy change of the system during this process. (c) Are your answers consistent with the second law of thermodynamics?

18P. A cylinder contains n moles of a monatomic ideal gas. If the gas undergoes a reversible isothermal expansion from initial volume V_i to final volume V_f along path I in Fig. 21-19, its change in entropy is $\Delta S = nR \ln (V_f/V_i)$. (See TE 21-2-1.) Now consider path II in Fig. 21-19, which takes the gas from the same initial state i to state x by a reversible adiabatic expansion, and then from that state x to the same final state f by a reversible constant volume process. (a) Describe how you would carry out the two reversible processes for path II. (b) Show that the temperature of the gas in state x is

$$T_x = T_i(V_i/V_f)^{2/3}.$$

(c) What are the energy Q_{I} transferred as heat along path I and the energy Q_{II} transferred as heat along path II? Are they equal? (d) What is the entropy change ΔS for path II? Is the entropy change for path I equal to it? (e) Evaluate T_x, Q_{I}, Q_{II}, and ΔS for $n = 1$, $T_i = 500$ K, and $V_f/V_i = 2$.

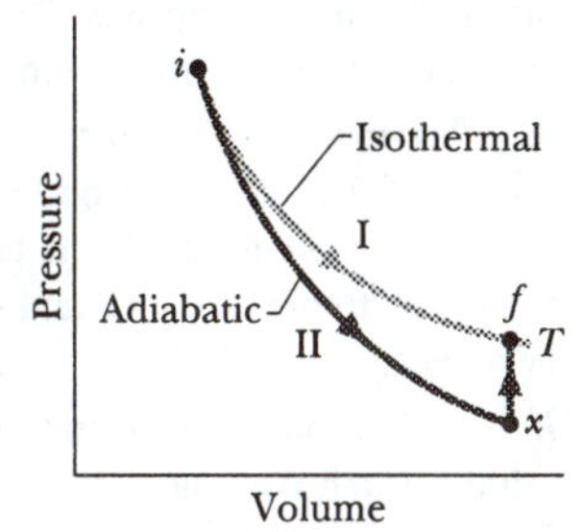

Fig. 21-19 Problem 18.

19P. One mole of an ideal monatomic gas is taken through the cycle in Fig. 21-20. (a) How much work is done by the gas in going from state a to state c along path abc? What are the changes in internal energy and entropy in going (b) from b to c and (c) through one complete cycle? Express all answers in terms of the pressure P_0, volume V_0, and temperature T_0 of state a.

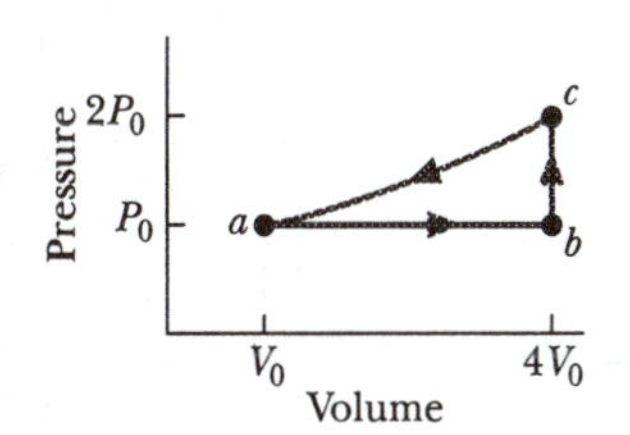

Fig. 21-20 Problem 19.

20P. One mole of an ideal monatomic gas, at an initial pressure of 5.00 kPa and initial temperature of 600 K, expands from initial volume $V_i = 1.00\ \text{m}^3$ to final volume $V_f = 2.00\ \text{m}^3$. During the expansion, the pressure P and volume V of the gas are related by $P = 5.00 \exp[(V_i - V)/a]$, where P is in kilopascals, V_i and V are in cubic meters, and $a = 1.00\ \text{m}^3$. What are (a) the final pressure and (b) the final temperature of the gas? (c) How much work is done by the gas during the expansion? (d) What is the change in entropy of the gas during the expansion? (*Hint:* Use two simple reversible processes to find the entropy change.)

SEC. 21-4 Entropy in the Real World: Engines

21E. A Carnot engine absorbs 52 kJ as heat and exhausts 36 kJ as heat in each cycle. Calculate (a) the engine's efficiency and (b) the work done per cycle in kilojoules.

22E. A Carnot engine whose low-temperature reservoir is at 17°C has an efficiency of 40%. By how much should the temperature of the high-temperature reservoir be increased to increase the efficiency to 50%?

23E. A Carnot engine operates between 235°C and 115°C, absorbing 6.30×10^4 J per cycle at the higher temperature. (a) What is the efficiency of the engine? (b) How much work per cycle is this engine capable of performing?

24E. In a hypothetical nuclear fusion reactor, the fuel is deuterium gas at a temperature of about 7×10^8 K. If this gas could be used to operate a Carnot engine with $T_L = 100$°C, what would be the engine's efficiency?

25E. A Carnot engine has an efficiency of 22.0%. It operates between constant-temperature reservoirs differing in temperature by 75.0 C°. What are the temperatures of the two reservoirs?

26P. A Carnot engine has a power of 500 W. It operates between constant-temperature reservoirs at 100°C and 60.0°C. What are (a) the rate of heat input and (b) the rate of exhaust heat output, in kilojoules per second?

27P. One mole of a monatomic ideal gas is taken through the reversible cycle shown in Fig. 21-21. Process bc is an adiabatic expansion, with $P_b = 10.0$ atm and $V_b = 1.00 \times 10^{-3}\ \text{m}^3$. Find (a) the energy added to the gas as heat, (b) the energy leaving the gas as heat, (c) the net work done by the gas, and (d) the efficiency of the cycle.

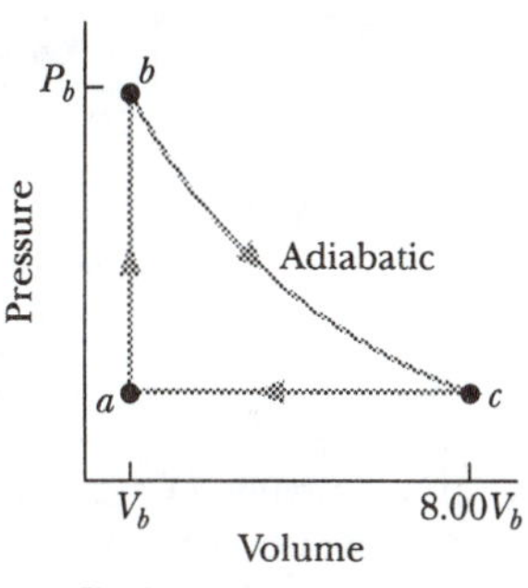

Fig. 21-21 Problem 27.

28P. Show that the area enclosed by the Carnot cycle on the temperature-entropy plot of Fig. 21-7 represents the net energy transfer per cycle as heat to the working substance.

29P. One mole of an ideal monatomic gas is taken through the cycle shown in Fig. 21-22. Assume that $P = 2P_0$, $V = 2V_0$, $P_0 = 1.01 \times 10^5$ Pa, and $V_0 = 0.0225\ \text{m}^3$. Calculate (a) the work done during the cycle, (b) the energy added as heat during stroke abc, and (c) the efficiency of the cycle. (d) What is the efficiency of a Carnot engine operating between the highest and lowest temperatures that occur in the cycle? How does this compare to the efficiency calculated in (c)?

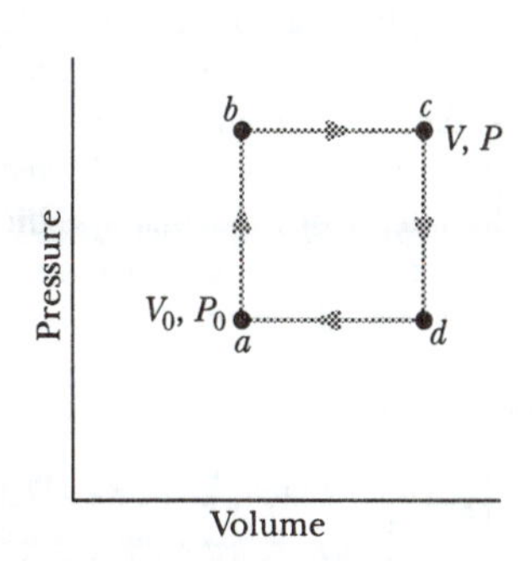

Fig. 21-22 Problem 29.

30P. In the first stage of a two-stage Carnot engine, energy Q_1 is absorbed as heat at temperature T_1, work W_1 is done, and energy

Q_2 is expelled as heat at a lower temperature T_2. The second stage absorbs that energy Q_2, does work W_2, and expels energy Q_3 at a still lower temperature T_3. Prove that the efficiency of the two-stage engine is $(T_1 - T_3)/T_1$.

31P. Suppose that a deep shaft were drilled in Earth's crust near one of the poles, where the surface temperature is $-40°C$, to a depth where the temperature is 800°C. (a) What is the theoretical limit to the efficiency of an engine operating between these temperatures? (b) If all the energy released as heat into the low-temperature reservoir were used to melt ice that was initially at $-40°C$, at what rate could liquid water at 0°C be produced by a 100 MW power plant (treat it as an engine)? The specific heat of ice is 2220 J/kg · K; water's heat of fusion is 333 kJ/kg. (Note that the engine can operate only between 0°C and 800°C in this case. Energy exhausted at $-40°C$ cannot be used to raise the temperature of anything above $-40°C$.)

32P. One mole of an ideal gas is used as the working substance of an engine that operates on the cycle shown in Fig. 21-23. *BC* and *DA* are reversible adiabatic processes. (a) Is the gas monatomic, diatomic, or polyatomic? (b) What is the efficiency of the engine?

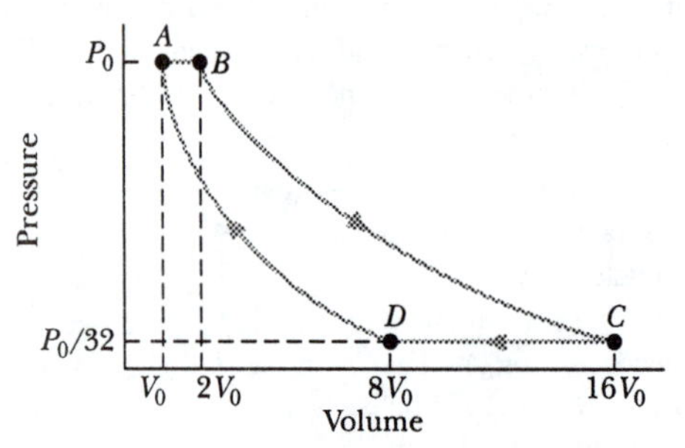

Fig. 21-23 Problem 32.

33P. The operation of a gasoline internal combustion engine is represented by the cycle in Fig. 21-24. Assume the gasoline–air intake mixture is an ideal gas and use a compression ratio of 4:1 ($V_4 = 4V_1$). Assume that $P_2 = 3P_1$. (a) Determine the pressure and temperature at each of the vertex points of the *P-V* diagram in terms of P_1, T_1, and the ratio γ of the molar specific heats of the gas. (b) What is the efficiency of the cycle?

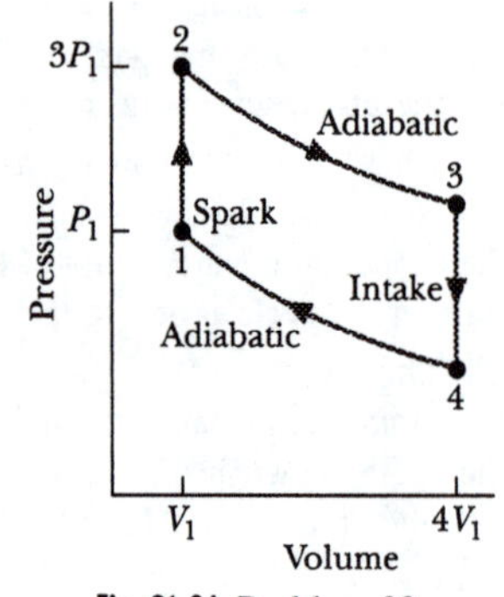

Fig. 21-24 Problem 33.

SEC. 21-5 Entropy in the Real World: Refrigerators

34E. A Carnot refrigerator does 200 J of work to remove 600 J from its cold compartment. (a) What is the refrigerator's coefficient of performance? (b) How much energy per cycle is exhausted to the kitchen as heat?

35E. A Carnot air conditioner takes energy from the thermal energy of a room at 70°F and transfers it to the outdoors, which is at 96°F. For each joule of electric energy required to operate the air conditioner, how many joules are removed from the room?

36E. The electric motor of a heat pump transfers energy as heat from the outdoors, which is at $-5.0°C$, to a room, which is at 17°C. If the heat pump were a Carnot heat pump (a Carnot engine working in reverse), how many joules of heat would be transferred to the thermal energy of the room for each joule of electric energy consumed?

37E. A heat pump is used to heat a building. The outside temperature is $-5.0°C$, and the temperature inside the building is to be maintained at 22°C. The pump's coefficient of performance is 3.8, and the heat pump delivers 7.54 MJ as heat to the building each hour. If the heat pump is a Carnot engine working in reverse, at what rate must work be done to run the heat pump?

38E. How much work must be done by a Carnot refrigerator to transfer 1.0 J as heat (a) from a reservoir at 7.0°C to one at 27°C, (b) from a reservoir at $-73°C$ to one at 27°C, (c) from a reservoir at $-173°C$ to one at 27°C, and (d) from a reservoir at $-223°C$ to one at 27°C?

39P. An air conditioner operating between 93°F and 70°F is rated at 4000 Btu/h cooling capacity. Its coefficient of performance is 27% of that of a Carnot refrigerator operating between the same two temperatures. What horsepower is required of the air conditioner motor?

40P. The motor in a refrigerator has a power of 200 W. If the freezing compartment is at 270 K and the outside air is at 300 K, and assuming the efficiency of a Carnot refrigerator, what is the maximum amount of energy that can be extracted as heat from the freezing compartment in 10.0 min?

41P. A Carnot engine works between temperatures T_1 and T_2. It drives a Carnot refrigerator that works between temperatures T_3 and T_4 (Fig. 21-25). Find the ratio Q_3/Q_1 in terms of T_1, T_2, T_3, and T_4.

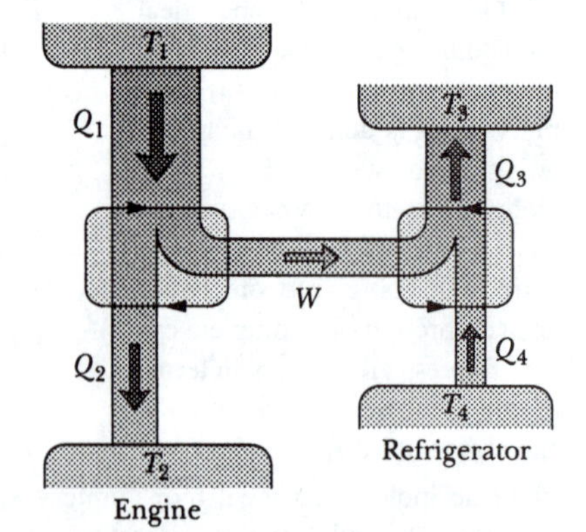

Fig. 21-25 Problem 41.

SEC. 21-7 A Statistical View of Entropy

42E. Construct a table like Table 21-1 for eight molecules.

43E. Show that for N molecules in a box, the number of possible microstates is 2^N when microstates are defined by whether a given molecule is in the left half of the box or the right half. Check this for the situation of Table 21-1.

44P. A box contains N gas molecules, equally divided between its two halves. For $N = 50$: (a) What is the multiplicity of this central configuration? (b) What is the total number of microstates for the system? (*Hint:* See Exercise 43.) (c) What percentage of the time does the system spend in its central configuration? (d) Repeat (a) through (c) for $N = 100$. (e) Repeat (a) through (c) for $N = 200$. (f) As N increases, you will find that the system spends *less* time (not more) in its central configuration. Explain why this is so.

45P. A box contains N gas molecules. Consider the box to be divided into three equal parts. (a) By extension of Eq. 21-18, write a formula for the multiplicity of any given configuration. (b) Consider two configurations: configuration A with equal numbers of molecules in all three thirds of the box, and configuration B with equal numbers of molecules in both halves of the box. What is the ratio W_A/W_B of the multiplicity of configuration A to that of configuration B? (c) Evaluate W_A/W_B for $N = 100$. (Because 100 is not evenly divisible by 3, put 34 molecules into one of the three box parts and 33 in each of the other parts for configuration A.)

APPENDIX A

The International System of Units (SI)*

1. SI Base Units

1. The SI Base Units

Quantity	Name	Symbol	Definition
length	meter	m	". . . the length of the path traveled by light in vacuum in 1/299,792,458 of a second." (1983)
mass	kilogram	kg	". . . this prototype [a certain platinum–iridium cylinder] shall henceforth be considered to be the unit of mass." (1889)
time	second	s	". . . the duration of 9,192,631,770 periods of the radiation corresponding to the transition between the two hyperfine levels of the ground state of the cesium-133 atom." (1967)
electric current	ampere	A	". . . that constant current which, if maintained in two straight parallel conductors of infinite length, of negligible circular cross section, and placed 1 meter apart in vacuum, would produce between these conductors a force equal to 2×10^{-7} newton per meter of length." (1946)
thermodynamic temperature	kelvin	K	". . . the fraction 1/273.16 of the thermodynamic temperature of the triple point of water." (1967)
amount of substance	mole	mol	". . , the amount of substance of a system which contains as many elementary entities as there are atoms in 0.012 kilogram of carbon-12." (1971)
luminous intensity	candela	cd	". . . the luminous intensity, in the perpendicular direction, of a surface of 1/600,000 square meter of a blackbody at the temperature of freezing platinum under a pressure of 101.325 newtons per square meter." (1967)

2. The SI Supplementary Units

2. The SI Supplementary Units

Quantity	Name of Unit	Symbol
plane angle	radian	rad
solid angle	steradian	sr

*Adapted from "The International System of Units (SI)," National Bureau of Standards Special Publication 330, 1972 edition. The definitions above were adopted by the General Conference of Weights and Measures, an international body, on the dates shown. In this book we do not use the candela.

3. Some SI Derivations

3. Some SI Derived Units

Quantity	Name of Unit	Symbol	
area	square meter	m^2	
volume	cubic meter	m^3	
frequency	hertz	Hz	s^{-1}
mass density (density)	kilogram per cubic meter	kg/m^3	
speed, velocity	meter per second	m/s	
angular velocity	radian per second	rad/s	
acceleration	meter per second per second	m/s^2	
angular acceleration	radian per second per second	rad/s^2	
force	newton	N	$kg \cdot m/s^2$
pressure	pascal	Pa	N/m^2
work, energy, quantity of heat	joule	J	$N \cdot m$
power	watt	W	J/s
quantity of electric charge	coulomb	C	$A \cdot s$
potential difference, electromotive force	volt	V	W/A
electric field strength	volt per meter (or newton per coulomb)	V/m	N/C
electric resistance	ohm	Ω	V/A
capacitance	farad	F	$A \cdot s/V$
magnetic flux	weber	Wb	$V \cdot s$
inductance	henry	H	$V \cdot s/A$
magnetic flux density	tesla	T	Wb/m^2
magnetic field strength	ampere per meter	A/m	
entropy	joule per kelvin	J/K	
specific heat	joule per kilogram kelvin	$J/(kg \cdot K)$	
thermal conductivity	watt per meter kelvin	$W/(m \cdot K)$	
radiant intensity	watt per steradian	W/sr	

4. Significant Figures and the Precision of Numerical Results

Quoting the result of a calculation or a measurement to the correct number of significant figures is merely a way of telling your reader how precise you believe the result to be. Quoting too many significant figures overstates the precision of your result while quoting too few implies less precision than the result may actually possess. So how many significant figures should you quote when reporting the result of a measurement or calculation?

Determining Significant Figures

Before answering the question of how many significant figures to quote, we need to have a clear method for determining how many significant figures a reported number has. The standard method is quite simple:

➤ **Method for Counting Significant Figures:** Read the number from left to right, and count the first non-zero digit and all the digits (zero or not) to the right of it as significant.

Using this rule, 350 mm, 0.000350 km, and 0.350 m each has *three* significant figures. In fact, each of these numbers merely represents the same distance, expressed in different units. As you can see from this example, the number of *decimal places* that a number has is *not* the same as its number of *significant figures.* The first of these distances has zero decimal places, the second has six decimal places, while the third has three, yet all three of these numbers have three significant figures.

One consequence of this method is especially worth noting. Trailing zeros count as significant figures. For example, 2700 m/s has four significant figures. If you really meant it to have only three significant figures, you would have to write it either as 2.70 km/s (changing the unit) or 2.70×10^3 m/s (using scientific notation.)

A Simple Rule for Reporting Significant Figures in a Calculated Result

Now that you know how to count significant figures, how many should the result of a calculation have? A simple rule that will work for each of the steps in most calculations is:

➤ **Significant Figures in a Calculated Result:** Unless addition is involved, no final result should be quoted to *more* significant figures than the original data from which it was derived.

In introductory physics you will only rarely encounter data that are known to better than two, three, or four significant figures. This simple rule then tells you that you can't go very far wrong if you round off all your final results to three significant figures. There are two situations in which the simple rule should not be applied to a calculation. One is when an exact number is involved in the calculation and another is when a calculation is done in parts so that intermediate results are used.

1. ***Using Exact Data*** There are some obvious situations in which a number used in a calculation is exact. Numbers based on counting items are exact. For example, if you are told that there are 5 people on an elevator, there are exactly 5 people, not 4.7 or 5.1. Another situation arises when a number is exact to a certain number of significant figures by definition. For example, the conversion factor 2.54 cm/inch does *not* have three significant figures because the inch is *defined* to be exactly 2.5400000 . . . cm. *Data that are known exactly should not be included when deciding which of the original data has the fewest significant figures.*
2. ***Significant Figures in Intermediate Results*** Only the final result that you quote at the end of your calculation should be rounded using the simple rule. Intermediate results should never be rounded at all. Modern spreadsheet software takes care of this for you, as does your calculator if you store your intermediate results in its memory rather than writing them down and then rekeying them. If you must write down intermediate results, always keep a few more significant figures than your final result will have.

Understanding and Refining the Simple Significant Figure Rule

Since quoting the result of a calculation or a measurement to the correct number of significant figures is merely a way of indicating its precision, you need to understand what limits the precision of data before you can acquire a better understanding of the simple rule and its exceptions.

Absolute Precision There are two ways of talking about precision. The first of these is *absolute precision*, which tells you explicitly the smallest scale division of the measurement. It's always quoted in the same units as the quantity being measured. For example, saying "I measured the length of the table to the nearest centimeter" states the absolute precision of the measurement. Knowing the absolute precision tells you how many *decimal places* the measurement has; it alone does not determine the number of significant figures. For example, if the table is 235 cm long, then 1 cm of absolute precision translates into three significant figures. On the other hand, if the table is for a doll's house and is only 8 cm long, then the same 1 cm of absolute precision yields only one significant figure.

Relative Precision Because of this disadvantage of absolute precision, scientists often prefer to describe the precision of data *relative* to the size of the quantity being measured. To use the previous examples, the *relative precision* of the length of the real table is 1 cm out of 235 cm. This is usually stated as a ratio (1 part in 235) or, more conveniently, as a percentage ($1/235 = 0.004255 \approx 0.4\%$). In the case of the toy table, the same 1 cm of absolute precision yields a relative precision of only 1 part in 8 or $1/8 = 0.125 = 12.5\%$.

Inconsistencies between Significant Figures and Relative Precision You may have noticed an inconsistency that goes with using a certain number of significant figures to express relative precision. Quoted to the same number of significant figures, the relative precision of results can be quite different. For example, 13 cm and 94 cm both have two significant figures. Yet the first is specified to only 1 part in 13 or $1/13 \approx 10\%$, while the second is known to 1 part in 94 or $1/94 \approx 1\%$. This bias toward greater relative precision for results with larger first significant figures is one weakness of using significant figures to track the precision of calculated results. To partially address this problem, you can include one additional significant figure than the simple rule suggests, when the final result of a calculation has a 1 as its first significant figure.

Multiplying and Dividing When multiplying or dividing numbers, the *relative* precision of the result can not exceed that of the least precise number used. Since the number of significant figures in the result tells us its relative precision, the simple rule is all that you need when you multiply or divide. For example, the area of a strip of paper whose measured size is 280 cm by 2.5 cm would be correctly reported, according to the simple rule, as 7.0×10^2 cm^2. This result has only two significant figures since the less precise mesurement, 2.5 cm, that went into the calculation had only two significant figures. Reporting this result as 700 cm^2 would not be correct since this reported result has three significant figures, exceeding the relative precision of the 2.5 cm measurement.

Addition and Subtraction When adding or subtracting, you line up the decimal points before you add or subtract. This means that it's the *absolute* precision of the least precise number that limits the precision of the sum or the difference. This can lead to some exceptions to the simple rule. For example, adding 957 cm and 878 cm yields 1835 cm. Here the result is reliable to an absolute precision of about 1 cm since both of the original distances had this reliability. But the result then has four significant figures while each of the original numbers had only three. If, on the other hand, you take the difference betwen these two distances you get 79 cm. The difference is still reliable to about 1 cm, but that absolute precision now translates into only two significant figures worth of relative precision. So, you should be careful when adding or subtracting, since addition can actually increase the relative precision of your result and, more importantly, subtraction can reduce it.

Evaluating Functions What about the evaluation of functions? For example, how many significant figures does the sin(88.2°) have? You can also take an empirical approach to answering this question. First use your calculator to note that sin(88.2°) = 0.999506. Now add 1 to the least significant decimal place of the argument of the function and evaluate it again. Here this gives sin(88.3°) = 0.999559. Take the last significant figure in the result to be *the first one from the left that changed* when you repeated the calculation. In this example the first digit that changed was the 0; it became a 5 (the second 5) in the recalculation. So, using the empirical approach gives you 5 significant figures.

APPENDIX B
Some Fundamental Constants of Physics*

Constant	Symbol	Computational Value	Best (1998) Value	
			Value[a]	Uncertainty[b]
Speed of light in a vacuum	c	3.00×10^{8} m/s	2.997 924 58	exact
Elementary charge	e	1.60×10^{-19} C	1.602 176 462	0.039
Gravitational constant	G	6.67×10^{-11} m³/s² · kg	6.673	1500
Universal gas constant	R	8.31 J/mol · K	8.314 472	1.7
Avogadro constant	N_A	6.02×10^{23} mol⁻¹	6.022 141 99	0.079
Boltzmann constant	k	1.38×10^{-23} J/K	1.380 650 3	1.7
Stefan–Boltzmann constant	σ	5.67×10^{-8} W/m² · K⁴	5.670 400	7.0
Molar volume of ideal gas at STP[d]	V_m	2.27×10^{-2} m³/mol	2.271 098 1	1.7
Permittivity constant	ϵ_0	8.85×10^{-12} F/m	8.854 187 817 62	exact
Permeability constant	μ_0	1.26×10^{-6} H/m	1.256 637 061 43	exact
Planck constant	h	6.63×10^{-34} J · s	6.626 068 76	0.078
Electron mass[c]	m_e	9.11×10^{-31} kg	9.109 381 88	0.079
		5.49×10^{-4} u	5.485 799 110	0.0021
Proton mass[c]	m_p	1.67×10^{-27} kg	1.672 621 58	0.079
		1.0073 u	1.007 276 466 88	1.3×10^{-4}
Ratio of proton mass to electron mass	m_p/m_e	1840	1836.152 667 5	0.0021
Electron charge-to-mass ratio	e/m_e	1.76×10^{11} C/kg	1.758 820 174	0.040
Neutron mass[c]	m_n	1.68×10^{-27} kg	1.674 927 16	0.079
		1.0087 u	1.008 664 915 78	5.4×10^{-4}
Hydrogen atom mass[c]	$m_{^1H}$	1.0078 u	1.007 825 031 6	0.0005
Deuterium atom mass[c]	$m_{^2H}$	2.0141 u	2.014 101 777 9	0.0005
Helium atom mass[c]	$m_{^4He}$	4.0026 u	4.002 603 2	0.067
Muon mass	m_μ	1.88×10^{-28} kg	1.883 531 09	0.084
Electron magnetic moment	μ_e	9.28×10^{-24} J/T	9.284 763 62	0.040
Proton magnetic moment	μ_p	1.41×10^{-26} J/T	1.410 606 663	0.041
Bohr magneton	μ_B	9.27×10^{-24} J/T	9.274 008 99	0.040
Nuclear magneton	μ_N	5.05×10^{-27} J/T	5.050 783 17	0.040
Bohr radius	r_B	5.29×10^{-11} m	5.291 772 083	0.0037
Rydberg constant	R	1.10×10^{7} m⁻¹	1.097 373 156 854 8	7.6×10^{-6}
Electron Compton wavelength	λ_C	2.43×10^{-12} m	2.426 310 215	0.0073

[a]Values given in this column should be given the same unit and power of 10 as the computational value.

[b]Parts per million.

[c]Masses given in u are in unified atomic mass units, where 1 u = $1.660\ 538\ 73 \times 10^{-27}$ kg.

[d]STP means standard temperature and pressure: 0°C and 1.0 atm (0.1 MPa).

*The values in this table were selected from the 1998 CODATA recommended values (www.physics.nist.gov).

APPENDIX C
Some Astronomical Data

Some Distances from Earth

To the Moon*	3.82×10^8 m	To the center of our galaxy	2.2×10^{20} m
To the Sun*	1.50×10^{11} m	To the Andromeda Galaxy	2.1×10^{22} m
To the nearest star (Proxima Centauri)	4.04×10^{16} m	To the edge of the observable universe	$\sim 10^{26}$ m

*Mean distance.

The Sun, Earth, and the Moon

Property	Unit	Sun	Earth	Moon
Mass	kg	1.99×10^{30}	5.98×10^{24}	7.36×10^{22}
Mean radius	m	6.96×10^8	6.37×10^6	1.74×10^6
Mean density	kg/m^3	1410	5520	3340
Free-fall acceleration at the surface	m/s^2	274	9.81	1.67
Escape velocity	km/s	618	11.2	2.38
Period of rotation[a]	—	37 d at poles[b] 26 d at equator[b]	23 h 56 min	27.3 d
Radiation power[c]	W	3.90×10^{26}		

[a]Measured with respect to the distant stars.
[b]The Sun, a ball of gas, does not rotate as a rigid body.
[c]Just outside Earth's atmosphere solar energy is received, assuming normal incidence, at the rate of 1340 W/m^2.

Some Properties of the Planets

	Mercury	Venus	Earth	Mars	Jupiter	Saturn	Uranus	Neptune	Pluto
Mean distance from Sun, 10^6 km	57.9	108	150	228	778	1430	2870	4500	5900
Period of revolution, y	0.241	0.615	1.00	1.88	11.9	29.5	84.0	165	248
Period of rotation,[a] d	58.7	−243[b]	0.997	1.03	0.409	0.426	−0.451[b]	0.658	6.39
Orbital speed, km/s	47.9	35.0	29.8	24.1	13.1	9.64	6.81	5.43	4.74
Inclination of axis to orbit	<28°	≈3°	23.4°	25.0°	3.08°	26.7°	97.9°	29.6°	57.5°
Inclination of orbit to Earth's orbit	7.00°	3.39°		1.85°	1.30°	2.49°	0.77°	1.77°	17.2°
Eccentricity of orbit	0.206	0.0068	0.0167	0.0934	0.0485	0.0556	0.0472	0.0086	0.250
Equatorial diameter, km	4880	12 100	12 800	6790	143 000	120 000	51 800	49 500	2300
Mass (Earth = 1)	0.0558	0.815	1.000	0.107	318	95.1	14.5	17.2	0.002
Density (water = 1)	5.60	5.20	5.52	3.95	1.31	0.704	1.21	1.67	2.03
Surface value of g,[c] m/s^2	3.78	8.60	9.78	3.72	22.9	9.05	7.77	11.0	0.5
Escape velocity,[c] km/s	4.3	10.3	11.2	5.0	59.5	35.6	21.2	23.6	1.1
Known satellites	0	0	1	2	16 + ring	18 + rings	17 + rings	8 + rings	1

[a]Measured with respect to the distant stars.
[b]Venus and Uranus rotate opposite their orbital motion.
[c]Gravitational acceleration measured at the planet's equator.

APPENDIX D

Conversion Factors

Conversion factors may be read directly from these tables. For example, 1 degree = 2.778×10^{-3} revolutions, so $16.7° = 16.7 \times 2.778 \times 10^{-3}$ rev. The SI units are fully capitalized. Adapted in part from G. Shortley and D. Williams, *Elements of Physics,* 1971, Prentice-Hall, Englewood Cliffs, NJ.

Plane Angle

	°	′	″	RADIAN	rev
1 degree =	1	60	3600	1.745×10^{-2}	2.778×10^{-3}
1 minute =	1.667×10^{-2}	1	60	2.909×10^{-4}	4.630×10^{-5}
1 second =	2.778×10^{-4}	1.667×10^{-2}	1	4.848×10^{-6}	7.716×10^{-7}
1 RADIAN =	57.30	3438	2.063×10^{5}	1	0.1592
1 revolution =	360	2.16×10^{4}	1.296×10^{6}	6.283	1

Solid Angle

1 sphere = 4π steradians = 12.57 steradians

Length

	cm	METER	km	in.	ft	mi
1 centimeter =	1	10^{-2}	10^{-5}	0.3937	3.281×10^{-2}	6.214×10^{-6}
1 METER =	100	1	10^{-3}	39.37	3.281	6.214×10^{-4}
1 kilometer =	10^{5}	1000	1	3.937×10^{4}	3281	0.6214
1 inch =	2.540	2.540×10^{-2}	2.540×10^{-5}	1	8.333×10^{-2}	1.578×10^{-5}
1 foot =	30.48	0.3048	3.048×10^{-4}	12	1	1.894×10^{-4}
1 mile =	1.609×10^{5}	1609	1.609	6.336×10^{4}	5280	1

1 angström = 10^{-10} m
1 nautical mile = 1852 m
= 1.151 miles = 6076 ft
1 fermi = 10^{-15} m
1 light-year = 9.460×10^{12} km
1 parsec = 3.084×10^{13} km
1 fathom = 6 ft
1 Bohr radius = 5.292×10^{-11} m
1 yard = 3 ft
1 rod = 16.5 ft
1 mil = 10^{-3} in.
1 nm = 10^{-9} m

Area

	METER²	cm²	ft²	in.²
1 SQUARE METER =	1	10^{4}	10.76	1550
1 square centimeter =	10^{-4}	1	1.076×10^{-3}	0.1550
1 square foot =	9.290×10^{-2}	929.0	1	144
1 square inch =	6.452×10^{-4}	6.452	6.944×10^{-3}	1

1 square mile = 2.788×10^{7} ft² = 640 acres
1 barn = 10^{-28} m²
1 acre = 43 560 ft²
1 hectare = 10^{4} m² = 2.471 acres

Volume

	METER^3	cm^3	L	ft^3	in.^{3}
1 CUBIC METER =	1	10^6	1000	35.31	6.102×10^4
1 cubic centimeter =	10^{-6}	1	1.000×10^{-3}	3.531×10^{-5}	6.102×10^{-2}
1 liter =	1.000×10^{-3}	1000	1	3.531×10^{-2}	61.02
1 cubic foot =	2.832×10^{-2}	2.832×10^4	28.32	1	1728
1 cubic inch =	1.639×10^{-5}	16.39	1.639×10^{-2}	5.787×10^{-4}	1

1 U.S. fluid gallon = 4 U.S. fluid quarts = 8 U.S. pints = 128 U.S. fluid ounces = 231 in.^{3}

1 British imperial gallon = 277.4 in.^{3} = 1.201 U.S. fluid gallons

Mass

Quantities in the colored areas are not mass units but are often used as such. When we write, for example, 1 kg "=" 2.205 lb, this means that a kilogram is a *mass* that *weighs* 2.205 pounds at a location where g has the standard value of 9.80665 m/s^2.

	g	KILOGRAM	slug	u	oz	lb	ton
1 gram =	1	0.001	6.852×10^{-5}	6.022×10^{23}	3.527×10^{-2}	2.205×10^{-3}	1.102×10^{-6}
1 KILOGRAM =	1000	1	6.852×10^{-2}	6.022×10^{26}	35.27	2.205	1.102×10^{-3}
1 slug =	1.459×10^4	14.59	1	8.786×10^{27}	514.8	32.17	1.609×10^{-2}
1 atomic mass unit =	1.661×10^{-24}	1.661×10^{-27}	1.138×10^{-28}	1	5.857×10^{-26}	3.662×10^{-27}	1.830×10^{-30}
1 ounce =	28.35	2.835×10^{-2}	1.943×10^{-3}	1.718×10^{25}	1	6.250×10^{-2}	3.125×10^{-5}
1 pound =	453.6	0.4536	3.108×10^{-2}	2.732×10^{26}	16	1	0.0005
1 ton =	9.072×10^5	907.2	62.16	5.463×10^{29}	3.2×10^4	2000	1

1 metric ton = 1000 kg

Density

Quantities in the colored areas are weight densities and, as such, are dimensionally different from mass densities. See note for mass table.

	slug/ft^3	KILOGRAM/ METER^3	g/cm^3	lb/ft^3	lb/in.^{3}
1 slug per foot^3 =	1	515.4	0.5154	32.17	1.862×10^{-2}
1 KILOGRAM per METER^3 =	1.940×10^{-3}	1	0.001	6.243×10^{-2}	3.613×10^{-5}
1 gram per centimeter^3 =	1.940	1000	1	62.43	3.613×10^{-2}
1 pound per foot^3 =	3.108×10^{-2}	16.02	16.02×10^{-2}	1	5.787×10^{-4}
1 pound per inch^3 =	53.71	2.768×10^4	27.68	1728	1

Time

	y	d	h	min	SECOND
1 year =	1	365.25	8.766×10^3	5.259×10^5	3.156×10^7
1 day =	2.738×10^{-3}	1	24	1440	8.640×10^4
1 hour =	1.141×10^{-4}	4.167×10^{-2}	1	60	3600
1 minute =	1.901×10^{-6}	6.944×10^{-4}	1.667×10^{-2}	1	60
1 SECOND =	3.169×10^{-8}	1.157×10^{-5}	2.778×10^{-4}	1.667×10^{-2}	1

Speed

	ft/s	km/h	METER/SECOND	mi/h	cm/s
1 foot per second =	1	1.097	0.3048	0.6818	30.48
1 kilometer per hour =	0.9113	1	0.2778	0.6214	27.78
1 METER per SECOND =	3.281	3.6	1	2.237	100
1 mile per hour =	1.467	1.609	0.4470	1	44.70
1 centimeter per second =	3.281×10^{-2}	3.6×10^{-2}	0.01	2.237×10^{-2}	1

1 knot = 1 nautical mi/h = 1.688 ft/s   1 mi/min = 88.00 ft/s = 60.00 mi/h

Force

Force units in the colored areas are now little used. To clarify: 1 gram-force (= 1 gf) is the force of gravity that would act on an object whose mass is 1 gram at a location where g has the standard value of 9.80665 m/s^2.

	dyne	NEWTON	lb	pdl	gf	kgf
1 dyne =	1	10^{-5}	2.248×10^{-6}	7.233×10^{-5}	1.020×10^{-3}	1.020×10^{-6}
1 NEWTON =	10^5	1	0.2248	7.233	102.0	0.1020
1 pound =	4.448×10^5	4.448	1	32.17	453.6	0.4536
1 poundal =	1.383×10^4	0.1383	3.108×10^{-2}	1	14.10	1.410×10^2
1 gram-force =	980.7	9.807×10^{-3}	2.205×10^{-3}	7.093×10^{-2}	1	0.001
1 kilogram-force =	9.807×10^5	9.807	2.205	70.93	1000	1

1 ton = 2000 lb

Pressure

	atm	dyne/cm^2	inch of water	cm Hg	PASCAL	lb/in.^{2}	lb/ft^2
1 atmosphere =	1	1.013×10^6	406.8	76	1.013×10^5	14.70	2116
1 dyne per centimeter^2 =	9.869×10^{-7}	1	4.015×10^{-4}	7.501×10^{-5}	0.1	1.405×10^{-5}	2.089×10^{-3}
1 inch of water[a] at 4°C =	2.458×10^{-3}	2491	1	0.1868	249.1	3.613×10^{-2}	5.202
1 centimeter of mercury[a] at 0°C =	1.316×10^{-2}	1.333×10^4	5.353	1	1333	0.1934	27.85
1 PASCAL =	9.869×10^{-6}	10	4.015×10^{-3}	7.501×10^{-4}	1	1.450×10^{-4}	2.089×10^{-2}
1 pound per inch^2 =	6.805×10^{-2}	6.895×10^4	27.68	5.171	6.895×10^3	1	144
1 pound per foot^2 =	4.725×10^{-4}	478.8	0.1922	3.591×10^{-2}	47.88	6.944×10^{-3}	1

[a]Where the acceleration of gravity has the standard value of 9.80665 m/s^2.

1 bar = 10^6 dyne/cm^2 = 0.1 MPa   1 millibar = 10^3 dyne/cm^2 = 10^2 Pa   1 torr = 1 mm Hg

Energy, Work, Heat

Quantities in the colored areas are not energy units but are included for convenience. They arise from the relativistic mass–energy equivalence formula $E = mc^2$ and represent the energy released if a kilogram or unified atomic mass unit (u) is completely converted to energy (bottom two rows) or the mass that would be completely converted to one unit of energy (rightmost two columns).

	Btu	erg	ft · lb	hp · h	JOULE	cal	kW · h	eV	MeV	kg	u
1 British thermal unit =	1	1.055×10^{10}	777.9	3.929×10^{-4}	1055	252.0	2.930×10^{-4}	6.585×10^{21}	6.585×10^{15}	1.174×10^{-14}	7.070×10^{12}
1 erg =	9.481×10^{-11}	1	7.376×10^{-8}	3.725×10^{-14}	10^{-7}	2.389×10^{-8}	2.778×10^{-14}	6.242×10^{11}	6.242×10^{5}	1.113×10^{-24}	670.2
1 foot-pound =	1.285×10^{-3}	1.356×10^{7}	1	5.051×10^{-7}	1.356	0.3238	3.766×10^{-7}	8.464×10^{18}	8.464×10^{12}	1.509×10^{-17}	9.037×10^{9}
1 horsepower-hour =	2545	2.685×10^{13}	1.980×10^{6}	1	2.685×10^{6}	6.413×10^{5}	0.7457	1.676×10^{25}	1.676×10^{19}	2.988×10^{-11}	1.799×10^{16}
1 JOULE =	9.481×10^{-4}	10^{7}	0.7376	3.725×10^{-7}	1	0.2389	2.778×10^{-7}	6.242×10^{18}	6.242×10^{12}	1.113×10^{-17}	6.702×10^{9}
1 calorie =	3.969×10^{-3}	4.186×10^{7}	3.088	1.560×10^{-6}	4.186	1	1.163×10^{-6}	2.613×10^{19}	2.613×10^{13}	4.660×10^{-17}	2.806×10^{10}
1 kilowatt-hour =	3413	3.600×10^{13}	2.655×10^{6}	1.341	3.600×10^{6}	8.600×10^{5}	1	2.247×10^{25}	2.247×10^{19}	4.007×10^{-11}	2.413×10^{16}
1 electron-volt =	1.519×10^{-22}	1.602×10^{-12}	1.182×10^{-19}	5.967×10^{-26}	1.602×10^{-19}	3.827×10^{-20}	4.450×10^{-26}	1	10^{-6}	1.783×10^{-36}	1.074×10^{-9}
1 million electron-volts =	1.519×10^{-16}	1.602×10^{-6}	1.182×10^{-13}	5.967×10^{-20}	1.602×10^{-13}	3.827×10^{-14}	4.450×10^{-20}	10^{-6}	1	1.783×10^{-30}	1.074×10^{-3}
1 kilogram =	8.521×10^{13}	8.987×10^{23}	6.629×10^{16}	3.348×10^{10}	8.987×10^{16}	2.146×10^{16}	2.497×10^{10}	5.610×10^{35}	5.610×10^{29}	1	6.022×10^{26}
1 unified atomic mass unit =	1.415×10^{-13}	1.492×10^{-3}	1.101×10^{-10}	5.559×10^{-17}	1.492×10^{-10}	3.564×10^{-11}	4.146×10^{-17}	9.320×10^{8}	932.0	1.661×10^{-27}	1

Power

	Btu/h	ft · lb/s	hp	cal/s	kW	WATT
1 British thermal unit per hour =	1	0.2161	3.929×10^{-4}	6.998×10^{-2}	2.930×10^{-4}	0.2930
1 foot-pound per second =	4.628	1	1.818×10^{-3}	0.3239	1.356×10^{-3}	1.356
1 horsepower =	2545	550	1	178.1	0.7457	745.7
1 calorie per second =	14.29	3.088	5.615×10^{-3}	1	4.186×10^{-3}	4.186
1 kilowatt =	3413	737.6	1.341	238.9	1	1000
1 WATT =	3.413	0.7376	1.341×10^{-3}	0.2389	0.001	1

Magnetic Field

	gauss	TESLA	milligauss
1 gauss =	1	10^{-4}	1000
1 TESLA =	10^{4}	1	10^{7}
1 milligauss =	0.001	10^{-7}	1

1 tesla = 1 weber/meter^2

Magnetic Flux

	maxwell	WEBER
1 maxwell =	1	10^{-8}
1 WEBER =	10^{8}	1

APPENDIX E

Mathematical Formulas

Geometry

Circle of radius r: circumference $= 2\pi r$; area $= \pi r^2$.

Sphere of radius r: area $= 4\pi r^2$; volume $= \frac{4}{3}\pi r^3$.

Right circular cylinder of radius r and height h:
area $= 2\pi r^2 + 2\pi rh$; volume $= \pi r^2 h$.

Triangle of base a and altitude h: area $= \frac{1}{2}ah$.

Quadratic Formula

If $ax^2 + bx + c = 0$, then $x = \dfrac{-b \pm \sqrt{b^2 - 4ac}}{2a}$.

Trigonometric Functions of Angle θ

$\sin\theta = \dfrac{y}{r}$ $\cos\theta = \dfrac{x}{r}$

$\tan\theta = \dfrac{y}{x}$ $\cot\theta = \dfrac{x}{y}$

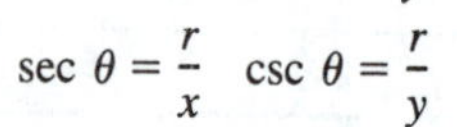

$\sec\theta = \dfrac{r}{x}$ $\csc\theta = \dfrac{r}{y}$

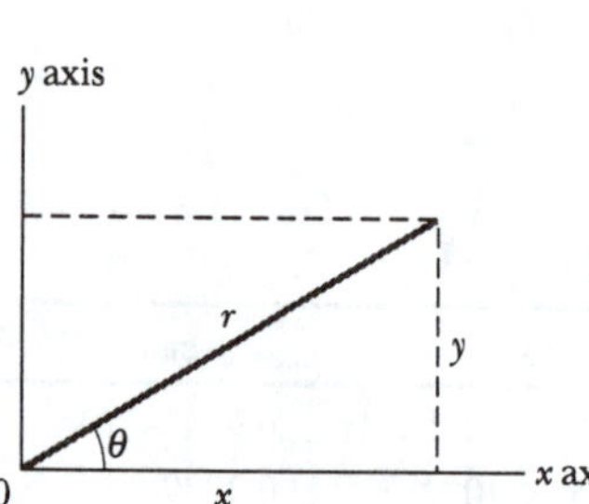

Pythagorean Theorem

In this right triangle,
$a^2 + b^2 = c^2$

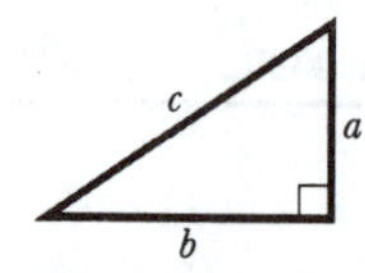

Triangles

Angles are A, B, C

Opposite sides are a, b, c

Angles $A + B + C = 180°$

$\dfrac{\sin A}{a} = \dfrac{\sin B}{b} = \dfrac{\sin C}{c}$

$c^2 = a^2 + b^2 - 2ab\cos C$

Exterior angle $D = A + C$

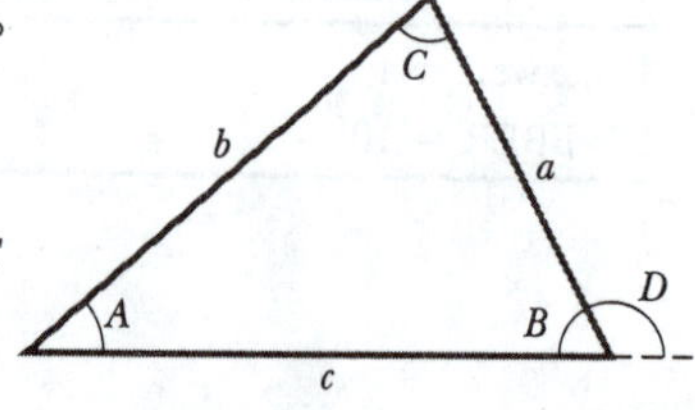

Mathematical Signs and Symbols

$=$ equals

$\approx$ equals approximately

$\sim$ is the order of magnitude of

$\neq$ is not equal to

$\equiv$ is identical to, is defined as

$>$ is greater than ($\gg$ is much greater than)

$<$ is less than ($\ll$ is much less than)

$\geq$ is greater than or equal to (or, is no less than)

$\leq$ is less than or equal to (or, is no more than)

$\pm$ plus or minus

$\propto$ is proportional to

Σ the sum of

x_{avg} the average value of x

Trigonometric Identities

$\sin(90° - \theta) = \cos\theta$

$\cos(90° - \theta) = \sin\theta$

$\sin\theta/\cos\theta = \tan\theta$

$\sin^2\theta + \cos^2\theta = 1$

$\sec^2\theta - \tan^2\theta = 1$

$\csc^2\theta - \cot^2\theta = 1$

$\sin 2\theta = 2\sin\theta\cos\theta$

$\cos 2\theta = \cos^2\theta - \sin^2\theta = 2\cos^2\theta - 1 = 1 - 2\sin^2\theta$

$\sin(\alpha \pm \beta) = \sin\alpha\cos\beta \pm \cos\alpha\sin\beta$

$\cos(\alpha \pm \beta) = \cos\alpha\cos\beta \mp \sin\alpha\sin\beta$

$\tan(\alpha \pm \beta) = \dfrac{\tan\alpha \pm \tan\beta}{1 \mp \tan\alpha\tan\beta}$

$\sin\alpha \pm \sin\beta = 2\sin\frac{1}{2}(\alpha \pm \beta)\cos\frac{1}{2}(\alpha \mp \beta)$

$\cos\alpha + \cos\beta = 2\cos\frac{1}{2}(\alpha + \beta)\cos\frac{1}{2}(\alpha - \beta)$

$\cos\alpha - \cos\beta = -2\sin\frac{1}{2}(\alpha + \beta)\sin\frac{1}{2}(\alpha - \beta)$

Binomial Theorem

$$(1 + x)^n = 1 + \frac{nx}{1!} + \frac{n(n-1)x^2}{2!} + \cdots \qquad (x^2 < 1)$$

Exponential Expansion

$$e^x = 1 + x + \frac{x^2}{2!} + \frac{x^3}{3!} + \cdots$$

Logarithmic Expansion

$$\ln(1 + x) = x - \tfrac{1}{2}x^2 + \tfrac{1}{3}x^3 - \cdots \qquad (|x| < 1)$$

Trigonometric Expansions (θ in radians)

$$\sin \theta = \theta - \frac{\theta^3}{3!} + \frac{\theta^5}{5!} - \cdots$$

$$\cos \theta = 1 - \frac{\theta^2}{2!} + \frac{\theta^4}{4!} - \cdots$$

$$\tan \theta = \theta + \frac{\theta^3}{3} + \frac{2\theta^5}{15} + \cdots$$

Cramer's Rule

Two simultaneous equations in unknowns x and y,

$$a_1x + b_1y = c_1 \quad \text{and} \quad a_2x + b_2y = c_2,$$

have the solutions

$$x = \frac{\begin{vmatrix} c_1 & b_1 \\ c_2 & b_2 \end{vmatrix}}{\begin{vmatrix} a_1 & b_1 \\ a_2 & b_2 \end{vmatrix}} = \frac{c_1b_2 - c_2b_1}{a_1b_2 - a_2b_1}$$

and

$$y = \frac{\begin{vmatrix} a_1 & c_1 \\ a_2 & c_2 \end{vmatrix}}{\begin{vmatrix} a_1 & b_1 \\ a_2 & b_2 \end{vmatrix}} = \frac{a_1c_2 - a_2c_1}{a_1b_2 - a_2b_1}.$$

Products of Vectors

Let $\hat{\mathrm{i}}$, $\hat{\mathrm{j}}$, and $\hat{\mathrm{k}}$ be unit vectors in the x, y, and z directions. Then

$$\hat{\mathrm{i}}\cdot\hat{\mathrm{i}} = \hat{\mathrm{j}}\cdot\hat{\mathrm{j}} = \hat{\mathrm{k}}\cdot\hat{\mathrm{k}} = 1, \qquad \hat{\mathrm{i}}\cdot\hat{\mathrm{j}} = \hat{\mathrm{j}}\cdot\hat{\mathrm{k}} = \hat{\mathrm{k}}\cdot\hat{\mathrm{i}} = 0,$$

$$\hat{\mathrm{i}}\times\hat{\mathrm{i}} = \hat{\mathrm{j}}\times\hat{\mathrm{j}} = \hat{\mathrm{k}}\times\hat{\mathrm{k}} = 0,$$

$$\hat{\mathrm{i}}\times\hat{\mathrm{j}} = \hat{\mathrm{k}}, \qquad \hat{\mathrm{j}}\times\hat{\mathrm{k}} = \hat{\mathrm{i}}, \qquad \hat{\mathrm{k}}\times\hat{\mathrm{i}} = \hat{\mathrm{j}}.$$

Any vector $\vec{a}$ with components a_x, a_y, and a_z along the x, y, and z axes can be written as

$$\vec{a} = a_x\hat{\mathrm{i}} + a_y\hat{\mathrm{j}} + a_z\hat{\mathrm{k}}.$$

Let $\vec{a}$, $\vec{b}$, and $\vec{c}$ be arbitrary vectors with magnitudes a, b, and c. Then

$$\vec{a}\times(\vec{b}+\vec{c}) = (\vec{a}\times\vec{b}) + (\vec{a}\times\vec{c})$$

$$(s\vec{a})\times\vec{b} = \vec{a}\times(s\vec{b}) = s(\vec{a}\times\vec{b}) \quad (s = \text{a scalar}).$$

Let θ be the smaller of the two angles between $\vec{a}$ and $\vec{b}$. Then

$$\vec{a}\cdot\vec{b} = \vec{b}\cdot\vec{a} = a_xb_x + a_yb_y + a_zb_z = ab\cos\theta$$

$$\vec{a}\times\vec{b} = -\vec{b}\times\vec{a} = \begin{vmatrix} \hat{\mathrm{i}} & \hat{\mathrm{j}} & \hat{\mathrm{k}} \\ a_x & a_y & a_z \\ b_x & b_y & b_z \end{vmatrix}$$

$$= \hat{\mathrm{i}}\begin{vmatrix} a_y & a_z \\ b_y & b_z \end{vmatrix} - \hat{\mathrm{j}}\begin{vmatrix} a_x & a_z \\ b_x & b_z \end{vmatrix} + \hat{\mathrm{k}}\begin{vmatrix} a_x & a_y \\ b_x & b_y \end{vmatrix}$$

$$= (a_yb_z - b_ya_z)\hat{\mathrm{i}} + (a_zb_x - b_za_x)\hat{\mathrm{j}} + (a_xb_y - b_xa_y)\hat{\mathrm{k}}$$

$$|\vec{a}\times\vec{b}| = ab\sin\theta$$

$$\vec{a}\cdot(\vec{b}\times\vec{c}) = \vec{b}\cdot(\vec{c}\times\vec{a}) = \vec{c}\cdot(\vec{a}\times\vec{b})$$

$$\vec{a}\times(\vec{b}\times\vec{c}) = (\vec{a}\cdot\vec{c})\vec{b} - (\vec{a}\cdot\vec{b})\vec{c}$$

Derivatives and Integrals

In what follows, the letters u and v stand for any functions of x, and a and m are constants. To each of the indefinite integrals should be added an arbitrary constant of integration. The *Handbook of Chemistry and Physics* (CRC Press Inc.) gives a more extensive tabulation.

1. $\frac{dx}{dx} = 1$
2. $\frac{d}{dx}(au) = a\frac{du}{dx}$
3. $\frac{d}{dx}(u + v) = \frac{du}{dx} + \frac{dv}{dx}$
4. $\frac{d}{dx}x^m = mx^{m-1}$
5. $\frac{d}{dx}\ln x = \frac{1}{x}$
6. $\frac{d}{dx}(uv) = u\frac{dv}{dx} + v\frac{du}{dx}$
7. $\frac{d}{dx}e^x = e^x$
8. $\frac{d}{dx}\sin x = \cos x$
9. $\frac{d}{dx}\cos x = -\sin x$
10. $\frac{d}{dx}\tan x = \sec^2 x$
11. $\frac{d}{dx}\cot x = -\csc^2 x$
12. $\frac{d}{dx}\sec x = \tan x \sec x$
13. $\frac{d}{dx}\csc x = -\cot x \csc x$
14. $\frac{d}{dx}e^u = e^u\frac{du}{dx}$
15. $\frac{d}{dx}\sin u = \cos u\frac{du}{dx}$
16. $\frac{d}{dx}\cos u = -\sin u\frac{du}{dx}$

1. $\int dx = x$
2. $\int au\,dx = a\int u\,dx$
3. $\int (u + v)\,dx = \int u\,dx + \int v\,dx$
4. $\int x^m\,dx = \frac{x^{m+1}}{m+1} \quad (m \neq -1)$
5. $\int \frac{dx}{x} = \ln|x|$
6. $\int u\frac{dv}{dx}\,dx = uv - \int v\frac{du}{dx}\,dx$
7. $\int e^x\,dx = e^x$
8. $\int \sin x\,dx = -\cos x$
9. $\int \cos x\,dx = \sin x$
10. $\int \tan x\,dx = \ln|\sec x|$
11. $\int \sin^2 x\,dx = \frac{1}{2}x - \frac{1}{4}\sin 2x$
12. $\int e^{-ax}\,dx = -\frac{1}{a}e^{-ax}$
13. $\int xe^{-ax}\,dx = -\frac{1}{a^2}(ax + 1)e^{-ax}$
14. $\int x^2e^{-ax}\,dx = -\frac{1}{a^3}(a^2x^2 + 2ax + 2)e^{-ax}$
15. $\int_0^\infty x^ne^{-ax}\,dx = \frac{n!}{a^{n+1}}$
16. $\int_0^\infty x^{2n}e^{-ax^2}\,dx = \frac{1\cdot 3\cdot 5 \cdots (2n-1)}{2^{n+1}a^n}\sqrt{\frac{\pi}{a}}$
17. $\int \frac{dx}{\sqrt{x^2 + a^2}} = \ln(x + \sqrt{x^2 + a^2})$
18. $\int \frac{x\,dx}{(x^2 + a^2)^{3/2}} = -\frac{1}{(x^2 + a^2)^{1/2}}$
19. $\int \frac{dx}{(x^2 + a^2)^{3/2}} = \frac{x}{a^2(x^2 + a^2)^{1/2}}$
20. $\int_0^\infty x^{2n+1}\,e^{-ax^2}\,dx = \frac{n!}{2a^{n+1}} \quad (a > 0)$
21. $\int \frac{x\,dx}{x + d} = x - d\ln(x + d)$

APPENDIX F

Properties of the Elements

All physical properties are for a pressure of 1 atm unless otherwise specified.

Element	Symbol	Atomic Number Z	Molar Mass, g/mol	Density, g/cm^3 at 20°C	Melting Point, °C	Boiling Point, °C	Specific Heat, J/(g · °C) at 25°C
Actinium	Ac	89	(227)	10.06	1323	(3473)	0.092
Aluminum	Al	13	26.9815	2.699	660	2450	0.900
Americium	Am	95	(243)	13.67	1541	—	—
Antimony	Sb	51	121.75	6.691	630.5	1380	0.205
Argon	Ar	18	39.948	1.6626×10^{-3}	−189.4	−185.8	0.523
Arsenic	As	33	74.9216	5.78	817 (28 atm)	613	0.331
Astatine	At	85	(210)	—	(302)	—	—
Barium	Ba	56	137.34	3.594	729	1640	0.205
Berkelium	Bk	97	(247)	14.79	—	—	—
Beryllium	Be	4	9.0122	1.848	1287	2770	1.83
Bismuth	Bi	83	208.980	9.747	271.37	1560	0.122
Bohrium	Bh	107	262.12	—	—	—	—
Boron	B	5	10.811	2.34	2030	—	1.11
Bromine	Br	35	79.909	3.12 (liquid)	−7.2	58	0.293
Cadmium	Cd	48	112.40	8.65	321.03	765	0.226
Calcium	Ca	20	40.08	1.55	838	1440	0.624
Californium	Cf	98	(251)	—	—	—	—
Carbon	C	6	12.01115	2.26	3727	4830	0.691
Cerium	Ce	58	140.12	6.768	804	3470	0.188
Cesium	Cs	55	132.905	1.873	28.40	690	0.243
Chlorine	Cl	17	35.453	3.214×10^{-3} (0°C)	−101	−34.7	0.486
Chromium	Cr	24	51.996	7.19	1857	2665	0.448
Cobalt	Co	27	58.9332	8.85	1495	2900	0.423
Copper	Cu	29	63.54	8.96	1083.40	2595	0.385
Curium	Cm	96	(247)	13.3	—	—	—
Dubnium	Db	105	262.114	—	—	—	—
Dysprosium	Dy	66	162.50	8.55	1409	2330	0.172
Einsteinium	Es	99	(254)	—	—	—	—
Erbium	Er	68	167.26	9.15	1522	2630	0.167
Europium	Eu	63	151.96	5.243	817	1490	0.163
Fermium	Fm	100	(237)	—	—	—	—
Fluorine	F	9	18.9984	1.696×10^{-3} (0°C)	−219.6	−188.2	0.753
Francium	Fr	87	(223)	—	(27)	—	—
Gadolinium	Gd	64	157.25	7.90	1312	2730	0.234
Gallium	Ga	31	69.72	5.907	29.75	2237	0.377

Element	Symbol	Atomic Number Z	Molar Mass, g/mol	Density, g/cm³ at 20°C	Melting Point, °C	Boiling Point, °C	Specific Heat, J/(g · °C) at 25°C
Germanium	Ge	32	72.59	5.323	937.25	2830	0.322
Gold	Au	79	196.967	19.32	1064.43	2970	0.131
Hafnium	Hf	72	178.49	13.31	2227	5400	0.144
Hassium	Hs	108	(265)	—	—	—	—
Helium	He	2	4.0026	0.1664×10^{-3}	−269.7	−268.9	5.23
Holmium	Ho	67	164.930	8.79	1470	2330	0.165
Hydrogen	H	1	1.00797	0.08375×10^{-3}	−259.19	−252.7	14.4
Indium	In	49	114.82	7.31	156.634	2000	0.233
Iodine	I	53	126.9044	4.93	113.7	183	0.218
Iridium	Ir	77	192.2	22.5	2447	(5300)	0.130
Iron	Fe	26	55.847	7.874	1536.5	3000	0.447
Krypton	Kr	36	83.80	3.488×10^{-3}	−157.37	−152	0.247
Lanthanum	La	57	138.91	6.189	920	3470	0.195
Lawrencium	Lr	103	(257)	—	—	—	—
Lead	Pb	82	207.19	11.35	327.45	1725	0.129
Lithium	Li	3	6.939	0.534	180.55	1300	3.58
Lutetium	Lu	71	174.97	9.849	1663	1930	0.155
Magnesium	Mg	12	24.312	1.738	650	1107	1.03
Manganese	Mn	25	54.9380	7.44	1244	2150	0.481
Meitnerium	Mt	109	(266)	—	—	—	—
Mendelevium	Md	101	(256)	—	—	—	—
Mercury	Hg	80	200.59	13.55	−38.87	357	0.138
Molybdenum	Mo	42	95.94	10.22	2617	5560	0.251
Neodymium	Nd	60	144.24	7.007	1016	3180	0.188
Neon	Ne	10	20.183	0.8387×10^{-3}	−248.597	−246.0	1.03
Neptunium	Np	93	(237)	20.25	637	—	1.26
Nickel	Ni	28	58.71	8.902	1453	2730	0.444
Niobium	Nb	41	92.906	8.57	2468	4927	0.264
Nitrogen	N	7	14.0067	1.1649×10^{-3}	−210	−195.8	1.03
Nobelium	No	102	(255)	—	—	—	—
Osmium	Os	76	190.2	22.59	3027	5500	0.130
Oxygen	O	8	15.9994	1.3318×10^{-3}	−218.80	−183.0	0.913
Palladium	Pd	46	106.4	12.02	1552	3980	0.243
Phosphorus	P	15	30.9738	1.83	44.25	280	0.741
Platinum	Pt	78	195.09	21.45	1769	4530	0.134
Plutonium	Pu	94	(244)	19.8	640	3235	0.130
Polonium	Po	84	(210)	9.32	254	—	—
Potassium	K	19	39.102	0.862	63.20	760	0.758
Praseodymium	Pr	59	140.907	6.773	931	3020	0.197
Promethium	Pm	61	(145)	7.22	(1027)	—	—
Protactinium	Pa	91	(231)	15.37 (estimated)	(1230)	—	—
Radium	Ra	88	(226)	5.0	700	—	—
Radon	Rn	86	(222)	9.96×10^{-3} (0°C)	(−71)	−61.8	0.092
Rhenium	Re	75	186.2	21.02	3180	5900	0.134

Element	Symbol	Atomic Number Z	Molar Mass, g/mol	Density, g/cm^3 at 20°C	Melting Point, °C	Boiling Point, °C	Specific Heat, J/(g · °C) at 25°C
Rhodium	Rh	45	102.905	12.41	1963	4500	0.243
Rubidium	Rb	37	85.47	1.532	39.49	688	0.364
Ruthenium	Ru	44	101.107	12.37	2250	4900	0.239
Rutherfordium	Rf	104	261.11	—	—	—	—
Samarium	Sm	62	150.35	7.52	1072	1630	0.197
Scandium	Sc	21	44.956	2.99	1539	2730	0.569
Seaborgium	Sg	106	263.118	—	—	—	—
Selenium	Se	34	78.96	4.79	221	685	0.318
Silicon	Si	14	28.086	2.33	1412	2680	0.712
Silver	Ag	47	107.870	10.49	960.8	2210	0.234
Sodium	Na	11	22.9898	0.9712	97.85	892	1.23
Strontium	Sr	38	87.62	2.54	768	1380	0.737
Sulfur	S	16	32.064	2.07	119.0	444.6	0.707
Tantalum	Ta	73	180.948	16.6	3014	5425	0.138
Technetium	Tc	43	(99)	11.46	2200	—	0.209
Tellurium	Te	52	127.60	6.24	449.5	990	0.201
Terbium	Tb	65	158.924	8.229	1357	2530	0.180
Thallium	Tl	81	204.37	11.85	304	1457	0.130
Thorium	Th	90	(232)	11.72	1755	(3850)	0.117
Thulium	Tm	69	168.934	9.32	1545	1720	0.159
Tin	Sn	50	118.69	7.2984	231.868	2270	0.226
Titanium	Ti	22	47.90	4.54	1670	3260	0.523
Tungsten	W	74	183.85	19.3	3380	5930	0.134
Un-named	Uun	110	(269)	—	—	—	—
Un-named	Uuu	111	(272)	—	—	—	—
Un-named	Uub	112	(264)	—	—	—	—
Un-named	Uut	113	—	—	—	—	—
Un-named	Unq	114	(285)	—	—	—	—
Un-named	Uup	115	—	—	—	—	—
Un-named	Uuh	116	(289)	—	—	—	—
Un-named	Uus	117	—	—	—	—	—
Un-named	Uuo	118	(293)	—	—	—	—
Uranium	U	92	(238)	18.95	1132	3818	0.117
Vanadium	V	23	50.942	6.11	1902	3400	0.490
Xenon	Xe	54	131.30	5.495×10^{-3}	−111.79	−108	0.159
Ytterbium	Yb	70	173.04	6.965	824	1530	0.155
Yttrium	Y	39	88.905	4.469	1526	3030	0.297
Zinc	Zn	30	65.37	7.133	419.58	906	0.389
Zirconium	Zr	40	91.22	6.506	1852	3580	0.276

The values in parentheses in the column of molar masses are the mass numbers of the longest-lived isotopes of those elements that are radioactive. Melting points and boiling points in parentheses are uncertain.

The data for gases are valid only when these are in their usual molecular state, such as H_2, He, O_2, Ne, etc. The specific heats of the gases are the values at constant pressure.

Source: Adapted from J. Emsley, *The Elements,* 3rd ed., 1998, Clarendon Press, Oxford. See also www.webelements.com for the latest values and newest elements.

APPENDIX G

Periodic Table of the Elements

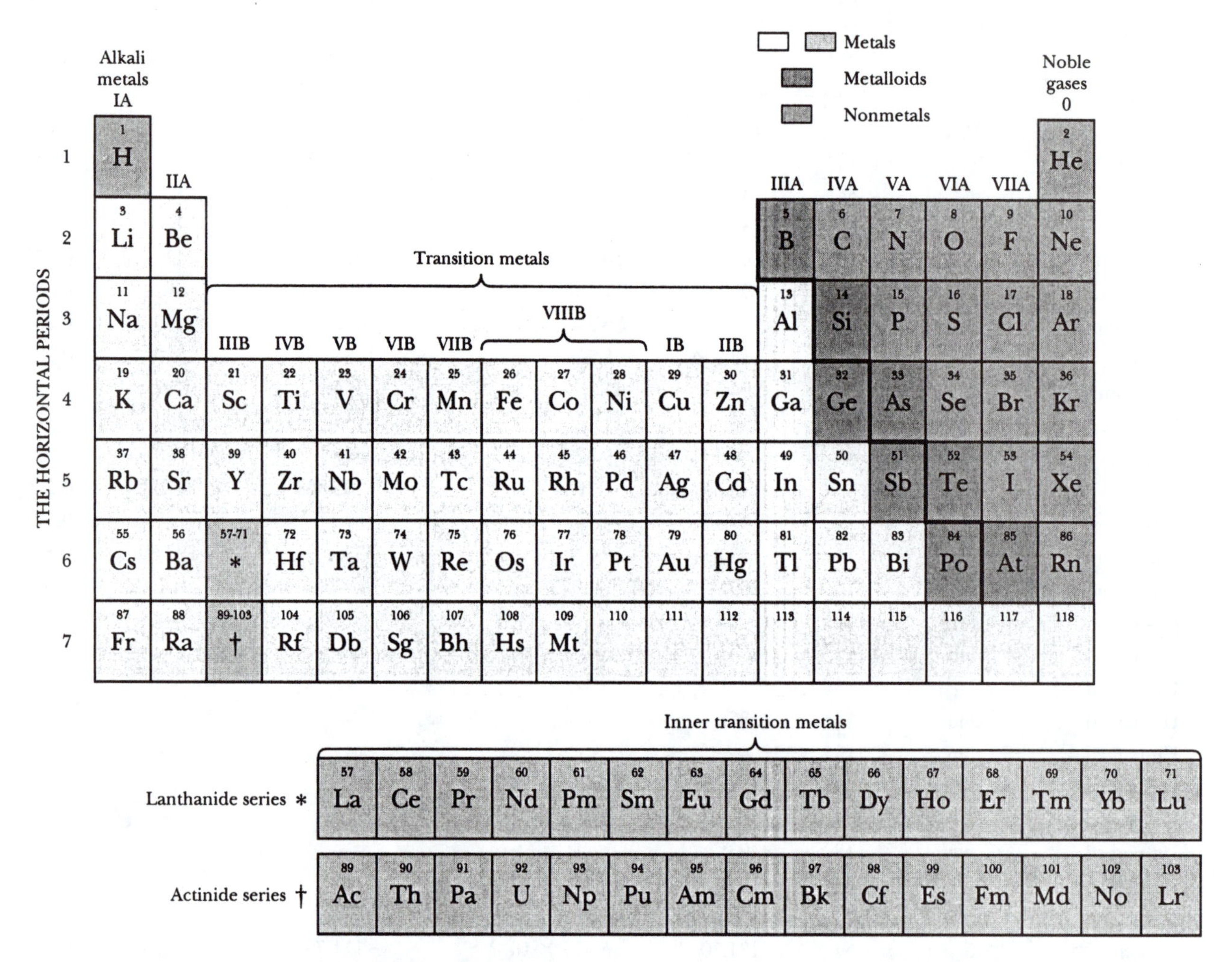

The names for elements 104 through 109 (Rutherfordium, Dubnium, Seaborgium, Bohrium, Hassium, and Meitnerium, respectively) were adopted by the International Union of Pure and Applied Chemistry (IUPAC) in 1997. Elements 110, 111, 112, 114, 116, and 118 have been discovered but, as of 2000, have not yet been named. See www.webelements.com for the latest information and newest elements.

PHOTO CREDITS

Chapter 13
Opener: Pascal Tournaire/Vandystadt/Allsport. Fig. 13-1: Fred Hirschmann/Allstock/Tony Stone Images/New York, Inc. Fig. 13-3: Andy Levin/Photo Researchers. Fig. 13-10: Courtesy Micro-Measurements Division, Measurements Group, Inc., Raleigh, North Carolina. Page 289: Hideo Kurihara/Tony, Stone Images/ New York, Inc.

Chapter 14
Opener: Courtesy Lund Observatory. Fig. 14-1 and page 14-19: Courtesy NASA. Fig. 14-17*b*: Courtesy National Radio Astronomy Observatory.

Chapter 15
Opener: Colin Prior/Tony Stone Images/New York, Inc. Fig. 15-8*a*: Courtesy Vernier Software and Technology. Fig. 15-8*b*: Courtesy PASCO scientific. Page 15-13: T. Orban/Sygma. Fig. 15-15: Will McIntyre/Photo Researchers. Fig. 15-16: Courtesy D.H. Peregrine, University of Bristol. Fig. 15-17: Courtesy Volvo North America.

Chapter 16
Opener: T Campion/Sygma. Figs 16-1 and 16-2: Courtesy Vernier Software and Technology. Figs. 16-4, 16-6, 16-10, and 16-16: Courtesy Priscilla Laws. Fig. 16-27: Courtesy NASA.

Chapter 17
Opener: John Visser/Bruce Coleman, Inc. Fig. 17-12: Courtesy Physical Science Study Committee. Fig. 17-20: Richard Megna/Fundamental Photographs. Fig. 17-22: Courtesy T.D. Rossing, Northern Illinois University.

Chapter 18
Opener: Stephen Dalton/Animals Animals. Fig. 18-1, 18-2, and 18-3: Courtesy Priscilla Laws. Page 18-7: Ben Rose/The Image Bank. Fig. 18-9: John Eastcott/Yva Momativk/DRK Photo. Fig. 18-20: U.S. Navy photo by Ensign John Gay. Fig. TE18-4: Bob Gruen/Star File.

Chapter 19
Opener: © Dr. Mosato Ono, Tamagawa University. Reproduced with permission. Fig. 19-1*a*: Courtesy Pocket Nurse Enterprises, Inc. Fig. 19-1*b*: Courtesy Electronic Temperature Instruments, Ltd. Fig. 19-1*c*: Courtesy Pocket Nurse Enterprises, Inc. Fig. 19-8: Courtesy Patrick J. Cooney. Fig. 19-9*a*: Courtesy Priscilla Laws. Fig. 19-21: AP/Wide World Photos. Fig. 19-25: Courtesy Daedalus Enterprises, Inc. Fig. TE19-3: © Dr. Mosato Ono, Tamagawa University. Reproduced with permission. Fig. 19-39: Kim Westerskov/Tony Stone Images/New York, Inc.

Chapter 20
Opener: Tom Branch.

Chapter 21
Opener: Steven Dalton/Photo Researchers. Fig. 21-9: Richard Ustinich/The Image Bank.Corporation.

INDEX

S

T